Introductory Quantum Optics
Second Edition

This established textbook provides an accessible but comprehensive introduction to the quantum nature of light and its interaction with matter. The field of quantum optics is covered with clarity and depth, from the underlying theoretical framework of field quantization, atom–field interactions, and quantum coherence theory, to important and modern applications at the forefront of current research such as quantum interferometry, squeezed light, quantum entanglement, cavity quantum electrodynamics, laser-cooled trapped ions, and quantum information processing. The text is suitable for advanced undergraduate and graduate students and would be an ideal main text for a course on quantum optics. This long-awaited second edition builds upon the success of the first edition, including many new developments in the field, particularly in the area of quantum state engineering. Additional homework problems have been added, and content from the first edition has been updated and clarified throughout.

Christopher C. Gerry is Professor of Physics at Lehman College, City University of New York. He was one of the first to exploit the use of group theoretical methods in quantum optics and is a highly regarded researcher and lecturer in the field. He has written well-regarded books, both for advanced students and researchers, and for a more general audience.

Sir Peter L. Knight FRS is Emeritus Professor at Imperial College London, a past President of the Institute of Physics (IOP), 2004 President of the Optical Society of America (OSA), Chair of the UK National Quantum Technology Programme Strategy Advisory Board, and Chair of the Quantum Metrology Institute at the National Physical Laboratory. His research centers on quantum technology and quantum optics and he has been the recipient of several prestigious awards, including the Thomas Young Medal and Glazebrook Medal of the Institute of Physics, Optica's Frederic Ives Medal and Herbert Walther Award, the Royal Medal of the Royal Society, and the Faraday Medal of the Institution of Engineering and Technology.

"Quantum technology is transitioning from the research laboratory into the commercial world. Scientists and engineers are learning new languages to understand how quantum will impact applications. Beyond the math, what I love about this book are the words highlighting both the intuitive and non-intuitive science, essential understanding for progressing the transition from quantum science to quantum technology."

Professor Miles Padgett FRS, The University of Glasgow, UK

"The book is absolutely a pleasure, with a wide coverage of the field including important developments such as optical tests of the foundational aspects of quantum mechanics, Heisenberg limited metrology, quantum gates, and decoherence. The simplicity with which deeper concepts are introduced is truly remarkable. It deserves to be on the shelf of everyone interested in the new quantum revolution of the twenty-first century."

Professor Girish S. Agarwal FRS, Texas A&M University, USA

Introductory Quantum Optics

Second Edition

CHRISTOPHER C. GERRY
Lehman College, City University of New York

PETER L. KNIGHT
Imperial College and UK National Physical Laboratory

CAMBRIDGE
UNIVERSITY PRESS

Shaftesbury Road, Cambridge CB2 8EA, United Kingdom

One Liberty Plaza, 20th Floor, New York, NY 10006, USA

477 Williamstown Road, Port Melbourne, VIC 3207, Australia

314–321, 3rd Floor, Plot 3, Splendor Forum, Jasola District Centre,
New Delhi – 110025, India

103 Penang Road, #05–06/07, Visioncrest Commercial, Singapore 238467

Cambridge University Press is part of Cambridge University Press & Assessment,
a department of the University of Cambridge.

We share the University's mission to contribute to society through the pursuit of
education, learning and research at the highest international levels of excellence.

www.cambridge.org
Information on this title: www.cambridge.org/highereducation/isbn/9781009415293

DOI: 10.1017/9781139151207

First edition © C. C. Gerry and P. L. Knight 2005
Second edition © Christopher C. Gerry and Peter L. Knight 2024

First published 2005
Third printing 2008
Second edition 2024

A catalogue record for this publication is available from the British Library

*A Cataloging-in-Publication data record for this book is available from the Library
of Congress*

ISBN 978-1-009-41529-3 Hardback

Additional resources for this publication at www.cambridge.org/gerry-knight2.

C. C. G. dedicates this book to his son, Eric.
P. L. K. dedicates this book to his wife, Chris.

Brief Contents

Contents

Preface

The first edition of this book, published in 2005, was intended to be a readable text for students interested in the nature of light fields and their interaction with atoms at the fundamental level, focusing on concepts and ideas. At that time we noted the vibrancy of the field, and in the subsequent 20 years it has grown and flourished, with major insights being uncovered on the nature of quantum light, on entanglement, and much else. The award of the 2022 Physics Nobel Prize to Alain Aspect, John Clauser, and Anton Zeilinger for their work on quantum optical entanglement highlights the central role the subject has continued to play in modern physics. Many universities now offer courses on quantum optics, some for final-year advanced undergraduates and many more for graduate students, and it is to these that we address our book. The text is designed for students taking courses in quantum optics who have already taken a course in quantum mechanics, although we do cover some key elements that they may not have encountered before, such as the Schmidt decomposition – so useful in describing entanglement, and some key ideas in quantum measurement theory.

The presentation continues to be very much concerned with the quantized electromagnetic field. As in the first edition, topics covered include single-mode field quantization in a cavity, quantization of multimode fields, quantum phase, coherent states, quasi-probability distribution in phase space, atom–field interactions, the Jaynes–Cummings model, quantum coherence theory, beam splitters and interferometers, nonclassical field states with squeezing and so on, tests of local realism with entangled photons from down-conversion, experimental realizations of cavity quantum electrodynamics, trapped ions, decoherence, and an introduction to applications to quantum information processing, particularly quantum cryptography. We have made many updates to the text and included new sections on the quantum phase operator in Chapter 2, a detailed discussion of the connection between the photon number parity operator and the Wigner function in Chapter 3, a major expansion on beam splitters and interferometers in Chapter 6 – to include more details on how to obtain beam splitter output states for any given number-state inputs. This chapter also includes a discussion of the use of the SU(2) (angular momentum) formalism in the Schwinger realization for the description of beam splitters and interferometers. The homework problems for this chapter have been significantly revised. In Chapter 7 we have added material on the distinction between Gaussian and non-Gaussian states and the significance of that distinction, and we have added a section on the so-called pair coherent states, a particular example of a two-mode non-Gaussian state. We have added a section on entanglement generation by beam splitting, and a section on quantum state engineering by photon-level operations on Gaussian states. In Chapter 8 we have added a section on modeling losses with fictious beam splitters, where we show how this method can be applied to number states. The section on decoherence has been reworked from this point of view. In Chapter 11 we have added sections on the no-signaling theorem and the no-cloning theorem. Also added is a section on an experiment in quantum optical interferometry where a squeezed vacuum state is mixed with a coherent state to generate a superposition of the so-called $N00N$ states. A section has been added on using photon number parity as the observable in an interferometric experiment with coherent light. In the section on quantum random number generation (QRNG) we have added a discussion on the use of measurements of the photon number parity operator in moderately intense laser light for generating random numbers and of the experiment that was performed in connection with this idea.

We took a decision to remain focused on quantum optics and resisted the temptation to extend substantially our coverage of quantum information processing, which has developed into a field of its own covered by excellent specialist texts. As theorists, we felt it would be inappropriate to go into the details of experiments, rather we concentrate on the basic theoretical ideas.

The book contains many homework problems, suggestions for further reading, and a comprehensive bibliography of the key papers that we feel students would benefit from accessing. Feedback from colleagues around the world has been invaluable as we worked on this second edition; we hope it will continue to be a useful guide to an exciting part of contemporary physics.

Acknowledgments

Acknowledgments to the First Edition

This book developed out of courses that we have given over the years at Imperial College London and the Graduate Center of the City University of New York, and we are grateful to the many students who have sat through our lectures and acted as guinea pigs for the material we have presented here.

We would like to thank our many colleagues who, over many years, have given us advice, ideas, and encouragement. We particularly thank Dr. Simon Capelin at Cambridge University Press who has had much more confidence than us that this would ever be completed. Over the years we have benefited from many discussions with our colleagues, especially Les Allen, Gabriel Barton, Janos Bergou, Keith Burnett, Vladimir Buzek, Richard Campos, Bryan Dalton, Joseph Eberly, Rainer Grobe, Edwin Hach III, Robert Hilborn, Mark Hillery, Ed Hinds, Rodney Loudon, Peter Milonni, Bill Munro, Geoffrey New, Edwin Power, George Series, Wolfgang Schleich, Bruce Shore, Carlos Stroud Jr., Stuart Swain, Dan Walls, and Krzysztof Wodkiewicz. We especially thank Adil Benmoussa for creating all the figures for this book using his expertise with Mathematica, Corel Draw, and Origin Graphics, for working through the homework problems, and for catching many errors in various drafts of the manuscript. We also thank Mrs. Ellen Calkins for typing the initial draft of several of the chapters.

We acknowledge our former students and postdocs, who have taught us so much and have gone on to become leaders themselves in this exciting subject, especially Stephen Barnett, Almut Beige, Artur Ekert, Barry Garraway, Christoph Keitel, Myungshik Kim, Gerard Milburn, Martin Plenio, Barry Sanders, Stefan Scheel, and Vlatko Vedral: they will recognize much that is here!

As this book is intended as an introduction to quantum optics, we have not attempted to be comprehensive in our citations. We apologize to those authors whose work is not cited.

C. C. G. wishes to thank the members of the Lehman College Department of Physics and Astronomy, and many other members of the

Lehman College community, for their encouragement during the writing of this book.

P. L. K. would especially like to acknowledge the support throughout of Chris Knight, who has patiently provided encouragement, chauffeuring and vast amounts of tea during the writing of this book.

Our work in quantum optics over the past four decades has been funded by many sources: for P. L. K. in particular the UK SRC, SERC, EPSRC, the Royal Society, the European Union, the Nuffield Foundation and the US Army are thanked for their support; for C. C. G. the National Science Foundation, the Research Corporation, Professional Staff Congress of the City University of New York (PSC-CUNY).

Acknowledgments to the Second Edition

We remain indebted to those named above, of course. But we would like to add our thanks to those who have given us invaluable insights since that first edition: especially Richard Birrittella, in particular for much help with graphics (Figures 3.9, 3.10, 6.6, 6.7, 6.8, 6.9, 6.10, 6.11, 6.12, 7.19, 7.20, 7.21, and 7.22) and with the student problems and Lior Cohen, for discussions on his experiments on the role of parity in quantum interferometry.

C. C. G. thanks former student and long-time collaborator Richard Birrittella for his help with various aspects of this edition of the book, and former student and long-term collaborator Ed Hach for many discussions on the contents of the book. He also thanks Lior Cohen for several discussions on the experiment he performed (with the group of H. Eisenberg) on the use of photon number parity in quantum optical interferometry. Also, he wishes to acknowledge the many conversations he has had on quantum optical interferometry over the past two decades and more with the late Jonathon Dowling. Finally, he wishes to acknowledge the summer funding he has received to support a long-term collaboration with Paul Alsing and his group at the U.S. Air Force Laboratory in Rome, New York.

P. L. K. thanks Terry Rudolph, the late Danny Segal and Richard Thompson, all at Imperial, and Miles Padgett in Glasgow, Jason Twamley in Okinawa for many discussions on the topics of this book.

1 Introduction

1.1 Scope and Aims of This Book

Quantum optics remains one of the liveliest fields in physics. While it has been a dominant research field for at least four decades, with much graduate activity, it has now impacted the undergraduate curriculum. This book developed from courses we have taught to final-year undergraduates and beginning graduate students at Imperial College London and City University of New York. There are plenty of good research monographs in this field, but we felt that there was a genuine need for a straightforward account for senior undergraduates and beginning postgraduates, which stresses basic concepts. This is a field which attracts some of the brightest students at present, in part because of the extraordinary progress in the field (e.g. the implementation of teleportation, quantum cryptography, Schrödinger-cat states, Bell violations of local realism and the like). The 2012 Physics Nobel Prize to Haroche and Wineland for their ground-breaking experimental work measuring and manipulating individual quantum systems, and the 2022 Physics Nobel Prize to Aspect, Clauser, and Zeilinger for their work on entanglement, are marks of the extraordinary achievements in quantum optics. We hope that this book provides an accessible introduction to this exciting subject.

Our aim was to write an elementary book on the essentials of quantum optics directed to an audience of upper-level undergraduates, assumed to have suffered through a course in quantum mechanics, and for first- or second-year graduate students interested in eventually pursuing research in this area. The material we introduce is not simple and will be a challenge for undergraduates and beginning graduate students, but we have tried to use the most straightforward approaches. Nevertheless, there are parts of the text that the reader will find more challenging than others. The problems at the end of each chapter similarly have a range of difficulty. The presentation is almost entirely concerned with the quantized electromagnetic field and its effects on atoms, and how nonclassical light behaves. A secondary aim of this book is to connect quantum optics with the subject of quantum information processing, which builds on much that we have learnt in quantum optics about the exploitation of quantum coherence. We will focus, however, on quantum optics but provide signposts to places that explore quantum information processing in more detail.

Topics covered are: single-mode field quantization in a cavity, quantization of multimode fields, the issue of the quantum phase, coherent states, quasi-probability distributions in phase space, atom–field interactions, the Jaynes–Cummings model, quantum coherence theory, beam splitters and interferometers, nonclassical field states with squeezing and so on, tests of local realism with entangled photons from down-conversion, experimental realizations of cavity quantum electrodynamics, trapped ions and so on, issues regarding decoherence, and some applications to quantum information processing, particularly quantum cryptography. The book includes many homework problems for each chapter and bibliographies for further reading. Many of the problems involve computational work, some more extensively than others.

1.2 History

In this chapter we briefly survey the historical development of our ideas of optics and photons. A detailed account can be found in the "Historical Introduction," for example, in the 7th edition of Born and Wolf [1]. A most readable account of the development of quantum ideas can be found in an accessible book by Whitaker [2]. An article by A. Muthukrishnan, M. O. Scully and M. S. Zubairy [3] ably surveys the historical development of our ideas on light and photons in a most readable manner.

The ancient world was already wrestling with the nature of light as rays. By the seventeenth century, the two rival concepts of waves and corpuscles were well established. Maxwell, in the second half of the nineteenth century, laid the foundations of modern field theory, with a detailed account of light as electromagnetic waves. At that point, classical physics seemed triumphant, with "minor" worries about the nature of black-body radiation and the photoelectric effect. These, of course, were the seeds of the quantum revolution. Planck, an inherently conservative theorist, was led rather reluctantly (it seems) to propose that thermal radiation was emitted and absorbed in discrete quanta in order to explain the spectra of thermal bodies. It was Einstein who generalized this idea so that these new quanta represented the light itself rather than the processes of absorption and emission, and was able to describe how matter and radiation could come into equilibrium (introducing on the way the idea of stimulated emission), and how the photoelectric effect could be explained. By 1913, Bohr applied the basic idea of quantization to atomic dynamics and was able to predict the positions of atomic spectral lines.

Gilbert Lewis, a chemist, coined the word "photon" well after the light quanta idea itself was introduced. In 1926, Lewis said:

It would seem appropriate to speak of one of these hypothetical entities as a particle of light, a corpuscle of light, a light quantum, or light quant, if we are to assume that it spends only a minute fraction of its existence as a carrier of radiant energy, while the rest of the time it remains as an important structural element within the atom ... I therefore take the liberty of proposing for this hypothetical new atom, which is not light but plays an important part in every process of radiation, the name photon. [4]

Clearly, Lewis's idea and ours are rather distantly connected!

De Broglie, in a remarkable leap of imagination, generalized what we knew about light quanta, exhibiting wave and particle properties to matter itself. Heisenberg, Born, Schrödinger, and Dirac laid the foundations of quantum mechanics in an amazingly short period from 1925 to 1926. They gave us the whole machinery we still use: representations, quantum-state evolution, unitary transformations, perturbation theory, and more. The intrinsic probabilistic nature of quantum mechanics was uncovered by Max Born, who proposed the idea of probability amplitudes, which allowed a fully quantum treatment of interference.

Fermi and Dirac, pioneers of quantum mechanics, were also among the first to address the question of how quantized light interacts with atomic sources and propagates. Fermi's *Reviews of Modern Physics* article in the 1930s, based on lectures he gave in Ann Arbor, summarizes what was known at that time within the context of nonrelativistic quantum electrodynamics in the Coulomb gauge. His treatment of interference (especially Lipmann fringes) still repays reading today. It is useful to quote Willis Lamb in this context:

Begin by deciding how much of the universe needs to be brought into the discussion. Decide what normal modes are needed for an adequate treatment. Decide how to model the light sources and work out how they drive the system. [5]

This statement sums up the approach we will take throughout this book.

Weisskopf and Wigner applied the newly developed ideas of nonrelativistic quantum mechanics to the dynamics of spontaneous emission and resonance fluorescence, predicting the exponential law for excited-state decay. This work already exhibited the self-energy problems, which were to plague quantum electrodynamics for the next 20 years until the development of the renormalization program by Schwinger, Feynman, Tomonaga, and Dyson. The observation of the anomalous magnetic moment of the electron by Kusch, and of radiative-level shifts of atoms by Lamb and Retherford, were the highlights of this era. The interested reader will find the history of this period very ably described by Schweber in his magisterial account of QED [6]. This period of research demonstrated the importance of considering the vacuum as a field, which had observable consequences. In a remarkable development in the late 1940s, triggered by the observation that colloids were more stable than expected (from considerations of van der

Waals interactions), Casimir showed that long-range intermolecular forces were intrinsically quantum electrodynamic. He linked them to the idea of zero-point motion of the field and showed that metal plates in vacuum attract as a consequence of such zero-point motion.

Einstein had continued his study of the basic nature of quantum mechanics and in 1935, in a remarkable paper with Podolsky and Rosen, was able to show how peculiar quantum correlations were. The ideas in this paper were to explode into one of the most active parts of modern physics, with the development by Bohm and Bell of concrete predictions of the nature of these correlations; this laid the foundations of what was to become the new subject of quantum information processing.

Optical coherence had been investigated for many years using amplitude interference: a first-order correlation. Hanbury Brown and Twiss in the 1950s worked on intensity correlations as a tool in stellar interferometry, and showed how thermal photon detection events were "bunched." This led to the development of the theory of photon statistics and photon counting, and to the beginnings of quantum optics as a separate subject. At the same time as ideas of photon statistics were being developed, researchers had begun to investigate coherence in light–matter interactions. Radio-frequency spectroscopy had already been initiated with atomic beams, in the work of Rabi, Ramsey, and others. Sensitive optical pumping probes of light interaction with atoms were developed in the 1950s and 1960s by Kastler, Brossel, Cohen-Tannoudji, Series, Dodd, and others.

By the early 1950s, Townes and his group, and Basov and Prokhorov, had developed molecular microwave sources of radiation: the new masers, based on precise initial-state preparation, population inversion, and stimulated emission. Ed Jaynes, in the 1950s onwards, played a major role in studies of whether quantization played a role in maser operation (and this set the stage for much later work on fully quantized atom–field coupling, in what became known as the Jaynes–Cummings model). Extending the maser idea to the optical regime and the development of lasers of course revolutionized modern physics and technology.

Glauber, Wolf, Sudarshan, Mandel, Klauder, and many others developed a quantum theory of coherence based on coherent states and photo-detection. Coherent states allowed us to describe the behavior of light in phase space, using the quasi-probabilities developed much earlier by Wigner and others.

For several years after the development of the laser, there were no tunable sources: researchers interested in the details of atom–light or molecule–light interactions had to rely on molecular chance resonances. Nevertheless, this led to the beginning of the study of coherent interactions and coherent transients such as photon echoes, self-induced transparency, optical nutation, and so on (well described in the standard monograph by Allen and Eberly). Tunable lasers became available in the early 1970s, and the dye laser in particular transformed precision studies in quantum optics and laser spectroscopy. Resonant interactions, coherent transients, and the like became much more straightforward to study and led to the beginnings of quantum optics proper, as we now understand it: for the first time we were able to study the dynamics of single atoms interacting with light in a nonperturbative manner. Stroud and his group initiated studies of resonance fluorescence with the observation of the splitting of resonance fluorescence spectral lines into component parts by the coherent driving predicted earlier by Mollow. Mandel, Kimble, and others demonstrated how the resonance fluorescence light was antibunched, a feature studied by a number of theorists including Walls, Carmichael, Cohen-Tannoudji, Mandel, and Kimble. The observation of antibunching, and the associated (but inequivalent) sub-Poissonian photon statistics, laid the foundation of the study of "nonclassical light." During the 1970s, several experiments explored the nature of photons: their indivisibility and the buildup of interference at the single photon level. Laser cooling rapidly developed in the 1980s and 1990s, and allowed the preparation of states of matter under precise control. Indeed, this has become a major subject in its own right and we have taken the decision here to exclude laser cooling from this book.

Following the development of high-intensity pulses of light from lasers, a whole set of nonlinear optical phenomena were investigated, starting with the pioneering work in Ann Arbor by Franken and co-workers. Harmonic generation, parametric down-conversion, and other phenomena were demonstrated. For the most part, none of this early work on nonlinear optics required field quantization and quantum optics proper for its description. But there were early signs that some could well do so: quantum nonlinear optics was really initiated by Burnham and Weinberg's study (see Chapter 9) of unusual nonclassical correlations in down-conversion. In the hands of Mandel, Zeilinger, and many others, these correlations in down-conversion became the fundamental tool used to uncover fundamental insights into quantum optics.

Until the 1980s, essentially all light fields investigated had phase-independent noise; this changed with the production of squeezed light sources with phase-sensitive noise. These squeezed light sources enabled us to investigate Heisenberg uncertainty relations for light fields. Again, parametric down-conversion proved to be the most effective tool to generate such unusual light fields.

Quantum opticians realized quite early that were atoms to be confined in resonators, then atomic radiative transition dynamics could be dramatically changed. Purcell, in a remarkable paper in 1946 within the context of magnetic resonance, had already predicted that spontaneous emission rates, previously thought of as pretty immutable, were in fact modified by enclosing the source atom within a cavity whose mode structure and densities are significantly different from those of free space. Putting atoms within resonators or close to mirrors became possible at the end of the 1960s. By the 1980s, the theorists' dream of studying single atoms interacting with single modes of the electromagnetic field became possible in the hands of Haroche, Walther, Kimble, and many others. At this point, the transition dynamics becomes wholly reversible, as the atom coherently exchanges excitation with the field, until coherence is eventually lost through a dissipative "decoherence" process. This dream is called the Jaynes–Cummings model, after its proposers, and forms a basic building block of quantum optics (and is discussed in detail in this book).

New fundamental concepts in information processing, leading to quantum cryptography and quantum computation, have been developed in recent years by Feynman, Benioff, Deutsch, Jozsa, Bennett, Brassard, Ekert, Shor, and others. Instead of using classical bits that can represent either the values 0 or 1, the basic unit of a quantum computer is a quantum mechanical two-level system (qubit) that can exist in coherent superpositions of the logical values 0 and 1. A set of n qubits can then be in a superposition of up to 2^n different states, each representing a binary number. Were we able to control and manipulate, say, 1500 qubits, we could access more states than there are particles in the visible universe. Computations are implemented by unitary transformations, which act on all states of a superposition simultaneously. Quantum gates form the basic units from which these unitary transformations are built up. In related developments, absolutely secure encryption can be guaranteed by using quantum sources of light.

The use of the quantum mechanical superpositions and entanglement results in a high degree of parallelism, which can increase the speed of computation exponentially. A number of problems which cannot feasibly be tackled on a classical computer can be solved efficiently on a quantum computer. In 1994, a quantum algorithm was discovered by Peter Shor that allows the solution of a practically important problem, namely factorization, with such an exponential increase of speed. Subsequently, possible experimental realizations of a quantum computer have been proposed, for example in linear ion traps and nuclear magnetic resonance schemes. Presently, we are at the stage where quantum gates have been demonstrated in a number of implementations. Quantum computation is closely related to quantum cryptography and quantum communication. Basic experiments demonstrating the in-principle possibility of these ideas have been carried out in various laboratories.

The linear ion trap is one of the most promising systems for quantum computation and is one we study in this book in detail. The quantum state preparation (laser cooling and optical pumping) in this system is a well-established technique, as is the state measurement by electron shelving and fluorescence. Singly charged ions of an atom such as calcium or beryllium are trapped and laser cooled to micro-Kelvin temperatures, where they form a string lying along the axis of a linear radio-frequency (r.f.) Paul trap. The internal state of any one ion can be exchanged with the quantum state of motion of the whole string. This can be achieved by illuminating the ion with a pulse of laser radiation at a frequency tuned below the ion's internal resonance by the vibrational frequency of one of the normal modes of oscillation of the string. This couples single phonons into and out of the vibrational mode. The motional state can then be coupled to the internal state of another ion by directing the laser onto the second ion and applying a similar laser pulse. In this way, general transformations of the quantum state of all the ions can be generated. The ion trap has several features to recommend it. It can achieve processing on quantum bits without the need for any new technological breakthroughs, such as micro-fabrication techniques or new cooling methods. The state of any ion can be measured and re-prepared many times without problem, which is an important feature for implementing quantum error-correction protocols.

Trapped atoms or ions can be strongly coupled to an electromagnetic field mode in a cavity, which permits the powerful combination of quantum processing and long-distance quantum communication. This suggests ways in which we may construct quantum memories. These systems can in principle realize a quantum processor larger than any which could be thoroughly simulated by classical computing, but the decoherence generated by dephasing and spontaneous emission is a formidable obstacle.

Entangled states are the key ingredient for certain forms of quantum cryptography and for quantum teleportation. Entanglement is also responsible for the power of quantum computing, which, under ideal conditions, can accomplish certain tasks exponentially faster than any classical computer. A deeper understanding of the role of quantum entanglement in quantum information theory will allow us to improve existing applications and to develop new methods of quantum information manipulation. These are all described in later chapters.

What then is the future of quantum optics? It underpins a great deal of laser science and novel atomic physics. Controlling and manipulating the quantum noise in large-scale interferometers is at the heart of the LIGO project, which detects gravitational waves. It may even be the vehicle by which we can realize a whole new technology, whereby quantum mechanics permits the processing and transmission of information in wholly novel ways. But, of course, whatever we may predict now to emerge will be confounded by the unexpected: the field remains an adventure repeatedly throwing up the unexpected.

1.3 Contents of This Book

The layout of this book is as follows. In Chapter 2, we show how the electromagnetic field can be quantized in terms of harmonic oscillators representing modes of the electromagnetic field, with states describing how many excitations (photons) are present in each normal mode. In Chapter 3, we introduce the coherent states, superposition states carrying phase information. In Chapter 4, we describe how light and matter interact. Chapter 5 quantifies our notions of coherence in terms of optical field correlation

functions. Chapter 6 introduces simple optical elements such as beam
splitters and interferometers, which manipulate the states of light.
Chapter 7 describes those nonclassical states whose basic properties are
dictated by their fundamental quantum nature. Spontaneous emission and
decay in an open environment are discussed in Chapter 8. Chapter 9
describes how quantum optical sources of radiation can be used to provide
tests of fundamental quantum mechanics, including tests of nonlocality
and Bell inequalities. Chapter 10 discusses how atoms confined in cavities
and trapped laser-cooled ions can be used to study basic interaction
phenomena. Chapter 11 applies what we have learnt to the newly emerging
problems of quantum information processing. Appendices set out some
mathematical ideas needed within the main body of the text. Throughout,
we have tried to illustrate the ideas we have been developing through
homework problems.

References

[1] M. Born and E. Wolf, *Principles of Optics*, 7th edition (Cambridge:
     Cambridge University Press, 1999).
[2] A. Whitaker, *Einstein, Bohr and the Quantum Dilemma*, 2nd edition
     (Cambridge: Cambridge University Press, 2006).
[3] A. Muthukrishnan, M. O. Scully, and M. S. Zubairy, *Opt. Photon.
     News Trends* **3**, No. 1 (October 2003).
[4] G. N. Lewis, *Nature* **118**, 874 (1926).
[5] W. E. Lamb Jr., *Appl. Phys. B* **66**, 77 (1995).
[6] S. S. Schweber, *QED and the Men Who Made It: Dyson, Feynman,
     Schwinger and Tomonaga* (Princeton, NJ: Princeton University Press,
     1994).

Suggestions for Further Reading

*Many books on quantum optics exist, most taking the story much further
than we do, in more specialized monographs.*

J. R. Klauder and E. C. G. Sudarshan, *Fundamentals of Quantum Optics*
     (New York: W. A. Benjamin, 1968).
W. H. Louisell, *Quantum Statistical Properties of Radiation* (New York:
     Wiley, 1973).
H. M. Nussenzveig, *Introduction to Quantum Optics* (London: Gordon &
     Breach, 1973).
M. Sargent III, M. O. Scully, and W. E. Lamb Jr., *Laser Physics* (Reading,
     MA: Addison-Wesley, 1974).
L. Allen and J. H. Eberly, *Optical Resonance and Two Level Atoms*
     (New York: Wiley, 1975 and Mineola, NY: Dover, 1987).

H. Haken, *Light, Volume I: Waves, Photons, and Atoms* (Amsterdam: North Holland, 1981).

C. Cohen-Tannoudji, J. Dupont-Roc, and G. Grynberg, *Photons and Atoms* (New York: Wiley-Interscience, 1989).

M. Weissbluth, *Photon–Atom Interactions* (New York: Academic Press, 1989).

B. W. Shore, *The Theory of Coherent Atomic Excitation* (New York: Wiley-Interscience, 1990).

P. Meystre and M. Sargent III, *Elements of Quantum Optics*, 2nd edition (Berlin: Springer-Verlag, 1991).

J. Peřina, *Quantum Statistics of Linear and Nonlinear Optical Phenomena*, 2nd edition (Dordrecht: Kluwer, 1991).

C. Cohen-Tannoudji, J. Dupont-Roc, and G. Grynberg, *Atom–Photon Interactions* (New York: Wiley-Interscience, 1992).

G. J. Milburn and D. F. Walls, *Quantum Optics* (Berlin: Springer-Verlag, 1994).

W. Vogel and D.-G. Welsch, *Lectures in Quantum Optics* (Berlin: Akademie Verlag, 1994).

L. Mandel and E. Wolf, *Optical Coherence and Quantum Optics* (Cambridge: Cambridge University Press, 1995).

S. M. Barnett and P. M. Radmore, *Methods in Theoretical Quantum Optics* (Oxford: Oxford University Press, 1997).

U. Leonhardt, *Measuring the Quantum State of Light* (Cambridge: Cambridge University Press, 1997).

M. O. Scully and M. S. Zubairy, *Quantum Optics* (Cambridge: Cambridge University Press, 1997).

V. Peřinová, A. Lukš, and J. Peřina, *Phase in Optics* (Singapore: World Scientific, 1998).

P. Ghosh, *Testing Quantum Mechanics on New Ground* (Cambridge: Cambridge University Press, 1999).

Y. Yamamoto and A. İmamoğlu, *Mesoscopic Quantum Optics* (New York: Wiley-Interscience, 1999).

R. Loudon, *The Quantum Theory of Light*, 3rd edition (Oxford: Oxford University Press, 2000).

M. Orszag, *Quantum Optics: Including Noise, Trapped Ions, Quantum Trajectories, and Decoherence* (Berlin: Springer, 2000).

R. R. Puri, *Mathematical Methods of Quantum Optics* (Berlin: Springer, 2001).

W. P. Schleich, *Quantum Optics in Phase Space* (Berlin: Wiley-VCH, 2001).

V. V. Dodonov and V. I. Man'ko (Eds.), *Theory of Nonclassical States of Light* (London: Taylor & Francis, 2003).

H. Bachor and T. H. Ralph, *A Guide to Experiments in Quantum Optics*, 2nd edition (Berlin: Wiley-VCH, 2004).

M. Fox, *Quantum Optics, An Introduction* (Oxford: Oxford University Press, 2006).

J. C. Garrison and R. Y. Chiao, *Quantum Optics* (Oxford: Oxford University Press, 2008).

A. B. Klimov and S. M. Chumkov, *A Group-Theoretical Approach to Quantum Optics* (Weinheim: Wiley-VCH, 2009).

G. Grynberg, A. Aspect, and C. Fabre, *Introduction to Quantum Optics* (Cambridge: Cambridge University Press, 2010).

P. Kok and B. W. Lovett, *Introduction to Quantum Optical Information Processing* (Cambridge: Cambridge University Press, 2010).

P. R. Berman and V. S. Malinovsky, *Laser Spectroscopy and Quantum Optics* (Princeton, NJ: Princeton University Press, 2011).

G. S. Agarwal, *Quantum Optics* (Cambridge: Cambridge University Press, 2013).

An undergraduate-level textbook with quantum optics orientation that includes a section of quantum optical-based laboratories for illustrating foundational aspects of quantum mechanics is

M. Beck, *Quantum Mechanics, Theory and Experiment* (Oxford: Oxford University Press, 2012).

A useful reprint collection of papers on coherent states, including the early work by Glauber, Klauder, and others, is the following:

J. R. Klauder and B.-S. Skagerstam (Eds.), *Coherent States* (Singapore: World Scientific, 1985).

The present book deals exclusively with the quantum optics of light, but there is a closely related field, quantum atom optics, that relies on some of the same concepts. A recent book on this topic (with many references) is

T. Byrnes and E. O. Ilo-Okeke, *Quantum Atom Optics, Theory and Applications to Quantum Technology* (Cambridge: Cambridge University Press, 2021).

2

Field Quantization

In this chapter we present a discussion of the quantization of the electromagnetic field and discuss some of its properties, with particular regard to the interpretation of the photon as an elementary excitation of a normal mode of the field. We start with the case of a single-mode field confined by conducting walls in a one-dimensional cavity and later generalize to multimode fields in free space. The photon number states are introduced and we discuss the fluctuations of the field observables with respect to these states. Finally, we discuss the problem of the quantum description of the phase of the quantized electromagnetic field.

2.1 Quantization of a Single-Mode Field

We begin with the rather simple but very important case of a radiation field confined to a one-dimensional cavity along the z-axis with perfectly conducting walls at $z = 0$ and $z = L$, as shown in Fig. 2.1.

The electric field must therefore vanish on the boundaries and will take the form of a standing wave. We assume there are no sources of radiation (i.e. no currents or charges) nor any dielectric media in the cavity. The field is assumed to be polarized along the x-direction, $\mathbf{E}(\mathbf{r}, t) = \mathbf{e}_x E_x(z, t)$, where $\mathbf{e}_x$ is a unit polarization vector. Maxwell's equations without sources are, in SI units,

$$\nabla \times \mathbf{E} = -\frac{\partial \mathbf{B}}{\partial t}, \tag{2.1}$$

$$\nabla \times \mathbf{B} = \mu_0 \varepsilon_0 \frac{\partial \mathbf{E}}{\partial t}, \tag{2.2}$$

$$\nabla \bullet \mathbf{B} = 0, \tag{2.3}$$

$$\nabla \bullet \mathbf{E} = 0. \tag{2.4}$$

A single-mode field satisfying Maxwell's equations and the boundary conditions is given by

$$E_x(z, t) = \left(\frac{2\omega^2}{V \varepsilon_0}\right)^{1/2} q(t) \sin(kz), \tag{2.5}$$

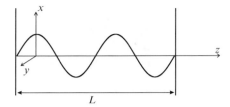

Fig. 2.1 Cavity with perfectly conducting walls located at $z = 0$ and $z = L$. The electric field is polarized along the x-direction.

where ω is the frequency of the mode and k is the wave number, which is related to the frequency according to $k = \omega/c$. The boundary condition at $z = L$ yields the allowed frequencies $\omega_m = c(m\pi/L)$, $m = 1, 2, \ldots$. We assume that ω in Eq. (2.5) is one of these frequencies and ignore the rest for now. V in Eq. (2.5) is the effective volume of the cavity and is a time dependent factor that will act as a canonical position. The magnetic field in the cavity, from Eq. (2.5) and Eq. (2.2), is $\mathbf{B}(\mathbf{r}, t) = \mathbf{e}_y B_y(z, t)$, where

$$B_y(z, t) = \left(\frac{\mu_0 \varepsilon_0}{k}\right) \left(\frac{2\omega^2}{V\varepsilon_0}\right)^{1/2} \dot{q}(t)\cos(kz). \tag{2.6}$$

Here, $\dot{q}(t)$ will play the role of a canonical momentum for a "particle" without the mass factor that is $p(t) = \dot{q}(t)$.

The classical field energy, or Hamiltonian H, of the single-mode field is given by

$$\begin{aligned} H &= \frac{1}{2}\int dV \left[\varepsilon_0 \mathbf{E}^2(\mathbf{r}, t) + \frac{1}{\mu_0}\mathbf{B}^2(\mathbf{r}, t)\right] \\ &= \frac{1}{2}\int dV \left[\varepsilon_0 E_x^2(z, t) + \frac{1}{\mu_0}B_y^2(z, t)\right]. \end{aligned} \tag{2.7}$$

Using Eqs. (2.5) and (2.6), it is straightforward to show (and is left as an exercise) that

$$H = \frac{1}{2}(p^2 + \omega^2 q^2), \tag{2.8}$$

from which it is apparent that a single-mode field is formally equivalent to a harmonic oscillator apart from the inclusion of mass, where the electric and magnetic fields, apart from some scale factors, play the roles of canonical position and momentum. The quantities p and q have the units momentum/(mass)$^{1/2}$ and length $\times$ (mass)$^{1/2}$, respectively. If we make the replacements $p \to p/\sqrt{m}$ and $q \to q\sqrt{m}$ where m is some mass, we trivially recover the usual Hamiltonian for a particle of said mass in a harmonic potential.

Every elementary textbook on quantum mechanics discusses the quantization of the one-dimensional harmonic oscillator. Here we take the approach that having identified the canonical variables q and p for the classical system, we simply use the correspondence rule to replace them by their operator equivalents $\hat{q}$ and $\hat{p}$, where operators will be distinguished from c-numbers by the caret. These operators must satisfy the canonical commutation relation

$$[\hat{q},\hat{p}] = i\hbar\hat{I} . \tag{2.9}$$

(Note that the commutation relation is invariant under the transformations $\hat{p} \rightarrow \hat{p}/\sqrt{m}$ and $\hat{q} \rightarrow \hat{q}\sqrt{m}$.) Henceforth we follow custom and drop the identity operator $\hat{I}$ and write $[\hat{q},\hat{p}] = i\hbar$. Then the electric and magnetic fields of the single mode become the operators

$$\hat{E}_x(z,t) = \left(\frac{2\omega^2}{V\varepsilon_0}\right)^{1/2} \hat{q}(t)\,\sin(kz) \tag{2.10}$$

and

$$\hat{B}_y(z,t) = \left(\frac{\mu_0\varepsilon_0}{k}\right)\left(\frac{2\omega^2}{V\varepsilon_0}\right)^{1/2} \hat{p}(t)\,\cos(kz) \tag{2.11}$$

respectively. The Hamiltonian becomes

$$\hat{H} = \frac{1}{2}(\hat{p}^2 + \omega^2\hat{q}^2). \tag{2.12}$$

The operators $\hat{q}$ and $\hat{p}$ are Hermitian and therefore correspond to observable quantities. However, it is convenient, and traditional, to introduce the non-Hermitian (and therefore non-observable) annihilation ($\hat{a}$) and creation ($\hat{a}^\dagger$) operators through the combinations

$$\hat{a} = (2\hbar\omega)^{-1/2}(\omega\hat{q} + i\hat{p}), \tag{2.13}$$

$$\hat{a}^\dagger = (2\hbar\omega)^{-1/2}(\omega\hat{q} - i\hat{p}). \tag{2.14}$$

The electric and magnetic field operators then become, respectively,

$$\hat{E}_x(z,t) = \mathcal{E}_0(\hat{a} + \hat{a}^\dagger)\sin(kz), \tag{2.15}$$

$$\hat{B}_y(z,t) = \mathcal{B}_0\frac{1}{i}(\hat{a} - \hat{a}^\dagger)\cos(kz), \tag{2.16}$$

where $\mathcal{E}_0 = (\hbar\omega/\varepsilon_0 V)^{1/2}$ and $\mathcal{B}_0 = (\mu_0/k)(\varepsilon_0\hbar\omega^3/V)^{1/2}$ represent, respectively, the electric and magnetic fields "per photon." The quotation marks indicate that this is not exactly correct since, as we shall show, the average of these fields for a definite number of photons is zero. Nevertheless, they are useful measures of the fluctuations of the quantized field. Operators $\hat{a}$ and $\hat{a}^\dagger$ satisfy the commutation relation

$$[\hat{a},\hat{a}^\dagger] = 1 \tag{2.17}$$

and, as a result, the Hamiltonian operator takes the form

$$\hat{H} = \hbar\omega\left(\hat{a}^\dagger\hat{a} + \frac{1}{2}\right). \tag{2.18}$$

So far, we have said nothing of the time dependence of the operators $\hat{a}$ and $\hat{a}^\dagger$. For an arbitrary operator $\hat{O}$ having no explicit time dependence, Heisenberg's equation reads

$$\frac{d\hat{O}}{dt} = \frac{i}{\hbar}[\hat{H}, \hat{O}]. \tag{2.19}$$

For the annihilation operator $\hat{a}$ this becomes

$$\begin{aligned}
\frac{d\hat{a}}{dt} &= \frac{i}{\hbar}[\hat{H}, \hat{a}] \\
&= \frac{i}{\hbar}\left[\hbar\omega\left(\hat{a}^\dagger\hat{a} + \frac{1}{2}\right), \hat{a}\right] \\
&= i\omega(\hat{a}^\dagger\hat{a}\hat{a} - \hat{a}\hat{a}^\dagger\hat{a}) \\
&= -i\omega[\hat{a}, \hat{a}^\dagger]\hat{a} = -i\omega\hat{a},
\end{aligned} \tag{2.20}$$

which has the solution

$$\hat{a}(t) = \hat{a}(0)e^{-i\omega t}. \tag{2.21}$$

By the same method, or simply by taking the Hermitian conjugate of Eq. (2.21), we have

$$\hat{a}^\dagger(t) = \hat{a}^\dagger(0)e^{i\omega t}. \tag{2.22}$$

An alternate way of obtaining these solutions is to write the formal solution to Eq. (2.19) in the form

$$\hat{O}(t) = e^{i\hat{H}t/\hbar}\hat{O}(0)e^{-i\hat{H}t/\hbar}, \tag{2.23}$$

and then to use the Baker–Hausdorff lemma [1], which for operators $\hat{A}$ and $\hat{B}$ is given by

$$e^{i\lambda\hat{A}}\hat{B}e^{-i\lambda\hat{A}} = \hat{B} + i\lambda[\hat{A}, \hat{B}] + \frac{(i\lambda)^2}{2!}[\hat{A}, [\hat{A}, \hat{B}]] + \frac{(i\lambda)^3}{3!}[\hat{A}, [\hat{A}, [\hat{A}, \hat{B}]]] + \cdots, \tag{2.24}$$

to obtain

$$\begin{aligned}
\hat{O}(t) = \ &\hat{O}(0) + \frac{it}{\hbar}[\hat{H}, \hat{O}(0)] \\
&+ \frac{1}{2!}\left(\frac{it}{\hbar}\right)^2[\hat{H}, [\hat{H}, \hat{O}(0)]] + \cdots \\
&+ \frac{1}{n!}\left(\frac{it}{\hbar}\right)^n[\hat{H}, [\hat{H}, [\hat{H}, \ldots[\hat{H}, \hat{O}(0)]]]] + \cdots.
\end{aligned} \tag{2.25}$$

For the operator $\hat{a}$, this results in

$$\hat{a}(t) = \hat{a}(0)\left[1 - i\omega t - \frac{\omega^2 t^2}{2!} + i\frac{\omega^3 t^3}{3!} + \cdots\right]$$

$$= \hat{a}(0)e^{-i\omega t}. \tag{2.26}$$

The use of this method of solution may seem analogous to the use of a sledgehammer to crack a nut, but will turn out to be quite useful later when we take up cases involving nonlinear interactions.

The operator product $\hat{a}^\dagger\hat{a}$ has a special significance and is called the number operator, which we denote as $\hat{n}$. We let $|n\rangle$ denote an energy eigenstate of the single-mode field with energy eigenvalue E_n such that

$$\hat{H}|n\rangle = \hbar\omega\left(\hat{a}^\dagger\hat{a} + \frac{1}{2}\right)|n\rangle = E_n|n\rangle. \tag{2.27}$$

If we multiply Eq. (2.27) by $\hat{a}^\dagger$, then we can generate a new eigenvalue equation

$$\hbar\omega\left(\hat{a}^\dagger\hat{a}^\dagger\hat{a} + \frac{1}{2}\hat{a}^\dagger\right)|n\rangle = E_n\hat{a}^\dagger|n\rangle. \tag{2.28}$$

Using the commutation relation of Eq. (2.17), we can rewrite this as

$$\hbar\omega\left[(\hat{a}^\dagger\hat{a}\hat{a}^\dagger - \hat{a}^\dagger) + \frac{1}{2}\hat{a}^\dagger\right]|n\rangle = E_n\hat{a}^\dagger|n\rangle, \tag{2.29}$$

or

$$\hbar\omega\left(\hat{a}^\dagger\hat{a} + \frac{1}{2}\right)(\hat{a}^\dagger|n\rangle) = (E_n + \hbar\omega)(\hat{a}^\dagger|n\rangle), \tag{2.30}$$

which is the eigenvalue problem for the eigenstate $(\hat{a}^\dagger|n\rangle)$ with energy eigenvalue $E_n + \hbar\omega$. It should be clear now why $\hat{a}^\dagger$ is called the creation operator: it creates a "quantum" of energy $\hbar\omega$. One could also say, rather loosely, that a "photon" of energy $\hbar\omega$ is created by $\hat{a}^\dagger$. Similarly, if we multiply Eq. (2.27) by the operator $\hat{a}$ and use the commutation relation, we obtain

$$\hat{H}(\hat{a}|n\rangle) = (E_n - \hbar\omega)(\hat{a}|n\rangle), \tag{2.31}$$

where it is evident that the operator $\hat{a}$ destroys or annihilates one quanta of energy or one photon, the eigenstate $(\hat{a}|n\rangle)$ possessing the energy eigenvalue $E_n - \hbar\omega$. Evidently, repeating the procedure on Eq. (2.30) would result in the lowering of the energy eigenvalue by integer multiples of $\hbar\omega$. But the energy of the harmonic oscillator must always be positive, so there must be a lowest-energy eigenvalue, $E_0 > 0$, with corresponding eigenstate $|0\rangle$ such that

$$\hat{H}(\hat{a}|0\rangle) = (E_0 - \hbar\omega)(\hat{a}|0\rangle) = 0, \tag{2.32}$$

because

$$\hat{a}|0\rangle = 0. \tag{2.33}$$

Thus, the eigenvalue problem for the ground state is

$$\hat{H}|0\rangle = \hbar\omega\left(\hat{a}^\dagger\hat{a} + \frac{1}{2}\right)|0\rangle = \frac{1}{2}\hbar\omega|0\rangle, \tag{2.34}$$

so that the lowest-energy eigenvalue is the so-called zero-point energy $\hbar\omega/2$. Since $E_{n+1} = E_n + \hbar\omega$, the energy eigenvalues are

$$E_n = \hbar\omega\left(n + \frac{1}{2}\right), \qquad n = 0, 1, 2, \ldots. \tag{2.35}$$

(These energy levels are pictured, against the harmonic oscillator potential, in Fig. 2.2.)

Thus, for the number operator $\hat{n} = \hat{a}^\dagger\hat{a}$, we have

$$\hat{n}|n\rangle = n|n\rangle. \tag{2.36}$$

These number states must be normalized according to $\langle n|n\rangle = 1$. For the state $\hat{a}|n\rangle$ we have

$$\hat{a}|n\rangle = c_n|n - 1\rangle, \tag{2.37}$$

where c_n is a constant to be determined. Then the inner product of $\hat{a}|n\rangle$ with itself is

$$\begin{aligned}((\langle n|\hat{a}^\dagger)(\hat{a}|n\rangle)) = \langle n|\hat{a}^\dagger\hat{a}|n\rangle &= n \\ &= \langle n - 1|c_n^*c_n|n - 1\rangle = |c_n^2|,\end{aligned} \tag{2.38}$$

and thus $|c_n^2| = n$, so we can take $c_n = \sqrt{n}$. Thus

$$\hat{a}|n\rangle = \sqrt{n}|n - 1\rangle. \tag{2.39}$$

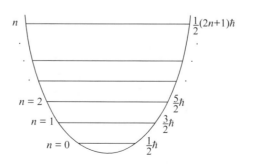

Fig 2.2
The energy levels of a harmonic oscillator of frequency ω.

Similarly, we can show that

$$\hat{a}^{\dagger}|n\rangle = \sqrt{n+1}|n+1\rangle. \tag{2.40}$$

From this last result it is straightforward to show that the number states $|n\rangle$ may be generated from the ground state $|n\rangle$ by the repeated action of the creation operator $\hat{a}^{\dagger}$:

$$|n\rangle = \frac{(\hat{a}^{\dagger})^{n}}{\sqrt{n!}}|0\rangle. \tag{2.41}$$

Because $\hat{H}$ and $\hat{n}$ are Hermitian operators, states of different number are orthogonal, that is $\langle n'|n\rangle = \delta_{nn'}$, and furthermore the number states form a complete set,

$$\sum_{n=0}^{\infty}|n\rangle\langle n| = 1. \tag{2.42}$$

The only nonvanishing matrix elements of the annihilation and creation operators are

$$\langle n-1|\hat{a}|n\rangle = \sqrt{n}\langle n-1|n-1\rangle = \sqrt{n}, \tag{2.43}$$

$$\langle n+1|\hat{a}^{\dagger}|n\rangle = \sqrt{n+1}\langle n+1|n+1\rangle = \sqrt{n+1}. \tag{2.44}$$

2.2 Quantum Fluctuations of a Single-Mode Field

The number state $|n\rangle$ is a state of well-defined energy but it is not a state of well-defined electric field, since

$$\langle n|\hat{E}_{x}(z,t)|n\rangle = \mathcal{E}_{0}\sin(kz)[\langle n|\hat{a}|n\rangle + \langle n|\hat{a}^{\dagger}|n\rangle] = 0, \tag{2.45}$$

that is the mean field is zero. However, the mean of the square of this field, which contributes to the energy density, is not zero:

$$\begin{aligned}
\langle n|\hat{E}_{x}^{2}(z,t)|n\rangle &= \mathcal{E}_{0}^{2}\sin^{2}(kz)\langle n|\hat{a}^{\dagger^{2}} + \hat{a}^{2} + \hat{a}^{\dagger}\hat{a} + \hat{a}\hat{a}^{\dagger}|n\rangle \\
&= \mathcal{E}_{0}^{2}\sin^{2}(kz)\langle n|\hat{a}^{\dagger^{2}} + \hat{a}^{2} + 2\hat{a}^{\dagger}\hat{a} + 1|n\rangle \\
&= 2\mathcal{E}_{0}^{2}\sin^{2}(kz)\left(n + \frac{1}{2}\right).
\end{aligned} \tag{2.46}$$

The fluctuations in the electric field may be characterized by the variance

$$\left\langle\left(\Delta\hat{E}_{x}(z,t)\right)^{2}\right\rangle = \left\langle\hat{E}_{x}^{2}(z,t)\right\rangle - \left\langle\hat{E}_{x}(z,t)\right\rangle^{2}, \tag{2.47}$$

or by the standard deviation $\Delta E_{x} = \langle(\Delta E_{x}(z,t))^{2}\rangle^{1/2}$, which is sometimes referred to as the uncertainty of the field. For the number state $|n\rangle$ we have

$$\Delta E_x = \sqrt{2}\mathcal{E}_0 \sin(kz)\left(n + \frac{1}{2}\right)^{1/2}. \tag{2.48}$$

Note that even when $n = 0$, the field has fluctuations, the so-called vacuum fluctuations. Now the number states $|n\rangle$ are taken to represent a state of the field containing n photons. Yet, as we have seen, the average field is zero. This is all in accordance with the uncertainty principle, because the number operator $\hat{n}$ does not commute with the electric field:

$$[\hat{n}, \hat{E}_x] = \mathcal{E}_0 \sin(kz)(\hat{a}^\dagger - \hat{a}). \tag{2.49}$$

Thus, $\hat{n}$ and $\hat{E}_x$ are complementary quantities for which their respective uncertainties obey the inequality[*]

$$\Delta n \Delta E_x \geq \frac{1}{2}\mathcal{E}_0 |\sin(kz)||\langle \hat{a}^\dagger - \hat{a}\rangle|. \tag{2.50}$$

For a number state $|n\rangle$, the right-hand side vanishes but $\Delta n = 0$ as well. If the field were accurately known, then the number of photons would be uncertain. There is a connection here to the notion of the phase of the electric field. In classical physics, the amplitude and phase of a field can be simultaneously well defined. Not so in quantum mechanics. In fact, the history of the concept of a quantum phase operator is long and contentious, and we shall deal with this issue at length later. For now, we simply take a heuristic point of view, for which the phase is in some sense complementary to number, much in the way that time is complementary to energy. By analogy to the time–energy uncertainty relation, there should be a number–phase uncertainty relation of the form

$$\Delta n \Delta \phi \geq 1. \tag{2.51}$$

From this, one could argue that for a well-defined (accurately known) phase, the photon number is uncertain, whereas for a well-defined photon number, the phase is uncertain and, in fact, the phase is randomly distributed over the range $0 < \phi < 2\pi$. We shall examine the issue of the quantum phase in more detail in Section 2.7.

2.3 Quadrature Operators for a Single-Mode Field

When we explicitly include the time dependence of the electric field operator, we have

$$\hat{E}_x(z, t) = \mathcal{E}_0(\hat{a}e^{-i\omega t} + \hat{a}^\dagger e^{i\omega t}) \sin(kz), \tag{2.52}$$

---

[*] Recall that for operators $\hat{A}$ and $\hat{B}$ satisfying $[\hat{A}, \hat{B}] = i\hat{C}$, $\Delta A \Delta B \geq \frac{1}{2}|\langle \hat{C}\rangle|$.

where $\hat{a}(0) \equiv \hat{a}$ and $\hat{a}^\dagger(0) \equiv \hat{a}^\dagger$. We now introduce the so-called quadrature operators

$$\hat{X}_1 = \frac{1}{2}(\hat{a} + \hat{a}^\dagger), \tag{2.53}$$

$$\hat{X}_2 = \frac{1}{2i}(\hat{a} - \hat{a}^\dagger), \tag{2.54}$$

in terms of which the field operator may be recast as

$$\hat{E}_x(z,t) = 2\mathcal{E}_0 \sin(kz)[\hat{X}_1\cos(\omega t) + \hat{X}_2\sin(\omega t)]. \tag{2.55}$$

It is evident that $\hat{X}_1$ and $\hat{X}_2$ are associated with field amplitudes oscillating out of phase with each other by 90° (and hence they are in quadrature). Note that $\hat{X}_1$ and $\hat{X}_2$ are essentially the position and momentum operators obtainable from Eqs. (2.13) and (2.14) but scaled to be dimensionless. They satisfy the commutation relation

$$[\hat{X}_1, \hat{X}_2] = \frac{i}{2}, \tag{2.56}$$

from which it follows that

$$\langle(\Delta\hat{X}_1)^2\rangle\langle(\Delta\hat{X}_2)^2\rangle \geq \frac{1}{16}. \tag{2.57}$$

For the number states,

$$\langle n|\hat{X}_1^2|n\rangle = \frac{1}{4}\langle n|\hat{a}^2 + \hat{a}^{\dagger 2} + \hat{a}^\dagger\hat{a} + \hat{a}\hat{a}^\dagger|n\rangle$$

$$= \frac{1}{4}\langle n|\hat{a}^2 + \hat{a}^{\dagger 2} + 2\hat{a}^\dagger\hat{a} + 1|n\rangle \tag{2.58}$$

$$= \frac{1}{4}(2n+1),$$

and similarly

$$\langle n|\hat{X}_2^2|n\rangle = \frac{1}{4}(2n+1). \tag{2.59}$$

Thus, for a number state, the uncertainties in both quadratures are the same and furthermore the vacuum state $n = 0$ minimizes the uncertainty product, since

$$\langle(\Delta\hat{X}_1)^2\rangle_{vac} = \frac{1}{4} = \langle(\Delta\hat{X}_2)^2\rangle_{vac}. \tag{2.60}$$

Before moving on to multimode fields, we want to stress that the quanta of the single-mode cavity field are the excitations of energy in discrete amounts of $\hbar\omega$. These quanta, universally referred to as photons, are not localized particles (in field theory, there is no position operator for photons) but rather are spread out over the entire mode volume. This is in sharp contrast to the view of photons as "corpuscles" of light, as in the old quantum theory.

2.4 Multimode Fields

The results for the single-mode field confined to a cavity can be generalized to multimode radiation fields. We shall consider these fields to be in free space, where it is assumed that there are no sources of radiation and no charges, so that Eqs. (2.1)–(2.4) still hold. The electric and magnetic radiation fields may be given in terms of the vector potential $\mathbf{A}(\mathbf{r}, t)$ which satisfies the wave equation

$$\nabla^2 \mathbf{A} - \frac{1}{c^2} \frac{\partial^2 \mathbf{A}}{\partial t^2} = 0, \tag{2.61}$$

and the Coulomb gauge condition

$$\nabla \cdot \mathbf{A}(\mathbf{r}, t) = 0, \tag{2.62}$$

where

$$\mathbf{E}(\mathbf{r}, t) = -\frac{\partial \mathbf{A}(\mathbf{r}, t)}{\partial t} \tag{2.63}$$

and

$$\mathbf{B}(\mathbf{r}, t) = \nabla \times \mathbf{A}(\mathbf{r}, t). \tag{2.64}$$

Note that SI units are being used here.

We now imagine that free space can be modeled as a cubic cavity of side length L with perfectly reflecting walls. The idea here is that L should be very large compared to the dimensions of anything inside the cube with which the radiation could interact (e.g. atoms). We also assume that L is much larger than the wavelengths of the field. All physical results obtained from such a model should be independent of the size of the cavity as, after all calculations are done, we take $L \to \infty$.

The purpose of the cubical cavity is to allow us to impose periodic boundary conditions on the faces of the cube. For example, in the x-direction we shall require that plane waves satisfy the condition

$$e^{ik_x x} = e^{ik_x(x+L)}, \tag{2.65}$$

from which it follows that

$$k_x = \left(\frac{2\pi}{L}\right) m_x, \qquad\qquad m_x = 0, \pm 1, \pm 2, \ldots. \tag{2.66}$$

Similarly for the y- and z-directions we have

$$k_y = \left(\frac{2\pi}{L}\right)m_y, \qquad m_y = 0, \pm1, \pm2, \ldots, \qquad (2.67)$$

$$k_z = \left(\frac{2\pi}{L}\right)m_z, \qquad m_z = 0, \pm1, \pm2, \ldots. \qquad (2.68)$$

The wave vector has the components

$$\mathbf{k} = \frac{2\pi}{L}(m_x, m_y, m_x), \qquad (2.69)$$

and its magnitude is related to the frequency ω_k according to $k = \omega_k/c$. A set of integers (m_x, m_y, m_z) specifies a normal mode of the field (apart from polarization), the number of modes being infinite but denumerable. This is mathematically simpler than dealing with the continuum of modes in free space. The total number of modes in the intervals $\Delta m_x, \Delta m_y, \Delta m_z$ is

$$\Delta m = \Delta m_x \Delta m_y \Delta m_z = 2\left(\frac{L}{2\pi}\right)^3 \Delta k_x \Delta k_y \Delta k_z, \qquad (2.70)$$

where the factor of 2 takes into account the two independent polarizations. In a quasi-continuous limit, wherein we assume that wavelengths are small compared to L, we shall have waves densely packed in k-space and may therefore approximate Δm by the differential

$$dm = 2\left(\frac{V}{8\pi^3}\right)dk_x dk_y dk_z, \qquad (2.71)$$

where we have set $V = L^3$. In k-space, spherical polar coordinates

$$\mathbf{k} = k(\sin\theta\cos\phi, \ \sin\theta\sin\phi, \ \cos\theta), \qquad (2.72)$$

and we have

$$dm = 2\left(\frac{V}{8\pi^3}\right)k^2 dk d\Omega, \qquad (2.73)$$

where $d\Omega = \sin\theta d\theta d\phi$ is the element of solid angle around the direction of $\mathbf{k}$. Using the relation $k = \omega_k/c$, we can transform Eq. (2.73) into

$$dm = 2\left(\frac{V}{8\pi^3}\right)\frac{\omega_k^2}{c^3}d\omega_k d\Omega. \qquad (2.74)$$

Integrating Eq. (2.73) over the solid angle gives us

$$\left.\begin{array}{l}\text{the numbers of modes}\\ \text{in all directions}\\ \text{in the range } k \text{ to}\\ k+dk\end{array}\right\} = V\frac{k^2}{\pi^2}dk = V\rho_k dk, \qquad (2.75)$$

where $\rho_k dk$ is the mode density (number of modes per unit volume) and obviously $\rho_k = k^2/\pi^2$. Integrating Eq. (2.74) in the same fashion yields

$$\left.\begin{array}{l}\text{the numbers of modes}\\ \text{in all directions}\\ \text{in the range } \omega_k \text{ to}\\ \omega_k + d\omega_k\end{array}\right\} = V\frac{\omega_k^2}{\pi^2 c^3}d\omega_k \equiv V\rho(\omega_k)d\omega_k, \qquad (2.76)$$

where $\rho(\omega_k)d\omega_k$ is also the mode density with $\rho(\omega_k) = \omega_k^2/(\pi^2 c^3)$.

The vector potential can be expressed as a superposition of plane waves in the form

$$\mathbf{A}(\mathbf{r},t) = \sum_{\mathbf{k},s}\mathbf{e}_{\mathbf{k}s}[A_{\mathbf{k}s}(t)e^{i\mathbf{k}\cdot\mathbf{r}} + A_{\mathbf{k}s}^*(t)e^{-i\mathbf{k}\cdot\mathbf{r}}], \qquad (2.77)$$

where $A_{\mathbf{k}s}$ is the complex amplitude of the field, where $\mathbf{e}_{\mathbf{k}s}$ is a real polarization vector. The sum over $\mathbf{k}$ simply means the sum over the set of integers (m_x, m_y, m_z) and the sum over s is the sum over the two independent polarizations. These polarizations must be orthogonal:

$$\mathbf{e}_{\mathbf{k}s}\cdot\mathbf{e}_{\mathbf{k}s'} = \delta_{ss'}, \qquad (2.78)$$

and from the gauge condition of Eq. (2.62) must satisfy

$$\mathbf{k}\cdot\mathbf{e}_{\mathbf{k}s} = 0, \qquad (2.79)$$

known as the *transversality* condition. The Coulomb gauge is sometimes known as the transverse gauge, wherein the polarization is orthogonal to the propagation direction. The polarization vectors $\mathbf{e}_{\mathbf{k}1}$ and $\mathbf{e}_{\mathbf{k}2}$ form a right-handed system such that

$$\mathbf{e}_{\mathbf{k}1} \times \mathbf{e}_{\mathbf{k}2} = \frac{\mathbf{k}}{|\mathbf{k}|} = \boldsymbol{\kappa}. \qquad (2.80)$$

In free space, the sum in Eq. (2.77) is replaced by the integral

$$\sum_k \longrightarrow \frac{V}{\pi^2}\int k^2 dk. \qquad (2.81)$$

Now from Eqs. (2.61) and (2.62) we obtain, for the complex amplitudes $A_{\mathbf{k},s}(t)$, the harmonic oscillator equation

$$\frac{d^2 A_{\mathbf{k}s}}{dt^2} + \omega_k^2 A_{\mathbf{k}s} = 0, \qquad (2.82)$$

where $\omega_k = ck$. The solution is

$$A_{\mathbf{k}s}(t) = A_{\mathbf{k}s}e^{-i\omega_k t}, \tag{2.83}$$

where we have set $A_{\mathbf{k}s}(0) \equiv A_{\mathbf{k}s}$. From Eqs. (2.63) and (2.64), the electric and magnetic fields, respectively, are

$$\mathbf{E}(\mathbf{r},t) = i\sum_{\mathbf{k},s}\omega_k\mathbf{e}_{\mathbf{k}s}\left[A_{\mathbf{k}s}e^{i(\mathbf{k}\cdot\mathbf{r}-\omega_k t)} - A_{\mathbf{k}s}^*e^{-i(\mathbf{k}\cdot\mathbf{r}-\omega_k t)}\right], \tag{2.84}$$

$$\mathbf{B}(\mathbf{r},t) = \frac{i}{c}\sum_{\mathbf{k},s}\omega_k(\boldsymbol{\kappa}\times\mathbf{e}_{\mathbf{k}s})[A_{\mathbf{k}s}e^{i(\mathbf{k}\cdot\mathbf{r}-\omega_k t)} - A_{\mathbf{k}s}^*e^{-i(\mathbf{k}\cdot\mathbf{r}-\omega_k t)}]. \tag{2.85}$$

The energy of the field is given by

$$H = \frac{1}{2}\int_V\left(\varepsilon_0\mathbf{E}\cdot\mathbf{E} + \frac{1}{\mu_0}\mathbf{B}\cdot\mathbf{B}\right)dV. \tag{2.86}$$

The periodic boundary condition results in

$$\int_0^L e^{\pm ik_x x}dx = \begin{cases} L & k_x = 0, \\ 0 & k_x \neq 0, \end{cases} \tag{2.87}$$

with similar results for the y- and z-directions. These may collectively be written as

$$\int_V e^{\pm\, i(\mathbf{k}-\mathbf{k}')\cdot\mathbf{r}}dV = \delta_{\mathbf{k}\mathbf{k}'}V. \tag{2.88}$$

From this we find that the contribution to H from the electric field is

$$\frac{1}{2}\int_V\varepsilon_0\mathbf{E}\cdot\mathbf{E}dV = \varepsilon_0 V\sum_{\mathbf{k}s}\omega_k^2 A_{\mathbf{k}s}(t)A_{\mathbf{k}s}^*(t) - R \tag{2.89}$$

where

$$R = \frac{1}{2}\varepsilon_0 V\sum_{\mathbf{k}ss'}\omega_k^2\mathbf{e}_{\mathbf{k}s}\cdot\mathbf{e}_{-\mathbf{k}s'}[A_{\mathbf{k}s}(t)A_{-\mathbf{k}s'}(t) + A_{\mathbf{k}s}^*(t)A_{-\mathbf{k}s'}^*(t)] \tag{2.90}$$

To obtain the magnetic contribution, we need the vector identity

$$(\mathbf{A}\times\mathbf{B})\cdot(\mathbf{C}\times\mathbf{D}) = (\mathbf{A}\cdot\mathbf{C})(\mathbf{B}\cdot\mathbf{D}) - (\mathbf{A}\cdot\mathbf{D})(\mathbf{B}\cdot\mathbf{C}), \tag{2.91}$$

from which we obtain

$$(\boldsymbol{\kappa}\times\mathbf{e}_{\mathbf{k}s})\cdot(\boldsymbol{\kappa}\times\mathbf{e}_{\mathbf{k}s'}) = \delta_{ss'}, \tag{2.92}$$

$$(\boldsymbol{\kappa}\times\mathbf{e}_{\mathbf{k}s})\cdot(-\boldsymbol{\kappa}\times\mathbf{e}_{-\mathbf{k}s'}) = -\mathbf{e}_{\mathbf{k}s}\cdot\mathbf{e}_{-\mathbf{k}s'}. \tag{2.93}$$

Using these results, we have

$$\frac{1}{2}\int \frac{1}{\mu_0}\mathbf{B}\cdot\mathbf{B}dV = \varepsilon_0 V\sum_{\mathbf{k}s}\omega_k^2 A_{\mathbf{k}s}(t)A_{\mathbf{k}s}^*(t) + R. \qquad (2.94)$$

Thus, adding Eqs. (2.89) and (2.94) we obtain the field energy

$$H = 2\varepsilon_0 V\sum_{\mathbf{k}s}\omega_k^2 A_{\mathbf{k}s}(t)A_{\mathbf{k}s}^*(t) \qquad (2.95)$$

$$= 2\varepsilon_0 V\sum_{\mathbf{k}s}\omega_k^2 A_{\mathbf{k}s}A_{\mathbf{k}s}^*, \qquad (2.96)$$

where we have used Eq. (2.83).

The energy of Eq. (2.95) has a very simple form in terms of the amplitudes $A_{\mathbf{k}s}$. In order to quantize the field, the canonical variables $p_{\mathbf{k}s}$ and $q_{\mathbf{k}s}$ must be introduced. We set

$$A_{\mathbf{k}s} = \frac{1}{2\omega_k(\varepsilon_0 V)^{1/2}}[\omega_k q_{\mathbf{k}s} + ip_{\mathbf{k}s}], \qquad (2.97)$$

$$A_{\mathbf{k}s}^* = \frac{1}{2\omega_k(\varepsilon_0 V)^{1/2}}[\omega_k q_{\mathbf{k}s} - ip_{\mathbf{k}s}], \qquad (2.98)$$

such that upon substitution into Eq. (2.95), we obtain

$$H = \frac{1}{2}\sum_{\mathbf{k}s}(p_{\mathbf{k}s}^2 + \omega_k^2 q_{\mathbf{k}s}^2), \qquad (2.99)$$

each term of which is the energy of a simple harmonic oscillator of unit mass. The quantization of the field proceeds by demanding that the canonical variables become operators satisfying the commutation relations

$$[\hat{q}_{\mathbf{k}s}, \hat{q}_{\mathbf{k}'s'}] = 0 = [\hat{p}_{\mathbf{k}s}, \hat{p}_{\mathbf{k}'s'}] \qquad (2.100)$$

$$[\hat{q}_{\mathbf{k}s}, \hat{p}_{\mathbf{k}'s'}] = i\hbar\delta_{\mathbf{k}\mathbf{k}'}\delta_{ss'}. \qquad (2.101)$$

As for the single-mode field, annihilation and creation operators may be defined as

$$\hat{a}_{\mathbf{k}s} = \frac{1}{(2\hbar\omega_k)^{1/2}}[\omega_k q_{\mathbf{k}s} + ip_{\mathbf{k}s}], \qquad (2.102)$$

$$\hat{a}_{\mathbf{k}s}^\dagger = \frac{1}{(2\hbar\omega_k)^{1/2}}[\omega_k q_{\mathbf{k}s} - ip_{\mathbf{k}s}], \qquad (2.103)$$

which satisfy

$$[\hat{a}_{\mathbf{k}s}, \hat{a}_{\mathbf{k}'s'}] = 0 = [\hat{a}_{\mathbf{k}s}^\dagger, \hat{a}_{\mathbf{k}'s'}^\dagger], \qquad (2.104)$$

$$[\hat{a}_{\mathbf{k}s}, \hat{a}_{\mathbf{k}'s'}^\dagger] = \delta_{\mathbf{k}\mathbf{k}'}\delta_{ss'}. \qquad (2.105)$$

The energy of the field becomes the Hamiltonian operator

$$\hat{H} = \sum_{\mathbf{k}s} \hbar\omega_k \left(\hat{a}^\dagger_{\mathbf{k}s} \hat{a}_{\mathbf{k}s} + \frac{1}{2} \right) \tag{2.106}$$

$$= \sum_{\mathbf{k}s} \hbar\omega_k \left(\hat{n}_{\mathbf{k}s} + \frac{1}{2} \right), \tag{2.107}$$

where

$$\hat{n}_{\mathbf{k}s} = \hat{a}^\dagger_{\mathbf{k}s} \hat{a}_{\mathbf{k}s} \tag{2.108}$$

is the number operator for the mode $\mathbf{k}s$. Each of these modes, being independent of all the others, has an associated set of number eigenstates $|n_{\mathbf{k}s}\rangle$. For the jth mode, let $n_{\mathbf{k}_j s_j} \equiv n_j$ and let $\hat{a}_{\mathbf{k}_j s_j} \equiv \hat{a}_j$, $\hat{a}^\dagger_{\mathbf{k}_j s_j} \equiv \hat{a}^\dagger_j$, and $\hat{n}_{\mathbf{k}_j s_j} \equiv \hat{n}_j$. The Hamiltonian for the field is then

$$\hat{H} = \sum_j \hbar\omega_j \left(\hat{n}_j + \frac{1}{2} \right), \tag{2.109}$$

and a multimode photon number state is just a product of the number states of all the modes, which we write as

$$|n_1\rangle |n_2\rangle |n_3\rangle \ldots \equiv |n_1, n_2, n_3 \ldots\rangle$$
$$= |\{n_j\}\rangle. \tag{2.110}$$

This is an eigenstate of $\hat{H}$ such that

$$\hat{H} |\{n_j\}\rangle = E |\{n_j\}\rangle, \tag{2.111}$$

where its eigenvalue E is

$$E = \sum_j \hbar\omega_j \left(n_j + \frac{1}{2} \right). \tag{2.112}$$

Of course, these number states are orthogonal according to

$$\langle n_1, n_2, \ldots | n'_1, n'_2, \ldots \rangle = \delta_{n_1 n'_1} \delta_{n_2 n'_2} \ldots . \tag{2.113}$$

The action of the annihilation operator of the jth mode on the multimode number state is

$$\hat{a}_j |n_1, n_2, \ldots, n_j, \ldots\rangle = \sqrt{n_j} \, |n_1, n_2, \ldots, n_j - 1, \ldots\rangle. \tag{2.114}$$

Similarly for the creation operator

$$\hat{a}^\dagger_j |n_1, n_2, \ldots, n_j, \ldots\rangle = \sqrt{n_j + 1} \, |n_1, n_2, \ldots, n_j + 1, \ldots\rangle. \tag{2.115}$$

The multimode vacuum state is denoted

$$|\{0\}\rangle = |0_1, 0_2, \ldots, 0_j, \ldots\rangle, \tag{2.116}$$

for which

$$\hat{a}_j|\{0\}\rangle = 0, \qquad (2.117)$$

for all j. All the number states can be generated from the vacuum according to

$$|\{n_j\}\rangle = \prod_j \frac{(\hat{a}_j^\dagger)^{n_j}}{\sqrt{n_j!}} \; |\{0\}\rangle \qquad (2.118)$$

Upon quantization of the field, the amplitudes $A_{\mathbf{k}s}$ become operators which, from Eqs. (2.99) and (2.102), have the form

$$\hat{A}_{\mathbf{k}s} = \left(\frac{\hbar}{2\omega_k \varepsilon_0 V}\right)^{\frac{1}{2}} \hat{a}_{\overline{\mathbf{k}}s}, \qquad (2.119)$$

and thus the quantized vector potential has the form

$$\hat{\mathbf{A}}(\mathbf{r}, t) = \sum_{\mathbf{k}s} \left(\frac{\hbar}{2\omega_k \varepsilon_0 V}\right)^{\frac{1}{2}} \mathbf{e}_{\mathbf{k}s}[\hat{a}_{\mathbf{k}s} e^{i(\mathbf{k}\cdot\mathbf{r}-\omega_k t)} + \hat{a}_{\mathbf{k}s}^\dagger e^{-i(\mathbf{k}\cdot\mathbf{r}-\omega_k t)}]. \qquad (2.120)$$

The electric field operator is then

$$\hat{\mathbf{E}}(\mathbf{r}, t) = i\sum_{\mathbf{k}s} \left(\frac{\hbar\omega_k}{2\varepsilon_0 V}\right)^{\frac{1}{2}} \mathbf{e}_{\mathbf{k}s}[\hat{a}_{\mathbf{k}s} e^{i(\mathbf{k}\cdot\mathbf{r}-\omega_k t)} - \hat{a}_{\mathbf{k}s}^\dagger e^{-i(\mathbf{k}\cdot\mathbf{r}-\omega_k t)}], \qquad (2.121)$$

while the magnetic field operator is

$$\hat{\mathbf{B}}(\mathbf{r}, t) = \frac{i}{c}\sum_{\mathbf{k}s} (\boldsymbol{\kappa} \times \mathbf{e}_{\mathbf{k}s}) \left(\frac{\hbar\omega_k}{2\varepsilon_0 V}\right)^{\frac{1}{2}} [\hat{a}_{\mathbf{k}s} e^{i(\mathbf{k}\cdot\mathbf{r}-\omega_k t)} - \hat{a}_{\mathbf{k}s}^\dagger e^{-i(\mathbf{k}\cdot\mathbf{r}-\omega_k t)}], \qquad (2.122)$$

where $\boldsymbol{\kappa} = \mathbf{k}/|\mathbf{k}|$. The annihilation and creation operators appearing in Eqs. (2.120) to (2.122) are to be understood as Heisenberg picture operators evaluated at time $t = 0$. As in the single-mode case, the time-dependent annihilation operator for a free field is given by

$$\hat{a}_{\mathbf{k}s}(t) = \hat{a}_{\mathbf{k}s}(0) e^{-i\omega_k t}. \qquad (2.123)$$

Thus, the electric field, for instance, can be written as

$$\hat{\mathbf{E}}(\mathbf{r}, t) = i\sum_{\mathbf{k}s} \left(\frac{\hbar\omega_k}{2\varepsilon_0 V}\right)^{\frac{1}{2}} \mathbf{e}_{\mathbf{k}s} \left(\hat{a}_{\mathbf{k}s}(t) e^{i\mathbf{k}\cdot\mathbf{r}} - \hat{a}_{\mathbf{k}s}^\dagger(t) e^{-i\mathbf{k}\cdot\mathbf{r}}\right). \qquad (2.124)$$

Sometimes this field is written as

$$\hat{\mathbf{E}}(\mathbf{r}, t) = \hat{\mathbf{E}}^{(+)}(\mathbf{r}, t) + \hat{\mathbf{E}}^{(-)}(\mathbf{r}, t), \qquad (2.125)$$

where

$$\hat{\mathbf{E}}^{(+)}(\mathbf{r}, t) = i \sum_{ks} \left(\frac{\hbar \omega_k}{2\varepsilon_0 V} \right)^{\frac{1}{2}} \mathbf{e}_{ks} \hat{a}_{ks}(t) e^{i\mathbf{k}\cdot\mathbf{r}}, \qquad (2.126)$$

and where

$$\hat{\mathbf{E}}^{(-)}(\mathbf{r}, t) = [\hat{\mathbf{E}}^{(+)}(\mathbf{r}, t)]^{\dagger}. \qquad (2.127)$$

$\hat{\mathbf{E}}^{(+)}$ is called the positive frequency part of the field as it contains all terms that oscillate as $e^{-i\omega t}$ for $\omega > 0$, while $\hat{\mathbf{E}}^{(-)}$ is called the negative frequency part. The former is essentially a collective annihilation operator while the latter is a collective creation operator. Similar expressions can be written for the magnetic field and for the vector potential.

In most quantum optical situations, the coupling of the field to matter is through the electric field interacting with a dipole moment, or through some nonlinear type of interaction involving powers of the electric field. Thus, we shall be mostly interested in the electric field throughout the rest of the book. Furthermore, note that the magnetic field is "weaker" than the electric field by a factor of $1/c$. The field couples to the spin magnetic moment of the electrons and we shall show that this interaction is negligible for essentially all aspects of quantum optics that we are concerned with.

For a single-mode plane-wave field, the electric field is

$$\hat{\mathbf{E}}(\mathbf{r}, t) = i \left(\frac{\hbar \omega}{2\varepsilon_0 V} \right)^{\frac{1}{2}} \mathbf{e}_x [\hat{a} e^{i\mathbf{k}\cdot\mathbf{r} - i\omega t} - \hat{a}^{\dagger} e^{-i\mathbf{k}\cdot\mathbf{r} + i\omega t}]. \qquad (2.128)$$

In much of quantum optics, the spatial variation of the field over the dimensions of the atomic system may be negligible. For optical radiation, λ is on the order of several thousand angstroms, so that

$$\frac{\lambda}{2\pi} = \frac{1}{|\mathbf{k}|} >> |\mathbf{r}_{atom}|, \qquad (2.129)$$

where $|\mathbf{r}_{atom}|$ is a characteristic length of the size of an atom. Under this condition

$$e^{\pm i\mathbf{k}\cdot\mathbf{r}} \approx 1 \pm i\mathbf{k}\cdot\mathbf{r}, \qquad (2.130)$$

and we can replace the exponential by unity to obtain

$$\hat{\mathbf{E}}(\mathbf{r}, t) \approx \hat{\mathbf{E}}(t)$$
$$= i \left(\frac{\hbar \omega}{2\varepsilon_0 V} \right)^{\frac{1}{2}} \mathbf{e}_x [\hat{a} e^{-i\omega t} - \hat{a}^{\dagger} e^{i\omega t}]. \qquad (2.131)$$

This approximation, which will be discussed again in Chapter 4, is called the "dipole" approximation.

2.5 Thermal Fields

As is well known, quantum theory originated with Planck's discovery of the radiation law that now bears his name. We refer, of course, to the law describing the radiation emitted by an ideal object known as a black body – a perfect emitter and absorber of radiation. A black body can be modeled as a cavity (or actually a small hole in the cavity) containing radiation at thermal equilibrium with its walls. The radiation is thus coupled to a heat bath and so is not, unlike in the preceding sections of this chapter, a truly free field. But assuming the coupling is weak, we can, according to the theory of statistical mechanics, treat the field as if it were an isolated system that can be described as a microcanonical ensemble.

We consider then, for the moment, a single-mode field in thermal equilibrium with the walls of a cavity at temperature T. Then, according to statistical mechanics, the probability P_n that the mode is thermally excited in the nth level is

$$P_n = \frac{\exp(-E_n/k_B T)}{\displaystyle\sum_m \exp(-E_m/k_B T)} \tag{2.132}$$

where the E_n are given in Eq. (2.34) and where k_B is the Boltzmann constant ($k_B = 1.38 \times 10^{-23} \mathrm{J/K}$). For later purposes, we introduce here the density operator for the thermal field:

$$\hat{\rho}_{Th} = \frac{\exp(-\hat{H}/k_B T)}{\mathrm{Tr}\left[\exp(-\hat{H}/k_B T)\right]}, \tag{2.133}$$

where $\hat{H} = \hbar\omega\left(\hat{a}^\dagger \hat{a} + \frac{1}{2}\right)$ and where

$$\begin{aligned}
\mathrm{Tr}\left[\exp(-\hat{H}/k_B T)\right] &= \sum_{n=0}^{\infty} \langle n|\exp(-\hat{H}/k_B T)|n\rangle \\
&= \sum_{n=0}^{\infty} \exp(-E_n/k_B T) \equiv Z
\end{aligned} \tag{2.134}$$

is the partition function. With $E_n = \hbar\omega\left(n + \frac{1}{2}\right)$,

$$Z = \exp(-\hbar\omega/2k_B T)\sum_{n=0}^{\infty} \exp(-\hbar\omega n/k_B T). \tag{2.135}$$

Since $\exp(-\hbar\omega/k_B T) < 1$, the sum is a geometric series and thus

$$\sum_{n=0}^{\infty} \exp(-\hbar\omega n/k_B T) = \frac{1}{1 - \exp(-\hbar\omega/k_B T)}, \tag{2.136}$$

so that

$$Z = \frac{\exp(-\hbar\omega/2k_BT)}{1 - \exp(-\hbar\omega/k_BT)}. \tag{2.137}$$

Evidently

$$P_n = \langle n|\hat{\rho}_{Th}|n\rangle = \frac{1}{Z}\exp(-E_n/k_BT). \tag{2.138}$$

Also note that the density operator itself can be written as

$$\hat{\rho}_{Th} = \sum_{n'=0}^{\infty}\sum_{n=0}^{\infty}|n'\rangle\langle n'|\hat{\rho}_{Th}|n\rangle\langle n|$$

$$= \frac{1}{Z}\sum_{n=0}^{\infty}\exp(-E_n/k_BT)|n\rangle\langle n| \tag{2.139}$$

$$= \sum_{n=0}^{\infty}P_n|n\rangle\langle n|.$$

The average photon number of the thermal field is calculated as

$$\bar{n} = \langle \hat{n} \rangle = \mathrm{Tr}(\hat{n}\hat{\rho}_{Th}) = \sum_{n=0}^{\infty}\langle n|\hat{n}\hat{\rho}_{Th}|n\rangle$$

$$= \sum_{n=0}^{\infty}nP_n = \exp(-\hbar\omega/2k_BT)\frac{1}{Z}\sum_{n=0}^{\infty}n\,\exp(-\hbar\omega n/k_BT) \tag{2.140}$$

Noting that with $x = \hbar\omega/k_BT$, we have

$$\sum_{n=0}^{\infty}ne^{-nx} = -\frac{d}{dx}\sum_{n=0}^{\infty}e^{-nx}$$

$$= -\frac{d}{dx}\left(\frac{1}{1 - e^{-x}}\right) \tag{2.141}$$

$$= \frac{e^{-x}}{(1 - e^{-x})^2}.$$

Thus we have

$$\bar{n} = \frac{\exp(-\hbar\omega/k_BT)}{1 - \exp(-\hbar\omega/k_BT)}$$

$$= \frac{1}{\exp(\hbar\omega/k_BT) - 1}. \tag{2.142}$$

Evidently

$$\bar{n} \approx \begin{cases} \dfrac{k_BT}{\hbar\omega} & (k_BT \gg \hbar\omega), \\ \exp(-\hbar\omega/k_BT) & (k_BT \ll \hbar\omega). \end{cases} \tag{2.143}$$

At room temperatures, the average number of photons at optical frequencies is very small (on the order of 10^{-40}). At the surface temperature of the sun (6000 K) and at the frequency of yellow light (6×10^{14} Hz, $\lambda = 500$ nm), the average photon number is about 10^{-2}. On the other hand, the average photon number rapidly increases with increasing wavelength. Again at room temperature, $\bar{n} \simeq 1$ for λ in the range $\lambda = 10 - 100$ μm. In the microwave part of the spectrum, $\bar{n} \gg 1$.

From Eq. (2.142) it follows that

$$\exp(-\hbar\omega/k_B T) = \frac{\bar{n}}{1+\bar{n}}, \tag{2.144}$$

and from Eqs. (2.138) and (2.139) it follows that $\hat{\rho}_{Th}$ can be written in terms of $\bar{n}$ as

$$\hat{\rho}_{Th} = \frac{1}{1+\bar{n}} \sum_{n=0}^{\infty} \left(\frac{\bar{n}}{1+\bar{n}}\right)^n |n\rangle\langle n|. \tag{2.145}$$

The probability of finding n photons in the field is given in terms of $\bar{n}$ as

$$P_n = \frac{\bar{n}^n}{(1+\bar{n})^{n+1}}. \tag{2.146}$$

In Fig. 2.3 we plot P_n versus n for two different values of $\bar{n}$. It is clear in both cases that the most probable photon number is the vacuum, P_n, decreasing monotonically with n. There is obviously nothing special about P_n for n near or at $\bar{n}$ (which need not be integer).

The fluctuations in the average photon number are given as

$$\langle(\Delta n)^2\rangle = \langle\hat{n}^2\rangle - \langle\hat{n}\rangle^2. \tag{2.147}$$

It can be shown, in a manner similar to the derivation of $\bar{n}$, that

$$\langle\hat{n}^2\rangle = \mathrm{Tr}(\hat{n}^2\hat{\rho}_{Th})$$
$$= \bar{n} + 2\bar{n}^2, \tag{2.148}$$

so that

$$\langle(\Delta n)^2\rangle = \bar{n} + \bar{n}^2, \tag{2.149}$$

from which it is apparent that the *fluctuations* of $\hat{n}$ are larger than the *average* $\bar{n}$. The root-mean-square deviation is

$$\Delta n = (\bar{n} + \bar{n}^2)^{1/2}, \tag{2.150}$$

which for $\bar{n} \gg 1$ is approximately

$$\Delta n \approx \bar{n} + \frac{1}{2}. \tag{2.151}$$

The relative uncertainty is given by the ratio $\Delta n/\bar{n}$, which is approximately 1 for $\bar{n} \gg 1$ and approximately $1/\sqrt{\bar{n}}$ for $\bar{n} \ll 1$. Obviously, $\Delta n/\bar{n} \to \infty$ as $\bar{n} \to 0$.

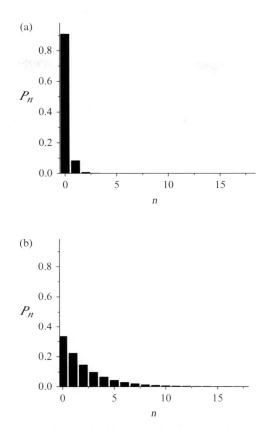

Fig. 2.3   Thermal photon number distributions for (a) $\bar{n} = 0.1$ and (b) $\bar{n} = 2$.

The average energy of the photons in the cavity is $\hbar\omega\,\bar{n}$. Planck's radiation law is obtained by multiplying the average energy of the photons by the density of modes per unit interval in ω in a unit volume, $\rho(\omega) = \omega^2/\pi^2c^3$ (where the two independent polarization directions have been taken into account), to obtain the average energy density per unit interval in ω as

$$\overline{U}(\omega) = \hbar\omega\bar{n}\rho(\omega)$$

$$= \frac{\hbar\omega^3}{\pi^2 c^3}\,\frac{1}{\exp(\hbar\omega/k_B T) - 1}. \tag{2.152}$$

For $k_B T \gg \hbar\omega$, this takes the simpler form

$$\overline{U}(\omega) \approx \frac{\omega^3 k_B T}{\pi^2 c^3}\quad (k_B T \gg \hbar\omega), \tag{2.153}$$

which is known as Rayleigh's law. This is sometimes called the "classical limit," obtained from Planck's law of Eq. (2.152) by setting $\hbar \to 0$. (We note, however, that setting $\hbar$ to zero is not a well-defined limit: $\hbar$ contains dimensional information.) On the other hand, for low temperatures where $k_B T \ll \hbar\omega$, we obtain

$$\overline{U}(\omega) \approx \frac{\hbar\omega^3}{\pi^2 c^3} \exp\left(-\frac{\hbar\omega}{k_B T}\right) \qquad (k_B T \ll \hbar\omega). \qquad (2.154)$$

This is Wien's law. By differentiation, it follows that $\overline{U}(\omega)$ has a maximum at

$$\omega_{\max} = \frac{2.8 k_B T}{\hbar} = \frac{2\pi c}{\lambda_{\max}}, \qquad (2.155)$$

which is Wien's displacement law.

The average energy per unit volume is obtained just by integrating over all frequencies:

$$\begin{aligned} \overline{U} &= \int\limits_0^\infty \overline{U}(\omega)\,d\omega \\ &= \frac{\hbar^2}{\pi^2 c^3} \int\limits_0^\infty \frac{\omega^3}{\exp(\hbar\omega/k_B T) - 1}\,d\omega \\ &= \frac{\pi^2 k_B^4 T^4}{15 c^3 \hbar^3}. \end{aligned} \qquad (2.156)$$

This is the Stefan–Boltzmann law.

2.6 Vacuum Fluctuations and the Zero-Point Energy

We have seen that the quantized radiation field fluctuates. For a single-mode field the fluctuations in the electric field strength are given by Eq. (2.48). With the field mode in the vacuum state $|0\rangle$, the r.m.s. fluctuation of the field strength is

$$\Delta E_x = \mathcal{E}_0 \sin(kz). \qquad (2.157)$$

These vacuum fluctuations and the zero-point energy have a common origin in the noncommutability of the operators $\hat{a}$ and $\hat{a}^\dagger$. The zero-point energy and the vacuum fluctuations actually present severe problems in quantum field theory. The most glaring of these comes about as follows: the universe contains an infinite number of radiation modes, each with a finite zero-point energy, $\hbar\omega/2$. The total zero-point energy (ZPE) of the universe is then

$$E_{ZPE} = \frac{\hbar}{2} \sum_\omega \omega \to \infty, \qquad (2.158)$$

unless somehow the high-frequency modes are excluded. It is frequently said that only energy differences are important, but this cannot quite be the whole story because according to general relativity, it is the total energy that counts, not just energy differences [2]. Other "infinites" appearing in the theory of quantum electrodynamics have been "swept under the rug" through the renormalization procedure, but this particular one still sticks out like a sore thumb. In fact, the vacuum energy and fluctuations actually give rise to observable effects. For example, spontaneous emission, which generates most of the visible light around us as thermal radiation, is a direct result of the vacuum fluctuations as we will show in Chapter 4. The ZPE gives rise to at least two other effects, one being the Lamb shift and the other being the Casimir effect.

The Lamb shift is a discrepancy between experiment and the Dirac relativistic theory of the hydrogen atom. The theory predicts that the $2^2S_{1/2}$ and $2^2P_{1/2}$ levels should be degenerate. Early optical work suggested that these states were not degenerate but separated by about 0.033 cm. Using an elegant combination of atomic beam and microwave techniques, Lamb and Retherford [3] showed that the $2^2S_{1/2}$ state has a higher energy than the $2^2P_{1/2}$ state by the equivalent of about 1000 MHz. In 1947, Bethe [4] explained this splitting as being due to the interaction of the bound electron with the ZPE. Here we present a simple intuitive interpretation originally given by Welton in 1948 [5].

Each mode contains ZPE $h\nu/2$, where $\nu = \omega/2\pi$. The number of modes in a cavity of volume V with frequency between ν and $\nu + d\nu$ is $(8\pi/c^3)\nu^2 d\nu$. Thus, the ZPE field energy is

$$\left(\frac{8\pi}{c^3}\nu^2 d\nu V\right)\frac{1}{2}h\nu = \frac{1}{8\pi}\int_V (E_\nu^2 + B_\nu^2)\, dV$$

$$= \frac{1}{8\pi}E_\nu^2 V, \tag{2.159}$$

where E_ν is the amplitude of the electric field component of frequency ν. Thus

$$E_\nu^2 = \frac{32\pi^2}{c^3}h\nu^3 d\nu. \tag{2.160}$$

The electron bound in the hydrogen atom interacts with the fluctuating zero-point electric field and with the Coulomb potential of the proton $-e^2/r$. If r represents the electron's "standard orbit" and Δr the fluctuations from this orbit, then the change in the potential energy is $\Delta V = V(r + \Delta r) - V(r)$, which by Taylor's theorem is

$$\Delta V = \Delta x \frac{\partial V}{\partial x} + \Delta y \frac{\partial V}{\partial y} + \Delta z \frac{\partial V}{\partial z}$$

$$+ \frac{1}{2}(\Delta x)^2 \frac{\partial^2 V}{\partial x^2} + \frac{1}{2}(\Delta y)^2 \frac{\partial^2 V}{\partial y^2} + \frac{1}{2}(\Delta z)^2 \frac{\partial^2 V}{\partial z^2} + \cdots. \tag{2.161}$$

Since the fluctuations are isotropic, $\langle \Delta x \rangle = \langle \Delta y \rangle = \langle \Delta z \rangle = 0$ and $\langle (\Delta x)^2 \rangle = \langle (\Delta y)^2 \rangle = \langle (\Delta z)^2 \rangle = \langle (\Delta r)^2 \rangle / 3$. Then

$$\langle \Delta V \rangle = \frac{1}{6} \langle (\Delta r)^2 \rangle \nabla^2 V. \tag{2.162}$$

For the atomic state $|nlm_l\rangle$ [2–5], the energy shift to first order is

$$\begin{aligned} \Delta E &= \langle nlm_l | \langle \Delta V \rangle | nlm_l \rangle \\ &= \frac{1}{6} \langle (\Delta r)^2 \rangle \langle nlm_l | \nabla^2 V | nlm_l \rangle. \end{aligned} \tag{2.163}$$

With $V = -e^2/r$ and $\nabla^2(1/r) = -4\pi\delta(r)$, we obtain

$$\langle nlm_l | \nabla^2 V | nlm_l \rangle = 4\pi e^2 |\psi_{nml_l}(r=0)|^2. \tag{2.164}$$

All the atomic wave functions of nonrelativistic quantum theory vanish at the origin except for the s-states with $l = 0$, where

$$|\psi_{n00}(r=0)|^2 = \frac{1}{\pi n^3 a_0^3}, \tag{2.165}$$

where a_0 is the Bohr radius. For p-states the wave function vanishes and therefore so does the energy shift. To obtain $\langle (\Delta r)^2 \rangle$ we assume that the important field frequencies greatly exceed the atomic resonance frequencies, the lower frequencies being shielded by the atomic binding and unable to influence the motion of the electrons. The displacement Δr_ν induced with frequency between ν and $\nu + d\nu$ is determined by

$$\frac{d^2 \Delta r_\nu}{dt^2} = \frac{eE_\nu}{m} \exp(2\pi\nu it). \tag{2.166}$$

The solution is

$$\Delta r_\nu = -\frac{e}{m} \frac{E_\nu}{4\pi^2\nu^2} \exp(2\pi\nu it). \tag{2.167}$$

The main square displacement induced by these modes is

$$\langle (\Delta r_\nu)^2 \rangle = -\frac{e^2}{m^2} \frac{E_\nu^2}{32\pi^4\nu^4} = \frac{e^2 h}{\pi^2 m^2 c^3} \frac{d\nu}{\nu}. \tag{2.168}$$

The s-state energy shift obtained by summing over all frequencies is

$$\Delta E = \frac{2}{3} \left(\frac{e^2}{\hbar c} \right)^2 \left(\frac{\hbar}{mc} \right)^2 \frac{hc}{\pi^2 n^3 a_0^3} \int \frac{d\nu}{\nu}, \tag{2.169}$$

where $e^2/\hbar c$ is the fine-structure constant and $\hbar/mc$ is the Compton wavelength of the electron. The integral is divergent but may be cut off at both high and low frequencies. At low frequencies, the atom does not respond to the fluctuating field, the frequency of the electron's orbit $\nu_0 = e^2/\hbar a_0^3 n^3$ being the natural cutoff. At high frequencies, relativistic effects show up in the electron's motion. But the preceding analysis is nonrelativistic, so that

$$\frac{v}{c} = \left(\frac{p/m}{c}\right) = \frac{pc}{mc^2} = \frac{\hbar k}{mc} < 1, \tag{2.170}$$

which restricts k to less than $mc/\hbar$ and angular frequencies to less than $mc^2/\hbar$ in the integral of Eq. (2.169). Thus, for the $2^2S_{1/2}$ state of hydrogen, with $a_0 = \hbar^2/me^2$, the energy shift is

$$\Delta E = \frac{1}{6\pi}\left(\frac{e^2}{\hbar c}\right)^3 \frac{me^4}{\hbar^2} \log\left(\frac{mc^2}{h\nu_0}\right), \tag{2.171}$$

which gives $\Delta E/h \sim 1000$ MHz. The $2^2P_{1/2}$ state is unaffected to this order.

The Casimir effect [6], in the simplest version, is actually the occurrence of a force between two parallel perfectly conducting plates due to a change in the ZPE resulting from the boundary conditions on the plates. We will follow the discussion of Milonni and Shih [7] to show how this force arises.

Consider a parallelepiped with perfectly conducting walls of length $L_x = L_y = L$ and $L_z = d$. The boundary conditions on the walls restrict the allowed frequencies to those given by

$$\omega_{lmn} = \pi c \left(\frac{l^2}{L^2} + \frac{m^2}{L^2} + \frac{n^2}{d^2}\right)^{\frac{1}{2}}, \tag{2.172}$$

where l, m, and n take on non-negative integer values. If there is no box, all frequencies are allowed. The ZPE in the box is

$$E_0(d) = {\sum_{l,m,n}}' (2)\frac{1}{2}\hbar\omega_{lmn}, \tag{2.173}$$

where the factor of 2 accounts for two independent polarizations and where the prime on the summation sign means that the 2 is to be removed if one of the integers l, m, or n is zero, as there is only one independent polarization for that case. We shall be interested only in the case where $L \gg d$, so that we can replace the sums of l and m by integrals to write

$$E_0(d) = \frac{\hbar c L^2}{\pi} \sum_{n=0}^{\infty} \int_0^{\infty} dx \int_0^{\infty} dy \left(x^2 + y^2 + \frac{\pi^2 n^2}{d^2}\right)^{\frac{1}{2}}. \tag{2.174}$$

On the other hand, if d is arbitrarily large, the sum over n can be replaced by an integral so that

$$E_0(\infty) = \frac{\hbar c L^2}{\pi^2} \frac{d}{\pi} \int_0^{\infty} dx \int_0^{\infty} dy \int_0^{\infty} dz (x^2 + y^2 + z^2)^{\frac{1}{2}}. \tag{2.175}$$

When the plates are separated by distance d, the potential energy of the system is just $U(d) = E_0(d) - E_0(\infty)$, which is the energy required to bring the plates from infinity to distance d. Thus

$$U(d) = \frac{L^2 \hbar c}{\pi} \left[\sum_{n=0}^{\infty} \int_0^{\infty} dx \int_0^{\infty} dy \left(x^2 + y^2 + \frac{\pi^2 n^2}{d^2} \right)^{\frac{1}{2}} \right.$$

$$\left. - \frac{d}{\pi} \int_0^{\infty} dx \int_0^{\infty} dy \int_0^{\infty} dz (x^2 + y^2 + z^2)^{\frac{1}{2}} \right]. \qquad (2.176)$$

Transforming to polar coordinates in the x–y plane, we have

$$U(d) = \frac{L^2 \hbar c}{\pi^2} \frac{\pi}{2} \left[\sum_{n=0}^{\infty} \int_0^{\infty} dr \, r \left(r^2 + \frac{n^2 \pi^2}{d^2} \right)^{\frac{1}{2}} \right.$$

$$\left. - \frac{d}{\pi} \int_0^{\infty} dz \int_0^{\infty} dr \, r (r^2 + z^2)^{\frac{1}{2}} \right]. \qquad (2.177)$$

Making the change of variable $w = r^2$,

$$U(d) = \frac{L^2 \hbar c}{4\pi^2} \frac{\pi^3}{d^3} \left[\sum_{n=0}^{\infty} \int_0^{\infty} dw \, (w + n^2)^{\frac{1}{2}} \right.$$

$$\left. - \int_0^{\infty} dz \int_0^{\infty} dw (w + z^2)^{\frac{1}{2}} \right]. \qquad (2.178)$$

Both ZPEs in Eq. (2.178) are infinite, but the difference is finite. We can write this as

$$U(d) = \frac{\pi^2 \hbar c}{4d^3} L^2 \left[\frac{1}{2} F(0) + \sum_{n=1}^{\infty} F(n) - \int_0^{\infty} dz F(z) \right] \qquad (2.179)$$

where

$$F(u) \equiv \int_0^{\infty} dw (w + u^2)^{\frac{1}{2}} \qquad (2.180)$$

with $u = n$ or z. The difference can be estimated by the Euler–Maclaurin formula

$$\sum_0^{\infty} F(n) - \int_0^{\infty} dz F(z) = -\frac{1}{2} F(0) - \frac{1}{12} F'(0) + \frac{1}{720} F'''(0)..., \qquad (2.181)$$

since $F'(z) = 2z^2$, $F'(0) = 0$, $F'''(0) = -4$, and all higher-order derivatives vanish. Thus

$$U(d) = \frac{\pi^2 \hbar c}{4d^3} L^3 \left(-\frac{4}{720} \right) = -\frac{\pi^2 \hbar c}{720 d^3} L^2. \qquad (2.182)$$

This implies that the force per unit area between the plates is given by

$$F(d) = -U'(d) = -\frac{\pi^2 \hbar c}{240 d^4},\qquad(2.183)$$

which is the Casimir force. The existence of this force was confirmed by experiments carried out by Sparnaay [8] in 1957.

Motivated by the development of nanotechnology, there has been a great deal of recent work on the Casimir effect, and we list some relevant reviews on this work in the bibliography for this chapter.

2.7 The Quantum Phase

Consider now a light wave as pictured in the classical electromagnetic theory. The electric field of a single mode can be written as

$$\begin{aligned}
\mathbf{E}(\mathbf{r}, t) &= \mathbf{e}_x E_0 \cos(\mathbf{k} \cdot \mathbf{r} - \omega t + \phi) \\
&= \mathbf{e}_x \frac{1}{2} E_0 \{ e^{i(\mathbf{k} \cdot \mathbf{r} - \omega t + \phi)} + e^{-i(\mathbf{k} \cdot \mathbf{r} - \omega t + \phi)} \}
\end{aligned}\qquad(2.184)$$

where E_0 is the amplitude of the field and ϕ is its phase. Compare this to Eq. (2.128). The equations would be quite similar if the operator $\hat{a}$ could be factored into polar form. The earliest attempt at such a decomposition appears to be due to Dirac [9], who factored the annihilation and creation operators according to

$$\hat{a} = e^{i\hat{\phi}} \sqrt{\hat{n}},\qquad(2.185)$$

$$\hat{a}^{\dagger} = \sqrt{\hat{n}} e^{-i\hat{\phi}},\qquad(2.186)$$

where $\hat{\phi}$ was to be interpreted as a Hermitian operator for phase. From the fundamental commutation relation $[\hat{a}, \hat{a}^{\dagger}] = 1$, it follows that

$$e^{i\hat{\phi}} \hat{n} e^{-i\hat{\phi}} - \hat{n} = 1,\qquad(2.187)$$

or

$$e^{i\hat{\phi}} \hat{n} - \hat{n} e^{i\hat{\phi}} = e^{i\hat{\phi}}.\qquad(2.188)$$

By expanding the exponentials, one can see that Eq. (2.187) is satisfied as long as

$$[\hat{n}, \hat{\phi}] = i.\qquad(2.189)$$

It thus appears that number and phase are complementary observables and therefore the fluctuations in these quantities should satisfy the uncertainty relation

$$\Delta n \, \Delta \phi \geq \frac{1}{2}.\qquad(2.190)$$

Unfortunately, things are not so simple. To see that something is quite wrong with the above, consider the matrix element of the commutator for the arbitrary number states $|n\rangle$ and $|n'\rangle$:

$$\langle n'|[\hat{n}, \hat{\phi}]|n\rangle = i\delta_{nn'}. \tag{2.191}$$

Expanding the left side results in

$$(n' - n)\langle n'|\hat{\varphi}|n\rangle = i\delta_{nn'}, \tag{2.192}$$

which contains an obvious contradiction in the case when $n' = n$ (giving $0 = i$). The Dirac approach fails because of the underlying assumption that a Hermitian phase operator $\hat{\phi}$ actually exists. The above contradiction seems to suggest otherwise. There are in fact two reasons for the failure of the Dirac approach. If $\hat{\phi}$ exists as a Hermitian operator, then $\exp(i\hat{\phi})$ should be a unitary operator. From Eqs. (2.185) and (2.186), we should have

$$e^{i\hat{\phi}} = \hat{a}(\hat{n})^{-\frac{1}{2}} \tag{2.193}$$

$$e^{-i\hat{\phi}} = (\hat{n})^{-\frac{1}{2}}\hat{a}^{\dagger} = (e^{i\hat{\phi}})^{\dagger}. \tag{2.194}$$

Now

$$(e^{i\hat{\phi}})^{\dagger}(e^{i\hat{\phi}}) = 1, \tag{2.195}$$

but

$$(e^{i\hat{\phi}})(e^{i\hat{\phi}})^{\dagger} = \hat{a}\frac{1}{\hat{n}}\hat{a}^{\dagger} \neq 1. \tag{2.196}$$

So, in fact, $\exp(i\hat{\phi})$ is not a unitary operator, from which it follows that $\hat{\phi}$ is not Hermitian. The root of the problem is that the operator $\hat{n}$ has a spectrum bounded from below; it does not include the negative integers. Furthermore, the number–phase uncertainty relation of Eq. (2.190) implies a phase uncertainty greater than 2π for a state of well-defined photon number. Ultimately, this is because the phase ϕ is a periodic variable, the nature of which is not accounted for in the commutation relation of Eq. (2.189).

Over the years, there have been many attempts to create a formalism for the description of the quantum phase that in one way or another overcomes the obstacles discussed above. These schemes have been reviewed extensively in the recent literature. For pedagogical reasons, and because it was one of the first schemes proposed against which the others may be gauged, we first introduce the approach due to Susskind and Glogower [11] and its improvements by Carruthers and Nieto [12], wherein a kind of one-sided "unitary" phase operator (actually an operator analogous to an exponential phase factor) is introduced.

The Susskind–Glogower (SG) operators are defined by the relations

$$\hat{E} \equiv (\hat{n} + 1)^{-\frac{1}{2}}\hat{a} = (\hat{a}\hat{a}^{\dagger})^{-\frac{1}{2}}\hat{a}, \tag{2.197}$$

$$\hat{E}^{\dagger} \equiv \hat{a}^{\dagger}(\hat{n} + 1)^{-\frac{1}{2}} = \hat{a}^{\dagger}(\hat{a}\hat{a}^{\dagger})^{-\frac{1}{2}}, \tag{2.198}$$

where $\hat{E}$ and $\hat{E}^{\dagger}$ are to be the analogs of the phase factors $\exp(\pm i\phi)$. These $\hat{E}$ operators are sometimes called "exponential" operators and, from context, should not be confused with the field operators. When applied to the number states $|n\rangle$, they yield

$$\begin{aligned} \hat{E}|n\rangle &= |n - 1\rangle && \text{for} \quad n \neq 0, \\ &= 0 && \text{for} \quad n = 0, \end{aligned} \tag{2.199}$$

$$\hat{E}^{\dagger}|n\rangle = |n + 1\rangle. \tag{2.200}$$

From this, it is easy to see that useful and equivalent expressions for these exponential operators are

$$\hat{E} = \sum_{n=0}^{\infty} |n\rangle\langle n + 1|, \quad \hat{E}^{\dagger} = \sum_{n=0}^{\infty} |n + 1\rangle\langle n|. \tag{2.201}$$

It is easy to show from these expressions that

$$\hat{E}\hat{E}^{\dagger} = \sum_{n=0}^{\infty}\sum_{n'=0}^{\infty} |n\rangle\langle n + 1|n' + 1\rangle\langle n'| = \sum_{n=0}^{\infty} |n\rangle\langle n| = 1, \tag{2.202}$$

but that

$$\hat{E}^{\dagger}\hat{E} = \sum_{n=0}^{\infty}\sum_{n'=0}^{\infty} |n + 1\rangle\langle n|n'\rangle\langle n' + 1| = \sum_{n=0}^{\infty} |n + 1\rangle\langle n + 1| = 1 - |0\rangle\langle 0|. \tag{2.203}$$

The presence of the vacuum state projection operator $|0\rangle\langle 0|$ spoils the unitarity of $\hat{E}$. But for a state with average photon number $\bar{n} \geq 1$, the contribution from the vacuum state will be small and $\hat{E}$ will be approximately unitary.

Of course, $\hat{E}$ and $\hat{E}^{\dagger}$ themselves are not observables, but the operators

$$\hat{C} \equiv \frac{1}{2}(\hat{E} + \hat{E}^{\dagger}), \quad \hat{S} \equiv \frac{1}{2}(\hat{E} - \hat{E}^{\dagger}) \tag{2.204}$$

are the obvious analogs of $\cos\phi$ and $\sin\phi$. These operators are Hermitian and satisfy the commutation relation

$$[\hat{C}, \hat{S}] = \frac{1}{2}i|0\rangle\langle 0|, \tag{2.205}$$

and thus commute for all states but the vacuum. Furthermore, the quantum form of the familiar trigonometric identity becomes

$$\hat{C}^2 + \hat{S}^2 = 1 - \frac{1}{2}|0\rangle\langle 0|, \tag{2.206}$$

where once again we see the spoiling effect of the vacuum. It can also be shown that

$$[\hat{C}, \hat{n}] = i\hat{S} \quad \text{and} \quad [\hat{S}, \hat{n}] = -i\hat{C}, \tag{2.207}$$

the first of which is similar to Eq. (2.49). The uncertainty relations obeyed in these cases are, respectively,

$$(\Delta n)(\Delta C) \geq \frac{1}{2}|\langle \hat{S}\rangle| \tag{2.208}$$

and

$$(\Delta n)(\Delta S) \geq \frac{1}{2}|\langle \hat{C}\rangle|, \tag{2.209}$$

where Δ means root-mean-square deviation.

In the case of number states $|n\rangle$, $\Delta n = 0$,

$$\langle n|\hat{C}|n\rangle = \langle n|\hat{S}|n\rangle = 0 \tag{2.210}$$

and

$$\langle n|\hat{C}^2|n\rangle = \langle n|\hat{S}^2|n\rangle = \begin{cases} \frac{1}{2} & n \geq 1, \\ \frac{1}{4} & n = 0. \end{cases} \tag{2.211}$$

Thus, for $n \geq 1$, the uncertainties in $\hat{C}$ and $\hat{S}$ are

$$\Delta C = \Delta S = \frac{1}{\sqrt{2}}, \tag{2.212}$$

which would seem to correspond to a phase angle equally likely to have any value in the range 0 to 2π. (Note that this will not be true for the vacuum state!) In any case, the right-hand sides of Eqs. (2.207) and (2.208) are zero for a number state, as required in order that the uncertainty relations be satisfied.

There exist eigenstates $|\phi\rangle$ of the exponential operator satisfying the eigenvalue equation

$$\hat{E}|\phi\rangle = e^{i\phi}|\phi\rangle, \tag{2.213}$$

given by

$$|\phi\rangle = \sum_{n=0}^{\infty} e^{in\phi}|n\rangle. \tag{2.214}$$

These states are not normalizable, nor are they orthogonal as the scalar product of $|\phi\rangle$ and $|\phi'\rangle$ is not the delta function $\delta(\phi - \phi')$. By virtue of the fact that

$$\int_0^{2\pi} e^{i(n-n')\phi} d\phi = 2\pi\delta_{nn'}, \tag{2.215}$$

it is easy to show that the phase eigenstates resolve unity according to

$$\frac{1}{2\pi} \int_0^{2\pi} d\phi |\phi\rangle\langle\phi| = 1. \tag{2.216}$$

The expectation values of the $\hat{C}$ and $\hat{S}$ operators obviously take the form $\cos\phi$ and $\sin\phi$, respectively.

Now an arbitrary state $|\psi\rangle$ of the field will be given as a superposition of all the number states, that is

$$|\psi\rangle = \sum_{n=0}^{\infty} C_n |n\rangle, \tag{2.217}$$

where the coefficients C_n must satisfy

$$\sum_{n=0}^{\infty} |C_n|^2 = 1 \tag{2.218}$$

in order that $|\psi\rangle$ be normalized. We may associate a phase distribution $\mathcal{P}(\phi)$ with the state $|\psi\rangle$ according to the prescription

$$\begin{aligned} \mathcal{P}(\phi) &\equiv \frac{1}{2\pi} |\langle\phi|\psi\rangle|^2 \\ &= \frac{1}{2\pi} \left| \sum_{n=0}^{\infty} e^{-in\phi} C_n \right|^2. \end{aligned} \tag{2.219}$$

Clearly, $\mathcal{P}(\phi)$ is always positive and can be shown to be normalized as follows: writing

$$\begin{aligned} \mathcal{P}(\phi) &= \frac{1}{2\pi} |\langle\phi|\psi\rangle|^2 = \frac{1}{2\pi} \langle\phi|\psi\rangle\langle\phi|\psi\rangle^* \\ &= \frac{1}{2\pi} \langle\phi|\psi\rangle\langle\psi|\phi\rangle = \frac{1}{2\pi} \langle\psi|\phi\rangle\langle\phi|\psi\rangle, \end{aligned} \tag{2.220}$$

and then integrating over ϕ using the resolution of unity given by Eq. (2.223), we have

$$\int_0^{2\pi} \mathcal{P}(\phi)d\phi = \langle\psi|\psi\rangle = 1. \tag{2.221}$$

More generally, for a state described by a density operator $\hat{\rho}$, we would have

$$\mathcal{P}(\phi) = \frac{1}{2\pi} \langle \phi | \hat{\rho} | \phi \rangle. \tag{2.222}$$

The distribution can be used to calculate the average of any function of ϕ, $f(\phi)$, according to

$$\langle f(\phi) \rangle = \int_0^{2\pi} f(\phi) \mathcal{P}(\phi) d\phi. \tag{2.223}$$

For the present, we consider only a number state $|n\rangle$ for which all the coefficients in Eq. (2.217) vanish but for $C_n = 1$. This leads to

$$\mathcal{P}(\phi) = \frac{1}{2\pi} \tag{2.224}$$

which is uniform, as expected. Furthermore, the average of ϕ is π and the average over ϕ^2 is $4\pi^2/3$ and thus the fluctuations in ϕ are

$$\Delta\phi = \sqrt{\langle \phi^2 \rangle - \langle \phi \rangle^2} = \frac{\pi}{\sqrt{3}}, \tag{2.225}$$

as we would expect for a uniform probability distribution over the range 0 to 2π. Note that this applies to all number states, including the vacuum. We shall apply this formalism to other field states, in particular the coherent states, in later chapters.

In the above discussion of the SG formalism, no Hermitian operator corresponding to the phase is presented. The space of number states is infinite-dimensional, making it impossible to introduce a Hermitian phase operator. Only the "exponential"-like operators $\hat{E}$ and $\hat{E}^\dagger$, which are one-way unitary, are invoked. On the other hand, in a finite-dimensional space, Hermitian phase operators are possible. The approach to the quantum mechanical phase for the quantized field taken by Pegg and Barnett [13] is to introduce a Hermitian phase operator in a space of finite dimensions, essentially a truncated space of number states, with which all calculations are performed, after which the dimension of the space is allowed to go to infinity. We outline the Pegg–Barnett (PB) approach next.

The PB approach to a Hermitian phase operator starts by considering a finite-dimensional Hilbert space $\mathcal{H}_s$ spanned by the $(s + 1)$ photon number states $|0\rangle, |1\rangle, ..., |s\rangle$. The expectation values of physical variables, that is of the Hermitian phase operator and functions of it, are calculated in this finite-dimensional space. These expectation values will parametrically depend on the value of s. The final stage of the calculation is to take the limit $s \to \infty$. Pegg and Barnett assume the existence of orthonormal number states $|n\rangle \in \mathcal{H}_s$ with the properties

$$\langle n | m \rangle = \delta_{n,m}, \quad \sum_{n=0}^{s} |n\rangle\langle n| = 1. \tag{2.226}$$

The annihilation and creation operators act as usual apart from the action of the creation operator on the state $|s\rangle$. For $|n\rangle \in \mathcal{H}_s$,

$$\hat{a}|n\rangle = \sqrt{n}|n-1\rangle,$$

$$\hat{a}^\dagger|n\rangle = \sqrt{n+1}|n+1\rangle, \quad \hat{a}^\dagger|s\rangle = 0. \tag{2.227}$$

The annihilation and creation operators in the truncated space can be rewritten, respectively, as

$$\hat{a} = \sum_{n=1}^{s} \sqrt{n}|n-1\rangle\langle n|, \quad \hat{a}^\dagger = \sum_{n=1}^{s} \sqrt{n}|n\rangle\langle n-1,| \tag{2.228}$$

and the commutation relation between these operators in $\mathcal{H}_s$ takes the form

$$[\hat{a}, \hat{a}^\dagger] = 1 - (s+1)|s\rangle\langle s|, \tag{2.229}$$

which means that if the dimension of the space is taken to be finite, the creation and annihilation operators no longer form the usual algebra of the harmonic oscillator, the so-called Heisenberg–Weyl algebra. The photon number operator $\hat{n}$ is defined in the obvious way according to

$$\hat{n} = \sum_{n=0}^{s} n|n\rangle\langle n| = \hat{a}^\dagger\hat{a}, \tag{2.230}$$

and the following additional commutation relations are satisfied:

$$[\hat{n}, \hat{a}^k] = -k\hat{a}^k, \quad [\hat{n}, (\hat{a}^\dagger)^k] = k(\hat{a}^\dagger)^k. \tag{2.231}$$

Pegg and Barnett [13] showed that the space $\mathcal{H}_s$ can also be spanned by the $s+1$ discrete *phase states* $|\theta_m\rangle$:

$$|\theta_m\rangle \equiv (s+1)^{-1/2} \sum_{n=0}^{s} \exp(i\theta_m n)|n\rangle, \tag{2.232}$$

with the properties

$$\langle\theta_m|\theta_l\rangle = \delta_{m,l}, \quad \sum_{m=0}^{s} |\theta_m\rangle\langle\theta_m| = 1. \tag{2.233}$$

The phase angle θ_m is defined as

$$\theta_m = \theta_0 + 2\pi\frac{m}{s+1}, \quad m = 0, 1, \ldots, s, \tag{2.234}$$

where the value of θ_0 is arbitrary and once chosen, it defines a particular basis set of phase states.

The Hermitian phase operator we denote as $\hat{\Phi}_\theta$, and define by

$$\hat{\Phi}_\theta = \sum_{m=0}^{s} \theta_m |\theta_m\rangle \langle \theta_m|, \tag{2.235}$$

from which it follows, by construction, that the phase states $|\theta_m\rangle$ are eigenstates of the Hermitian phase operator:

$$\hat{\Phi}_\theta |\theta_m\rangle = \theta_m |\theta_m\rangle. \tag{2.236}$$

The subscript θ on the symbol for the phase operator is there to remind us that the decomposition of Eq. (2.235) depends on the choice of θ_0 used in Eq. (2.234). Using Eq. (2.232), the phase operator in the number state basis is written as

$$\hat{\Phi}_\theta = \theta_0 + 2\pi \frac{s}{s+1}$$

$$+ \frac{2\pi}{(s+1)} \sum_{\substack{k,n \\ k \neq n}}^{s} \frac{\exp[-i(k-n)\theta_0]}{\exp[-i(k-n)2\pi/(s+1)] - 1} |n\rangle \langle k|. \tag{2.237}$$

The commutation relation between the number operator and the phase operator can be worked out, to yield

$$[\hat{n}, \hat{\Phi}_\theta] = \frac{2\pi}{(s+1)} \sum_{\substack{k,n=0 \\ k \neq n}}^{s} \frac{(n-k)\exp[-i(k-n)\theta_0]}{\exp[-i(k-n)2\pi/(s+1)] - 1} |n\rangle \langle k|, \tag{2.238}$$

which clearly differs from the proposed commutation relation of Dirac as given in Eq. (2.189).

Consider now an arbitrary (normalized) state $|\psi_s\rangle$ given in the $(s+1)$-dimensional space $\mathcal{H}_s$ as

$$|\psi_s\rangle = \sum_{n=0}^{s} C_n^{(s)} |n\rangle, \quad \sum_{n=0}^{s} |C_n^{(s)}|^2 = 1. \tag{2.239}$$

We can form a phase probability distribution using the phase states of Eq. (2.232) according to

$$P^{(s)}(\theta_m) = |\langle \theta_m | \psi_s \rangle|^2, \tag{2.240}$$

which is normalized according to

$$\sum_{m=0}^{s} P^{(s)}(\theta_m) = 1. \tag{2.241}$$

In the limit $s \to \infty$, we see from Eq. (2.234) that the discrete phases θ_m go over to a continuum of phases which we label θ. We can then take into account the discrete spacing of the phases by defining a new, continuous, phase distribution as

$$\mathcal{P}(\theta) \equiv \lim_{s\to\infty} \frac{s+1}{2\pi} P^{(s)}(\theta_m) = \lim_{s\to\infty} \frac{s+1}{2\pi} \left| \langle \theta_m | \psi_s \rangle \right|^2, \qquad (2.242)$$

which, upon using Eq. (2.232), in the limit goes over to

$$\mathcal{P}(\theta) = \frac{1}{2\pi} \left| \sum_{n=0}^{\infty} C_n e^{-i\theta n} \right|^2, \qquad (2.243)$$

where we have taken $\lim_{s\to\infty} C_n^{(s)} \to C_n$. This is the same result as obtained in Eq. (2.219) with the SG formalism. It is important to note, though, that the operator for the phase of Eq. (2.237) does not converge to a Hermitian operator in the limit $s \to \infty$. That, of course, is to be expected as the lack of such an operator was the motivation to propose a Hermitian phase operator in the truncated, finite-dimensional Hilbert space in the first place. Further analysis of the phase operator in the finite-dimensional space can be found in a paper by Bužek et al. [14].

Other approaches, in fact many alternative approaches to the quantum phase problem, have been put forward. Agarwal et al. [16], noting that the phase distributions obtained by either the SG or the PB theories are the same, have taken the view that it is sensible and computationally advantageous to describe the quantum-optical phase by a phase distribution rather than a quantum phase operator, a view which they claim is supported by the work of Shapiro and collaborators [16] who have shown, using quantum estimation theory [17], that the phase states $|\phi\rangle$ of Eq. (2.214) generate the probability operator measure for maximum likelihood phase estimation. On the other hand, Mandel and co-workers [18] have taken an operational approach to phase operators based on classical phase measurements. They show that phase measurements are difficult even in classical optics and that these difficulties carry over into quantum optics, and further, different measurement procedures lead to different phase operators. It does not appear, at least from the operational point of view, that there exists one correct phase operator. In the balance of the book, whenever we need to address the issue of the phase for a quantized state of light, we shall have in mind the phase distributions given by either Eq. (2.219) or Eq. (2.243).

A short review of approaches to the quantum phase problem has been published by Lynch [19], and an extensive volume of reprints on the topic has been published by Barnett and Vaccaro [20].

Problems

1. For the single-mode field given by Eq. (2.5), use Maxwell's equations to obtain the corresponding magnetic field given by Eq. (2.6).
2. For the single-mode field of the previous problem, obtain the Hamiltonian and show that it has the form of a simple harmonic oscillator.

3. Give a proof of the Baker–Hausdorff lemma of Eq. (2.24) by expanding $e^{i\lambda\hat{A}}\hat{B}\,e^{-i\lambda\hat{A}}$ in a Maclaurin series with respect to λ and then setting $\lambda = 1$.

4. In the special case where $[\hat{A},\hat{B}] \neq 0$, but where $[\hat{A},[\hat{A},\hat{B}]] = 0 = [\hat{B},[\hat{A},\hat{B}]]$, show that

$$e^{\hat{A}+\hat{B}} = \exp\left(-\tfrac{1}{2}[\hat{A},\hat{B}]\right)e^{\hat{A}}e^{\hat{B}} = \exp\left(\tfrac{1}{2}[\hat{A},\hat{B}]\right)e^{\hat{B}}e^{\hat{A}}.$$

This is known as the Baker–Campbell–Hausdorff theorem.

5. Suppose the state of a single-mode cavity field is given at time $t = 0$ by

$$|\psi(0)\rangle = \frac{1}{\sqrt{2}}\left(|n\rangle + e^{i\varphi}|n+1\rangle\right),$$

where φ is some phase. Find the state $|\psi(t)\rangle$ at times $t > 0$. For this time-evolved state, verify the uncertainty relation of Eq. (2.50).

6. Consider the superposition state $|\psi_{01}\rangle = \alpha|0\rangle + \beta|1\rangle$, where α and β are complex and satisfy $|\alpha|^2 + |\beta|^2 = 1$. Calculate the variances of the quadrature operators $\hat{X}_1$ and $\hat{X}_2$. Are there any values of the parameters α and β for which either of the quadrature variances become *less* than for a vacuum state? If so, check to see if the uncertainty principle is violated. Repeat with the state $|\psi_{02}\rangle = \alpha|0\rangle + \beta|2\rangle$.

7. Many processes involve the absorption of single photons from a quantum field state, the process of absorption being represented by the action of the annihilation operator $\hat{a}$. For an arbitrary field state $|\psi\rangle$, the absorption of a single photon yields the state $|\psi'\rangle \sim \hat{a}|\psi\rangle$. Normalize this state. Compare the average photon numbers $\bar{n}$ of the state $|\psi\rangle$ and $\bar{n}'$ of $|\psi'\rangle$. Do you find that $\bar{n}' = \bar{n} - 1$?

8. Consider the superposition of the vacuum and 10 photon number state

$$|\psi\rangle = \frac{1}{\sqrt{2}}\left(|0\rangle + |10\rangle\right).$$

Calculate the average photon number for this state. Next assume that a single photon is absorbed and recalculate the average photon number. Does your result seem sensible in comparison with your answer to the previous question?

9. Consider the multimode expressions for the electric and magnetic fields of Eqs. (2.84) and (2.85), respectively. (a) Show that they satisfy the free-space Maxwell's equations. (b) Use these fields and follow through the derivation of the field energy given by Eq. (2.96).

10. It is sometimes useful to characterize the photon number probability distribution P_n by its factorial moments. The rth factorial moment is defined as

$$\langle \hat{n}(\hat{n}-1)(\hat{n}-2)...(\hat{n}-r+1) \rangle = \sum_n n(n-1)(n-2)...(n-r+1)P_n.$$

Show that for a thermal field, the right-hand side has the value $r!\bar{n}^r$.

11. Work out the commutator of the cosine and sine operators, $\hat{C}$ and $\hat{S}$, respectively. Calculate the matrix elements of the commutator and show that only the diagonal ones are nonzero.

12. Consider the mixed state (see Appendix A) described by the density operator

$$\hat{\rho} = \frac{1}{2}(|0\rangle\langle 0| + |1\rangle\langle 1|)$$

and the pure superposition state

$$|\psi\rangle = \frac{1}{\sqrt{2}}(|0\rangle + e^{i\theta}|1\rangle).$$

Calculate the corresponding phase distributions $\mathcal{P}(\phi)$ and compare them.

13. Show that for the thermal state, $\mathcal{P}(\phi) = 1/2\pi$.

14. Work out expressions for the operators $\exp(i\hat{\Phi}_\theta)$, $\cos\hat{\Phi}_\theta$, and $\sin\hat{\Phi}_\theta$, where $\hat{\Phi}_\theta$ is the Pegg–Barnett Hermitian phase operator, in terms of (a) the phase state projectors and (b) the number states of the truncated basis. (c) Is the relation $\sin^2\hat{\Phi}_\theta + \cos^2\hat{\Phi}_\theta = 1$ satisfied by these operators? (d) Using the results of part (b), work out the commutation relations between these operators and the photon number operator.

References

[1] See, for example, J. J. Sakurai, *Modern Quantum Mechanics*, revised edition (Reading, MA: Addison-Wesley, 1994), p. 96.
[2] See the discussion in I. J. K. Aitchison, *Contemp. Phys.* **26**, 333 (1985).
[3] W. E. Lamb Jr. and R. C. Retherford, *Phys. Rev.* **72**, 241 (1947).
[4] H. A. Bethe, *Phys. Rev.* **72**, 339 (1947).
[5] T. A. Welton, *Phys. Rev.* **74**, 1157 (1948).
[6] H. B. G. Casimir, *Proc. K. Ned. Akad. Wet.* **51**, 793 (1948); *Physica* **19**, 846 (1953).
[7] P. W. Milonni and M.-L. Shih, *Contemp. Phys.* **33**, 313 (1992).
[8] M. J. Sparnaay, *Nature* **180**, 334 (1957).
[9] P. A. M. Dirac, *Proc. R. Soc. London, Series A* **114**, 243 (1927).
[10] S. M. Barnett and D. T. Pegg, *J. Phys. A* **19**, 3849 (1986).
[11] L. Susskind and J. Glogower, *Physics* **1**, 49 (1964).
[12] P. Carruthers and M. M. Nieto, *Rev. Mod. Phys.* **40**, 411 (1968).

[13] D. T. Pegg and S. M. Barnett, *Europhys. Lett.* **6**, 483 (1988); *Phys. Rev. A* **39**, 1665 (1989); **43**, 2579 (1991).

[14] V. Bužek, A. D. Wilson-Gordon, P. L. Knight, and W. K. Lai, *Phys. Rev. A* **45**, 8079 (1992).

[15] G. S. Agarwal, S. Chaturvedi, K. Tara, and V. Srinivasan, *Phys. Rev. A* **45**, 4904 (1992).

[16] J. H. Shapiro and S. R. Shepard, *Phys. Rev. A* **43**, 3795 (1991).

[17] C. W. Helstrom, *Quantum Detection and Estimation Theory* (New York: Academic Press, 1976).

[18] J. W. Noh, S. Fougères, and L. Mandel, *Phys. Rev. A* **45**, 424 (1992); **46**, 2840 (1992).

[19] R. Lynch, *Phys. Rep.* 256, 367 (1995).

[20] S. M. Barnett and J. A. Vaccaro, The *Quantum Phase Operator, A Review* (London: Taylor & Francis, 2007).

Bibliography

Quantum Mechanics

Virtually every quantum mechanics textbook covers the harmonic oscillator through the operator method, sometimes known as the algebraic, or factorization, method. Some good undergraduate textbooks are

J. S. Townsend, *A Modern Approach to Quantum Mechanics* (New York: McGraw-Hill, 1992).

D. H. McIntyre, *Quantum Mechanics, A Paradigms Approach* (New York: Pearson, 2012).

An excellent graduate-level text is

J. J. Sakurai and J. Napolitano, *Modern Quantum Mechanics*, 2nd edition (Boston, MA: Addison-Wesley, 2011).

Field Theory

E. G. Harris, *A Pedestrian Approach to Quantum Field Theory* (New York: Wiley, 1972).

L. H. Ryder, *Quantum Field Theory*, 2nd edition (Cambridge: Cambridge University Press, 1996).

T. Lancaster and S. J. Blundell, *Quantum Field Theory for the Gifted Amateur* (Oxford: Oxford University Press, 2014).

For an extensive set of references on the Casimir effect, see

S. K. Lamoreaux, "Resource Letter CF-1: Casimir force," *Am. J. Phys.* **67**, 850 (1999).

For recent developments on the Casimir effect, see

M. Bordag, G. L. Klimchitskaya, G. L. Mohideen, and V. M. Mostepanenko, *Advances in the Casimir Effect* (Oxford: Oxford University Press, 2009).

K. A. Milton, "Recent developments in the Casimir effect," *J. Phys. Conf. Ser.* **161**, 012001 (2009).

The Quantum Phase

V. Peřinova, A. Lukš, and J. Peřina, *Phase in Optics* (Singapore: World Scientific, 1998).

A. Luis and L. L. Sanchez-Soto, "Quantum phase difference, phase measurements, and Stokes operators," in E. Wolf (Ed.), *Progress in Optics*, Vol. 41 (Amsterdam: Elsevier, 2000), p. 423.

3

Coherent States

At the end of Chapter 2, we showed that the photon number states $|n\rangle$ have a uniform phase distribution over the range 0 to 2π. Essentially, then, there is no well-defined phase for these states and, as we have already shown, the expectation value of the field operator for a number state vanishes. It is sometimes suggested (see e.g. Sakurai [1]) that the classical limit of the quantized field is the limit in which the number of photons becomes very large, such that the number operator becomes a continuous variable. However, this cannot be the whole story. For a single-mode field as given by Eq. (2.128), the average of the field in a number state yields $\langle n|\hat{\mathbf{E}}(\mathbf{r},t)|n\rangle = 0$ no matter how large the value of n. We know that at a fixed point in space, a classical field oscillates sinusoidally in time. Clearly this does not happen for the expectation value of the field operator for a number state. In this chapter, we present a set of states, the coherent states [2], which do give rise to a sensible classical limit and, in fact, these states are the "most classical" quantum states of a harmonic oscillator, as we shall see.

3.1 Eigenstates of the Annihilation Operator and Minimum Uncertainty States

In order to have a non-zero expectation value of the electric field operator, or equivalently, of the annihilation and creation operators, we are required to have a superposition of number states differing only by +1. For example, this could contain only the states $|n\rangle$ and $|n+1\rangle$:

$$|\psi\rangle = C_n|n\rangle + C_{n+1}|n+1\rangle, \tag{3.1}$$

where $|C_n|^2 + |C_{n+1}|^2 = 1$. Generally, a superposition of all the number states will have the property that the expectation value of, say, $\hat{a}$ will not vanish. Clearly, by inspecting Eqs. (2.15) and (2.16), the replacement of $\hat{a}$ and $\hat{a}^\dagger$ by continuous variables produces a classical field. A unique way to make this replacement is to seek eigenstates of the annihilation operator. These states are denoted as $|\alpha\rangle$ and satisfy the relation

$$\hat{a}|\alpha\rangle = \alpha|\alpha\rangle, \tag{3.2}$$

where α is a complex number, otherwise arbitrary. (Note that complex eigenvalues are allowed here as $\hat{a}$ is non-Hermitian.) The states $|\alpha\rangle$ are "right" eigenstates of $\hat{a}$ and $\langle\alpha|$ are "left" eigenstates of $\hat{a}^\dagger$ with eigenvalue α^*:

$$\langle\alpha|\hat{a}^\dagger = \alpha^*\langle\alpha|. \tag{3.3}$$

Since the number states $|n\rangle$ form a complete set, we may expand $|\alpha\rangle$ according to

$$|\alpha\rangle = \sum_{n=0}^{\infty} C_n|n\rangle. \tag{3.4}$$

Acting with $\hat{a}$ on each term of the expansion, Eq. (3.2) becomes

$$\hat{a}|\alpha\rangle = \sum_{n=1}^{\infty} C_n\sqrt{n}|n-1\rangle = \alpha\sum_{n=0}^{\infty} C_n|n\rangle. \tag{3.5}$$

Equating coefficients of $|n\rangle$ on both sides leads to

$$C_n\sqrt{n} = \alpha C_{n-1}, \tag{3.6}$$

or

$$\begin{aligned} C_n &= \frac{\alpha}{\sqrt{n}}C_{n-1} = \frac{\alpha^2}{\sqrt{n(n-1)}}C_{n-2} = \cdots \\ &= \frac{\alpha^n}{\sqrt{n!}}C_0, \end{aligned} \tag{3.7}$$

and thus

$$|\alpha\rangle = C_0\sum_{n=0}^{\infty} \frac{\alpha^n}{\sqrt{n!}}|n\rangle. \tag{3.8}$$

From the normalization requirement, we determine C_0:

$$\begin{aligned} \langle\alpha|\alpha\rangle = 1 &= |C_0|^2\sum_n\sum_{n'} \frac{\alpha^{*n}\alpha^{n'}}{\sqrt{n!n'!}} \langle n|n'\rangle, \\ &= |C_0|^2\sum_{n=0}^{\infty} \frac{|\alpha|^{2n}}{n!} = |C_0|^2 e^{|\alpha|^2}, \end{aligned} \tag{3.9}$$

which implies that $C_0 = \exp\left(-\frac{1}{2}|\alpha|^2\right)$. Thus, our normalized coherent states are

$$|\alpha\rangle = \exp\left(-\frac{1}{2}|\alpha|^2\right)\sum_{n=0}^{\infty} \frac{\alpha^n}{\sqrt{n!}}|n\rangle. \tag{3.10}$$

Let us now consider the expectation value of the single-mode electric field operator $\hat{\mathbf{E}}(\mathbf{r},t) = \mathbf{e}_x\hat{E}_x(\mathbf{r},t)$ where

$$\hat{E}_x(\mathbf{r},t) = i\left(\frac{\hbar\omega}{2\varepsilon_0 V}\right)^{\frac{1}{2}}\left[\hat{a}e^{i(\mathbf{k}\cdot\mathbf{r}-\omega t)} - \hat{a}^\dagger e^{-i(\mathbf{k}\cdot\mathbf{r}-\omega t)}\right]. \tag{3.11}$$

We obtain

$$\langle\alpha|\hat{E}_x(\mathbf{r},t)|\alpha\rangle = i\left(\frac{\hbar\omega}{2\varepsilon_0 V}\right)^{\frac{1}{2}}\left[\alpha e^{i(\mathbf{k}\cdot\mathbf{r}-\omega t)} - \alpha^* e^{-i(\mathbf{k}\cdot\mathbf{r}-\omega t)}\right]. \tag{3.12}$$

Writing α in polar form, $\alpha = |\alpha|e^{i\theta}$, we have

$$\langle\alpha|\hat{E}_x(\mathbf{r},t)|\alpha\rangle = 2|\alpha|\left(\frac{\hbar\omega}{2\varepsilon_0 V}\right)^{\frac{1}{2}}\sin(\omega t - \mathbf{k}\cdot\mathbf{r} - \theta) , \qquad (3.13)$$

which looks like a classical field. Furthermore, we can show that

$$\langle\alpha|\hat{E}_x^2(\mathbf{r},t)|\alpha\rangle = \frac{\hbar\omega}{2\varepsilon_0 V}[1 + 4|\alpha|^2 \sin^2(\omega t - \mathbf{k}\cdot\mathbf{r} - \theta)] . \qquad (3.14)$$

Thus, the fluctuations in $\hat{E}_x(\mathbf{r},t)$,

$$\Delta E_x \equiv \left\langle(\Delta\hat{E}_x)^2\right\rangle^{\frac{1}{2}} = \left(\frac{\hbar\omega}{2\varepsilon_0 V}\right)^{\frac{1}{2}} , \qquad (3.15)$$

are identical to those for a vacuum state. The coherent state is nearly a classical state because it not only yields the correct form for the field expectation values, but also contains only the noise of the vacuum. Indeed, using the quadrature operators of Eqs. (2.53) and (2.54), it is an easy exercise to show that for the coherent states

$$\left\langle(\Delta\hat{X}_1)^2\right\rangle_\alpha = \frac{1}{4} = \langle(\Delta\hat{X}_2)^2\rangle_\alpha, \qquad (3.16)$$

which again shows that these states have the fluctuations of the vacuum. Thus, with respect to the field quadrature operators, the coherent states both minimize (actually equalize) the uncertainty product and exhibit equal uncertainties, those of the vacuum, in each quadrature. As a matter of fact, this property can be used as an alternate definition for the coherent states. Consider the three Hermitian operators $\hat{A}$, $\hat{B}$, and $\hat{C}$ satisfying

$$[\hat{A},\hat{B}] = i\hat{C}, \qquad (3.17)$$

which implies the uncertainty relation

$$\langle(\Delta\hat{A})^2\rangle\langle(\Delta\hat{B})^2\rangle \geq \frac{1}{4}\langle\hat{C}^2\rangle. \qquad (3.18)$$

States that equalize this relation are those that satisfy the eigenvalue equation [3]

$$\left[\hat{A} + \frac{i\langle\hat{C}\rangle}{2\langle(\Delta\hat{B})^2\rangle}\hat{B}\right]|\psi\rangle = \left[\langle\hat{A}\rangle + \frac{i\langle\hat{C}\rangle}{2\langle(\Delta\hat{B})^2\rangle}\langle\hat{B}\rangle\right]|\psi\rangle . \qquad (3.19)$$

The states $|\psi\rangle$ satisfying Eq. (3.19) sometimes go by the name "intelligent" states [4]. For the case where

$$\langle(\Delta\hat{A})^2\rangle = \langle(\Delta\hat{B})^2\rangle = \frac{1}{4}\langle\hat{C}\rangle, \qquad (3.20)$$

the eigenvalue equation becomes

$$[\hat{A} + i\hat{B}] \, |\psi\rangle = [\langle\hat{A}\rangle + i\langle\hat{B}\rangle] \, |\psi\rangle. \tag{3.21}$$

For $\hat{A} = \hat{X}_1$ and $\hat{B} = \hat{X}_2$, this is equivalent to Eq. (3.2) with $\alpha = \langle\hat{X}_1\rangle + i\langle\hat{X}_2\rangle$.

What is the physical meaning of the complex parameter α? From Eq. (3.13) it is apparent that $|\alpha|$ is related to the amplitude of the field. The average of the photon number operator $\hat{n} = \hat{a}^\dagger\hat{a}$ is

$$\bar{n} = \langle\alpha|\hat{n}|\alpha\rangle = |\alpha|^2, \tag{3.22}$$

and thus $|\alpha|^2$ is just the average photon number of the field. To calculate the fluctuations of the photon number, we need to calculate

$$\begin{aligned}\langle\alpha|\hat{n}^2|\alpha\rangle &= \langle\alpha|\hat{a}^\dagger\hat{a}\hat{a}^\dagger\hat{a}|\alpha\rangle \\ &= \langle\alpha|(\hat{a}^\dagger\hat{a}^\dagger\hat{a}\hat{a} + \hat{a}^\dagger\hat{a})|\alpha\rangle \\ &= |\alpha|^4 + |\alpha|^2 = \bar{n}^2 + \bar{n},\end{aligned} \tag{3.23}$$

and thus

$$\Delta n = \sqrt{\langle\hat{n}^2\rangle - \langle\hat{n}\rangle^2} = \bar{n}^{1/2}, \tag{3.24}$$

which is characteristic of a Poisson process. In fact, for a measurement of the number of photons in the field, the probability of detecting n photons is

$$\begin{aligned}P_n &= |\langle n|\alpha\rangle|^2 = e^{-|\alpha|^2}\frac{|\alpha|^{2n}}{n!} \\ &= e^{-\bar{n}}\frac{\bar{n}^n}{n!},\end{aligned} \tag{3.25}$$

which is a Poisson distribution with mean $\bar{n}$. Note that the fractional uncertainty in the photon number is

$$\frac{\Delta n}{\bar{n}} = \frac{1}{\sqrt{\bar{n}}}, \tag{3.26}$$

which decreases with increasing $\bar{n}$. In Fig. 3.1 we plot a couple of examples of the photon number probability distribution for different $\bar{n}$.

Let us now look at the phase distribution of the coherent states. For a coherent state $|\alpha\rangle$, with $\alpha = |\alpha| \, e^{i\theta}$, the corresponding phase distribution is

$$\begin{aligned}\mathcal{P}(\varphi) &= \frac{1}{2\pi}|\langle\varphi|\alpha\rangle|^2 \\ &= \frac{1}{2\pi} e^{-|\alpha|^2}\left|\sum_{n=0}^{\infty} e^{in(\varphi-\theta)}\frac{|\alpha|^n}{\sqrt{n!}}\right|^2.\end{aligned} \tag{3.27}$$

For large $|\alpha|^2$, the Poisson distribution may be approximated as a Gaussian [5]:

(a)

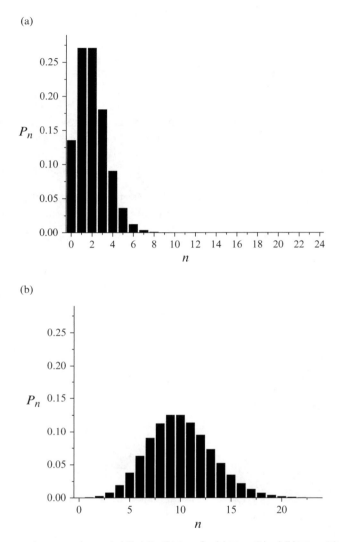

(b)

Fig. 3.1   Coherent state photon number probability distributions for (a) $\bar{n} = 2$ and (b) $\bar{n} = 10$.

$$e^{-|\alpha|^2} \frac{|\alpha|^{2n}}{n!} \approx (2\pi|\alpha|^2)^{-\frac{1}{2}}\exp\left[-\frac{(n - |\alpha|^2)^2}{2|\alpha|^2}\right], \tag{3.28}$$

so that the sum in Eq. (3.27) may be evaluated to obtain the approximate form for $\mathcal{P}(\varphi)$ as

$$\mathcal{P}(\varphi) \approx \left(\frac{2|\alpha|^2}{\pi}\right)^{\frac{1}{2}} \exp[-2|\alpha|^2(\varphi - \theta)^2] . \tag{3.29}$$

This is a Gaussian peaked at $\varphi = \theta$. Furthermore, the peak becomes narrower with increasing $\bar{n} = |\alpha|^2$, as illustrated in Fig. 3.2.

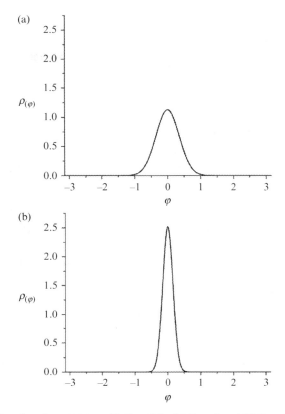

Fig. 3.2   Phase distributions for coherent states with $\theta = 0$ for (a) $\bar{n} = 2$ and (b) $\bar{n} = 10$.

The coherent states $|\alpha\rangle$ are quantum states very close to classical states because (i) the expectation value of the electric field has the form of the classical expression, (ii) the fluctuations in the electric field variables are the same as for a vacuum, (iii) the fluctuations in the fractional uncertainty for the photon number decrease with the increasing average photon number, and (iv) the states become well localized in phase with increasing average photon number. However, in spite of their near classical properties, they are still quantum states. In Fig. 3.3 we illustrate this with a sketch, where the expectation of the field operator, and its fluctuations, are plotted against time. The field clearly has the classical form of a sine wave but with the quantum fluctuations superimposed, indicating that it does have quantum features. In fact, *all* states of light must have some quantum features as the quantum theory of light is more fundamental than the classical theory. However, the quantum features of light are generally difficult to observe. We shall deal with some of these features in Chapter 7.

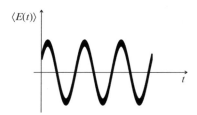

$\langle E(t) \rangle$

Fig. 3.3 Coherent state expectation value of the electric field as a function of time for a fixed position showing the quantum fluctuations. The fluctuations of the field are the same at all times, such that the field is as close to a classical field as is possible for any quantum state.

3.2 Displaced Vacuum States

We have discussed two ways in which the coherent states may be defined: as right eigenstates of the annihilation operator and as states that minimize the uncertainty relation for the two orthogonal field quadratures with equal uncertainties (identical to those of a vacuum state) in each quadrature. There is, in fact, a third definition that leads to equivalent states. This involves the displacement of the vacuum. As we will show, this is closely related to a mechanism for generating the coherent states from classical currents.

The displacement operator $\hat{D}(\alpha)$ is defined as [2]

$$\hat{D}(\alpha) = \exp(\alpha \hat{a}^\dagger - \alpha^* \hat{a}), \tag{3.30}$$

and the coherent states are given as

$$|\alpha\rangle = \hat{D}(\alpha)|0\rangle. \tag{3.31}$$

To see this, consider the identity (known as the disentangling theorem)

$$\begin{aligned} e^{\hat{A}+\hat{B}} &= e^{\hat{A}} \, e^{\hat{B}} e^{-\frac{1}{2}[\hat{A},\hat{B}]} \\ &= e^{\hat{B}} \, e^{\hat{A}} e^{\frac{1}{2}[\hat{A},\hat{B}]}, \end{aligned} \tag{3.32}$$

valid if $[\hat{A}, \hat{B}] \neq 0$ but where

$$[\hat{A}, [\hat{A}, \hat{B}]] = [\hat{B}, [\hat{A}, \hat{B}]] = 0. \tag{3.33}$$

With $\hat{A} = \alpha \, \hat{a}^\dagger$ and $\hat{B} = -\alpha^* \hat{a}$, then $[\hat{A}, \hat{B}] = |\alpha|^2$ and Eq. (3.33) holds. Thus

$$\hat{D}(\alpha) = e^{\alpha \hat{a}^\dagger - \alpha^* \hat{a}} = e^{-\frac{1}{2}|\alpha|^2} e^{\alpha \hat{a}^\dagger} e^{-\alpha^* \hat{a}}. \tag{3.34}$$

Expanding $\exp(-\alpha^* \hat{a})$ makes it clear that

$$e^{-\alpha^* \hat{a}}|0\rangle = \sum_{l=0}^{\infty} \frac{(-\alpha^* \hat{a})^l}{l!}|0\rangle = |0\rangle, \tag{3.35}$$

since $\hat{a}^l|0\rangle = 0$ except for $l = 0$. But

$$
\begin{aligned}
e^{\alpha \hat{a}^\dagger}|0\rangle &= \sum_{n=0}^{\infty} \frac{\alpha^n}{n!}(\hat{a}^\dagger)^n|0\rangle \\
&= \sum_{n=0}^{\infty} \frac{\alpha^n}{\sqrt{n!}}|n\rangle,
\end{aligned}
\tag{3.36}
$$

where we have used the fact that $(\hat{a}^\dagger)^n|0\rangle = \sqrt{n!}|n\rangle$. Thus, we have

$$
\begin{aligned}
|\alpha\rangle &= \hat{D}(\alpha)|0\rangle \\
&= e^{-\frac{1}{2}|\alpha|^2} \sum_{n=0}^{\infty} \frac{\alpha^n}{\sqrt{n!}}|n\rangle,
\end{aligned}
\tag{3.37}
$$

in agreement with our previous definitions.

The displacement operator $\hat{D}$ is, of course, a unitary operator. It can be shown that

$$
\begin{aligned}
\hat{D}^\dagger(\alpha) &= \hat{D}(-\alpha) \\
&= e^{-\frac{1}{2}|\alpha|^2} e^{-\alpha \hat{a}^\dagger} e^{\alpha^* \hat{a}}.
\end{aligned}
\tag{3.38}
$$

An alternate representation of $\hat{D}(\alpha)$ is

$$
\hat{D}(\alpha) = e^{\frac{1}{2}|\alpha|^2} e^{-\alpha^* \hat{a}} e^{\alpha \hat{a}^\dagger}.
\tag{3.39}
$$

Evidently

$$
\hat{D}(\alpha)\hat{D}^\dagger(\alpha) = \hat{D}^\dagger(\alpha)\hat{D}(\alpha) = 1,
\tag{3.40}
$$

as required for unitarity.

The displacement operator obeys the semigroup relation: the product of two displacement operators, say of $\hat{D}(\alpha)$ and $\hat{D}(\beta)$, is, up to an overall phase factor, the displacement operator $\hat{D}(\alpha + \beta)$. To see this, let $\hat{A} = \alpha \hat{a}^\dagger - \alpha^* \hat{a}, \hat{B} = \beta \hat{a}^\dagger - \beta^* \hat{a}$, where

$$
[\hat{A}, \hat{B}] = \alpha \beta^* - \alpha^* \beta = 2i \mathrm{Im}(\alpha \beta^*).
\tag{3.41}
$$

Then, using Eq. (3.32),

$$
\begin{aligned}
\hat{D}(\alpha)\hat{D}(\beta) &= e^{\hat{A}} e^{\hat{B}} \\
&= \exp[i \mathrm{Im}(\alpha \beta^*)] \\
&\quad \times \exp[(\alpha + \beta)\hat{a}^\dagger - (\alpha^* + \beta^*)\hat{a}] \\
&= \exp[i \mathrm{Im}(\alpha \beta^*)]\hat{D}(\alpha + \beta).
\end{aligned}
\tag{3.42}
$$

Applied to the vacuum state, we have

$$
\begin{aligned}
\hat{D}(\alpha)\hat{D}(\beta)|0\rangle &= \hat{D}(\alpha)|\beta\rangle \\
&= \exp[i \, \mathrm{Im}(\alpha \beta^*)] \, |\alpha + \beta\rangle \ .
\end{aligned}
\tag{3.43}
$$

The phase factor $\exp[i \, \mathrm{Im}(\alpha\beta^*)]$ is an overall phase factor and is physically irrelevant.

For future reference and applications, we now work out the matrix elements of the displacement operator in the photon number basis. That is, we provide an expression for the quantities $\langle m|\hat{D}(\alpha)|n\rangle$.

Using the decomposition of the displacement operator given in Eq. (3.34), we have

$$\langle m|\hat{D}(\alpha)|n\rangle = e^{-\frac{1}{2}|\alpha|^2}\langle m|e^{\alpha\hat{a}^\dagger}e^{-\alpha^*\hat{a}}|n\rangle. \tag{3.44}$$

Now we expand the right-most exponential to obtain

$$e^{-\alpha^*\hat{a}}|n\rangle = \sum_{p=0}^{\infty}\frac{(-\alpha^*\hat{a})^p}{p!}|n\rangle = \sum_{p=0}^{n}\frac{(-\alpha^*)^p}{p!}\sqrt{\frac{n!}{(n-p)!}}|n-p\rangle. \tag{3.45}$$

Setting $n - p = l$, we can rewrite this result as

$$e^{-\alpha^*\hat{a}}|n\rangle = \sum_{l=0}^{n}\frac{(-\alpha^*)^{n-l}}{(n-l)!}\sqrt{\frac{n!}{l!}}|l\rangle. \tag{3.46}$$

In the same manner we find that

$$\langle m|e^{\alpha\hat{a}^\dagger} = \sum_{k=0}^{m}\frac{\alpha^{m-k}}{(m-k)!}\sqrt{\frac{m!}{k!}}\langle k|. \tag{3.47}$$

It follows that

$$\begin{aligned}\langle m|\hat{D}(\alpha)|n\rangle &= e^{-\frac{1}{2}|\alpha|^2}\sum_{k=0}^{m}\sum_{l=0}^{n}\sqrt{\frac{m!n!}{k!l!}}\frac{\alpha^{m-k}}{(m-k)!}\frac{(-\alpha^*)^{n-l}}{(n-l)!}\delta_{kl} \\ &= (m!n!)^{1/2}e^{-\frac{1}{2}|\alpha|^2}\sum_{l=0}^{\min(m,n)}\frac{\alpha^{m-l}(-\alpha^*)^{n-l}}{l!(m-l)!(n-l)!}.\end{aligned} \tag{3.48}$$

3.3 Wave Packets and Time Evolution

From Eqs. (2.13) and (2.14), we obtain the "position" operator

$$\hat{q} = \sqrt{\frac{\hbar}{2\omega}}(\hat{a}+\hat{a}^\dagger) = \sqrt{\frac{2\hbar}{\omega}}\hat{X}_1, \tag{3.49}$$

where $\hat{X}_1$ is the quadrature operator of Eq. (2.53). The eigenstates of the operator $\hat{q}$ we denote as $|q\rangle$, where $\hat{q}|q\rangle = q|q\rangle$. The corresponding wave functions for the number states are [6]

$$\psi_n(q) = \langle q|n \rangle = (2^n n!)^{-1/2}\left(\frac{\omega}{\pi\hbar}\right)^{1/4}\exp(-\xi^2/2)H_n(\xi), \tag{3.50}$$

where $\xi = q\sqrt{\omega/\hbar}$ and the $H_n(\xi)$ are Hermite polynomials. The corresponding wave function for the coherent state is then

$$\psi_\alpha(q) \equiv \langle q|\alpha \rangle = \left(\frac{\omega}{\pi\hbar}\right)^{1/4}e^{-|\alpha|^2/2}\sum_{n=0}^{\infty}\frac{(\alpha/\sqrt{2})^n}{n!}H_n(\xi). \tag{3.51}$$

Using the generating function for Hermite polynomials, we can obtain the closed-form expression for the coherent state wave function

$$\psi_\alpha(q) = \left(\frac{\omega}{\pi\hbar}\right)^{1/4}e^{-|\alpha|^2/2}e^{\xi^2/2}e^{-(\xi-\alpha/\sqrt{2})^2}, \tag{3.52}$$

a Gaussian wave function. Of course, the probability distribution over the "position" variable q,

$$P(q) = |\psi_\alpha(q)|^2, \tag{3.53}$$

is also Gaussian.

We now consider the time evolution of a coherent state for a single-mode free field where the Hamiltonian is given by Eq. (2.18). The time-evolving coherent state is given by

$$\begin{aligned} |\alpha, t\rangle &\equiv \exp(-i\hat{H}t/\hbar)|\alpha\rangle = e^{-i\omega t/2}e^{-i\omega\hat{n}}|\alpha\rangle \\ &= e^{-i\omega t/2}|\alpha e^{-i\omega t}\rangle, \end{aligned} \tag{3.54}$$

and so the coherent state remains a coherent state under free-field evolution. The corresponding time-evolving wave function is

$$\psi_\alpha(q,t) = \left(\frac{\omega}{\pi\hbar}\right)^{1/4}e^{-|\alpha|^2/2}e^{\xi^2/2}e^{-(\xi-\alpha e^{-i\omega t}/\sqrt{2})^2}, \tag{3.55}$$

a Gaussian whose shape does not change with time and whose centroid follows the motion of a classical point particle in a harmonic oscillator potential (the demonstration of this is left as an exercise, e.g. see problem 2 at the end of the chapter). In Fig. 3.4 we illustrate the motion of a coherent state wave packet in the harmonic oscillator potential. The motional states of laser-cooled trapped atoms or ions can be engineered to have this minimum uncertainty character, as we shall see in a later chapter.

Ultimately, the stability of the wave packet is due to the fact that, for the harmonic oscillator, the energy levels are integer spaced. For quantum systems where the energy levels are not integer spaced, such as for the Coulomb problem, formulating coherent states is a bit of a challenge and no such states are truly stable for all times. Consideration of such states is beyond the scope of this book, but see Ref. [7].

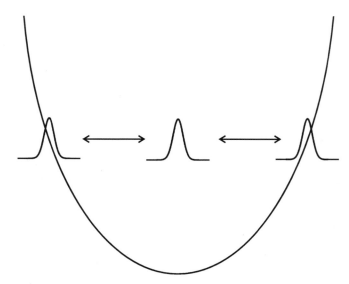

Fig. 3.4 A coherent state wave function moves through the harmonic oscillator potential, between the classical turning points, without dispersion.

3.4 Generation of Coherent States

A coherent state may be generated by a classical oscillating current. Let the quantum electromagnetic vector potential be $\hat{\mathbf{A}}(\mathbf{r}, t)$ for a field interacting with a classical current described by the current density $\mathbf{j}(\mathbf{r}, t)$. According to classical electromagnetic theory, the interaction energy $\hat{V}(t)$ is given by

$$\hat{V}(t) = -\int d^3\mathbf{r}\, \mathbf{j}(\mathbf{r}, t) \cdot \hat{\mathbf{A}}(\mathbf{r}, t). \tag{3.56}$$

For a single-mode field, $\hat{\mathbf{A}}(\mathbf{r}, t)$ is given, in the interaction picture, by

$$\hat{\mathbf{A}}(\mathbf{r}, t) = \mathbf{e}\left(\frac{\hbar}{2\omega\varepsilon_0 V}\right)^{\frac{1}{2}}[\hat{a}e^{i(\mathbf{k}\cdot\mathbf{r}-\omega t)} + \hat{a}^\dagger e^{-i(\mathbf{k}\cdot\mathbf{r}-\omega t)}], \tag{3.57}$$

where $\hat{a}$ means $\hat{a}(0)$. Substituting this into Eq. (3.56), we have

$$\hat{V}(t) = -\left(\frac{\hbar}{2\omega\varepsilon_0 V}\right)^{\frac{1}{2}}[\hat{a}\mathbf{e} \cdot \mathbf{J}(\mathbf{k}, t)e^{-i\omega t} + \hat{a}^\dagger\mathbf{e} \cdot \mathbf{J}^*(\mathbf{k}, t)e^{i\omega t}], \tag{3.58}$$

where

$$\mathbf{J}(\mathbf{k}, t) = \int d^3\mathbf{r}\, \mathbf{j}(\mathbf{r}, t)e^{i\mathbf{k}\cdot\mathbf{r}}. \tag{3.59}$$

Since $\hat{V}(t)$ depends on time, the associated evolution operator is a time-ordered product [8]. But for an infinitesimally short time interval from t to $t + \delta t$, the evolution operator is

$$
\begin{aligned}
\hat{U}(t + \delta t, t) &\cong \exp[-i\hat{V}(t)\delta t/\hbar] \\
&= \exp\{\delta t[u(t)\hat{a} - u^*(t)\hat{a}^\dagger]\} \\
&= \hat{D}[u(t)\delta t],
\end{aligned}
\tag{3.60}
$$

where

$$
u(t) = -i\left(\frac{1}{2\hbar\omega\varepsilon_0 V}\right)^{\frac{1}{2}} \mathbf{e} \cdot \mathbf{J}^*(\mathbf{k}, t)e^{i\omega t}.
\tag{3.61}
$$

For a finite time interval, say from 0 to T, the evolution operator may be written

$$
\hat{U}(T, 0) = \lim_{\delta t \to 0} \hat{T} \prod_{l=0}^{T/\delta t} \hat{D}[u(t_l)\delta t],
\tag{3.62}
$$

where $\hat{T}$ is the time-ordering operator and where $t_l = l\delta t$. Equation (3.49) becomes, using Eq. (3.42),

$$
\begin{aligned}
\hat{U}(T, 0) &= \lim_{\delta t \to 0} e^{i\Phi}\hat{D}\left[\sum_{l=0}^{T/\delta t} u(t_l)\delta t\right] \\
&= e^{i\Phi}\hat{D}[\alpha(T)],
\end{aligned}
\tag{3.63}
$$

where

$$
\alpha(T) = \lim_{\delta t \to 0} \sum_{l=0}^{T/\delta t} u(t_l)\,\delta t = \int_0^T u(t')dt',
\tag{3.64}
$$

and where Φ is the accumulated overall phase. With initial state the vacuum, the state at time T is just the coherent state $|\alpha(T)\rangle$, with $\alpha(T)$ given by Eq. (3.59), apart from the irrelevant overall phase.

3.5 More on the Properties of Coherent States

The number states are orthonormal, $\langle n|n'\rangle = \delta_{nn'}$, and complete, $\sum_{n=0}^{\infty} |n\rangle\langle n| = \hat{I}$, and as such can be used as a basis for the expansion of an arbitrary state vector of the field, that is for a given $|\psi\rangle$,

$$
|\psi\rangle = \sum_n C_n|n\rangle,
\tag{3.65}
$$

where $C_n = \langle n|\psi\rangle$, the coherent states being a particular example. But the coherent states themselves are not orthogonal: for $|\alpha\rangle$ and $|\beta\rangle$,

$$\langle \beta | \alpha \rangle = e^{-\frac{1}{2}|\alpha|^2 - \frac{1}{2}|\beta|^2}$$

$$\times \sum_{n=0}^{\infty} \sum_{m=0}^{\infty} \frac{(\beta^*)^n \alpha^m}{\sqrt{n! m!}} \langle n | m \rangle$$

$$= e^{-\frac{1}{2}|\alpha|^2 - \frac{1}{2}|\beta|^2} \sum_{n=0}^{\infty} \frac{(\beta^* \alpha)^n}{n!} \tag{3.66}$$

$$= e^{-\frac{1}{2}|\alpha|^2 - \frac{1}{2}|\beta|^2 + \beta^* \alpha}$$

$$= \exp\left[\frac{1}{2}(\beta^* \alpha - \beta \alpha^*)\right] \exp\left[-\frac{1}{2}|\beta - \alpha|^2\right].$$

The first term is just a complex phase, so that

$$|\langle \beta | \alpha \rangle|^2 = e^{-|\beta - \alpha|^2} \neq 0. \tag{3.67}$$

Thus, the coherent states are not orthogonal. Of course, if $|\beta - \alpha|^2$ is large, they are nearly orthogonal.

The completeness relation for the coherent states is given as an integral over the complex α-plane according to

$$\int |\alpha\rangle\langle\alpha| \frac{d^2\alpha}{\pi} = 1, \tag{3.68}$$

where $d^2\alpha = d \operatorname{Re}(\alpha) d \operatorname{Im}(\alpha)$. The proof of this goes as follows: writing

$$\int |\alpha\rangle\langle\alpha| d^2\alpha = \int e^{-|\alpha|^2} \sum_n \sum_m \frac{\alpha^n \alpha^{*m}}{\sqrt{n! m!}} |n\rangle\langle m| d^2\alpha, \tag{3.69}$$

we transform to polar coordinates, setting $\alpha = r e^{i\theta}, d^2\alpha = r \, dr \, d\theta$ so that

$$\int |\alpha\rangle\langle\alpha| d^2\alpha = \sum_n \sum_m \frac{|n\rangle\langle m|}{\sqrt{n! m!}} \int_0^{\infty} dr e^{-r^2} r^{n+m+1} \int_0^{2\pi} d\theta e^{i(n-m)\theta} . \tag{3.70}$$

But

$$\int_0^{2\pi} d\theta e^{i(n-m)\theta} = 2\pi \delta_{nm}, \tag{3.71}$$

and, with a further change of variables, $r^2 = y, 2r\,dr = dy$, we thus have

$$\int |\alpha\rangle\langle\alpha| d^2\alpha = \pi \sum_{n=0}^{\infty} \frac{|n\rangle\langle n|}{n!} \int_0^{\infty} dy e^{-y} y^n. \tag{3.72}$$

Since

$$\int_0^\infty dy\, e^{-y} y^n = n!, \tag{3.73}$$

we have

$$\int |\alpha\rangle\langle\alpha| d^2\alpha = \pi \sum_{n=0}^\infty |n\rangle\langle n| = \pi, \tag{3.74}$$

which completes the proof.

Any state vector $|\psi\rangle$ in the Hilbert space of the quantized single-mode field can be expressed in terms of the coherent states as

$$|\psi\rangle = \int \frac{d^2\alpha}{\pi} |\alpha\rangle\langle\alpha|\psi\rangle. \tag{3.75}$$

But suppose the state $|\psi\rangle$ itself is the coherent state $|\beta\rangle$. Then

$$\begin{aligned}
|\beta\rangle &= \int \frac{d^2\alpha}{\pi} |\alpha\rangle\langle\alpha|\beta\rangle \\
&= \int \frac{d^2\alpha}{\pi} |\alpha\rangle \exp\left[-\frac{1}{2}|\alpha|^2 - \frac{1}{2}|\beta|^2 + \alpha^*\beta\right].
\end{aligned} \tag{3.76}$$

This last equation shows that the coherent states are not linearly independent. The coherent states are said to be "overcomplete," there being more than enough states available to express any state in terms of the coherent states. Note that in Eq. (3.71) the quantity $\langle\alpha|\beta\rangle = \exp(-\frac{1}{2}|\alpha|^2 - \frac{1}{2}|\beta|^2 + \alpha^*\beta)$ plays the role of a Dirac delta function. It is often referred to as a reproducing kernel.

For arbitrary state $|\psi\rangle$, we may write

$$\begin{aligned}
\langle\alpha|\psi\rangle &= \exp\left(-\frac{1}{2}|\alpha|^2\right) \sum_{n=0}^\infty \psi_n \frac{(\alpha^*)^n}{\sqrt{n!}} \\
&= \exp\left(-\frac{1}{2}|\alpha|^2\right) \psi(\alpha^*),
\end{aligned} \tag{3.77}$$

where $\psi_n = \langle n|\psi\rangle$ and

$$\psi(z) = \sum_{n=0}^\infty \psi_n \frac{z^n}{\sqrt{n!}} \tag{3.78}$$

is absolutely convergent anywhere on the complex z-plane, hence $\psi(z)$ is an entire function, since $\langle\psi|\psi\rangle = \sum_n |\langle n|\psi\rangle|^2 = 1$. The functions $\psi(z)$ constitute the Segal–Bargmann [9] space of entire functions. If $|\psi\rangle$ is a number state, $|\psi\rangle = |n\rangle$, then $\psi_n(z) = z^n/\sqrt{n!}$. These functions form an orthonormal basis on the Segal–Bargmann space.

Let $\hat{F}$ be an operator given as a function of $\hat{a}$ and $\hat{a}^\dagger$, $\hat{F} = F(\hat{a}, \hat{a}^\dagger)$. In terms of the number states, $\hat{F}$ may be decomposed as

$$\hat{F} = \sum_n \sum_m |m\rangle \langle m|\hat{F}|n\rangle \langle n|$$

$$= \sum_n \sum_m |m\rangle \hat{F}_{mn} \langle n|, \tag{3.79}$$

where the F_{mn} are the matrix elements $\langle m|\hat{F}|n\rangle$.

With coherent states,

$$\hat{F} = \frac{1}{\pi^2} \int d^2\beta \int d^2\alpha |\beta\rangle \langle \beta|\hat{F}|\alpha\rangle \langle \alpha|. \tag{3.80}$$

But

$$\langle \beta|\hat{F}|\alpha\rangle = \sum_n \sum_m F_{mn} \langle \beta|m\rangle \langle n|\alpha\rangle$$

$$= \exp\left[-\frac{1}{2}(|\beta|^2 + |\alpha|^2)\right] F(\beta^*, \alpha), \tag{3.81}$$

where

$$F(\beta^*, \alpha) = \sum_m \sum_n F_{mn} \frac{(\beta^*)^m (\alpha)^n}{\sqrt{m!n!}} . \tag{3.82}$$

Thus

$$\hat{F} = \frac{1}{\pi^2} \int d^2\beta \int d^2\alpha \, \exp\left[-\frac{1}{2}(|\beta|^2 + |\alpha|^2)\right]$$

$$\times F(\beta^*, \alpha)|\beta\rangle \langle \alpha|. \tag{3.83}$$

Suppose now that $\hat{F}$ is a Hermitian operator with eigenstates $|\lambda\rangle$ such that

$$\hat{F} = \sum_\lambda \lambda|\lambda\rangle \langle \lambda|. \tag{3.84}$$

Then

$$\langle m|\hat{F}|n\rangle = \sum_\lambda \lambda\langle m|\lambda\rangle \langle \lambda|n\rangle. \tag{3.85}$$

But

$$|\langle m|\hat{F}|n\rangle| \le \sum_\lambda \lambda|\langle m|\lambda\rangle \langle \lambda|n\rangle|$$

$$\le \sum_\lambda \lambda = \mathrm{Tr} F, \tag{3.86}$$

which implies that $|\langle m|\hat{F}|n\rangle|$ has an upper bound. This being the case, it follows that the function $F(\beta^*, \alpha)$ is an entire function in both β^* and α.

The diagonal elements of an operator $\hat{F}$ in a coherent state basis completely determine the operator. From Eqs. (3.76) and (3.77), we have

$$\langle \alpha|\hat{F}|\alpha\rangle e^{\alpha^*\alpha} = \sum_n \sum_m \frac{\alpha^{*m} \alpha^n}{\sqrt{m!n!}} \langle m|\hat{F}|n\rangle. \tag{3.87}$$

Treating α and α^* as independent variables, it is apparent that

$$\frac{1}{\sqrt{m!n!}} \left[\frac{\partial^{n+m} \left(\langle \alpha | \hat{F} | \alpha \rangle e^{\alpha^* \alpha} \right)}{\partial \alpha^{*m} \partial \alpha^n} \right] \Bigg|_{\substack{\alpha=0 \\ \alpha^*=0}} = \langle m | \hat{F} | n \rangle. \qquad (3.88)$$

Thus, from "diagonal" coherent state matrix elements of $\hat{F}$ we can obtain all the matrix elements of the operator in the number basis.

3.6 Phase-Space Pictures of Coherent States

It is well known that the concept of phase space in quantum mechanics is problematic due to the fact that the canonical variables $\hat{x}$ and $\hat{p}$ are incompatible, that is they do not commute. Thus, the state of a system is not well localized as a point in phase space as it is in classical statistical mechanics. Nevertheless, we have shown that the coherent states minimize the uncertainty relation for the two orthogonal quadrature operators and that the uncertainties of the two quadratures are equal. Recall that $\hat{X}_1 = (\hat{a} + \hat{a}^\dagger)/2$ and $\hat{X}_2 = (\hat{a} - \hat{a}^\dagger)/2i$. These operators are dimensionless scaled position and momentum operators, respectively. Their coherent state expectation values are $\langle \hat{X}_1 \rangle_\alpha = \frac{1}{2}(\alpha + \alpha^*) = \text{Re}\,\alpha$, $\langle \hat{X}_2 \rangle_\alpha = \frac{1}{2i}(\alpha - \alpha^*) = \text{Im}\,\alpha$. Thus, the complex α-plane plays the role of phase space where, up to scale factors, the real and imaginary parts of α are coordinate and momentum variables, respectively. A coherent state $|\alpha\rangle$ with $\alpha = |\alpha|e^{i\theta}$ may be represented pictorially then as in Fig. 3.5, where the shaded circle represents the "area of uncertainty" of the coherent state, the fluctuations being equal in all directions of the phase space, with the center of the circle located at distance $|\alpha| = \langle \hat{n} \rangle^{1/2}$ from the origin and at angle θ above the position axis.

Further, $\Delta\theta$, in a qualitative sense, represents the phase uncertainty of the coherent state and it should be clear that $\Delta\theta$ diminishes for increasing $|\alpha|$, the fluctuations in X_1 and X_2 being independent of α and identical to

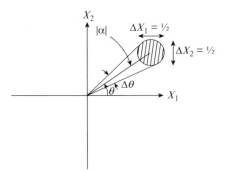

Fig. 3.5   Phase-space portrait of a coherent state of amplitude $|\alpha|$ and phase angle θ.

Note the error circle is the same for all coherent states. Note that as $|\alpha|$ increases, the phase uncertainty $\Delta\theta$ decreases, as would be expected in the "classical limit."

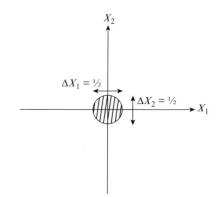

Fig. 3.6   Phase-space portrait of the quantum vacuum state.

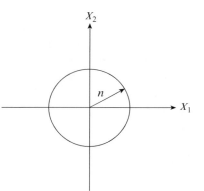

Fig. 3.7   Phase-space portrait of the number state $|n\rangle$. The uncertainty in the photon number is $\Delta n = 0$ while the phase is entirely random.

those of the vacuum. In fact, for the vacuum $|\alpha| = 0$, the phase-space representation is given in Fig. 3.6, where it is evident that uncertainty in the phase is as large as possible, that is $\Delta\theta = 2\pi$.

A number state $|n\rangle$ can be represented in phase space as a circle of radius n, the uncertainty in n being zero and the uncertainty in phase again being 2π, as in Fig. 3.7.

It must be understood that these pictures are qualitative in nature but are useful as a graphical way of visualizing the distribution of noise in various quantum states of the field. Since most quantum states have no classical analog (e.g. the number state does not), the corresponding phase-space portraits should not be taken too literally. Yet these representations will be quite useful when we discuss the nature of the squeezed states of light in Chapter 7.

Finally in this section, we make one more use of the phase-space diagrams, namely to illustrate the time evolution of quantum states for a non-interacting field. We have seen that for a non-interacting field, a coherent state $|\alpha\rangle$ evolves to the coherent state $|\alpha e^{-i\omega t}\rangle$. This can be pictured as a clockwise rotation of the error circle in phase, since $\langle \alpha e^{-i\omega t}|\hat{X}_1|\alpha e^{-i\omega t}\rangle = \alpha \cos \omega t$, $\langle \alpha e^{-i\omega t}|\hat{X}_2|\alpha e^{-i\omega t}\rangle = -\alpha \sin \omega t$, assuming α real. Since in the Schrödinger picture, the electric field operator is given, from Eqs. (2.15) and (2.53), as

$$\hat{E}_x = 2\mathcal{E}_0 \sin(kz)\hat{X}_1, \qquad (3.89)$$

the expectation value for the coherent state $|\alpha e^{-i\omega t}\rangle$ is

$$\langle \alpha e^{-i\omega t}|\hat{E}_x|\alpha e^{-i\omega t}\rangle = 2\mathcal{E}_0 \sin(kz)\alpha \cos (\omega t). \qquad (3.90)$$

Thus, apart from the scale factor $2\mathcal{E}_0 \sin(kz)$, the time evolution of the electric field is given by the projection of the $\langle \hat{X}_1 \rangle$ axis, and points within the error circle as a function of time, as indicated in Fig. 3.8.

The evolution of representative points within the error circle is shown, indicating the uncertainty of the electric field – the "quantum flesh" on the "classical bones," so to speak. Note that the greater the excitation of the field (i.e. α), the more classical the field appears since the fluctuations are independent of α. But the coherent state is the most classical of all the quantum states, so it is apparent that for the field in something other than a coherent state, the expectation value of the field may in no way appear classical-like. A number state is a very nonclassical state and, using representative points from its phase-space portrait, it is easy to see from Fig. 3.7 that the expectation value of the field is zero. But it is possible to imagine other kinds of states having no vanishing expectation values of the field where fluctuations may be less than those of a coherent state in one part of the field. These are the squeezed states, to be taken up in Chapter 7.

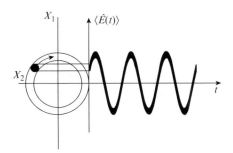

Fig. 3.8   The error circle of a coherent state (black dot) revolves about the origin of phase space at the oscillator angular frequency ω and the expectation value of the electric field is the projection onto an axis parallel to X_1.

3.7 Density Operators and Phase-Space Probability Distributions

Recall (see Appendix A) that for a mixture (or ensemble) of quantum states $|\psi_1\rangle$, $|\psi_2\rangle$, ..., the density operator is given as

$$\hat{\rho} = \sum_i p_i |\psi_i\rangle\langle\psi_i|, \tag{3.91}$$

where the p_i are the probabilities of finding the system in the ith member of the ensemble and

$$\mathrm{Tr}(\hat{\rho}) = \sum_i p_i = 1. \tag{3.92}$$

(For a pure state, $\hat{\rho} = |\psi\rangle\langle\psi|$.) The expectation value of an operator $\hat{O}$ is given by

$$\langle\hat{O}\rangle = \mathrm{Tr}(\hat{O}\hat{\rho}) = \sum_i p_i \langle\psi_i|\hat{O}|\psi_i\rangle. \tag{3.93}$$

As in Eq. (3.74), we can resolve unity in terms of the number states on both sides of the density operator to obtain

$$\hat{\rho} = \sum_n \sum_m |m\rangle \rho_{mn} \langle n|. \tag{3.94}$$

All the matrix elements $\rho_{mn} = \langle m|\hat{\rho}|n\rangle$ are required to completely determine the operator $\hat{\rho}$. The diagonal elements, $P_n = \rho_{nn}$, are just the probabilities of finding n photons in the field.

On the other hand, resolving unity with coherent states on both sides of $\hat{\rho}$, as in Eq. (3.75), results in

$$\hat{\rho} = \int\int \langle\alpha'|\hat{\rho}|\alpha''\rangle |\alpha'\rangle\langle\alpha''| \frac{d^2\alpha' d^2\alpha''}{\pi^2}. \tag{3.95}$$

But there is yet another way to represent $\hat{\rho}$ in terms of coherent states, namely

$$\hat{\rho} = \int |\alpha\rangle P(\alpha)\langle\alpha| d^2\alpha, \tag{3.96}$$

where $P(\alpha)$ is a weight function sometimes known as the Glauber–Sudarshan P function [2, 10]. The right-hand side of Eq. (3.91) is the "diagonal" form of the density operator and the P function is analogous to the phase-space distributions of statistical mechanics. Here, the real and imaginary parts of α are the variables of the phase space. $P(\alpha)$ must be real as $\hat{\rho}$ is a Hermitian operator. Also

$$\begin{aligned}
\text{Tr}\hat{\rho} &= \text{Tr}\int |\alpha\rangle P(\alpha)\langle\alpha| d^2\alpha \\
&= \int \sum_n \langle n|\alpha\rangle P(\alpha)\langle\alpha|n\rangle d^2\alpha \\
&= \int P(\alpha)\sum_n \langle\alpha|n\rangle\langle n|\alpha\rangle d^2\alpha \\
&= \int P(\alpha)\langle\alpha|\alpha\rangle d^2\alpha = \int P(\alpha) d^2\alpha = 1,
\end{aligned} \qquad (3.97)$$

just as we would expect of a phase-space probability distribution.

But for some quantum states of the field, $P(\alpha)$ can have properties quite unlike those of any *true* probability distribution where one would expect to have $P(\alpha) \geq 0$. There are quantum states for which $P(\alpha)$ is negative or highly singular. In these cases, the corresponding quantum states are called "nonclassical." In fact, we may define a nonclassical state as a state for which the corresponding $P(\alpha)$ is negative in some regions of phase space (the α-plane), or is more singular than a delta function (the reason for the second criterion will become clear shortly). To speak of nonclassical states of light might seem an oxymoron. After all, aren't *all* states of light quantum mechanical? Well, yes, *all* states of light *are* quantum mechanical, but, waxing Orwellian, it turns out that some states are more quantum mechanical than others. States for which $P(\alpha)$ is positive, or no more singular than a delta function, are, in this sense, classical. Coherent states, which we have already shown to be quasi-classical in that they describe states of the field having properties close to what we would expect for classical oscillating coherent fields, have P functions given as delta functions, as we shall show, and are therefore classical in the sense described here. Certain effects, among them being quadrature and amplitude (or number) squeezing, can occur only for states for which the P functions are negative or highly singular. For that reason, the various forms of squeezing are known as distinctly nonclassical effects. These effects will be discussed in Chapter 7.

But how do we calculate $P(\alpha)$? A rather general procedure has been given by Mehta [11], which we follow here. Starting with Eq. (3.91) and using coherent state $|u\rangle$ and $|-u\rangle$, we have

$$\begin{aligned}
\langle -u|\hat{\rho}|u\rangle &= \int \langle -u|\alpha\rangle P(\alpha)\langle\alpha|u\rangle\ d^2\alpha \\
&= \int P(\alpha)\exp\left[-\frac{1}{2}|u|^2 - \frac{1}{2}|\alpha|^2 - u^*\alpha\right] \\
&\quad \times \exp\left[-\frac{1}{2}|\alpha|^2 - \frac{1}{2}|u|^2 + \alpha^* u\right]\ d^2\alpha \\
&= e^{-|u|^2}\int P(\alpha)\ e^{-|\alpha|^2} e^{\alpha^* u - \alpha u^*} d^2\alpha.
\end{aligned} \qquad (3.98)$$

Now let $\alpha = x + iy$ and $u = x' + iy'$ so that $\alpha^* u - \alpha u^* = 2i(x'y - xy')$. We then define the Fourier transforms in the complex plane:

$$g(u) = \int f(\alpha) e^{\alpha^* u - \alpha u^*} d^2\alpha, \tag{3.99}$$

$$f(\alpha) = \frac{1}{\pi^2} \int g(u) e^{u^*\alpha - u\alpha^*} d^2u. \tag{3.100}$$

With the identifications

$$g(u) = e^{|u|^2} \langle -u|\hat{\rho}|u\rangle,$$
$$f(\alpha) = P(\alpha) e^{-|u|^2}, \tag{3.101}$$

we then obtain from Eq. (3.100),

$$P(\alpha) = \frac{e^{|\alpha|^2}}{\pi^2} \int e^{|u|^2} \langle -u|\hat{\rho}|u\rangle e^{u^*\alpha - u\alpha^*} d^2u. \tag{3.102}$$

Care must be taken in regard to the convergence of the integral, since $e^{|u|^2} \to \infty$ as $|u| \to \infty$.

Now let us consider some examples. First, we consider the pure coherent state $|\beta\rangle$, where $\hat{\rho} = |\beta\rangle\langle\beta|$. With

$$\begin{aligned} \langle -u|\hat{\rho}|u\rangle &= \langle -u|\beta\rangle\langle\beta|u\rangle \\ &= e^{-|\beta|^2} e^{-|u|^2} e^{-u^*\beta + \beta^* u}, \end{aligned} \tag{3.103}$$

then

$$P(\alpha) = e^{|\alpha|^2} e^{-|\beta|^2} \frac{1}{\pi^2} \int e^{u^*(\alpha-\beta) - u(\alpha^* - \beta^*)} d^2u. \tag{3.104}$$

But the Fourier integral is just the two-dimensional form of the Dirac delta function:

$$\begin{aligned} \delta^2(\alpha - \beta) &= \delta[\text{Re}(\alpha) - \text{Re}(\beta)]\delta[\text{Im}(\alpha) - \text{Im}(\beta)] \\ &= \frac{1}{\pi^2} \int e^{u^*(\alpha-\beta) - u(\alpha^* - \beta^*)} d^2u, \end{aligned} \tag{3.105}$$

so that

$$P(\alpha) = \delta^2(\alpha - \beta). \tag{3.106}$$

This is the same as the distribution for a classical harmonic oscillator.

But if a coherent state is a classical-like state, then the number state $|n\rangle$ is at the other extreme, a state that cannot at all be described classically. For a pure number state $|n\rangle$, $\hat{\rho} = |n\rangle\langle n|$ and

$$\langle -u|\hat{\rho}|u\rangle = \langle -u|n\rangle\langle n|u\rangle = e^{-|u|^2} \frac{(-u^* u)^n}{n!}. \tag{3.107}$$

Thus

$$P(\alpha) = \frac{e^{|\alpha|^2}}{n!} \frac{1}{\pi^2} \int (-u^*u)^n e^{u^*\alpha - u\alpha^*} d^2u. \tag{3.108}$$

The integral does not exist in terms of ordinary functions. Formally, we can write

$$
\begin{aligned}
P(\alpha) &= \frac{e^{|\alpha|^2}}{n!} \frac{\partial^{2n}}{\partial \alpha^n \partial \alpha^{*n}} \frac{1}{\pi^2} \int e^{u^*\alpha - u\alpha^*} d^2u \\
&= \frac{e^{|\alpha|^2}}{n!} \frac{\partial^{2n}}{\partial \alpha^n \partial \alpha^{*n}} \delta^{(2)}(\alpha).
\end{aligned}
\tag{3.109}
$$

The derivative of the delta function, called a tempered distribution, is more singular than the delta function, and has meaning only under the integral sign, for example for some function $F(\alpha, \alpha^*)$,

$$
\begin{aligned}
\int F(\alpha, \alpha^*) & \frac{\partial^{2n}}{\partial \alpha^n \partial \alpha^{*n}} \delta^{(2)}(\alpha) d^2\alpha \\
&= \left[\frac{\partial^{2n} F(\alpha, \alpha^*)}{\partial \alpha^n \partial \alpha^{*n}} \right]_{\substack{\alpha=0 \\ \alpha^*=0}} .
\end{aligned}
\tag{3.110}
$$

At this point we introduce the optical equivalence theorem of Sudarshan [10]. Suppose we have a "normally ordered" function of the operators $\hat{a}$ and $\hat{a}^\dagger$, $G^{(N)}(\hat{a}, \hat{a}^\dagger)$, where the annihilation operators stand to the right of the creation operators:

$$\hat{G}^{(N)}(\hat{a}, \hat{a}^\dagger) = \sum_n \sum_m C_{nm} (\hat{a}^\dagger)^n \hat{a}^m. \tag{3.111}$$

(It should not be hard to guess what an *anti*-normally ordered operator looks like!) The average of this function is

$$
\begin{aligned}
\langle G^{(N)}(\hat{a}, \hat{a}^\dagger) \rangle &= \mathrm{Tr}\left[\hat{G}^{(N)}(\hat{a}, \hat{a}^\dagger) \hat{\rho} \right] \\
&= \mathrm{Tr} \int P(\alpha) \sum_n \sum_m C_{nm} (\hat{a}^\dagger)^n \hat{a}^m |\alpha\rangle \langle\alpha| d^2\alpha \\
&= \int P(\alpha) \sum_n \sum_m C_{nm} \langle\alpha| (\hat{a}^\dagger)^n \hat{a}^m |\alpha\rangle d^2\alpha \\
&= \int P(\alpha) \sum_n \sum_m C_{nm} (\alpha^*)^n \alpha^m d^2\alpha \\
&= \int P(\alpha) G^{(N)}(\alpha, \alpha^*) d^2\alpha.
\end{aligned}
\tag{3.112}
$$

The last line is the **optical equivalence theorem**: the expectation value of a normally ordered operator is just the P function weighted average of the function obtained from the operator by the replacements $\hat{a} \to \alpha$ and $\hat{a}^\dagger \to \alpha^*$.

Normally ordered operators will be important later, particularly in the discussion of the photoelectric detection of light resulting from the absorption of photons. It is useful to introduce the normal ordering operator denoted $::$. For an arbitrary function of $\hat{a}$ and $\hat{a}^\dagger$, $O(\hat{a}, \hat{a}^\dagger)$, we have

$$:O(\hat{a}, \hat{a}^\dagger): \equiv O^{(N)}(\hat{a}, \hat{a}^\dagger), \tag{3.113}$$

where the commutation relations are to be *disregarded*. The number operator $\hat{n} = \hat{a}^\dagger \hat{a}$ is already normally ordered, so

$$\langle \hat{n} \rangle = \langle \hat{a}^\dagger \hat{a} \rangle = \int P(\alpha) |\alpha|^2 d^2\alpha. \tag{3.114}$$

But $\hat{n}^2 = \hat{a}^\dagger \hat{a} \hat{a}^\dagger \hat{a}$ is not normally ordered. Thus

$$:\hat{n}^2: = (\hat{a}^\dagger)^2 \hat{a}^2 \tag{3.115}$$

and

$$\langle : \hat{n}^2 : \rangle = \langle (\hat{a}^\dagger)^2 \hat{a}^2 \rangle = \int P(\alpha) |\alpha|^4 d^2\alpha. \tag{3.116}$$

The utility of the normal ordering operator and the optical equivalence theorem will be evident in a later chapter.

Consider now some operator $\hat{B}$ defined in the Hilbert space of the single-mode quantized field. We can expand this operator in terms of the number states by twice using the completeness relation for those states to obtain

$$\hat{B} = \sum_{m=0}^{\infty} \sum_{n=0}^{\infty} |m\rangle \langle m|\hat{B}|n\rangle \langle n| = \sum_{m=0}^{\infty} \sum_{n=0}^{\infty} |m\rangle B_{mn} \langle n|, \tag{3.117}$$

where $B_{mn} = \langle m|\hat{B}|n\rangle$. But we can also expand the operator using twice the closure relation for the coherent states to arrive at

$$\hat{B} = \int \int |\alpha'\rangle \langle \alpha'|\hat{B}|\alpha\rangle \langle \alpha| \frac{d^2\alpha' d^2\alpha}{\pi^2},$$
$$= \int \int |\alpha'\rangle B(\alpha'^*, \alpha) \langle \alpha| \frac{d^2\alpha' d^2\alpha}{\pi^2}, \tag{3.118}$$

where

$$B(\alpha'^*, \alpha) = \langle \alpha'|\hat{B}|\alpha\rangle$$
$$= \exp[-(|\alpha'|^2 + |\alpha|^2)/2] \sum_{m=0}^{\infty} \sum_{n=0}^{\infty} B_{mn} \frac{(\alpha'^*)^m \alpha^n}{\sqrt{m!n!}}. \tag{3.119}$$

Yet another way to represent the operator is, in analogy to the representation of the density operator in Eq. (3.96), to introduce the "diagonal" coherent state form

$$\hat{B} = \int |\alpha\rangle B_P(\alpha, \alpha^*) \langle \alpha| \frac{d^2\alpha}{\pi}, \tag{3.120}$$

where $B_P(\alpha, \alpha^*)$ is the *P*-representative of $\hat{B}$. The average of $\hat{B}$ is given by

$$\langle \hat{B} \rangle = \text{Tr}(\hat{B}\hat{\rho})$$

$$= \sum_n \langle n| \int B_p(\alpha, \alpha^*) \, |\alpha\rangle\langle\alpha|\hat{\rho}|n\rangle \frac{d^2\alpha}{\pi}$$

$$= \int B_p(\alpha, \alpha^*)\langle\alpha|\hat{\rho}|\alpha\rangle \frac{d^2\alpha}{\pi}. \tag{3.121}$$

Evidently, the expectation value of the density operator with respect to the coherent state also plays the role of a phase-space probability distribution. This is usually called the Q, or Husimi, function [12]:

$$Q(\alpha) = \frac{\langle\alpha|\hat{\rho}|\alpha\rangle}{\pi}. \tag{3.122}$$

For $\hat{B} = \hat{I}$ we obtain the normalization condition

$$\int Q(\alpha) \, d^2\alpha = 1. \tag{3.123}$$

Unlike the P function, the Q function is positive for all quantum states.

But now let us calculate the expectation value of $\hat{B}$ using the P-representation of $\hat{\rho}$:

$$\langle \hat{B} \rangle = \text{Tr}(\hat{B}\hat{\rho})$$

$$= \sum_{n=0}^{\infty} \int \langle n|\hat{B}|\alpha\rangle P(\alpha)\langle\alpha|n\rangle d^2\alpha \tag{3.124}$$

$$= \int B_Q(\alpha, \alpha^*) P(\alpha) d^2\alpha,$$

where

$$B_Q(\alpha, \alpha^*) \equiv \langle\alpha|\hat{B}|\alpha\rangle$$

$$= e^{-|\alpha|^2} \sum_m \sum_n \frac{B_{mn}}{(m!n!)^{1/2}} \, (\alpha^*)^m \alpha^n, \tag{3.125}$$

which constitutes the Q-representation of $\hat{B}$. Thus, in calculating expectation values, if we use the P-representation of $\hat{\rho}$, we need the Q-representation of $\hat{B}$ [Eq. (3.125)], or if we use the P-representation of $\hat{B}$, we need the Q-representation of $\hat{\rho}$, that is the Q function of Eq. (3.122).

The Q function has the character of a probability distribution in the sense of positivity, whereas the P function is really a quasi-probability distribution. There is, in fact, one other important quasi-probability distribution over phase space, namely the Wigner function. The Wigner function seems to be the earliest introduced of the phase-space quasi-probability distributions, making its debut in 1932 [13]. It is defined, for an arbitrary density operator $\hat{\rho}$, according to

$$F_W(q,p) \equiv \frac{1}{2\pi\hbar} \int_{-\infty}^{\infty} \left\langle q + \frac{1}{2}x|\hat{\rho}|q - \frac{1}{2}x\right\rangle e^{-ipx/\hbar}dx, \qquad (3.126)$$

where $|q \pm \frac{1}{2}x\rangle$ are the eigenkets of the position operator. If the state in question is a pure state $\hat{\rho} = |\psi\rangle\langle\psi|$, then

$$F_W(q,p) \equiv \frac{1}{2\pi\hbar} \int_{-\infty}^{\infty} \psi^* \left(q - \frac{1}{2}x\right) \psi\left(q + \frac{1}{2}x\right) e^{-ipx/\hbar}dx, \qquad (3.127)$$

where $\langle q + \frac{1}{2}x|\psi\rangle = \psi(q + \frac{1}{2}x)$, and so on. Integrating over the momentum, we obtain

$$\begin{aligned}
\int_{-\infty}^{\infty} F_W(q,p)\, dp &= \frac{1}{2\pi\hbar} \int_{-\infty}^{\infty} \psi^*\left(q - \frac{1}{2}x\right) \psi\left(q + \frac{1}{2}x\right) \int_{-\infty}^{\infty} e^{-ipx/\hbar}dpdx \\
&= \int_{-\infty}^{\infty} \psi^*\left(q - \frac{1}{2}x\right) \psi\left(q + \frac{1}{2}x\right) \delta(x)dx \\
&= |\psi(q)|^2,
\end{aligned} \qquad (3.128)$$

which is the probability density for position variable q. Likewise, integrating over q, we obtain

$$\int_{-\infty}^{\infty} F_W(q,p)\, dq = |\varphi(p)|^2, \qquad (3.129)$$

where $\varphi(p)$ is the momentum space wave related to the coordinate space wave function $\psi(q)$ by a Fourier transform. The right-hand side of Eq. (3.129) is, of course, just the probability density in momentum space. But $F_W(q,p)$ itself is not a true probability distribution since it can take on negative values for nonclassical states, as we shall show. Like the other distributions, the Wigner function can be used to calculate averages. However, functions of the operators $\hat{q}$ and $\hat{p}$ to be averaged must be Weyl, or symmetrically, ordered in terms of these operators. For example, the classical function qp must be replaced by $(\hat{q}\hat{p} + \hat{p}\hat{q})/2$ and

$$\langle \hat{q}\hat{p} + \hat{p}\hat{q}\rangle = \int (qp + pq)\, F_W(q,p)dqdp\,. \qquad (3.130)$$

In general, if $\{G(\hat{q},\hat{p})\}_W$ is a Weyl-ordered function, where the bracket $\{\quad\}_W$ means Weyl, or symmetric, ordering, then

$$\langle\{G(\hat{q},\hat{p})\}_W\rangle = \int \{G(q,p)\}_W\, F_W(q,p)dqdp \qquad (3.131)$$

is the corresponding phase-space average.

For use in Section 3.8, an alternative expression for the Wigner function of Eq. (3.126) is given by

$$F_W(q,p) \equiv \frac{1}{\pi\hbar} \int_{-\infty}^{\infty} \langle q + s|\hat{\rho}|q - s\rangle e^{-2ips/\hbar} ds, \qquad (3.132)$$

where we have simply made the change of variable $x = 2s$.

3.8 The Photon Number Parity Operator and the Wigner Function

It turns out that the Wigner function can be expressed as the expectation value of the displaced photon number parity operator. The operator in question has many equivalent representations, but the simplest one is as follows. We define this operator, denoted $\hat{\Pi}$, as

$$\hat{\Pi} = (-1)^{\hat{a}^\dagger \hat{a}} = \sum_{n=0}^{\infty} (-1)^n |n\rangle\langle n|. \qquad (3.133)$$

Obviously this is a dichotomic observable (it is represented by a Hermitian operator) having eigenvalues $+1$ for even photon numbers and -1 for odd photon numbers. It is an unusual quantum observable as it has no classical analogue. An alternative representation of the parity operator is the integral form

$$\hat{\Pi} = \int_{-\infty}^{\infty} dx |-x\rangle\langle x|, \qquad (3.134)$$

where the ket vector $|x\rangle$ is an eigenstate of the position operator $\hat{x}$ according to $\hat{x}|x\rangle = x|x\rangle$, $-\infty < x < \infty$. Furthermore, these states satisfy the completeness relation

$$\hat{\mathbf{I}} = \int_{-\infty}^{\infty} dx |x\rangle\langle x|. \qquad (3.135)$$

To see the equivalence between these two expressions of the parity operator, we begin by taking the expectation value of Eq. (3.134) with respect to the number state $|n\rangle$ such that

$$\langle n|\hat{\Pi}|n\rangle = (-1)^n = \int_{-\infty}^{\infty} dx \langle n|-x\rangle\langle x|n\rangle. \qquad (3.136)$$

The amplitudes $\langle x|n\rangle$ are just the well-known energy eigenstates of the quantum harmonic oscillator problem of a particle with unit mass, which are given by

$$\langle x|n\rangle = \psi_n(x) = \frac{1}{\sqrt{2^n n!}} \left(\frac{\omega}{\pi\hbar}\right)^{1/4} e^{-\xi^2} H_n(\xi), \ \xi = x\left(\frac{\omega}{\hbar}\right)^{1/2}, \qquad (3.137)$$

where the $H_n(\xi)$ are the Hermite polynomials. Because of the property $H_n(-\xi) = (-1)^n H_n(\xi)$, we have that $\langle -x|n\rangle = (-1)^n \langle x|n\rangle$. Substituting this result into Eq. (3.136), we have

$$\langle n|\hat{\Pi}|n\rangle = (-1)^n = \int_{-\infty}^{\infty} dx (-1)^n \langle n|x\rangle \langle x|n\rangle$$

$$= (-1)^n \int_{-\infty}^{\infty} dx \langle n|x\rangle \langle x|n\rangle, \qquad (3.138)$$

where we have used the fact that $\langle x|n\rangle = \langle n|x\rangle$ as these are real functions. We can now use the completeness relation of Eq. (3.135) and the normalization condition $\langle n|n\rangle = 1$ to arrive at

$$\int_{-\infty}^{\infty} dx \langle n|x\rangle \langle x|n\rangle = 1, \qquad (3.139)$$

which completes the demonstration that the parity operator can be represented as given by Eq. (3.134).

Now we consider the displacement operator $\hat{D}(\alpha) = \exp(\alpha \hat{a}^{\dagger} - \alpha^* \hat{a})$ and set

$$\alpha = (2\hbar\omega)^{-1/2}(\omega q + ip) \quad \text{and} \quad \alpha^* = (2\hbar\omega)^{-1/2}(\omega q - ip), \qquad (3.140)$$

where q and p are c-number phase-space variables, not to be confused with the operators $\hat{q}$ and $\hat{p}$. Then, using Eqs. (2.13) and (2.14), our displacement operator takes the form

$$\hat{D}(\alpha) = \exp\left\{\frac{i}{\hbar}(p\hat{q} - q\hat{p})\right\}. \qquad (3.141)$$

Using the first line of the disentangling theorem of Eq. (3.32), the displacement operator can be decomposed according to

$$\hat{D}(\alpha) = e^{-\frac{i}{2\hbar}pq} e^{\frac{i}{\hbar}p\hat{q}} e^{-\frac{i}{\hbar}q\hat{p}}. \qquad (3.142)$$

The factor on the far right of this operator is the translation operator [14] given in one dimension as $\hat{T}(q) = \exp(-iq\hat{p}/\hbar)$, $\hat{p}$ being the generator of translations. This operator has the action $\hat{T}(q)|x\rangle = |x + q\rangle$.

Now apply the displacement operator in the form of Eq. (3.142) to the states $|\mp s\rangle$. This produces the result

$$\hat{D}(\alpha)|-s\rangle = e^{-ipq/2\hbar}e^{ip(q-s)/\hbar}|q-s\rangle,$$
$$\hat{D}(\alpha)|+s\rangle = e^{-ipq/2\hbar}e^{ip(q+s)/\hbar}|q+s\rangle. \tag{3.143}$$

From these we find that

$$\langle q+s|\hat{\rho}|q-s\rangle = e^{2ips/\hbar}\langle s|\hat{D}^{\dagger}(\alpha)\hat{\rho}\hat{D}(\alpha)|-s\rangle, \tag{3.144}$$

such that Eq. (3.132) now becomes

$$F_W(q,p) = \frac{1}{\pi\hbar}\int_{-\infty}^{\infty}\langle s|\hat{D}^{\dagger}(\alpha)\hat{\rho}\hat{D}(\alpha)|-s\rangle ds. \tag{3.145}$$

Taking the general density operator to have the form $\hat{\rho} = \sum_{i,j}c_{ij}|\psi_i\rangle\langle\psi_j|$, we have

$$F_W(q,p) = \frac{1}{\pi\hbar}\sum_{i,j}c_{ij}\int_{-\infty}^{\infty}\langle s|\hat{D}^{\dagger}(\alpha)|\psi_i\rangle\langle\psi_j|\hat{D}(\alpha)|-s\rangle ds$$

$$= \frac{1}{\pi\hbar}\sum_{i,j}c_{ij}\int_{-\infty}^{\infty}\langle\psi_j|\hat{D}(\alpha)|-s\rangle\langle s|\hat{D}^{\dagger}(\alpha)|\psi_i\rangle ds. \tag{3.146}$$

Integrating over s and then using the results of Eq. (3.134) and then of Eq. (3.133), we arrive at

$$F_W(q,p) = \frac{1}{\pi\hbar}\sum_{i,j}c_{ij}\sum_{n=0}^{\infty}\langle\psi_j|\hat{D}(\alpha)(-1)^n|n\rangle\langle n|\hat{D}^{\dagger}(\alpha)|\psi_i\rangle$$

$$= \frac{1}{\pi\hbar}\langle\hat{D}(\alpha)\hat{\Pi}\hat{D}^{\dagger}(\alpha)\rangle, \tag{3.147}$$

where $\hat{D}^{\dagger}(\alpha) = \hat{D}(-\alpha)$. Thus the Wigner function can be expressed as the expectation value of the displaced parity operator. That the Wigner function can be expressed in this way was explicitly noted by Royer [15], but it appears even earlier in the work of Cahill and Glauber [16].

That the right-hand side of Eq. (3.147) is expressed as a function of α must be taken into account in the normalization condition. From Eqs. (3.140) we find that $dqdp = 2\hbar d\,\mathrm{Re}(\alpha)d\,\mathrm{Im}(\alpha) = 2\hbar d^2\alpha$, so that the normalization condition becomes

$$\iint F_W(q,p)dqdp = \int W(\alpha)d^2\alpha, \tag{3.148}$$

where we define

$$W(\alpha) \equiv \frac{2}{\pi}\langle\hat{D}(\alpha)\hat{\Pi}\hat{D}^{\dagger}(\alpha)\rangle, \tag{3.149}$$

which we refer to in what follows also as the Wigner function. It is essentially the "quantum optical" Wigner function.

As was shown by Cahill and Glauber [16], the above expression for the Wigner function can be further simplified by taking note of the fact that the parity operator can be written as $\hat{\Pi} = (-1)^{\hat{a}^\dagger \hat{a}} = \exp(i\pi\hat{a}^\dagger\hat{a})$, which means we can write

$$\hat{D}(\alpha)\hat{\Pi}\hat{D}^\dagger(\alpha) = \hat{D}(\alpha)e^{i\pi\hat{a}^\dagger\hat{a}}\hat{D}(-\alpha)e^{-i\pi\hat{a}^\dagger\hat{a}}e^{i\pi\hat{a}^\dagger\hat{a}}$$
$$= \hat{D}(2\alpha)\hat{\Pi}, \tag{3.150}$$

where we used the fact that the parity operator as a transformation operator performs the reflection

$$e^{i\pi\hat{a}^\dagger\hat{a}}\begin{pmatrix} \hat{a} \\ \hat{a}^\dagger \end{pmatrix}e^{-i\pi\hat{a}^\dagger\hat{a}} = \begin{pmatrix} -\hat{a} \\ -\hat{a}^\dagger \end{pmatrix}, \tag{3.151}$$

such that

$$e^{i\pi\hat{a}^\dagger\hat{a}}\hat{D}(-\alpha)e^{-i\pi\hat{a}^\dagger\hat{a}} = \hat{D}(\alpha). \tag{3.152}$$

We then used the fact that $\hat{D}(\alpha)\hat{D}(\alpha) = \hat{D}(2\alpha)$. Thus we finally have

$$W(\alpha) = \frac{2}{\pi}\langle\hat{D}(2\alpha)\hat{\Pi}\rangle. \tag{3.153}$$

This is a very useful form for obtaining the Wigner function in the context of quantum optics. For a pure state of the form $|\psi\rangle = \sum_{n=0}^{\infty} C_n|n\rangle$, we have

$$W(\alpha) = \frac{2}{\pi}\sum_{n'=0}^{\infty}\sum_{n=0}^{\infty} C_{n'}^* C_n (-1)^n \langle n'|\hat{D}(2\alpha)|n\rangle, \tag{3.154}$$

where the matrix elements of the displacement operator are given by Eq. (3.48).

3.9 Characteristic Functions

Consider for a moment a classical random variable that we denote x. Suppose that $\rho(x)$ is a classical probability density associated with the variable x. Then it follows that

$$\rho(x) \geq 0 \tag{3.155}$$

and

$$\int \rho(x)dx = 1. \tag{3.156}$$

The nth moment of x is defined as

$$\langle x^n \rangle = \int dx\ x^n \rho(x). \tag{3.157}$$

If all moments $\langle x^n \rangle$ are known, then $\rho(x)$ is completely specified. This can be seen by introducing the characteristic function

$$C(k) = \langle e^{ikx} \rangle = \int dx\ e^{ikx} \rho(x)$$
$$= \sum_{n=0}^{\infty} \frac{(ik)^n}{n!} \langle x^n \rangle. \tag{3.158}$$

Evidently, the probability density is just the Fourier transform of the characteristic function:

$$\rho(x) = \frac{1}{2\pi} \int dk e^{-ikx} C(k). \tag{3.159}$$

Thus, if all the moments $\langle x^n \rangle$ are known and if $C(k)$ is known, then $\rho(x)$ is known. On the other hand, if we are given the characteristic function, we can calculate the moments according to

$$\langle x^n \rangle = \frac{1}{i^n} \frac{d^n C(k)}{dk^n} \bigg|_{k=0}. \tag{3.160}$$

We now proceed to introduce quantum mechanical characteristic functions. There are, in fact, three such functions:

$$C_W(\lambda) = \mathrm{Tr}[\hat{\rho} e^{\lambda \hat{a}^\dagger - \lambda^* \hat{a}}] = \mathrm{Tr}[\hat{\rho} \hat{D}(\lambda)] \qquad \text{(Wigner)}, \tag{3.161}$$

$$C_N(\lambda) = \mathrm{Tr}[\hat{\rho} e^{\lambda \hat{a}^\dagger} e^{-\lambda^* \hat{a}}] \qquad \text{(normally ordered)}, \tag{3.162}$$

$$C_A(\lambda) = \mathrm{Tr}[\hat{\rho} e^{-\lambda^* \hat{a}} e^{\lambda \hat{a}^\dagger}] \qquad \text{(anti-normally ordered)}. \tag{3.163}$$

These functions are related through the disentangling theorem of Eq. (3.32) according to

$$C_W(\lambda) = C_N(\lambda)\ e^{-\frac{1}{2}|\lambda|^2} = C_A(\lambda) e^{\frac{1}{2}|\lambda|^2}. \tag{3.164}$$

Furthermore, it is easy to show that

$$\langle (\hat{a}^\dagger)^m \hat{a}^n \rangle = \mathrm{Tr}[\hat{\rho} (\hat{a}^\dagger)^m \hat{a}^n] = \frac{\partial^{(m+n)}}{\partial \lambda^m \partial(-\lambda^*)^n} C_N(\lambda) \bigg|_{\lambda=0}, \tag{3.165}$$

$$\langle \hat{a}^m (\hat{a}^\dagger)^n \rangle = \mathrm{Tr}[\hat{\rho} \hat{a}^m (\hat{a}^\dagger)^n] = \frac{\partial^{(m+n)}}{\partial \lambda^n \partial(-\lambda^*)^m} C_A(\lambda) \bigg|_{\lambda=0}, \tag{3.166}$$

$$\langle \{(\hat{a}^\dagger)^m \hat{a}^n\}_W \rangle = \text{Tr}[\hat{\rho}\{(\hat{a}^\dagger)^m \hat{a}^n\}_W]$$

$$= \frac{\partial^{(m+n)}}{\partial \lambda^m \partial(-\lambda^*)^n} C_W(\lambda)\Big|_{\lambda=0}. \tag{3.167}$$

Instead of the three different characteristic functions, we can introduce the s-parameterized function of Cahill and Glauber [16]:

$$C(\lambda, s) = \text{Tr}[\hat{\rho} \exp(\lambda \hat{a}^\dagger - \lambda^* \hat{a} + s|\lambda|^2/2)], \tag{3.168}$$

such that $C(\lambda, 0) = C_W(\lambda), C(\lambda, 1) = C_N(\lambda)$, and $C(\lambda, -1) = C_A(\lambda)$.

The connections between these characteristic functions and the various quasi-probability distributions will now be made. For example, the anti-normally ordered characteristic function may be written as

$$C_A(\lambda) = \text{Tr}[\hat{\rho} e^{-\lambda^* \hat{a}} e^{\lambda \hat{a}^\dagger}]$$

$$= \text{Tr}[e^{\lambda \hat{a}^\dagger} \hat{\rho} e^{-\lambda^* \hat{a}}]$$

$$= \frac{1}{\pi} \int d^2\alpha \langle \alpha | e^{\lambda \hat{a}^\dagger} \hat{\rho} e^{-\lambda^* \hat{a}} | \alpha \rangle \tag{3.169}$$

$$= \int d^2\alpha \, Q(\alpha) e^{\lambda \alpha^* - \lambda^* \alpha},$$

which is just a two-dimensional Fourier transform of the Q function. The inverse Fourier transform yields

$$Q(\alpha) = \frac{1}{\pi^2} \int C_A(\lambda) e^{\lambda^* \alpha - \lambda \alpha^*} d^2\lambda. \tag{3.170}$$

Now consider the normally ordered characteristic function. Writing $\hat{\rho}$ in the P-representation, we have

$$C_N(\lambda) = \text{Tr}[\hat{\rho} e^{\lambda \hat{a}^\dagger} e^{-\lambda^* \hat{a}}]$$

$$= \int P(\alpha) \langle \alpha | e^{\lambda \hat{a}^\dagger} e^{-\lambda^* \hat{a}} | \alpha \rangle d^2\alpha \tag{3.171}$$

$$= \int P(\alpha) \, e^{\lambda \alpha^* - \lambda^* \alpha} d^2\alpha,$$

which is just the Fourier transform of the P function. The inverse transform yields

$$P(\alpha) = \frac{1}{\pi^2} \int e^{\lambda^* \alpha - \lambda \, \alpha^*} C_N(\lambda) d^2\lambda. \tag{3.172}$$

Finally, and perhaps to no great surprise by now, it turns out that the Wigner function may be obtained as the Fourier transform of the Weyl-ordered characteristic function:

$$W(\alpha) \equiv \frac{1}{\pi^2} \int \exp(\lambda^* \alpha - \lambda \alpha^*) \; C_W(\lambda) d^2\lambda$$

$$= \frac{1}{\pi^2} \int \exp(\lambda^* \alpha - \lambda \alpha^*) \; C_N(\lambda) e^{-|\lambda|^2/2} \; d^2\lambda. \tag{3.173}$$

This definition is equivalent to the previous one with the proper interpretation of variables.

As an application of the characteristic function, we now derive the P function corresponding to the thermal, or chaotic, state of the field. Recall that the field in this case is a mixed state given by the density operator $\hat{\rho}_{Th}$ of Eq. (2.145). We first calculate the Q function for this density operator according to

$$Q(\alpha) = \langle \alpha | \hat{\rho}_{Th} | \alpha \rangle / \pi$$

$$= \frac{1}{\pi} \; e^{-|\alpha|^2} \sum_n \sum_m \langle m | \hat{\rho}_{Th} | n \rangle \; \frac{(\alpha^*)^m \alpha^n}{(m!n!)^{1/2}}$$

$$= \frac{e^{-|\alpha|^2}}{\pi(1+\bar{n})} \; \sum_m \left(\frac{\bar{n}}{1+\bar{n}}\right)^n \frac{(\alpha^*\alpha)^n}{n!}$$

$$= \frac{1}{\pi(1+\bar{n})} \exp\left(-\frac{|\alpha|^2}{1+\bar{n}}\right) \;, \tag{3.174}$$

where $\bar{n}$ for the thermal state is given by Eq. (2.142). Then from Eq. (3.169) we have

$$C_A(\lambda) = \frac{1}{\pi(1+\bar{n})} \int d^2\alpha \, \exp\left(-\frac{|\alpha|^2}{1+\bar{n}}\right) e^{\lambda\alpha^* - \lambda^*\alpha}. \tag{3.175}$$

Now letting $\alpha = (q+ip)/\sqrt{2}, \quad \lambda = (x+iy)/\sqrt{2}$ where $d^2\alpha = dqdp/2$, we have

$$C_A(x,y) = \frac{1}{2\pi(1+\bar{n})} \int \exp\left[-\frac{(q^2+p^2)}{2(1+\bar{n})}\right]$$

$$\times \exp[i(yq - xp)] \; dqdp. \tag{3.176}$$

Using the standard Gaussian integral

$$\int e^{-as^2} e^{\pm\beta s} ds = \sqrt{\frac{\pi}{a}} \, e^{-\beta^2/4a}, \tag{3.177}$$

we straightforwardly obtain

$$C_A(\lambda) = \exp[-(1+\bar{n})|\lambda|^2] \;. \tag{3.178}$$

But from Eq. (3.164) we have $C_N(\lambda) = C_A(\lambda)\exp(|\lambda|^2)$ and thus from Eq. (3.172) we finally obtain

$$P(\alpha) = \frac{1}{\pi^2} \int \exp(-\overline{n} \, |\lambda|^2) \; e^{\lambda^* \alpha - \lambda \alpha^*} d^2\lambda$$

$$= \frac{1}{\pi \overline{n}} \; \exp\left(-\frac{|\alpha|^2}{\overline{n}}\right). \tag{3.179}$$

This is Gaussian so it may be interpreted as a true probability distribution.

Finally, to conclude this chapter, we examine the Q functions and Wigner functions for the most classical quantum state, the coherent state, and the most quantum mechanical state, the number state. For the coherent state $\hat{\rho} = |\beta\rangle \langle\beta|$, one easily has the Q function

$$Q(\alpha) = \frac{1}{\pi} |\langle\alpha|\beta\rangle|^2 = \frac{1}{\pi} \exp(-|\alpha - \beta|^2), \tag{3.180}$$

whereas for the number state $\hat{\rho} = |n\rangle \langle n|$,

$$Q(\alpha) = \frac{1}{\pi} |\langle\alpha|n\rangle|^2 = \frac{1}{\pi} \exp(-|\alpha|^2) \frac{|\alpha|^{2n}}{n!}. \tag{3.181}$$

Setting $\alpha = x + iy$, we plot these as functions of x and y in Fig. 3.9.

The Q function of the coherent state is just a Gaussian centered at β, while the Q function for a number state is essentially annular and of radius $r \sim n$. Note how these functions seem to correlate with the phase-space figures introduced in Section 3.6. The corresponding Wigner functions, obtained from Eq. (3.174), are

$$W(\alpha) = \frac{2}{\pi} \exp(-2|\alpha - \beta|^2) \tag{3.182}$$

for the coherent state $|\beta\rangle$ and

$$W(\alpha) = \frac{2}{\pi} (-1)^n L_n(4|\alpha|^2) \exp(-2|\alpha|^2), \tag{3.183}$$

where $L_n(\zeta)$ is a Laguerre polynomial. (The derivation of these functions is left as an exercise. See problem 12 at the end of the chapter.) We plot these functions in Fig. 3.10, again with $\alpha = x + iy$. Evidently, the Q and Wigner functions for the coherent state are identical apart from an overall scale factor. But for the number state we see that the Wigner function oscillates and becomes negative over a wide region of phase space. The Q function, of course, can never become negative. It is always a probability distribution. But the Wigner function is not always positive, the Wigner function of the number state being a case in point, in which case it is not a probability distribution. A state whose Wigner function takes on negative values over some region of phase space is *nonclassical*. However, the converse is not necessarily true. A state can be nonclassical yet have a non-negative Wigner

(a)

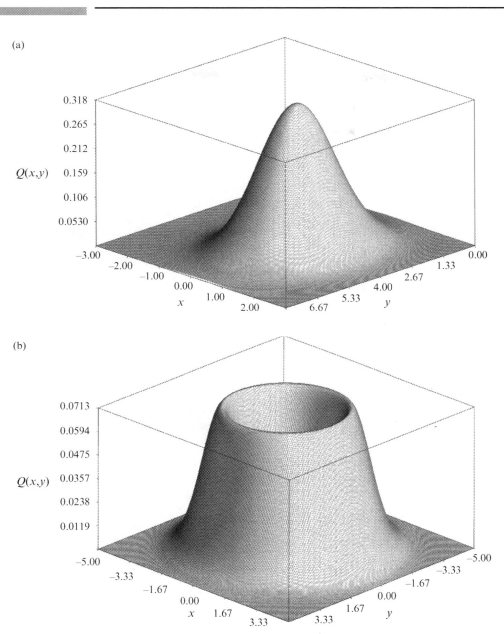

(b)

Fig. 3.9
  Q function for (a) a coherent state with $\bar{n} = 10$ and (b) a number state with $n = 3$.

function. As we said earlier, for nonclassical states, the P function becomes negative or more singular than a delta function over some region of phase space. The squeezed states are strongly nonclassical in this sense, as we shall discuss in Chapter 7, yet their Wigner functions are always positive. Nevertheless, for these and other nonclassical

(a)

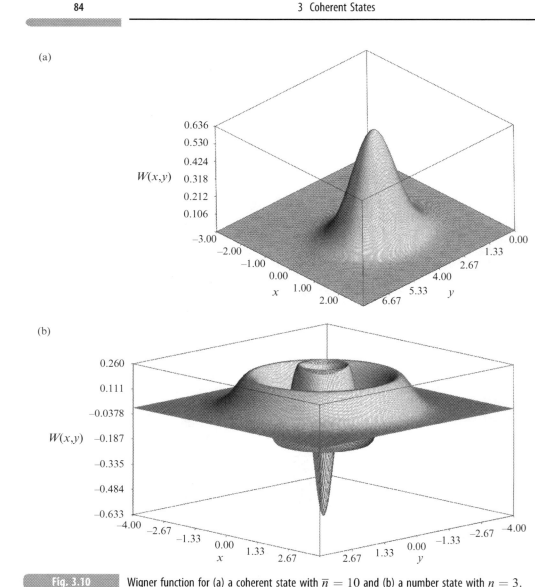

(b)

Wigner function for (a) a coherent state with $\bar{n} = 10$ and (b) a number state with $n = 3$.

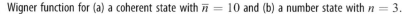

states, the Wigner function is still of primary interest as the P function generally cannot be written down as a function in the usual sense, whereas this can always be done for the Wigner function. Furthermore, the Wigner function can be more sensitive to the quantum nature of some states than can the Q function, as we have seen for the number state. More importantly, as we shall see in Chapter 7 (and below in problem 12), the Wigner function can display the *totality* of interference effects associated with a quantum state [17]. Finally, it happens that it is possible to reconstruct the Wigner function from experimental data, a procedure known as quantum state tomography [18]. A discussion of this procedure would take us beyond the scope of this book, but we refer the reader to the references and bibliography for this chapter.

Problems

1. Investigate the possible existence of *right* eigenstates of the creation operator $\hat{a}^\dagger$.
2. Show for a coherent state $|\alpha\rangle$, where $\alpha = |\alpha|e^{i\theta}$ and where $\bar{n} = |\alpha|^2$, that the phase uncertainty for the state is $\Delta\varphi = \frac{1}{2\sqrt{\bar{n}}}$ if $\bar{n} \gg 1$.
3. Complete the derivation of the wave function for the coherent state given by Eq. (3.52).
4. Prove the following identities:

$$\hat{a}^\dagger |\alpha\rangle\langle\alpha| = \left(\alpha^* + \frac{\partial}{\partial\alpha}\right)|\alpha\rangle\langle\alpha|,$$

$$|\alpha\rangle\langle\alpha|\hat{a} = \left(\alpha + \frac{\partial}{\partial\alpha^*}\right)|\alpha\rangle\langle\alpha|.$$

5. Verify Eq. (3.16), that the quantum fluctuations of the field quadrature operators are the same as for the vacuum when the field is in a coherent state.
6. For a coherent state $|\alpha\rangle$, evaluate the factorial moments

$$\langle\hat{n}(\hat{n}-1)(\hat{n}-2)...(\hat{n}-r+1)\rangle.$$

7. For a coherent state $|\alpha\rangle$, evaluate the sine and cosine operators of Eqs. (2.204) and their squares. (You will not get closed forms.) Examine the limit where the average photon number $\bar{n} = |\alpha|^2 \gg 1$. Then examine the uncertainty products of Eqs. (2.208) and (2.209) in this limit. Hint: See Ref. [19] for some useful mathematical results.
8. Consider again the exponential phase operator $\hat{E} = (\hat{n}+1)^{-1/2}\hat{a}$. (a) Seek normalized right eigenstates $|z\rangle$ of this operator satisfying the equation $\hat{E}|z\rangle = z|z\rangle$ where z is a complex number. Is there a restriction on the range of $|z|$? (b) Can one resolve unity with these states? (c) Examine the states for the case where $|z| \to 1$. How is this special case related to the phase states $|\phi\rangle$ of Eq. (2.214)? Is there a conflict with Eq. (2.216)? (d) Obtain the photon number distribution for this state and express it in terms of the average photon number $\bar{n}$ for the state. Does it look like anything you may have already seen in this book? (e) Obtain the phase distribution for $|z\rangle$.

9. Work out the normally ordered variance of the photon number operator, $\langle :(\Delta\hat{n})^2: \rangle \equiv \langle :\hat{n}^2: \rangle - \langle :\hat{n}: \rangle^2$, where $:\hat{n}: = \hat{n} = \hat{a}^\dagger\hat{a}$ is already normally ordered, in terms of the P function. Show that for a coherent state $\langle :(\Delta\hat{n})^2: \rangle = 0$. Suppose now that for some quantum state we find that $\langle :(\Delta\hat{n})^2: \rangle < 0$. What does this say about the P function for the state? Can such a state be considered a classical state?

10. Work out the normally ordered variances $\langle :(\Delta\hat{X})_i^2: \rangle$, $i = 1, 2$, of the quadrature operators in terms of the P function. Show that these variances vanish for a coherent state. Examine the conditions under which some quantum state might yield $\langle :(\Delta\hat{X})_i^2: \rangle < 0$ for either $i = 1$ or 2. Can this condition hold simultaneously for both quadratures?

11. Carry through the derivation of the Wigner functions of Eqs. (3.182) and (3.183) for the coherent state and the number state, respectively.

12. Show that the photon number parity operator can be represented according to

$$\hat{\Pi} = \int \frac{d^2\beta}{\pi} |\beta\rangle\langle-\beta|,$$

where $|\pm\beta\rangle$ are coherent states.

13. Show numerically that the matrix elements of the displacement operator in the photon number state basis given by Eq. (3.48) satisfy the condition

$$\sum_{m=0}^{\infty} |\langle m|D(\alpha)|n\rangle|^2 = 1, \quad \text{for } n = 0, 1, 2, \ldots, \tag{3.184}$$

for arbitrary α, as required for unitarity.

References

[1] J. J. Sakurai, *Advanced Quantum Mechanics* (Reading, MA: Addison-Wesley, 1967), p. 35.

[2] R. J. Glauber, *Phys. Rev.* **131**, 2766 (1963); J. R. Klauder, *J. Math. Phys.* **4**, 1055 (1963).

[3] E. Merzbacher, *Quantum Mechanics*, 2nd edition (New York: Wiley, 1970), pp. 158–161.

[4] See C. Aragone, E. Chalbaud, and S. Salamó, *J. Math. Phys.* **17**, 1963 (1976).

[5] See S. M. Barnett and D. T. Pegg. *J. Mod. Opt.* **36**, 7 (1989).

[6] The wave functions for the energy eigenstates for the one-dimensional harmonic oscillator can be found in any standard quantum mechanics textbook.

[7] See, for example, M. Nauenberg, *Phys. Rev. A* **40**, 1133 (1989); Z. Dačić Gaeta and C. R. Stroud Jr., *Phys. Rev. A* **42**, 6308 (1990).

[8] See L. W. Ryder, *Quantum Field Theory*, 2nd edition (Cambridge: Cambridge University Press, 1996), p. 178.

[9] I. Segal, *Illinois J. Math.* **6**, 500 (1962); V. Bargmann, *Commun. Pure Appl. Math.* **14**, 187 (1961).

[10] E. C. G. Sudarshan, *Phys. Rev. Lett.* **10**, 277 (1963).

[11] C. L. Mehta, *Phys. Rev. Lett.* **18**, 752 (1967).

[12] Y. Kano, *J. Math. Phys.* **6**, 1913 (1965); C. L. Mehta and E. C. G. Sudarshan, *Phys. Rev. B* **138**, 274 (1965); R. J. Glauber, in C. Dewitt, A. Blandin, and C. Cohen-Tannoudji (Eds.), *Quantum Optics and Electronics* (New York: Gordon & Breach, 1965), p. 65.

[13] E. P. Wigner, *Phys. Rev.* **40**, 794 (1932).

[14] See Ref. [3], pp. 327–330.

[15] A. Royer, *Phys. Rev. A* **15**, 449 (1977).

[16] K. E. Cahill and R. J. Glauber, *Phys. Rev.* **177**, 1882 (1969).

[17] G. S. Agarwal, *Found. Phys.* **25**, 219 (1995).

[18] See K. Vogel and H. Risken, *Phys. Rev. A* **40**, 2847 (1989); D. T. Smithey, M. Beck, M. G. Raymer, and A. Faridani, *Phys. Rev. Lett.* **70**, 1244 (1993); K. Banaszek, C. Radzewicz, K. Wodkiewicz, and J. S. Krasinski, *Phys. Rev. A* **60**, 674 (1999).

[19] P. Carruthers and M.M. Nieto, *Phys. Rev. Lett.* **14**, 387 (1965).

Bibliography

Coherent States

J. R. Klauder and B.-S. Skagerstam (Eds.), *Coherent States: Applications in Physics and Mathematical Physics* (Singapore: World Scientific, 1985).

W.-M. Zhang, D. H. Feng, and R. Gilmore, *Rev. Mod. Phys.* **62**, 867 (1990).

J.-P. Gazeau, *Coherent States in Quantum Physics* (Weinheim: Wiley-VCH, 2009).

Quasi-probability Distributions

G. S. Agarwal and E. Wolf, *Phys. Rev. D* **2**, 2161 (1970).

M. Hillery, R. F. O'Connell, M. O. Scully, and E. P. Wigner, *Phys. Rep.* **106**, 121 (1984).

Quantum State Tomography

U. Leonhardt, *Measuring the Quantum State of Light* (Cambridge: Cambridge University Press, 1997).

M. G. Raymer, *Contemp. Phys.* **38**, 343 (1997).

In 2005, Roy Glauber was awarded the Nobel Prize for his work on coherent states and the quantum theory of optical coherence. His Nobel Lecture is

R. J. Glauber, "Nobel Lecture: One hundred years of light quanta," *Rev. Mod. Phys.* **78**, 1267 (2006).

4   Emission and Absorption of Radiation by Atoms

In this chapter, we first discuss atom–field interactions using quantum mechanical perturbation theory for a classical driving field and then for a quantum mechanical field. In the latter case, spontaneous emission appears as a fully quantum mechanical phenomenon. We then examine the so-called Rabi model, a model of "two-level" atom interacting with a strong near-resonant classical field, introduce the rotating wave approximation, and then introduce the fully quantum mechanical Rabi model, better known as the Jaynes–Cummings model. Differences in the evolution predicted by the two models are drawn out, where we show that the Jaynes–Cummings model predicts behavior which has no semiclassical analog and depends entirely on the discreteness of photons. Finally, we study an extension of the Jaynes–Cummings model, the dispersive model, for the case where the quantized field is far out-of-resonance with the atomic transition frequency. This will eventually allow us to describe how superpositions of different coherent states of the field, states known as Schrödinger-cat states, can be generated from atom–field interactions.

4.1 Atom–Field Interactions

To begin, let us suppose that the Hamiltonian of an electron bound to an atom in the absence of external fields, in the configuration representation, is given by

$$\hat{H}_0 = \frac{1}{2m}\hat{\mathbf{P}}^2 + V(r), \qquad (4.1)$$

where $V(r)$ is the usual Coulomb interaction binding the electron to the nucleus and $r = |\mathbf{r}|$. In the configuration space representation, $\hat{\mathbf{P}} = -i\hbar\nabla$, with $\hat{\mathbf{r}}|\mathbf{r}\rangle = \mathbf{r}|\mathbf{r}\rangle$, the wave functions are given by $\psi(\mathbf{r}) = \langle\mathbf{r}|\psi\rangle$. We assume that energy eigenstates $|k\rangle$ of $\hat{H}_0$, satisfying the time-independent Schrödinger equation

$$\hat{H}_0\psi_k^{(0)}(\mathbf{r}) = E_k\psi_k^{(0)}(\mathbf{r}), \qquad (4.2)$$

where $\langle\mathbf{r}|k\rangle = \psi_k^{(0)}(\mathbf{r})$, are known. In the presence of external fields, the Hamiltonian is modified to

$$\hat{H}(\mathbf{r}, t) = \frac{1}{2m}[\hat{\mathbf{P}} + e\mathbf{A}(\mathbf{r}, t)]^2 - e\Phi(\mathbf{r}, t) + V(\hat{r}), \qquad (4.3)$$

where $\mathbf{A}(\mathbf{r}, t)$ and $\Phi(\mathbf{r}, t)$ are the vector and scalar potentials, respectively, of the external field. The fields themselves are given by

$$\mathbf{E}(\mathbf{r}, t) = -\nabla\Phi(\mathbf{r}, t) - \frac{\partial\mathbf{A}(\mathbf{r}, t)}{\partial t},$$
$$\mathbf{B}(\mathbf{r}, t) = \nabla \times \mathbf{A}(\mathbf{r}, t), \qquad (4.4)$$

and are invariant under the gauge transformations

$$\Phi'(\mathbf{r}, t) = \Phi(\mathbf{r}, t) - \frac{\partial\chi(\mathbf{r}, t)}{\partial t},$$
$$\mathbf{A}'(\mathbf{r}, t) = \mathbf{A}(\mathbf{r}, t) + \nabla\chi(\mathbf{r}, t). \qquad (4.5)$$

The time-dependent Schrödinger equation is

$$\hat{H}(\mathbf{r}, t)\Psi(\mathbf{r}, t) = i\hbar\frac{\partial\Psi(\mathbf{r}, t)}{\partial t}. \qquad (4.6)$$

In order to eventually simplify the form of the atom–field interaction, we define a unitary $\hat{R}$ such that with $\Psi'(\mathbf{r}, t) \equiv \hat{R}\,\Psi(\mathbf{r}, t)$ we have

$$\hat{H}'\Psi'(\mathbf{r}, t) = i\hbar\frac{\partial\Psi'(\mathbf{r}, t)}{\partial t}, \qquad (4.7)$$

where

$$\hat{H}' = \hat{R}\hat{H}\hat{R}^\dagger + i\hbar\frac{\partial\hat{R}}{\partial t}\hat{R}^\dagger. \qquad (4.8)$$

We now choose $\hat{R} = \exp(-ie\chi(\mathbf{r}, t)/\hbar)$ so that (using $\hat{\mathbf{P}} = -i\hbar\nabla$)

$$\hat{H}' = \frac{1}{2m}[\hat{\mathbf{P}} + e\mathbf{A}']^2 - e\Phi' + V(r), \qquad (4.9)$$

where $\mathbf{A}'$ and Φ' are given by Eq. (4.5). At this point we make a definite choice of gauge, namely the Coulomb (or radiation) gauge, for which $\Phi = 0$ and $\mathbf{A}$ satisfies the transversality condition $\nabla \cdot \mathbf{A} = 0$. The vector potential $\mathbf{A}$, for no sources near the atom, satisfies the wave equation

$$\nabla^2\mathbf{A} - \frac{1}{c^2}\frac{\partial^2\mathbf{A}}{\partial t^2} = 0. \qquad (4.10)$$

This choice of gauge is not relativistically invariant, in contrast to the Lorentz gauge, but the domain of quantum optics is, for the most part, nonrelativistic so that no inconsistency will be introduced. The Coulomb gauge has the advantage that the radiation field is completely described by the vector potential, as is obvious from Eq. (4.3), which in this gauge reads

$$\hat{H}(\mathbf{r}, t) = \frac{1}{2m}[\hat{\mathbf{P}} + e\mathbf{A}(\mathbf{r}, t)]^2 + V(r)$$

$$= \frac{\hat{\mathbf{P}}^2}{2m} + \frac{e}{m}\mathbf{A} \cdot \hat{\mathbf{P}} + \frac{e^2}{2m}\mathbf{A}^2 + V(r). \qquad (4.11)$$

Equation (4.9) now reads

$$\hat{H}'(\mathbf{r},t) = \frac{1}{2m}[\hat{\mathbf{P}} + e(\mathbf{A} + \boldsymbol{\nabla}\chi)]^2 + e\frac{\partial\chi}{\partial t} + V(r). \qquad (4.12)$$

The solution of the wave equation (4.10) has the form

$$\mathbf{A} = \mathbf{A}_0\, e^{i(\mathbf{k}\cdot\mathbf{r}-\omega t)} + c.c., \qquad (4.13)$$

where $|\mathbf{k}| = 2\pi/\lambda$ is the wave vector of the radiation. For $|\mathbf{r}|$ of typical atomic dimensions (a few Ångstroms) and λ of typical optical wavelengths (a few hundred nanometers in the range 400–700 nm), $\mathbf{k}\cdot\mathbf{r} \ll 1$, so that over the extent of an atom, the vector potential is spatially uniform, $\mathbf{A}(\mathbf{r},t) \simeq \mathbf{A}(t)$. This is the so-called dipole approximation. We now choose the gauge function $\chi(\mathbf{r},t)$ as $\chi(\mathbf{r},t) = -\mathbf{A}(t)\cdot\mathbf{r}$. With this choice,

$$\boldsymbol{\nabla}\chi(\mathbf{r},t) = -\mathbf{A}(t),$$
$$\frac{\partial\chi}{\partial t}(\mathbf{r},t) = -\mathbf{r}\cdot\frac{\partial\mathbf{A}}{\partial t} = -\mathbf{r}\cdot\mathbf{E}(t), \qquad (4.14)$$

where thus

$$\hat{H}' = \frac{\hat{\mathbf{P}}^2}{2m} + V(r) + e\mathbf{r}\cdot\mathbf{E}(t). \qquad (4.15)$$

This equation contains only one interaction term (within the dipole approximation) as opposed to the two terms of Eq. (4.11). We shall work with $\hat{H}'$ in all that follows, the interaction being in what is sometimes called the "length" gauge. The quantity $-e\mathbf{r}$ is the dipole moment: $\mathbf{d} = -e\mathbf{r}$. In general, that is for an unspecified representation, the dipole moment is an operator, $\hat{\mathbf{d}}$. We shall denote it as such in what follows. We write

$$\hat{H}' = \hat{H}_0 - \hat{\mathbf{d}}\cdot\mathbf{E}(t), \qquad (4.16)$$

where $\hat{H}_0$ is given by Eq. (4.1).

4.2 Interaction of an Atom with a Classical Field

So far, we have not specified the nature of the interacting field and have not even stated whether, or not, we consider the field to be classical or quantum mechanical. The derivation leading to Eq. (4.16) is valid for both classical and quantum fields. But eventually we want to be able to demonstrate differences in the way an atom behaves when interacting with classical or quantum fields. With that in mind, we first turn to the case where an atom is driven by a classical sinusoidal electric field.

We assume that the field has the form $\mathbf{E}(t) = \mathbf{E}_0 \cos(\omega t)$, ω being the frequency of the radiation, and that this field is abruptly turned on at $t = 0$.

The dipole approximation, where we assume that $\mathbf{k} \cdot \mathbf{r} << 1$ over the atom, has already been made. We further assume that the initial state of the atom is $|i\rangle$, where $\hat{H}_0|i\rangle = E_i|i\rangle$. For times $t > 0$ we expand the state vector $|\psi(t)\rangle$ in terms of the complete set of uncoupled atomic states $|k\rangle$:

$$|\psi(t)\rangle = \sum_k C_k(t)e^{-iE_kt/\hbar}|k\rangle, \qquad (4.17)$$

where the time-dependent amplitudes $C_k(t)$ satisfy the normalization requirement

$$\sum_k |C_k(t)|^2 = 1. \qquad (4.18)$$

Substituting this expansion into the time-dependent Schrödinger equation

$$i\hbar \frac{\partial |\psi(t)\rangle}{\partial t} = \left(\hat{H}_0 + \hat{H}^{(I)}\right)|\psi(t)\rangle, \qquad (4.19)$$

where $\hat{H}^{(I)} = -\hat{\mathbf{d}} \cdot \mathbf{E}(t)$, then multiplying from the left by $\langle l|e^{iE_lt/\hbar}$ leads to the set of coupled first-order differential equations for the amplitudes

$$\dot{C}_l(t) = -\frac{i}{\hbar} \sum_k C_k(t)\langle l|\hat{H}^{(I)}|k\rangle e^{i\omega_{lk}t} \qquad (4.20)$$

where $\omega_{lk} = (E_\ell - E_k)/\hbar$ are the transition frequencies between levels l and k. These equations are, so far, exact and need to be solved subject to the initial condition $C_i(0) = 1$, only state $|i\rangle$ being initially populated. As time goes forward, population will be lost from state $|i\rangle$ and increased in some initially unpopulated state $|f\rangle$, that is the amplitude $C_f(t)$ increases. The probability of the atom making a transition from state $|i\rangle$ to state $|f\rangle$ in time t is given by

$$P_{i \to f}(t) = C_f^*(t)C_f(t) = |C_f(t)|^2. \qquad (4.21)$$

The equations for the amplitudes are solvable only for very simple cases. These days, of course, one might solve the set of differential equations numerically, but in the case of a driving field that in some sense is "weak," we can use a time-dependent perturbation theory [1] approach to the problem. "Weak" in this case means that $|\mathbf{E}_0|$ is small, or actually that $|\langle f|\hat{\mathbf{d}} \cdot \mathbf{E}_0|i\rangle|$ is small. As a matter of bookkeeping, we write the interaction Hamiltonian as $\lambda\hat{H}^{(I)}$, where λ is treated as a number in the range $0 \leq \lambda \leq 1$. (At the end of the calculations we will always take $\lambda \to 1$.) We then expand the probability amplitude for, say, state $|l\rangle$ in the power series

$$C_l(t) = C_l^{(0)}(t) + \lambda C_l^{(1)}(t) + \lambda^2 C_l^{(2)}(t) + \cdots. \qquad (4.22)$$

Inserting such expansions into Eq. (4.19) and equating like powers of λ, we obtain, up to second order,

$$\dot{C}_l^{(0)} = 0, \tag{4.23}$$

$$\dot{C}_l^{(1)} = -\frac{i}{\hbar} \sum_k C_k^{(0)} H_{lk}^{(I)}(t) e^{i\omega_{lk}t}, \tag{4.24}$$

$$\dot{C}_l^{(2)} = -\frac{i}{\hbar} \sum_k C_k^{(1)} H_{lk}^{(I)}(t) e^{i\omega_{lk}t}, \tag{4.25}$$

where $H_{lk}^{(I)}(t) \equiv \langle l | \hat{H}^{(I)}(t) | k \rangle$. Note the general pattern that relates the nth order to the $(n-1)$th order:

$$\dot{C}_l^{(n)} = -\frac{i}{\hbar} \sum_k C_k^{(n-1)}(t) H_{lk}^{(I)}(t) e^{i\omega_{lk}t}. \tag{4.26}$$

The essential assumption underlying the perturbation theory approach is that the driving field is so weak that the atomic populations change very little. That is, if $C_i(0) = 1$, $C_f(0) = 0$ $(f \neq i)$ then for $t > 0$, to a good approximation, $C_i(t) \approx 1, |C_f(t)| << 1$ $(f \neq i)$. Thus, in the first-order equation (4.24), the only term surviving the sum on the right-hand side is for $k = i$, yielding

$$\dot{C}_f^{(1)}(t) = -\frac{i}{\hbar} H_{fi}^{(I)}(t) e^{i\omega_{fi}t} C_i^{(0)}(t) \tag{4.27}$$

or

$$C_f^{(1)}(t) = -\frac{i}{\hbar} \int_0^t dt' H_{fi}^{(I)}(t') e^{i\omega_{fi}t'} C_i^{(0)}(t'). \tag{4.28}$$

Inserting this result into the second-order equation (4.25), we obtain

$$C_f^{(2)}(t) = -\frac{i}{\hbar} \sum_l \int_0^t dt' \, H_{fl}^{(I)}(t') e^{i\omega_{fl}t'} C_l^{(1)}(t')$$

$$= \left(-\frac{i}{\hbar}\right)^2 \sum_l \int_0^t dt' \int_0^{t'} dt'' \, H_{fl}^{(I)}(t') e^{i\omega_{fl}t'}$$

$$\times H_{li}^{(I)}(t'') e^{i\omega_{li}t''} C_i^{(0)}(t'') \tag{4.29}$$

Equation (4.28) gives the amplitude for a transition from state $|i\rangle$ to state $|f\rangle$, while Eq. (4.29) gives the amplitude for a transition from state $|i\rangle$ to states $\{|l\rangle\}$ then to state $|f\rangle$. The total transaction probability from state $|i\rangle$ to state $|f\rangle$ is

$$P_{i \to f}(t) = |C_f^{(0)}(t) + C_f^{(1)}(t) + C_f^{(2)}(t) + \cdots|^2. \tag{4.30}$$

Now the dipole moment operator $\hat{\mathbf{d}}$ has nonvanishing matrix elements only between states of opposite parity. Thus, the first-order correction to the amplitude of the initial state vanishes:

$$C_i^{(1)}(t) = -\frac{i}{\hbar} \int_0^t dt' H_{ii}^{(I)}(t') C_i^{(0)}(t') = 0 \tag{4.31}$$

since $H_{ii}^{(I)}(t) = 0$. Therefore, to first order, $C_i(t) = C_i^{(0)}(t) = 1$ so that

$$C_f^{(1)}(t) = -\frac{i}{\hbar} \int_0^t dt' H_{fi}^{(I)}(t') e^{i\omega_{fi}t'}. \tag{4.32}$$

With $H^{(I)} = -\hat{\mathbf{d}} \cdot \mathbf{E}_0 \cos \omega t$, and expanding the cosine in terms of exponentials, this integrates to

$$C_f^{(1)}(t) = \frac{1}{2\hbar} (\hat{\mathbf{d}} \cdot \mathbf{E}_0)_{fi}$$
$$\times \left\{ \frac{\left(e^{i(\omega+\omega_{fi})t} - 1\right)}{(\omega + \omega_{fi})} - \frac{\left(e^{-i(\omega-\omega_{fi})t} - 1\right)}{(\omega - \omega_{fi})} \right\}, \tag{4.33}$$

where $(\hat{\mathbf{d}} \cdot \mathbf{E}_0)_{fi} = \langle f|\hat{\mathbf{d}} \cdot \mathbf{E}_0|i\rangle$. If the frequency of the radiation, ω, is near resonance with the atomic transition frequency ω_{fi}, the second term clearly dominates the first, assuming $\omega_{fi} > 0$. Therefore we drop the "anti-resonant" first term, making the so-called "rotating wave approximation" (RWA) familiar in the context of magnetic resonance [2]. Upon making this approximation, we have, to first order, the transition probability

$$P_{i \to f}^{(1)}(t) = \left|C_f^{(1)}(t)\right|^2 = \frac{\left|(\hat{\mathbf{d}} \cdot \mathbf{E}_0)_{fi}\right|^2}{\hbar^2} \frac{\sin^2(\Delta t/2)}{\Delta^2}, \tag{4.34}$$

where $\Delta = \omega - \omega_{fi}$ is the "detuning" between the radiation field and the atomic transition. When $\Delta \neq 0$, $P_{i \to f}^{(1)}(t)$ maximizes at

$$\left(P_{i \to f}^{(1)}\right)_{\max} = \frac{\left|(\hat{\mathbf{d}} \cdot \mathbf{E}_0)_{fi}\right|^2}{\hbar^2} \frac{1}{\Delta^2}. \tag{4.35}$$

For the case of exact resonance, $\Delta = 0$,

$$\left(P_{i \to f}^{(1)}\right)_{\max} = \frac{\left|(\hat{\mathbf{d}} \cdot \mathbf{E}_0)_{fi}\right|^2}{4\hbar^2} t^2. \tag{4.36}$$

For the perturbation expansion to be valid, we must have $\left(P_{i \to f}^{(1)}\right)_{\max} \ll 1$. For the off-resonance case, this places conditions on both $|(\hat{\mathbf{d}} \cdot \mathbf{E}_0)_{fi}|$ and Δ. For the resonant case, Eq. (4.36) is valid only for very short times. In Fig. 4.1 we plot the evolution of the probability distribution $P_{i \to f}^{(1)}(t)$ for both small detuning ($\Delta \approx 0$) and large detuning. The latter is periodic.

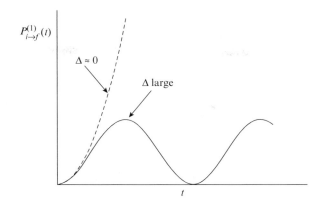

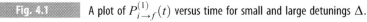

A plot of $P^{(1)}_{i \to f}(t)$ versus time for small and large detunings Δ.

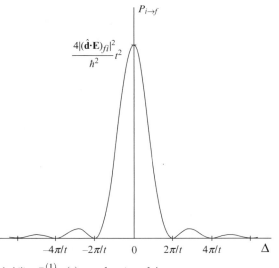

The transition probability $P^{(1)}_{i \to f}(t)$ as a function of Δ.

The transition probability $P^{(1)}_{i \to f}(t)$ is a sharply peaked function at $\Delta = 0$, as shown in Fig. 4.2. The width of the peak is proportional to t^{-1} while the height is proportional to t^2. Thus the area under the peak is proportional to t.

In fact

$$\int_{-\infty}^{\infty} \frac{\sin^2(\Delta t/2)}{\Delta^2} \, d(\Delta) = \frac{\pi}{2} t. \tag{4.37}$$

Furthermore, in the limit when $\Delta \approx 0$ and $t \gg 2\pi/\omega_{fi}$, the function in the integrand of Eq. (4.37) may be approximated by a Dirac delta function:

$$\lim_{t \to \infty} \frac{\sin^2(\Delta t/2)}{\Delta^2} = \frac{\pi}{2} t\, \delta(\Delta), \tag{4.38}$$

although the limit $t \to \infty$ is, in fact, actually constrained by the requirement that the right-hand side of Eq. (4.36) be $\ll 1$. In this case, the transition probability is

$$P_{i \to f}^{(1)}(t) = \frac{\pi}{2} \frac{\left|(\hat{\mathbf{d}} \cdot \mathbf{E}_0)_{fi}\right|^2}{\hbar^2} t \delta(\omega - \omega_{fi}). \tag{4.39}$$

We define the time-independent transition probability *rate* as

$$W_{i \to f} = \frac{P_{i \to f}^{(1)}}{t} = \frac{\pi}{2} \frac{\left|(\hat{\mathbf{d}} \cdot \mathbf{E}_0)_{fi}\right|^2}{\hbar^2} \delta(\omega - \omega_{fi}). \tag{4.40}$$

In practice, there will be a broad range of final states $|f\rangle$ accessible from the states, and the driving field will not be monochromatic so that a range of frequencies will have to be summed or integrate over to obtain the total transition rate. If $[f]$ represents a set of accessible final states, then the transition rate for a monochromatic field is

$$W_{i \to [f]} = \frac{\pi}{2} \sum_f \frac{\left|(\hat{\mathbf{d}} \cdot \mathbf{E}_0)_{fi}\right|^2}{\hbar^2} \delta(\omega - \omega_{fi}). \tag{4.41}$$

This expression is often famously called Fermi's Golden Rule [3].

Now suppose that the light irradiating the atom is from a lamp emitting a broad range of frequencies where there is no phase relationship between the different frequency components. The amplitude of the light will be frequency dependent, so that the transition probability rate induced by all the frequency components must be

$$\frac{P_{i \to f}^{(1)}(t)}{t} = \frac{1}{\hbar^2} \int d\omega\, \frac{\sin^2(\Delta t/2)}{\Delta^2} F(\omega), \tag{4.42}$$

where

$$F(\omega) \equiv |\langle f|\hat{\mathbf{d}} \cdot \mathbf{E}_0(\omega)|i\rangle|^2. \tag{4.43}$$

If the function $F(\omega)$ is broadband and varies slowly with ω in comparison to $(\sin^2(\Delta t/2)/\Delta^2)$, then $F(\omega)$ can be replaced by its resonance value to the peak $\Delta = 0$, so that

$$P_{i \to f}^{(1)}(t) = \frac{\pi}{2\hbar^2} F(\omega_{fi}) t, \tag{4.44}$$

and the transition rate is

$$W_{i \to f} = \frac{\pi}{2\hbar^2} F(\omega_{fi}). \tag{4.45}$$

The spread of frequencies results in the washing out, or dephasing, of the oscillations seen in Fig. 4.1b. This happens because of the lack of phase relations between the various frequency components – the light is incoherent. If the atom is driven by a coherent light field, such as from a laser, dephasing does not occur and the perturbative time-independent transition rates above generally do not adequately describe the dynamics. We postpone a discussion on driving the atom by a coherent laser field until Section 4.4.

4.3 Interaction of an Atom with a Quantized Field

In the preceding discussion, we made no assumption regarding the relative positions of the energy levels i and f. Transitions between the levels occur with some nonzero probability as long as $\mathbf{E}_0 \neq 0$, whether $E_i < E_f$ or $E_i > E_f$. As we will show, when the field is quantized, transitions will occur for the case $E_i > E_f$ even when no photons are present – the so-called spontaneous emission. This is only one of several differences that will appear in the atom–field dynamics in the comparison between cases when the field is quantized and when it is not.

We consider a single free-space field mode of the form given by Eq. (2.131):

$$\hat{\mathbf{E}}(t) = i \left(\frac{\hbar\omega}{2\varepsilon_0 V} \right)^{1/2} \mathbf{e}[\hat{a} e^{-i\omega t} - \hat{a}^\dagger e^{i\omega t}], \tag{4.46}$$

where the dipole approximation has been made. This operator is in the Heisenberg picture but we are going to be working in the Schrödinger picture where the field operator is

$$\hat{\mathbf{E}} = i \left(\frac{\hbar\omega}{2\varepsilon_0 V} \right)^{1/2} \mathbf{e}(\hat{a} - \hat{a}^\dagger). \tag{4.47}$$

The free Hamiltonian $\hat{H}_0$ now must be

$$\hat{H}_0 = \hat{H}_{\text{atom}} + \hat{H}_{\text{field}}, \tag{4.48}$$

where $\hat{H}_{\text{atom}}$ is just the free-atom Hamiltonian as before and $\hat{H}_{\text{field}}$ is the free-field Hamiltonian $\hbar\omega \hat{a}^\dagger \hat{a}$, where the zero-point term has been dropped as it does not contribute to the dynamics. The interaction Hamiltonian becomes

$$\hat{H}^{(I)} = -\hat{\mathbf{d}} \cdot \hat{\mathbf{E}} = -i \left(\frac{\hbar\omega}{2\varepsilon_0 V} \right)^{1/2} (\hat{\mathbf{d}} \cdot \mathbf{e})(\hat{a} - \hat{a}^\dagger)$$

$$= -\hat{\mathbf{d}} \cdot \mathbf{E}_0 (\hat{a} - \hat{a}^\dagger), \tag{4.49}$$

where $\mathbf{E}_0 = i(\hbar\omega/2\varepsilon_0 V)^{1/2}\mathbf{e}$.

Since both atomic and field systems are now quantized, the states of the combined system will involve products of states of both systems. Suppose the initial state of the atom–field system is $|i\rangle = |a\rangle|n\rangle$, where $|a\rangle$ is the initial state of the atom and the field contains n photons. The perturbation interaction of the quantized field causes a transition to the state $|f_1\rangle = |b\rangle|n-1\rangle$ (where $|b\rangle$ is another atomic state) by the absorption of a photon, and to the state $|f_2\rangle = |b\rangle|n+1\rangle$ by the emission of a photon. The energies of these states are

$$\text{for} \qquad |i\rangle = |a\rangle|n\rangle, \; E_i = E_a + n\hbar\omega, \qquad\qquad (4.50a)$$

$$\text{for} \qquad |f_1\rangle = |b\rangle|n-1\rangle, \; E_{f_1} = E_b + (n-1)\hbar\omega, \qquad (4.50b)$$

$$\text{for} \qquad |f_2\rangle = |b\rangle|n+1\rangle, \; E_{f_2} = E_b + (n+1)\hbar\omega, \qquad (4.50c)$$

where E_a and E_b are the energies of the atomic states $|a\rangle$ and $|b\rangle$, respectively.

The perturbation given by Eq. (4.46) is now time independent. The matrix elements of the interaction are as follows:

$$\langle f_1|\hat{H}^{(I)}|i\rangle = \langle b, n-1|\hat{H}^{(I)}|a, n\rangle$$
$$= -(\hat{\mathbf{d}}\cdot\vec{\mathbf{E}}_0)_{ba}\sqrt{n} \qquad \text{(absorption)} \qquad (4.51)$$

and

$$\langle f_2|\hat{H}^{(I)}|i\rangle = \langle b, n+1|\hat{H}^{(I)}|a, n\rangle$$
$$= (\hat{\mathbf{d}}\cdot\vec{\mathbf{E}}_0)_{ba}\sqrt{n+1} \qquad \text{(emission)} \qquad (4.52)$$

where

$$(\hat{\mathbf{d}}\cdot\vec{\mathbf{E}}_0)_{ab} = \langle a|\hat{\mathbf{d}}|b\rangle \cdot \vec{\mathbf{E}}_0 \equiv \mathbf{d}_{ab} \cdot \vec{\mathbf{E}}_0, \qquad (4.53)$$

the factor $\langle a|\hat{\mathbf{d}}|b\rangle = \mathbf{d}_{ab}$ being the dipole matrix element between states $|a\rangle$ and $|b\rangle$. In comparison to the semiclassical case, two things are noteworthy here. The absence of photons ($n = 0$) precludes absorption, just as one might expect. This is obviously in agreement with the case of a classical driving field – no field, no transitions. But in the case of emission, according to Eq. (4.49), transitions may occur even when no photons are present. This is spontaneous emission and it has no semiclassical counterpart. If $n > 0$, the emission of an additional photon is called stimulated emission, this process being the essential one for the operation of the laser (or LASER: Light Amplification by Stimulated Emission of Radiation). The rates of emission and absorption are proportional to the moduli squared of the above matrix elements for the respective processes. The ratio of these rates is

$$\frac{|\langle f_2|\hat{H}^{(I)}|i\rangle|^2}{|\langle f_1|\hat{H}^{(I)}|i\rangle|^2} = \frac{n+1}{n}, \tag{4.54}$$

a result to be used shortly.

The perturbation method developed previously can still be used with appropriate modifications to accommodate the fact that the field is now quantized. The Schrödinger equation still has the form of Eq. (4.19), but where now $\hat{H}_0$ is given by Eq. (4.48) and $\hat{H}^{(I)}$ by Eq. (4.49). Ignoring all other atomic states except $|a\rangle$ and $|b\rangle$, the state vector can be written as

$$\begin{aligned}
|\psi(t)\rangle = {} & C_i(t)|a\rangle|n\rangle e^{-iE_a t/\hbar} e^{-in\omega t} \\
& + C_{f_1}(t)|b\rangle|n-1\rangle e^{-iE_b t/\hbar} e^{-i(n-1)\omega t} \\
& + C_{f_2}(t)|b\rangle|n+1\rangle e^{-iE_b t/\hbar} e^{-i(n+1)\omega t},
\end{aligned} \tag{4.55}$$

where, assuming that $|\psi(0)\rangle = |a\rangle|n\rangle$, $C_i(0) = 1$ and $C_{f1}(0) = C_{f2}(0) = 0$. Following the perturbative method used before, we obtain the first-order correction for the amplitudes C_{f1} and C_{f2} associated with the atom being in state $|b\rangle$:

$$\begin{aligned}
C_{f1}^{(1)}(t) = {} & -\frac{i}{\hbar}\int_0^t dt' \langle f_1|\hat{H}^{(I)}|i\rangle e^{i(E_{f1}-E_i)t/\hbar}, \\
C_{f2}^{(1)}(t) = {} & -\frac{i}{\hbar}\int_0^t dt' \langle f_2|\hat{H}^{(I)}|i\rangle e^{i(E_{f2}-E_i)t/\hbar},
\end{aligned} \tag{4.56}$$

where the former is associated with absorption and the latter with emission. The amplitude associated with the atom being in the state $|b\rangle$, regardless of how it got there, is just the sum of the amplitudes in Eq. (4.56), that is $C_f^{(1)} = C_{f1}^{(1)} + C_{f2}^{(1)}$. Using Eqs. (4.47)–(4.49), we obtain

$$C_f^{(1)}(t) = \frac{i}{\hbar}(\hat{\mathbf{d}}\cdot\mathbf{E}_0)_{ab}\left\{(n+1)^{1/2}\frac{[e^{i(\omega+\omega_{ba})t}-1]}{(\omega+\omega_{ba})} - n^{1/2}\frac{[e^{i(\omega-\omega_{ba})t}-1]}{(\omega-\omega_{ba})}\right\}, \tag{4.57}$$

where $\omega_{ba} = (E_b - E_a)/\hbar$ and where the first term is due to emission and the second to absorption. If the number of photons is large, $n \gg 1$ then we can replace $\sqrt{n+1}$ by $\sqrt{n}$ in the first term and then Eqs. (4.54) and (4.30) are essentially the same, there being the correspondence between the classical and quantum field amplitudes $(\mathbf{E}_0)_{\text{classical}} \leftrightarrow (2i\mathbf{E}_0\sqrt{n})_{\text{quantum}}$, where the quantum mechanical $\mathbf{E}_0$ is given at the top of page 98. This correspondence between quantum and classical fields has its limits, one being the case when $n = 0$, as already discussed.

If $|b\rangle$ is the excited state, then $\omega_{ba} > 0$, so if $\omega \sim \omega_{ba}$ then the first term of Eq. (4.57) can be dropped, again the rotating wave approximation. Of course, if $|a\rangle$ is the excited state, then $\omega_{ba} < 0$ and if $\omega \sim -\omega_{ba}$, then the second term of Eq. (4.57) can be dropped, and we notice that the remaining term does not vanish even when $n = 0$, the transition between $|a\rangle$ and $|b\rangle$ taking place by spontaneous emission. Thus, the rotating wave approximation carries over into the case where the field and the atom are quantized. It can be shown that Fermi's Golden Rule carries over in a similar fashion.

We conclude this section with a field-theoretic derivation of the Planck distribution law. Suppose we have a collection of atoms interacting resonantly with a quantized field of frequency $\omega = (E_a - E_b)/\hbar$, where $|a\rangle$ and $|b\rangle$ are the atomic states with $E_a > E_b$. We let N_a and N_b represent the populations of atoms in states $|a\rangle$ and $|b\rangle$, respectively. Further, we let W_{emis} represent the transition rate due to photon emission and W_{abs} the transition rate due to photon absorption. Because the atoms are constantly emitting and absorbing photons, the atomic populations change with time according to

$$
\begin{aligned}
\frac{dN_a}{dt} &= -N_a W_{emis} + N_b W_{abs}, \\
\frac{dN_b}{dt} &= -N_b W_{abs} + N_a W_{emis}.
\end{aligned}
\tag{4.58}
$$

At thermal equilibrium, we have

$$
\frac{dN_a}{dt} = 0 = \frac{dN_b}{dt},
\tag{4.59}
$$

and thus we obtain

$$
N_a W_{emis} = N_b W_{abs}.
\tag{4.60}
$$

But, according to Boltzmann,

$$
\frac{N_b}{N_a} = \exp[(E_a - E_b)/kT] = \exp(\hbar\omega/kT),
\tag{4.61}
$$

and from Eq. (4.54) we have that

$$
\frac{N_b}{N_a} = \frac{W_{emis}}{W_{abs}} = \frac{n+1}{n}.
\tag{4.62}
$$

Thus, from Eqs. (4.61) and (4.62) it follows that

$$
n = \frac{1}{\exp(\hbar\omega/kT) - 1},
\tag{4.63}
$$

in agreement with Eq. (2.142) if we replace n by $\bar{n}$ in Eq. (4.63) to incorporate the fact that we cannot assume a definite number of photons.

Let's compare this derivation to the one given by Einstein [4] before quantum electrodynamics was even invented. His derivation is similar to the above but explicitly makes the distinction between spontaneous and stimulated emission. He introduced the coefficients A, B, and C having the following meanings: AN_a is the growth rate of the population in state $|b\rangle$

due to spontaneous emission from state $|b\rangle$ (A being the rate of spontaneous emission); $BU(\omega)N_a$ is the growth rate of the population of state $|b\rangle$ due to stimulated emission from $|a\rangle$, $U(\omega)$ being the spectral energy density; and $CU(\omega)N_b$ is the rate of growth of state $|a\rangle$ as a result of absorption by atoms in state $|b\rangle$. Note that the spontaneous emission term is independent of $U(\omega)$. The rate equations for the populations are now

$$\frac{dN_a}{dt} = -[A + BU(\omega)]N_a + CU(\omega)N_b,$$
$$\frac{dN_b}{dt} = -CU(\omega)N_b + [A + BU(\omega)]N_a. \tag{4.64}$$

At long times, the populations reach a steady state and the derivatives on the left-hand sides vanish to yield

$$[A + BU(\omega)]N_a = CU(\omega)N_b. \tag{4.65}$$

Using once again the relation in Eq. (4.61), we are led to

$$U(\omega) = \frac{A}{C \exp(\hbar\omega/kT) - B}. \tag{4.66}$$

But for a thermal field we must have, by comparison to Eq. (2.152), that $C = B$ and

$$\frac{A}{B} = \frac{\hbar\omega^3}{\pi^2 c^3}. \tag{4.67}$$

It is worth comparing the rates of spontaneous emission A and stimulated emission $BU(\omega)$:

$$\frac{A}{BU(\omega)} = \exp(\hbar\omega/kT) - 1. \tag{4.68}$$

For a natural source such as the sun, which we approximate as a black body with surface temperature $T \approx 6000$ K, this ratio is about 400 for $\lambda = 400$ nm and about 30 for $\lambda = 700$ nm [5]. Thus, at both ends of the visible spectrum, spontaneous emission dominates stimulated emission. In the range of visible light, it is only in "unnatural" sources, there exists a population inversion (i.e. more atoms in the excited state than in the ground), such as in lasers, where stimulated emission dominates spontaneous emission.

Spontaneous emission is a complex phenomenon and we have discussed only its most essential features. For atoms in free space where there is an infinity of modes into which the atoms may radiate. Spontaneous emission in this case is well described by the Weisskopf–Wigner theory [6] as an irreversible decay process. A discussion of that theory is outside the scope of this book. However, under certain circumstances, where there may be only one mode into which an atom can radiate, such as in the case of an atom in a cavity, spontaneous emission can be reversible, that is the atom can reabsorb the emitted photon. Such behavior is discussed in Section 4.5.

4.4 The Rabi Model

The perturbation theory approach to atom–field interactions assumes that the initial atomic state population is essentially unchanged, that is the probability amplitude for the atom being in any other state remains small. On the other hand, a strong laser field of near-resonant frequency with a pair of atomic levels (assumed of opposite parity) will cause a large population transfer to the near-resonant state but not to any other. In such a case, perturbation theory must be abandoned. Only the two dominant states will be retained and the problem will be solved more "exactly." This is the Rabi model [7], so named because of its original setting in magnetic resonance as studied by Rabi long ago. We study the semiclassical case first.

For definiteness and to follow convention, we label our two atomic states $|g\rangle$ (for ground) and $|e\rangle$ (for excited). The energy difference between these states is characterized by the transition frequency $\omega_0 = (E_e - E_g)/\hbar$. This frequency is close to the frequency ω of the driving laser field as shown in Fig. 4.3.

The interaction Hamiltonian we write as

$$\hat{H}^{(I)}(t) = \hat{V}_0 \cos \omega t, \tag{4.69}$$

where $\hat{V}_0 = -\hat{\mathbf{d}} \cdot \mathbf{E}_0$. We write the state vector as

$$|\psi(t)\rangle = C_g(t) e^{-iE_g t/\hbar}|g\rangle + C_e(t) e^{-iE_e t/\hbar}|e\rangle. \tag{4.70}$$

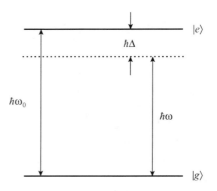

 Fig. 4.3 Energy-level diagram for a two-level atom acting with a near-resonant classical driving field of frequency ω. The resonant frequency between the two atomic levels is ω_0 and the detuning $\Delta = \omega_0 - \omega$.

From the Schrödinger equation

$$i\hbar \frac{\partial |\psi(t)\rangle}{\partial t} = \hat{H}(t)|\psi(t)\rangle, \qquad (4.71)$$

where

$$\hat{H} = \hat{H}_0 + \hat{V}_0 \cos \omega t, \qquad (4.72)$$

we arrive at the coupled set of equations for the amplitudes C_g and C_e:

$$\dot{C}_g = -\frac{i}{\hbar} \mathcal{V}^* \cos \omega t e^{-i\omega_0 t} \ C_e,$$

$$\dot{C}_e = -\frac{i}{\hbar} \mathcal{V} \cos \omega t e^{i\omega_0 t} \ C_g, \qquad (4.73)$$

where $\mathcal{V} = \langle e|\hat{V}_0|g\rangle = -\hat{\mathbf{d}}_{eg} \cdot \mathbf{E}_0$, which in general is a complex number. As an initial condition, we take all the population to be in the ground state: $C_g(0) = 1$ and $C_e(0) = 0$. In Eqs. (4.73) we expand $\cos \omega t$ as $\cos \omega t = (e^{i\omega t} + e^{-i\omega t})/2$ and retain only those terms oscillating at the frequency $\omega_0 - \omega$ to obtain

$$\dot{C}_g = -\frac{i}{2\hbar} \mathcal{V}^* \exp[i(\omega - \omega_0)t] \ C_e,$$

$$\dot{C}_e = -\frac{i}{2\hbar} \mathcal{V} \exp[-i(\omega - \omega_0)t] \ C_g. \qquad (4.74)$$

Dropping the terms oscillating at $\omega_0 + \omega$, of course, constitutes the RWA. Eliminating C_g we have for C_e:

$$\ddot{C}_e + i(\omega - \omega_0)\dot{C}_e + \frac{1}{4} \frac{|\mathcal{V}|^2}{\hbar^2} C_e = 0. \qquad (4.75)$$

As a trial solution we set

$$C_e(t) = e^{i\lambda t}, \qquad (4.76)$$

which leads to the two roots

$$\lambda_{\pm} = \frac{1}{2}\{-\Delta \pm \Omega\}, \quad \Omega = [\Delta^2 + \Omega_R^2]^{1/2}, \ \Omega_R = |\mathcal{V}|/\hbar, \qquad (4.77)$$

where $\Delta = \omega_0 - \omega$ is the detuning of atomic transition frequency and the laser field, Ω_R is the Rabi frequency, and Ω is the generalized Rabi frequency. Thus, the general solution is of the form

$$C_e(t) = A_+ e^{i\lambda_+ t} + A_- e^{i\lambda_- t}, \qquad (4.78)$$

where from the initial conditions we must have

$$A_{\pm} = \mp \frac{\mathcal{V}}{2\hbar\Omega} = \mp \frac{\Omega_R}{2\Omega}. \qquad (4.79)$$

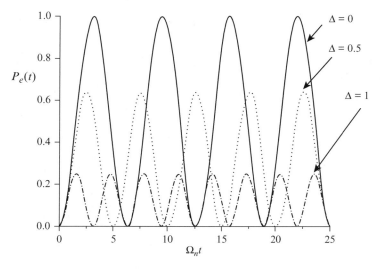

Fig. 4.4 Plots of $P_e(t)$ versus t for various detunings Δ.

Finally, our solution is

$$C_e(t) = i\frac{\Omega_R}{\Omega} \; e^{-i\Delta t/2} \; \sin(\Omega t/2),$$

$$C_g(t) = e^{-i\Delta t/2}\left\{ \cos(\Omega t/2) + i\frac{\Delta}{\Omega}\sin(\Omega t/2) \right\}. \quad (4.80)$$

The probability that the atom is in state $|e\rangle$ is

$$P_e(t) = |C_e(t)|^2$$

$$= \frac{\Omega_R{}^2}{\Omega^2} \; \sin^2(\Omega t/2), \quad (4.81)$$

which is plotted in Fig. 4.4 for various values of Δ. For the case of exact resonance, $\Delta = 0$, we have

$$P_e(t) = \sin^2(\Omega_R t/2), \quad (4.82)$$

and at the time $t = \pi/\Omega_R$ all the atomic population has been transferred to the excited state.

It is frequently convenient to consider the quantity known as the atomic inversion $W(t)$, defined as the difference in the excited and ground-state populations:

$$W(t) = P_e(t) - P_g(t), \quad (4.83)$$

which, for the resonant case and with the atom initially in the ground state, is

$$W(t) = \sin^2\left(\frac{\Omega_R t}{2}\right) - \cos^2\left(\frac{\Omega_R t}{2}\right) \qquad (4.84)$$
$$= -\cos(\Omega_R t).$$

We note that $W(\pi/\Omega_R) = 1$. In the parlance of nuclear magnetic resonance (NMR) experiments [8], such transfers are called π pulses. On the other hand, if $t = \pi/2\Omega_R$ then $W(\pi/2\Omega_R) = 0$ and the population is shared coherently between the excited and ground states with

$$C_e(\pi/2\Omega_R) = \frac{i}{\sqrt{2}},$$
$$\qquad (4.85)$$
$$C_g(\pi/2\Omega_R) = \frac{1}{\sqrt{2}},$$

so that

$$|\psi(\pi/2\Omega_R)\rangle = \frac{1}{\sqrt{2}}(|g\rangle + i|e\rangle). \qquad (4.86)$$

For obvious reasons, the transfer of population from the ground state to that of Eq. (4.84) is called a $\pi/2$-pulse. Such transfers of population by π- or $\pi/2$-pulses are standard procedures for manipulating not only spin states in NMR or electron paramagnetic resonance (EPR) experiments [2], but have become routine for manipulating atomic or ionic states in laser spectroscopy experiments [9].

The results of the perturbation theory may be recovered from the Rabi model either when the size of $\Omega_R/2$ is so small compared to the detuning Δ that it can be neglected in Ω, *or* if the radiation field acts only for such a short time that the term $\sin^2(\Omega_R t/2)$ can legitimately be represented by the first term in its expansion. In both cases the depletion of the initial atomic population is small and the perturbation theory approach remains valid.

4.5 Fully Quantum Mechanical Model: The Jaynes–Cummings Model

We now turn to the quantum electrodynamic version of the Rabi model. In our previous perturbation discussion of an atom interacting with a quantized electromagnetic field, we assumed the field to be a single-mode free field (plane wave). As we just discussed above, a free atom interacts with an infinite number of modes and thus the dynamics is not well described assuming only a single-mode field. On the other hand, it has recently become possible to manufacture environments where the density of modes is significantly different than in free space. We have in mind here small microwave cavities, or in some cases optical cavities, capable of supporting only a single mode or maybe a few widely spaced (in frequency) modes. Thus, in some cases, the ideal single-mode interaction can be realized in the laboratory. We shall discuss some specific examples later in Chapter 10, but for now we consider an atom with levels $|g\rangle$ and $|e\rangle$ as before, interacting with a single-mode cavity field of the form

$$\hat{\mathbf{E}} = \mathbf{e}\left(\frac{\hbar\omega}{\varepsilon_0 V}\right)^{1/2}(\hat{a} + \hat{a}^\dagger)\sin(kz), \tag{4.87}$$

where $\hat{e}$ is an arbitrarily oriented polarization vector.

The interaction Hamiltonian is now

$$\begin{aligned}\hat{H}^{(I)} &= -\hat{\mathbf{d}}\cdot\hat{\mathbf{E}} \\ &= \hat{d}g(\hat{a} + \hat{a}^\dagger),\end{aligned} \tag{4.88}$$

where

$$g = -\left(\frac{\hbar\omega}{\varepsilon_0 V}\right)^{1/2}\sin(kz), \tag{4.89}$$

and where $\hat{d} = \hat{\mathbf{d}}\cdot\mathbf{e}$.

At this point it is convenient to introduce the so-called atomic transition operators

$$\hat{\sigma}_+ = |e\rangle\langle g|, \qquad \hat{\sigma}_- = |g\rangle\langle e| = \hat{\sigma}_+^\dagger, \tag{4.90}$$

and the inversion operator

$$\hat{\sigma}_3 = |e\rangle\langle e| - |g\rangle\langle g|. \tag{4.91}$$

These operators obey the Pauli spin algebra

$$\begin{aligned}[\hat{\sigma}_+, \hat{\sigma}_-] &= \hat{\sigma}_3, \\ [\hat{\sigma}_3, \hat{\sigma}_\pm] &= \pm 2\hat{\sigma}_\pm.\end{aligned} \tag{4.92}$$

Only the off-diagonal elements of the dipole operator are nonzero, since by parity considerations $\langle e|\hat{d}|e\rangle = 0 = \langle g|\hat{d}|g\rangle$, so we may write

$$\begin{aligned}\hat{d} &= d|g\rangle\langle e| + d^*|e\rangle\langle g| \\ &= d\hat{\sigma}_- + d^*\hat{\sigma}_+ = d(\hat{\sigma}_+ + \hat{\sigma}_-),\end{aligned} \tag{4.93}$$

where we have set $\langle e|\hat{d}|g\rangle = d$ and assumed, without loss of generality, that d is real. Thus, the interaction Hamiltonian is

$$\hat{H}^{(I)} = \hbar\lambda(\hat{\sigma}_+ + \hat{\sigma}_-)(\hat{a} + \hat{a}^\dagger), \tag{4.94}$$

where $\lambda = dg/\hbar$.

If we define the level of the energy to be zero halfway between the states $|g\rangle$ and $|e\rangle$ as in Fig. 4.5, then the free-atom Hamiltonian may be written as

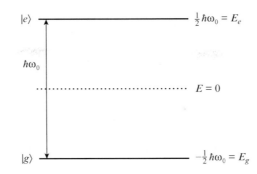

Fig. 4.5   Atomic energy-level diagram where $E = 0$ is taken halfway between the two levels.

$$\hat{H}_A = \frac{1}{2}(E_e - E_g)\hat{\sigma}_3 = \frac{1}{2}\hbar\omega_0\hat{\sigma}_3, \qquad (4.95)$$

where $E_e = -E_g = \frac{1}{2}\hbar\omega_0$.

The free-field Hamiltonian is, after dropping the zero-point energy term,

$$\hat{H}_F = \hbar\omega\hat{a}^\dagger\hat{a}. \qquad (4.96)$$

Thus, the total Hamiltonian is

$$\begin{aligned}
\hat{H} &= \hat{H}_A + \hat{H}_F + \hat{H}^{(I)} \\
&= \frac{1}{2}\hbar\omega_0\hat{\sigma}_3 + \hbar\omega\hat{a}^\dagger\hat{a} + \hbar\lambda(\hat{\sigma}_+ + \hat{\sigma}_-)(\hat{a} + \hat{a}^\dagger).
\end{aligned} \qquad (4.97)$$

In the free-field case, as we have already shown, the operators $\hat{a}$ and $\hat{a}^+$ evolve as

$$\hat{a}(t) = \hat{a}(0)e^{-i\omega t}, \quad \hat{a}^\dagger(t) = \hat{a}^\dagger(0)e^{i\omega t}. \qquad (4.98)$$

One can show similarly that for the free-atom case,

$$\hat{\sigma}_\pm(t) = \hat{\sigma}_\pm(0)e^{\pm\, i\omega_0 t}. \qquad (4.99)$$

Thus we can see that the approximate time dependences of the operator products in Eq. (4.82) are as follows:

$$\begin{aligned}
\hat{\sigma}_+\hat{a} &\sim e^{i(\omega_0 - \omega)t}, \\
\hat{\sigma}_-\hat{a}^\dagger &\sim e^{-i(\omega_0 - \omega)t}, \\
\hat{\sigma}_+\hat{a}^\dagger &\sim e^{i(\omega + \omega_0)t}, \\
\hat{\sigma}_-\hat{a} &\sim e^{-i(\omega + \omega_0)t}.
\end{aligned} \qquad (4.100)$$

For $\omega_0 \approx \omega$, the last two terms vary much more rapidly than the first two. Furthermore, the last two terms do not conserve energy in contrast to the first two. The term $\hat{\sigma}_+\hat{a}^\dagger$ corresponds to the emission of a photon as the atom goes from the ground to the excited state, whereas $\hat{\sigma}_-\hat{a}$ corresponds to the absorption of a photon as the atom goes from the excited to the ground state. Integrating the time-dependent Schrödinger equation, as in the perturbative case, will lead, for the last two terms, to denominators containing $\omega_0 + \omega$ compared to $\omega_0 - \omega$ for the first two terms. The reader will not be surprised to learn that we are going to drop the non-energy-conserving terms, the RWA again, so that our Hamiltonian in this approximation is

$$\hat{H} = \frac{1}{2}\hbar\omega_0\hat{\sigma}_3 + \hbar\omega\hat{a}^\dagger\hat{a} + \hbar\lambda(\hat{\sigma}_+\hat{a} + \hat{\sigma}_-\hat{a}^\dagger). \tag{4.101}$$

The interaction described by this Hamiltonian is widely referred to as the Jaynes–Cummings model [10].

Before solving for the dynamics for any specific cases, we take note of certain constants of the motion. An obvious one is the electron "number"

$$\hat{P}_E = |e\rangle\langle e| + |g\rangle\langle g| = 1, \quad [\hat{H}, \hat{P}_E] = 0, \tag{4.102}$$

valid when no other atomic states can become populated. Another is the excitation number

$$\hat{N}_e = \hat{a}^\dagger\hat{a} + |e\rangle\langle e|, [\hat{H}, \hat{N}_e] = 0. \tag{4.103}$$

Using these constants of the motion, we may break Eq. (4.101) into two commuting parts:

$$\hat{H} = \hat{H}_I + \hat{H}_{II}, \tag{4.104}$$

where

$$\hat{H}_I = \hbar\omega\hat{N}_e + \hbar\left(\frac{\omega_0}{2} - \omega\right)\hat{P}_E,$$
$$\hat{H}_{II} = -\hbar\Delta + \hbar\lambda(\hat{\sigma}_+\hat{a} + \hat{\sigma}_-\hat{a}^\dagger), \tag{4.105}$$

such that $[\hat{H}_I, \hat{H}_{II}] = 0$. Clearly, all the essential dynamics is contained in $\hat{H}_{II}$, whereas $\hat{H}_I$ contributes only overall irrelevant phase factors.

Let us now consider a simple example, with $\Delta = 0$, where the atom is initially in the excited state $|e\rangle$ and the field is initially in the number state $|n\rangle$. The initial state of the atom–field system is then $|i\rangle = |e\rangle|n\rangle$ and is of energy $E_i = \frac{1}{2}\hbar\omega + n\hbar\omega$. State $|i\rangle$ is coupled to (and only to) the state $|f\rangle = |g\rangle|n+1\rangle$ with energy $E_f = -\frac{1}{2}\hbar\omega + (n+1)\hbar\omega$. Note that $E_i = E_f$. We write the state vector as

$$|\psi(t)\rangle = C_i(t)e^{-iE_it/\hbar}|i\rangle + C_f(t)e^{-iE_ft/\hbar}|f\rangle, \tag{4.106}$$

where $C_i(0) = 1$ and $C_f(0) = 0$. Following standard procedures we obtain, from the interaction picture of the Schrödinger equation $i\hbar d|\psi(t)\rangle/dt = \hat{H}_{II}|\psi(t)\rangle$, the equations for the coefficients

$$\dot{C}_i = -i\lambda\sqrt{n+1}\ C_f,$$

$$\dot{C}_f = -i\lambda\sqrt{n+1}\ C_i. \tag{4.107}$$

Eliminating C_f, we obtain

$$\ddot{C}_i + \lambda^2(n+1)C_i = 0. \tag{4.108}$$

The solution matching the initial conditions is

$$C_i(t) = \cos(\lambda t\sqrt{n+1}). \tag{4.109}$$

From Eq. (4.107) we obtain

$$C_f(t) = -\sin(\lambda t\sqrt{n+1}). \tag{4.110}$$

Thus, our solution is

$$|\psi(t)\rangle = \cos(\lambda t\sqrt{n+1})|e\rangle|n\rangle$$
$$-i\sin(\lambda t\sqrt{n+1})|g\rangle|n+1\rangle. \tag{4.111}$$

The probability that the system remains in the initial state is

$$P_i(t) = |C_i(t)|^2 = \cos^2(\lambda t\sqrt{n+1}), \tag{4.112}$$

while the probability that it makes a transition to the state $|f\rangle$ is

$$P_f(t) = |C_f(t)|^2 = \sin^2(\lambda t\sqrt{n+1}). \tag{4.113}$$

The atomic inversion is given by

$$W(t) = \langle\psi(t)|\hat{\sigma}_3|\psi(t)\rangle$$
$$= P_i(t) - P_f(t) \tag{4.114}$$
$$= \cos(2\lambda t\sqrt{n+1}).$$

We may define a quantum electrodynamic Rabi frequency $\Omega(n) = 2\lambda\sqrt{n+1}$ so that

$$W(t) = \cos[\Omega(n)t]. \tag{4.115}$$

Clearly, the atomic inversion for the field initially in a number state is strictly periodic (Fig. 4.6), just as in the semiclassical case of Eq. (4.82) (apart from the minus sign due only to a different initial atomic state) and except for the fact that in the classical case there must always be a field present initially. But in the quantum mechanical case there are Rabi oscillations even for the case when $n = 0$. These are the vacuum field Rabi oscillations [11] and, of course, they have no classical counterpart. They are due to the atom spontaneously emitting a photon then

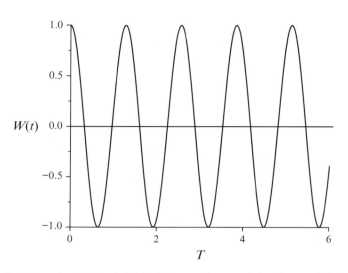

Fig. 4.6 Periodic atomic inversion with the field initially in a number state $|n\rangle$ with $n = 5$ photons.

reabsorbing it, re-emitting it, and so on: an example of reversible spontaneous emission. Such effects can be observed if atoms interact with very high
Q cavities. But aside from this, overall, the behavior of the atomic dynamics for a definite number of photons is very much like the semiclassical Rabi
model (i.e. periodic and regular). Perhaps this is a bit counterintuitive,
since a number state is the most nonclassical of all the field states. Intuition
might suggest that when the field is initially in a coherent state, we should
recover the semiclassical (periodic and regular) Rabi oscillations. As we are
about to demonstrate, intuition, in this case, is wrong.

Let us now consider a more general (pure-state) solution of the dynamics.
We assume the atom is initially in a superposition of states $|e\rangle$ and $|g\rangle$:

$$|\psi(0)\rangle_{\text{atom}} = C_g|g\rangle + C_e|e\rangle, \tag{4.116}$$

and the field is initially in the state

$$|\psi(0)\rangle_{\text{field}} = \sum_{n=0}^{\infty} C_n|n\rangle, \tag{4.117}$$

such that the initial atom–field state is

$$|\psi(0)\rangle = |\psi(0)\rangle_{\text{atom}} \otimes |\psi(0)\rangle_{\text{field}}. \tag{4.118}$$

The solution of the Schrödinger equation is now

$$|\psi(t)\rangle = \sum_{n=0}^{\infty} \left\{ \left[C_e C_n \cos(\lambda t\sqrt{n+1}) - iC_g C_{n+1}\sin(\lambda t\sqrt{n+1}) \right]|e\rangle \right.$$

$$\left. + \left[-iC_e C_{n-1}\sin(\lambda t\sqrt{n}) + C_g C_n\cos(\lambda t\sqrt{n}) \right]|g\rangle \right\}|n\rangle$$

$$(C_{-1} = 0). \tag{4.119}$$

Note that, in general, this is an entangled state.

For the case of the atom initially in the excited state, $C_e = 1$ and $C_g = 0$, so we may write the solution as

$$|\psi(t)\rangle = |\psi_g(t)\rangle|g\rangle + |\psi_e(t)\rangle|e\rangle, \tag{4.120}$$

where $|\psi_g(t)\rangle$ and $|\psi_e(t)\rangle$ are the field components of $|\psi(t)\rangle$ given by

$$\begin{aligned}
|\psi_g(t)\rangle &= -i\sum_{n=0}^{\infty} C_n \sin(\lambda t\sqrt{n+1})|n+1\rangle, \\
|\psi_e(t)\rangle &= \sum_{n=0}^{\infty} C_n \cos(\lambda t\sqrt{n+1})|n\rangle.
\end{aligned} \tag{4.121}$$

The atomic inversion is

$$\begin{aligned}
W(t) &= \langle\psi(t)|\hat{\sigma}_3|\psi(t)\rangle \\
&= \langle\psi_e(t)|\psi_e(t)\rangle - \langle\psi_g(t)|\psi_g(t)\rangle \\
&= \sum_{n=0}^{\infty}|C_n|^2 \cos(2\lambda t\sqrt{n+1}).
\end{aligned} \tag{4.122}$$

The result is just the n-photon inversion of Eq. (4.114) weighted with the photon number distribution of the initial field state.

For the coherent state, again that most classical of all quantum states, we have

$$C_n = e^{-|\alpha|^2/2}\frac{\alpha^n}{\sqrt{n!}}, \tag{4.123}$$

and thus the inversion is

$$W(t) = e^{-\bar{n}}\sum_{n=0}^{\infty}\frac{\bar{n}^n}{n!}\cos(2\lambda t\sqrt{n+1}). \tag{4.124}$$

A plot of $W(t)$ versus the scaled time λt in Fig. 4.7 reveals significant discrepancies between the fully quantized and semiclassical Rabi oscillations. We note first that the Rabi oscillations initially appear to damp out, or collapse. The collapse of the Rabi oscillations was noted fairly early in the study of this "idealized" model interaction [12]. Several years later, perhaps by executing longer runs of a computer program, it was found that after a period of quiescence following the collapse, the Rabi oscillations start to revive [13], though not completely. At longer times one finds a sequence of collapses and revivals, the revivals becoming less distinct as time increases. This collapse and revival behavior of the Rabi oscillations in the fully quantized model is strikingly different than in the semiclassical case, where the oscillations have constant amplitude. We must now explain this difference.

First, we consider the collapse. The average photon number is $\bar{n} = |\alpha|^2$, so the dominant Rabi frequency is

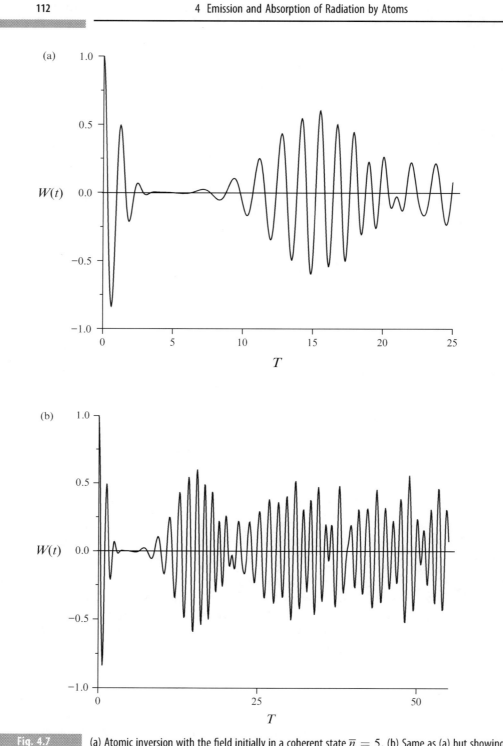

Fig. 4.7 (a) Atomic inversion with the field initially in a coherent state $\bar{n} = 5$. (b) Same as (a) but showing the evolution for a longer time, beyond the first revival.

$$\Omega(\bar{n}) = 2\lambda\sqrt{n+1} \approx 2\lambda\sqrt{n}, \quad \bar{n} \gg 1. \tag{4.125}$$

But there will be a range of "important" frequencies as a result of the spread of the probabilities $|C_n|^2$ about $\bar{n}$, $\bar{n} \pm \Delta n$; that is the frequencies in the range $\Omega(\bar{n} - \Delta n)$ to $\Omega(\bar{n} + \Delta n)$. The collapse time t_c may be estimated from the time–frequency "uncertainty" relation

$$t_c[\Omega(\bar{n} + \Delta n) - \Omega(\bar{n} - \Delta n)] \simeq 1, \tag{4.126}$$

where the spread of frequencies is responsible for the "dephasing" of the Rabi oscillations. For the coherent state, $\Delta n = \bar{n}^{1/2}$, and with

$$
\begin{aligned}
\Omega(\bar{n} \pm \bar{n}^{1/2}) &\simeq 2\lambda[\bar{n} \pm \bar{n}^{1/2}]^{1/2} \\
&= 2\lambda\bar{n}^{1/2}\left[1 \pm \frac{1}{\bar{n}^{1/2}}\right]^{1/2} \\
&\simeq 2\lambda\bar{n}^{1/2}\left(1 \pm \frac{1}{2\bar{n}^{1/2}}\right) \\
&= 2\lambda\bar{n}^{1/2} \pm \lambda,
\end{aligned}
\tag{4.127}
$$

it follows that

$$t_c[\Omega(\bar{n} + \bar{n}^{1/2}) - \Omega(\bar{n} - \bar{n}^{1/2})] \simeq t_c 2\lambda \simeq 1 \tag{4.128}$$

and thus $t_c \simeq (2\lambda)^{-1}$, which is independent of $\bar{n}$.

The preceding "derivation" of the collapse time is not very rigorous. We shall now give a more rigorous derivation. We expand $[n+1]^{1/2}$ about $\bar{n}$ as

$$[n+1]^{1/2} = [\bar{n}+1]^{1/2} + \frac{1}{2(\bar{n}+1)^{1/2}}(n - \bar{n}) + \cdots \tag{4.129}$$

so that we may approximate the inversion as

$$W(t) \simeq e^{-\bar{n}} \mathrm{Re}\left\{\sum_{n=0}^{\infty} \frac{\bar{n}^n}{n!}[e^{2i\lambda t(\bar{n}+1)^{1/2}} e^{i\lambda t n/(\bar{n}+1)^{1/2}} e^{-i\lambda t \bar{n}/(\bar{n}+1)^{1/2}}]\right\}. \tag{4.130}$$

Note that

$$\sum_{n=0}^{\infty} \frac{\bar{n}^n}{n!} e^{in\lambda t/(\bar{n}+1)^{1/2}} = \exp\left[\bar{n}e^{i\lambda t/(\bar{n}+1)^{1/2}}\right]. \tag{4.131}$$

For short times t,

$$e^{i\lambda t/(\bar{n}+1)^{1/2}} \simeq 1 + i\lambda t/(\bar{n}+1)^{1/2} - \frac{\lambda^2 t^2}{2(\bar{n}+1)}, \tag{4.132}$$

and thus

$$e^{\bar{n}}e^{i\lambda t/(\bar{n}+1)^{1/2}} \simeq e^{\bar{n}}e^{i\lambda t\bar{n}/(\bar{n}+1)}e^{-\frac{1}{2}\frac{\lambda^2 t^2}{(\bar{n}+1)^3}\bar{n}}. \tag{4.133}$$

Putting all of this together, we arrive at

$$W(t) \simeq \cos\{2\lambda t(\bar{n}+1)^{1/2}\}\exp\left\{-\frac{1}{2}\frac{\lambda^2 t^2 \bar{n}}{\bar{n}+1}\right\}, \tag{4.134}$$

valid for a short time. The inversion evidently exhibits a Gaussian decay law with a decay time given by

$$t_c = \frac{\sqrt{2}}{\lambda}\frac{\sqrt{\bar{n}+1}}{\bar{n}} \simeq \frac{\sqrt{2}}{\lambda}, \quad \bar{n} \gg 1 \tag{4.135}$$

which, apart from a numerical constant on the order of unity, agrees with our previous estimate.

Now let us examine the phenomenon of the revivals. $W(t)$ obviously consists of a sum of oscillating terms, each term oscillating at a particular Rabi frequency $\Omega(n) = 2\lambda\sqrt{n+1}$. If two neighboring terms are oscillating 180° out of phase with each other, we expect at least an approximate cancellation of these terms. On the other hand, if the neighboring terms are in phase with each other, we expect a constructive interference. In fact, this should be so whenever neighboring phases differ by some multiple of 2π. Since only those important frequencies around $\bar{n}$ will contribute, revivals should occur for times $t = t_R$ such that

$$[\Omega(\bar{n}+1) - \Omega(\bar{n})]t_R = 2\pi k, \quad k = 0, 1, 2, \ldots \tag{4.136}$$

holds. Expanding $\Omega(\bar{n})$ and $\Omega(\bar{n}+1)$, we easily arrive at $t_R = (2\pi/\lambda)\bar{n}^{1/2}k$ ($\bar{n} \gg 1$).

More rigorously, using Eq. (4.130) and the result in Eq. (4.131) we have

$$W(t) \simeq \cos\left[2\lambda t(\bar{n}+1)^{1/2} + \lambda t\bar{n}/(\bar{n}+1)^{-1/2} - \bar{n}\sin\left(\frac{\lambda t}{(\bar{n}+1)^{1/2}}\right)\right]$$

$$\times \exp\left\{-\bar{n}\left[1 - \cos\left(\frac{\lambda t}{(\bar{n}+1)^{1/2}}\right)\right]\right\}. \tag{4.137}$$

Obviously, the amplitude will be a maximum wherever the time $t = t_R = (2\pi/\lambda)(\bar{n}+1)^{1/2}k \approx (2\pi/\lambda)\bar{n}^{1/2}k$ ($\bar{n} \gg 1$), in agreement with the previous analysis. In Chapter 10 we discuss two experiments, one in the context of cavity quantum electrodynamics and the other in the context of the center of mass motion of a trapped ion, where the predicted collapse and revival of the Rabi oscillations have been observed.

Before we leave this section, we point out that the Jaynes–Cummings model is a simple fully quantum mechanical model of a field–matter interaction wherein entanglement between the quantized field and the two levels of the atom is generated. The initial state considered in Eq.

(4.118) is a factorizable state (i.e. it is a product of the initial atom and field states). The system evolves into the state of Eq. (4.120), a state that generally cannot be factored into a product of atomic and field states. In this case, the state is said to be entangled.

Intuition might suggest that the atom and field states become fully (or maximally) entangled in the middle of the quiescent period following the initial collapse of the Rabi oscillations. However, intuition would be quite wrong in this case. To study the evolution of the entanglement, we turn to the density operator of the atom–field system. The reader should review Appendix A at this juncture.

We first construct the atom–field density operator as follows, assuming an arbitrary initial state of the atom. That is, we write the atom–field density operator as

$$\hat{\rho}_{af}(t) = |\psi(t)\rangle\langle\psi(t)|, \tag{4.138}$$

where $|\psi(t)\rangle$ is given by Eq. (4.119). The density operator for the atom is given by

$$\hat{\rho}_a(t) = \text{Tr}_f \hat{\rho}_{af}(t) = \sum_{n=0}^{\infty} \langle n|\psi(t)\rangle\langle\psi(t)|n\rangle. \tag{4.139}$$

Then we trace over the square of this reduced density operator of the atom according to

$$\text{Tr}\hat{\rho}_a^2(t) = \langle g|\hat{\rho}_a^2(t)|g\rangle + \langle e|\hat{\rho}_a^2(t)|e\rangle. \tag{4.140}$$

Figure 4.8 is a plot of $\text{Tr}\hat{\rho}_a^2(t)$ versus the scaled time λt along with the atomic inversion, for temporal comparison, for the atom initially prepared in the ground state and for the coherent state of the field prepared with $\alpha = 7.0$. We see that the atom and field become highly entangled during the period of the collapse of the Rabi oscillations as $\text{Tr}\hat{\rho}_a^2(t)$ collapses to 0.5. However, $\text{Tr}\hat{\rho}_a^2(t)$ slowly rises to near unity at about half-way between the time of the collapse of the Rabi oscillations and the time of their first revival. This means that during the middle of the time of quiescence in the Rabi oscillations, the states of the atom–field system are factorizable. This unexpected behavior in the resonant Jaynes–Cummings model was revealed by Gea-Banacloche [14] in 1990. Problem 10 at the end of this chapter requests the reader to go through the details of this calculation and reproduce the results displayed in Fig. 4.8.

4.6 The Dressed States

There are many ways to solve for the dynamics of the Jaynes–Cummings model (see the reviews of Ref. [15]). In Section 4.5 we solved the time-dependent Schrödinger equation first for a field containing n photons and then, by simple extrapolation, for the case of a field in a superposition of

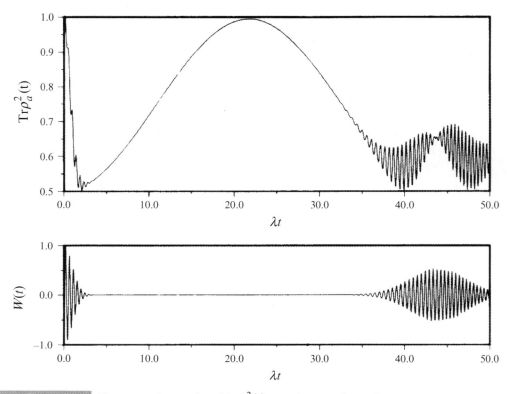

 The top graph is a plot of $\mathrm{Tr}\hat{\rho}_a^2(t)$ as a function of time for the atom initially in the ground state with the field initially in a coherent state with $\bar{n} = |\alpha|^2 = 49$. The bottom graph is the atomic inversion plotted against time for the same initial conditions. (Reprinted figure with permission from [14]. Copyright © 1990 by the American Physical Society. DOI: 10.1103/PhysRevLett.65.3385.)

the number states. Another important way to obtain the dynamics is to first find the stationary states of the Jaynes–Cummings Hamiltonian. For reasons that should become clear shortly, these eigenstates are called the "dressed" states [16].

Consider once again the Jaynes–Cummings model Hamiltonian

$$\hat{H} = \frac{1}{2}\hbar\omega_0\hat{\sigma}_3 + \hbar\omega\hat{a}^\dagger\hat{a} + \hbar\lambda(\hat{a}\hat{\sigma}_+ + \hat{a}^\dagger\hat{\sigma}_-), \qquad (4.141)$$

where we have *not* assumed the resonance condition $\omega = \omega_0$ at this point. In terms of the field number states, the interaction term in $\hat{H}$ causes only transitions of the type

$$|e\rangle|n\rangle \leftrightarrow |g\rangle|n + 1\rangle, \qquad (4.142)$$

or

$$|e\rangle|n-1\rangle \leftrightarrow |g\rangle|n\rangle. \tag{4.143}$$

The product states $|e\rangle|n-1\rangle$, $|g\rangle|n\rangle$, and so on are sometimes referred to as the "bare" states of the Jaynes–Cummings model; they are product states of the unperturbed atom and field. For a fixed n, the dynamics is completely confined to the two-dimensional space of product states, either $(|e\rangle|n-1\rangle, |g\rangle|n\rangle)$ or $(|e\rangle|n\rangle, |g\rangle|n+1\rangle)$. We define the following product states for a given n:

$$\begin{aligned}|\psi_{1n}\rangle &= |e\rangle|n\rangle, \\ |\psi_{2n}\rangle &= |g\rangle|n+1\rangle.\end{aligned} \tag{4.144}$$

Obviously $\langle\psi_{1n}|\psi_{2n}\rangle = 0$. Using this basis we obtain the matrix elements of $\hat{H}$, $H_{ij}^{(n)} = \langle\psi_{in}|\hat{H}|\psi_{jn}\rangle$, which are

$$H_{11}^{(n)} = \hbar\left[n\omega + \frac{1}{2}\omega_0\right],$$

$$H_{22}^{(n)} = \hbar\left[(n+1)\omega - \frac{1}{2}\omega_0\right], \tag{4.145}$$

$$H_{12}^{(n)} = \hbar\lambda\sqrt{n+1} = H_{21}^{(n)}.$$

Thus in the 2×2 subspace of Eq. (4.128), we obtain the matrix representation of $\hat{H}$:

$$\boldsymbol{H}^{(n)} = \hbar \begin{bmatrix} n\omega + \dfrac{1}{2}\omega_0 & \lambda\sqrt{n+1} \\ \lambda\sqrt{n+1} & (n+1)\omega - \dfrac{1}{2}\omega_0 \end{bmatrix}. \tag{4.146}$$

This matrix is "self-contained" since, as we have said, the dynamics connects only those states for which the photon number changes by ± 1. For a given n, the energy eigenvalues of $\boldsymbol{H}^{(n)}$ are as follows:

$$E_{\pm}(n) = \left(n+\frac{1}{2}\right)\hbar\omega \pm \frac{1}{2}\hbar\Omega_n(\Delta), \tag{4.147}$$

where

$$\Omega_n(\Delta) = [\Delta^2 + 4\lambda^2(n+1)]^{1/2} \quad (\Delta = \omega_0 - \omega) \tag{4.148}$$

is the Rabi frequency which now includes the effects of the detuning Δ. Obviously, for $\Delta = 0$ we obtain $\Omega_n(0) = 2\lambda\sqrt{n+1}$, the same quantum electrodynamic Rabi frequencies seen earlier. The eigenstates $|n, \pm\rangle$ associated with the energy eigenvalues are given by

$$\begin{aligned}|n, +\rangle &= \cos(\Phi_n/2)|\psi_{1n}\rangle + \sin(\Phi_n/2)|\psi_{2n}\rangle, \\ |n, -\rangle &= -\sin(\Phi_n/2)|\psi_{1n}\rangle + \cos(\Phi_n/2)|\psi_{2n}\rangle,\end{aligned} \tag{4.149}$$

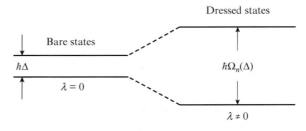

Fig. 4.9 Energy-level splitting due to the interaction of the atom with a quantized field. The split levels on the right are the energy levels of the dressed states.

where the angle Φ_n is defined through

$$\Phi_n = \tan^{-1}\left(\frac{2\lambda\sqrt{n+1}}{\Delta}\right) = \tan^{-1}\left(\frac{\Omega_n(0)}{\Delta}\right), \tag{4.150}$$

and where

$$\sin(\Phi_n/2) = \frac{1}{\sqrt{2}}\left[\frac{\Omega_n(\Delta) - \Delta}{\Omega_n(\Delta)}\right]^{1/2},$$

$$\cos(\Phi_n/2) = \frac{1}{\sqrt{2}}\left[\frac{\Omega_n(\Delta) + \Delta}{\Omega_n(\Delta)}\right]^{1/2}. \tag{4.151}$$

The states $|n, \pm\rangle$ are often referred to as "dressed states" or as the Jaynes–Cummings doublet. The bare states $|\psi_{1n}\rangle$ and $|\psi_{2n}\rangle$, of energies $E_{1n} = \hbar(\omega_0/2 + n\omega)$ and $E_{2n} = \hbar[-\omega_0/2 + (n+1)\omega]$, are each further split in energy due to the interaction, as indicated in Fig. 4.9. The splitting of the bare states into the dressed states is a kind of Stark shift, often called the AC, or dynamic, Stark shift. Note that in the limit of exact resonance, $\Delta = 0$, the bare states are degenerate but the splitting, of course, remains. In this limit, the dressed states are related to the bare states according to

$$|n, +\rangle = \frac{1}{\sqrt{2}}(|e\rangle|n\rangle + |g\rangle|n+1\rangle),$$

$$|n, -\rangle = \frac{1}{\sqrt{2}}(-|e\rangle|n\rangle + |g\rangle|n+1\rangle). \tag{4.152}$$

To see how the dressed states can be used to obtain the dynamics for rather general initial states, let us consider the specific case of a field prepared in some superposition of number states

$$|\psi_f(0)\rangle = \sum_n C_n|n\rangle, \tag{4.153}$$

where an atom, prepared in state $|e\rangle$, is injected into the field. Thus, the initial state of the atom–field system is

$$|\psi_{af}(0)\rangle = |\psi_f(0)\rangle|e\rangle$$

$$= \sum_n C_n |n\rangle|e\rangle = \sum_n C_n|\psi_{1n}\rangle, \qquad (4.154)$$

where, from Eqs. (4.149), we obtain $|\psi_{1n}\rangle$ in terms of the dressed states $|n, \pm\rangle$ as

$$|\psi_{1n}\rangle = \cos(\Phi_n/2)|n, +\rangle - \sin(\Phi_n/2)|n, -\rangle, \qquad (4.155)$$

and thus

$$|\psi_{af}(0)\rangle = \sum_n C_n[\cos(\Phi_n/2)|n, +\rangle - \sin(\Phi_n/2)|n, -\rangle]. \qquad (4.156)$$

Since the dressed states $|n, \pm\rangle$ are stationary states of the atom–field system, the state vector for times $t > 0$ is just given by

$$|\psi_{af}(t)\rangle = \exp\left[-\frac{i}{\hbar}\hat{H}t\right]|\psi_{af}(0)\rangle$$

$$= \sum_n C_n\left[\cos(\Phi_n/2)|n, +\rangle e^{-iE_+(n)t/\hbar} - \sin(\Phi_n/2)|n, -\rangle e^{-iE_-(n)t/\hbar}\right]. \qquad (4.157)$$

Of course, the entire result may now be recast back into the more familiar "bare" state basis by simply substituting $|n, \pm\rangle$ from Eqs. (4.144). In the limit $\Delta = 0$ we will recover the previous result of Eq. (4.117). The demonstration of this is left as an exercise.

4.7 Density Operator Approach: Application to Thermal States

So far, we have considered only cases where the field and the atom are initially in pure states. In general, though, one or both of the subsystems may initially be in mixed states, which requires us to seek a solution in terms of the density operator. For example, the field may initially be in a thermal state described by the density operator of Eq. (2.145). In studying the case of a two-level atom interacting with a thermal state described by a density operator, we are afforded yet another way to solve for the dynamics of the Jaynes–Cummings model.

We shall work in the interaction picture and once again assume the resonance condition so that the dynamics is driven by

$$\hat{H}_I = \hbar\lambda(\hat{a}\hat{\sigma}_+ + \hat{a}^\dagger\hat{\sigma}_-). \qquad (4.158)$$

If $\hat{\rho}(t)$ is the density operator of the atom–field system at time t, the evolution of this operator is given by

$$\frac{d\hat{\rho}}{dt} = -\frac{i}{\hbar}[\hat{H}_I, \hat{\rho}], \tag{4.159}$$

whose solution may be written as

$$\hat{\rho}(t) = \hat{U}_I(t)\hat{\rho}(0)\hat{U}_I^\dagger(t) \tag{4.160}$$

where

$$\begin{aligned}\hat{U}_I(t) &= \exp[-i\hat{H}_I t/\hbar] \\ &= \exp[-i\lambda t(\hat{a}\hat{\sigma}_+ + \hat{a}^\dagger\hat{\sigma}_-)].\end{aligned} \tag{4.161}$$

In the two-dimensional atomic subspace, the operators $\hat{\sigma}_\pm$ and $\hat{\sigma}_3$ have the matrix representations

$$\sigma_+ = \begin{pmatrix} 0 & 1 \\ 0 & 0 \end{pmatrix}, \sigma_- = \begin{pmatrix} 0 & 0 \\ 1 & 0 \end{pmatrix}, \sigma_3 = \begin{pmatrix} 1 & 0 \\ 0 & -1 \end{pmatrix} \tag{4.162}$$

where we have used the convention

$$\sigma_j = \begin{pmatrix} \langle e|\hat{\sigma}_j|e\rangle & \langle e|\hat{\sigma}_j|g\rangle \\ \langle g|\hat{\sigma}_j|e\rangle & \langle g|\hat{\sigma}_j|g\rangle \end{pmatrix} \quad j = \pm, 3. \tag{4.163}$$

In this two-dimensional subspace, the evolution operator $\hat{U}_I(t)$ may be expanded as

$$\hat{U}_I(t) = \begin{pmatrix} \hat{C}(t) & \hat{S}'(t) \\ \hat{S}(t) & \hat{C}'(t) \end{pmatrix} \tag{4.164}$$

where

$$\hat{C}(t) = \cos(\lambda t\sqrt{\hat{a}\hat{a}^\dagger}), \tag{4.165}$$

$$\hat{S}(t) = -i\hat{a}^\dagger \frac{\sin(\lambda t\sqrt{\hat{a}\hat{a}^\dagger})}{\sqrt{\hat{a}\hat{a}^\dagger}}, \tag{4.166}$$

$$\hat{C}'(t) = \cos(\lambda t\sqrt{\hat{a}^\dagger\hat{a}}), \tag{4.167}$$

$$\hat{S}'(t) = -i\hat{a} \frac{\sin(\lambda t\sqrt{\hat{a}^\dagger\hat{a}})}{\sqrt{\hat{a}^\dagger\hat{a}}}. \tag{4.168}$$

(The operators $\hat{C}, \hat{S}$, etc. here are not to be confused with the cosine and sine operators of the phase introduced in Chapter 2.) The Hermitian adjoint of $\hat{U}_I(t)$ given by Eq. (4.164) is just

$$\hat{U}_I^{\dagger}(t) = \hat{U}_I(-t) = \begin{pmatrix} \hat{C}(t) & -\hat{S}'(t) \\ -\hat{S}(t) & \hat{C}'(t) \end{pmatrix}. \tag{4.169}$$

We now suppose that at $t = 0$, the density operator for the atom–field system factors into field and atomic parts:

$$\hat{\rho}(0) = \hat{\rho}^F(0) \otimes \hat{\rho}^A(0). \tag{4.170}$$

We further suppose (to work out a particular example) that the atom is initially in the excited state $|e\rangle$ such that $\hat{\rho}^A(0) = |e\rangle\langle e|$. The corresponding density matrix for the atom is (using the convention of Eq. (4.163))

$$\rho^A(0) = \begin{pmatrix} 1 & 0 \\ 0 & 0 \end{pmatrix} \tag{4.171}$$

and thus we may write for the system

$$\hat{\rho}(0) = \begin{pmatrix} \hat{\rho}^F(0) & 0 \\ 0 & 0 \end{pmatrix} = \hat{\rho}^F(0) \otimes \begin{pmatrix} 1 & 0 \\ 0 & 0 \end{pmatrix}. \tag{4.172}$$

Using Eqs. (4.167) and (4.159)–(4.163) in Eq. (4.155), we find that

$$\hat{\rho}(t) = \begin{pmatrix} \hat{C}(t)\hat{\rho}(0)\hat{C}(t) & -\hat{C}(t)\hat{\rho}^F(0)\hat{S}'(t) \\ \hat{S}(t)\hat{\rho}^F(0)\hat{C}(t) & -\hat{S}(t)\hat{\rho}^F(0)\hat{S}'(t) \end{pmatrix}. \tag{4.173}$$

The reduced density operator of the field is found by tracing over the atomic states and thus

$$\hat{\rho}^F(t) = \mathrm{Tr}_A \hat{\rho}(t) = \hat{C}(t)\hat{\rho}^F(0)\hat{C}(t) - \hat{S}(t)\hat{\rho}^F(0)\hat{S}'(t). \tag{4.174}$$

The density matrix elements for the field are

$$\begin{aligned} \hat{\rho}_{nm}^F(t) &\equiv \langle n|\hat{\rho}^F(t)|m\rangle \\ &= \langle n|\hat{C}(t)\hat{\rho}^F(0)\hat{C}(t)|m\rangle \\ &\quad - \langle n|\hat{S}(t)\hat{\rho}^F(0)\hat{S}'(t)|m\rangle. \end{aligned} \tag{4.175}$$

On the other hand, tracing over the field states we obtain the reduced density operator of the atom:

$$\hat{\rho}^A(t) = \mathrm{Tr}_F \hat{\rho}(t) = \sum_{n=0}^{\infty} \langle n|\hat{\rho}(t)|n\rangle. \tag{4.176}$$

The density matrix elements are given by

$$\langle i|\hat{\rho}^A(t)|j\rangle = \sum_{n=0}^{\infty} \langle i, n|\hat{\rho}(t)|j, n\rangle = \rho_{ij}^A(t), \tag{4.177}$$

where $i, j = e, g$. The diagonal elements $\rho_{ee}^A(t)$ and $\rho_{gg}^A(t)$ are the populations of the excited and ground states, respectively, and satisfy the condition

$$\rho_{gg}^A(t) + \rho_{ee}^A(t) = 1. \tag{4.178}$$

The atomic inversion is given by

$$W(t) = \rho_{ee}^A(t) - \rho_{gg}^A(t) = 2\rho_{ee}^A(t) - 1. \tag{4.179}$$

From Eqs. (4.171) and (4.175) we find that

$$\rho_{ee}^A(t) \equiv \sum_{n=0}^{\infty} \langle n|\hat{C}(t)\hat{\rho}^F(0)\hat{C}(t)|n\rangle$$

$$= \sum_{n=0}^{\infty} \langle n|\,\hat{\rho}^F(0)|n\rangle \,\cos^2(\lambda t\sqrt{n+1}). \tag{4.180}$$

If the field is initially in a pure state

$$|\psi_F\rangle = \sum_{n=0}^{\infty} C_n|n\rangle, \tag{4.181}$$

then

$$\hat{\rho}^F(0) = |\psi_F\rangle\langle\psi_F| \tag{4.182}$$

and thus

$$\rho_{ee}^A(t) = \sum_{n=0}^{\infty} |C_n|^2 \cos^2(\lambda t\sqrt{n+1}) \tag{4.183}$$

which, through Eq. (4.158), yields the atomic inversion found in Eq. (4.122). But suppose the field is initially in a thermal state (a mixed state), where

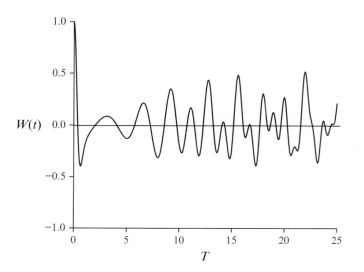

Fig. 4.10   The atomic inversion versus time for the atom initially in the excited state and the field initially in a thermal state with $\bar{n} = 2$.

$$\hat{\rho}^F(0) = \hat{\rho}_T = \sum_n P_n |n\rangle \langle n|, \qquad (4.184)$$

with P_n given by Eq. (2.146). From Eq. (4.174) we ultimately obtain the atomic inversion for an atom resonantly interacting with a thermal field as [17]

$$W(t) = \sum_{n=0}^{\infty} P_n \cos(2\lambda t \sqrt{n+1}). \qquad (4.185)$$

In Fig. 4.10 we plot $W(t)$ versus λt for a thermal field containing an average photon number of $\bar{n} = 2$. We leave to the reader an analysis of the observed behavior.

4.8 The Jaynes–Cummings Model with Large Detuning: A Dispersive Interaction

In the foregoing, we have mostly assumed that the tuning $\Delta = \omega_0 - \omega = 0$. An important variation on the original Jaynes–Cummings model is the situation in which the detuning is large enough such that direct atomic transitions do not occur but where nevertheless "dispersive" interactions between a single atom and a cavity field do occur [18]. This version of the Jaynes–Cummings model is important in a number of applications related to fundamental testing of quantum mechanics, some of which are discussed in Chapter 10.

The arrangement of atomic levels that we consider is given in Fig. 4.11. As shown in Appendix C, the effective atom–field interaction Hamiltonian is given by

$$\hat{H}_{eff} = \hbar \chi [\hat{\sigma}_+ \hat{\sigma}_- + \hat{a}^\dagger \hat{a} \hat{\sigma}_3], \qquad (4.186)$$

where $\chi = \lambda^2/\Delta$. Note that $\hat{\sigma}_+ \hat{\sigma}_- = |e\rangle\langle e|$. Suppose the initial state of the atom–field system is $|\psi(0)\rangle = |g\rangle|n\rangle$, that is the atom is in the ground state and the field is in a number state. Then according to the interaction Hamiltonian of Eq. (4.186), the state at time $t > 0$ is

$$|\psi(t)\rangle = e^{-i\hat{H}_{eff}t/\hbar}|\psi(0)\rangle = e^{i\chi nt}|g\rangle|n\rangle, \qquad (4.187)$$

while for the initial state $|\psi(0)\rangle = |e\rangle|n\rangle$ we have

$$|\psi(t)\rangle = e^{-i\hat{H}_{eff}t/\hbar}|\psi(0)\rangle = e^{-i\chi(n+1)t}|e\rangle|n\rangle. \qquad (4.188)$$

Evidently, nothing very interesting happens, just the production of unmeasurable phase factors. On the other hand, for initial coherent states of the field we have, for $|\psi(0)\rangle = |g\rangle|\alpha\rangle$,

$$|\psi(t)\rangle = e^{-i\hat{H}_{\mathit{eff}}t/\hbar}|\psi(0)\rangle = |g\rangle|\alpha e^{i\chi t}\rangle, \qquad (4.189)$$

and for $|\psi(0)\rangle = |e\rangle|\alpha\rangle$ we have

$$|\psi(t)\rangle = e^{-i\hat{H}_{\mathit{eff}}t/\hbar}|\psi(0)\rangle = e^{-i\chi t}|e\rangle|\alpha e^{-i\chi t}\rangle. \qquad (4.190)$$

We notice now that the coherent state amplitude is rotated in phase space by the angle χt but that the direction of the rotation depends on the state of the atom. Suppose now that the atom is prepared in a superposition of the ground and excited states. For simplicity, we take this to be a "balanced" state of the form $|\psi_{\mathrm{atom}}\rangle = (|g\rangle + e^{i\phi}|e\rangle)/\sqrt{2}$, where ϕ is some phase. With the initial state $|\psi(0)\rangle = |\psi_{\mathrm{atom}}\rangle|\alpha\rangle$ we have

$$|\psi(t)\rangle = e^{-i\hat{H}_{\mathit{eff}}t/\hbar}|\psi(0)\rangle = \frac{1}{\sqrt{2}}\left(|g\rangle|\alpha\,e^{i\chi t}\rangle + e^{-i(\chi t-\phi)}|e\rangle|\alpha\,e^{-i\chi t}\rangle\right). \quad (4.191)$$

This state is a bit more interesting, in fact much more interesting as in general there is entanglement between the atom and the field. If we take $\chi t = \pi/2$, we then have

$$\left|\psi\left(\frac{\pi}{2\chi}\right)\right\rangle = \frac{1}{\sqrt{2}}\left(|g\rangle|i\alpha\rangle - ie^{i\phi}|e\rangle|-i\alpha\rangle\right). \qquad (4.192)$$

Notice that, in terms of our phase-space pictures, the two coherent states in Eq. (4.192) are separated by 180°, the maximal separation. Coherent states differing in phase by 180° are also maximally *distinguishable*, in the sense that there is essentially no overlap between the two states, at least if $|\alpha|$ is large enough. In fact, this is the case even with $|\alpha|$ as low as $\sqrt{2}$. With very large values of $|\alpha|$, the two coherent states are said to be *macroscopically* distinguishable, and for moderate values *mesoscopically* distinguishable. The entangled state of Eq. (4.192) might bring to mind the tale of Schrödinger's ill-fated cat [19], suspended in a state of limbo, suspended in an entanglement between life and death and a nondecayed or decayed radioactive *microscopic* atom. Qualitatively speaking, the entangled state in Schrödinger's famous "paradox" is thus

$$|\psi_{\mathrm{atom\text{-}cat}}\rangle = \frac{1}{\sqrt{2}}\left[|\text{atom not decayed}\rangle|\text{cat alive}\rangle + |\text{atom decayed}\rangle|\text{cat dead}\rangle\right].$$

$$(4.193)$$

The parallel between this state and that of Eq. (4.192) is obvious; the states of the two-level atom playing the role of the radioactive atom and the coherent field states that of the cat.

Finally, there is another initial atomic state that is often considered. Suppose there is another atomic state, that we denote $|f\rangle$, of energy $E_f \ll E_g$ (as pictured in Fig. 4.11) and of parity opposite that of $|g\rangle$. The cavity is assumed to support no mode resonant with the $f \leftrightarrow g$ transition and we further assume that the state $|f\rangle$ is so far out of resonance with the

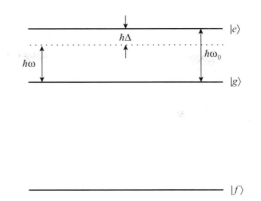

Fig. 4.11
The levels $|e\rangle$ and $|g\rangle$ are out of resonance with the field enough so that there are no direct transitions between them and only the dispersive interaction occurs. The state $|f\rangle$ is so far out of resonance with state $|g\rangle$ and the field that not even a dispersive interaction is present.

available cavity field mode that there is no discernable dispersive interaction either. Thus, with the atom initially prepared in state $|f\rangle$ and with the field initially in the coherent state $|\alpha\rangle$, the initial product state $|f\rangle|\alpha\rangle$ remains as such (i.e. it does not evolve). Now suppose the atom is prepared in a superposition of the form $|\psi_{\text{atom}}\rangle = (|g\rangle + e^{i\phi}|f\rangle)/\sqrt{2}$. The initial atom–field state $|\psi(0)\rangle = |\psi_{\text{atom}}\rangle|\alpha\rangle$ is easily seen to evolve into

$$\left|\psi(t)\right\rangle = \frac{1}{\sqrt{2}}\left(|g\rangle|\alpha e^{i\chi t}\rangle + e^{i\phi}|f\rangle|\alpha\rangle\right). \tag{4.194}$$

In this case, only one component of the initial atomic superposition causes a phase shift in the coherent, an advantage in certain applications. For $\chi t = \pi$ we have

$$\left|\psi\left(\frac{\pi}{\chi}\right)\right\rangle = \frac{1}{\sqrt{2}}(|g\rangle| - \alpha\rangle + e^{i\phi}|f\rangle|\alpha\rangle), \tag{4.195}$$

another version of the Schrödinger-cat state.

In Chapters 7, 8 and 10 we shall further elaborate on the issues of Schrödinger's cat.

4.9 Extensions of the Jaynes–Cummings Model

There are many extensions of the original Jaynes–Cummings model involving various types of alternative interactions. Among them are models involving two-photon transitions, multimode and multilevel models, Raman coupled models, two-channel models, and so on. We shall not discuss these models here, but we do refer to the various review articles that have appeared (see Refs. [16, 20] and references cited therein). Some of these extensions will appear in the context of homework problems. Further, it turns out that Jaynes–Cummings-type interactions also occur in the context of the vibrational motion of an ion in an electromagnetic trap. The simplest of these will be discussed in Chapter 10.

4.10 Schmidt Decomposition and Von Neumann Entropy for the Jaynes–Cummings Model

We finish this chapter with a discussion of the Schmidt decomposition and the related von Neumann entropy as they pertain to the Jaynes–Cummings model. As this system is bipartite, a Schmidt decomposition is assured, as discussed in Appendix A. We have already presented the solution of the time-dependent Schrödinger equation in Eq. (4.119), which we rewrite here as

$$|\Psi(t)\rangle = \sum_{n=0}^{\infty}[a_n(t)|g\rangle|n\rangle + b_n(t)|e\rangle|n\rangle], \qquad (4.196)$$

where

$$a_n(t) = C_g C_n \cos(\lambda t \sqrt{n}) - i C_e C_{n-1} \sin(\lambda t \sqrt{n}),$$
$$b_n(t) = C_e C_n \cos(\lambda t \sqrt{n+1}) - i C_g C_{n+1} \sin(\lambda t \sqrt{n+1}). \qquad (4.197)$$

But according to the Schmidt decomposition, for any instant in time t we can always find bases $\{|u_i(t)\rangle\}$ for the atom and $\{v_i(t)\}$ for the field such that the pure state of the system can be written as

$$|\Psi(t)\rangle = g_1(t)|u_1(t)\rangle|v_1(t)\rangle + g_2(t)|u_2(t)\rangle|v_2(t)\rangle. \qquad (4.198)$$

The reduced density matrices of the atom and field in these bases are

$$\rho_u(t) = \begin{pmatrix} |g_1(t)|^2 & 0 \\ 0 & |g_2(t)|^2 \end{pmatrix},$$

$$\rho_v(t) = \begin{pmatrix} |g_1(t)|^2 & 0 \\ 0 & |g_2(t)|^2 \end{pmatrix}. \qquad (4.199)$$

In order to find the coefficients $g_1(t)$ and $g_2(t)$ and the eigenvectors $|u_i\rangle$ and $|v_i\rangle$, we first calculate the reduced density operator of the atom in the bare basis specified by $|e\rangle$ and $|g\rangle$ and obtain

$$\rho_u = \begin{pmatrix} \sum_{n=0}^{\infty} |a_n(t)|^2 & \sum_{n=0}^{\infty} a_n(t)b_n^*(t) \\ \sum_{n=0}^{\infty} b_n(t)a_n^*(t) & \sum_{n=0}^{\infty} |b_n(t)|^2 \end{pmatrix}. \qquad (4.200)$$

Then we can use the parameterization of the density matrix in terms of the components of the Bloch vector, as in Eq. (A.25), to obtain

$$
\begin{aligned}
s_1(t) &= \sum_{n=0}^{\infty}[a_n(t)b_n^*(t) + b_n(t)a_n^*(t)], \\
s_2(t) &= -i\sum_{n=0}^{\infty}[a_n(t)b_n^*(t) - b_n(t)a_n^*(t)], \\
s_3(t) &= \sum_{n=0}^{\infty}[|a_n(t)|^2 - |b_n(t)|^2].
\end{aligned}
\tag{4.201}
$$

Then, as discussed in Appendix A, the coefficients $g_1(t)$ and $g_2(t)$ can be expressed in terms of the length of the Bloch vector according to

$$
g_1(t) = \frac{1}{2}[1 + |\mathbf{s}(t)|], \quad g_2(t) = \frac{1}{2}[1 - |\mathbf{s}(t)|].
\tag{4.202}
$$

Notice that the field mode, previously described in an infinite-dimensional Fock-state basis, is reduced to a "two-level" system in the Schmidt basis. The length of the Bloch vector is a measure of the purity of the atom–field system. For the atom and field in pure states, the length of the Bloch vector is unity.

As discussed in Appendix A, the utility of Schmidt decomposition is that it is easy to obtain an expression for von Neumann entropy, which for each of the subsystems of the Jaynes–Cummings model is

$$
S(\hat{\rho}_u) = -g_1(t)\ln g_1(t) - g_2(t)\ln g_2(t) = S(\hat{\rho}_v).
\tag{4.203}
$$

Whenever $|\mathbf{s}| = 1$ we have $S(\hat{\rho}_u) = 0 = S(\hat{\rho}_v)$. In Fig. 4.12, for an atom initially in the excited state and for the field initially in a coherent state with $\alpha = \sqrt{30}$, we plot (a) the s_3 component of the Bloch vector, (b) the s_2 component, and (c) the von Neumann entropy $S(\hat{\rho}_u)$, all against the scale time λt. With our particular initial condition $s_1 = 0$ for all times. We notice that s_3 undergoes the collapse and revival of the atomic inversion and that s_2 goes close to unity at one point during the quiescent period of the collapse of the former, indicating that the atom and field are nearly in pure states at that time. The von Neumann entropy, of course, is at a minimum at that point. This result is perhaps a bit surprising and counterintuitive. This behavior was first noticed, through a different type of calculation, by Gea-Banacloche [14], as already mentioned, and was further examined by Phoenix and Knight [21].

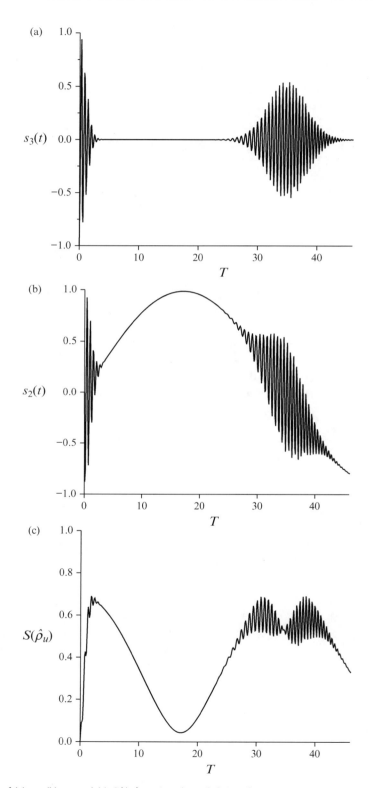

Fig. 4.12 Plots of (a) s_3, (b) s_2, and (c) $S(\hat{\rho}_u)$ against the scaled time λt.

Problems

1. Consider the semiclassical Rabi model (i.e. a two-level atom with a prescribed classical field) in the RWA as described by Eqs. (4.74). Obtain the solution assuming the atom initially in the excited state. Calculate the atomic inversion as a function of time.

2. Using the result of the previous problem, obtain the time-dependent expectation value of the atomic dipole moment operator $\hat{d} = d(\hat{\sigma}_+ + \hat{\sigma}_-)$ for the case of exact resonance. Compare the evolution of the dipole moment with the atomic inversion for the same resonance condition and comment on any similarities or differences.

3. In the fully quantized model of a two-level atom interacting with a quantized field within the RWA, the Jaynes–Cummings model, we have obtained the exact resonance solution for the initial state where the atom is excited and where the field is in a number state $|n\rangle$ (see Eqs. (4.106) to (4.115)). Recall that the Rabi oscillations of the atomic inversion are periodic for this case, just as in the semiclassical model. Use that solution to obtain the expectation value of the atomic dipole moment operator and compare with the result of the previous question. Is the time evolution of the atomic dipole moment in any way similar to that obtained in the semiclassical case? Comment on the result.

4. Obtain the expectation value of the atomic dipole moment as given by the Jaynes–Cummings model in the case where the field is initially in a coherent state. How does the result compare with the two previous cases? You should make a plot of the expectation value of the dipole moment as a function of time.

5. In the text, we obtained the dynamics of the Jaynes–Cummings model assuming exact resonance, $\Delta = 0$. Reconsider the problem for the case where $\Delta \neq 0$. Obtain plots of the atomic inversion and note the effect of the nonzero detuning on the collapse and revivals of the Rabi oscillations. Perform an analysis to obtain the effect of the nonzero detuning on the collapse and revival times.

6. Consider the resonant Jaynes–Cummings model for the initial thermal state as in Section 4.7. Assume the atom is initially in the excited state. Analyze the collapse of the Rabi oscillations and determine the dependence of the collapse time on the average photon number of the thermal field.

7. Consider a simple model of degenerate Raman scattering where $E_g = E_e$, as pictured in Fig. 4.13, described by the interaction Hamiltonian $\hat{H}_I = \hbar\lambda\hat{a}^\dagger\hat{a}(\sigma_+ + \sigma_-)$, where, as usual, $\sigma_+ = |e\rangle\langle g|$ and $\sigma_- = |g\rangle\langle e|$.
   (a) Obtain the dressed states for this model.
   (b) Assuming the field to be initially in a coherent state and the atom in the ground state, obtain the atomic inversion and show that the revivals of the Rabi oscillations are regular and complete.
   (c) Obtain the atomic inversion for an initial thermal state.

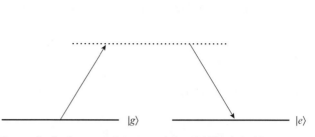

Fig. 4.13 Energy-level diagram for the degenerate Raman coupled model. The dashed line represents a "virtual" intermediate state, too far off-resonance from a real level (upper solid line) to become populated.

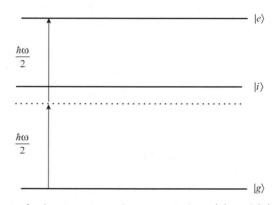

Fig. 4.14 Energy-level diagram for the resonant two-photon process. States $|e\rangle$ and $|g\rangle$ are of like parity, whereas the intermediate state $|i\rangle$ is of opposite parity. The dashed line represents a virtual atomic level, detuned from state $|i\rangle$.

8. A resonant two-photon extension of the Jaynes–Cummings model is described by the effective Hamiltonian $H_{eff} = \hbar\eta(\hat{a}^2\hat{\sigma}_+ + \hat{a}^{2\dagger}\hat{\sigma}_-)$, where, for the sake of simplicity, a small Stark shift term has been ignored. This Hamiltonian represents two-photon absorption and emission between atomic levels of like parity. The process is represented by Fig. 4.14, where the dashed line represents a virtual intermediate state of opposite parity.
    (a) Obtain the dressed states for this system.
    (b) Obtain the atomic inversion for this model assuming the atom to be initially in the ground state and the field initially in a number state. Repeat for a coherent state. Comment on the nature of the collapse and revival phenomena for these states.
    (c) Obtain the atomic inversion for an initial thermal state.

9. A two-mode variation on the two-photon model of the previous problem is described by the Hamiltonian $H_{eff} = \hbar\eta(\hat{a}\hat{b}\hat{\sigma}_+ + \hat{a}^\dagger\hat{b}^\dagger\hat{\sigma}_-)$, that is

a photon from each mode is absorbed or emitted. Obtain the atomic inversion for the case where both modes are initially in coherent states. Analyze the collapse and revival phenomena.

10. Consider the resonant Jaynes–Cummings model with the atom initially in the ground state and the field initially in a coherent state with $\alpha = 7$.

(a) Obtain the results presented in Fig. 4.8.

(b) Obtain plots (either contour or 3D) of the Q function of the field for various times between $t = 0$ and the time of the first revival of the Rabi oscillations.

11. Consider the atom–field state of Eq. (4.195) and assume that $\phi = 0$. Suppose that there is some way to determine the state of the atom (such is possible via field ionization, as will be discussed in Chapter 10). For example, if the atom is detected in state $|g\rangle$ then the field state will be reduced to $|-\alpha\rangle = \langle g|\psi(\pi)\rangle/|\langle g|\psi(\pi)\rangle|$ with a similar result for detection of the atomic state $|f\rangle$. (See Appendix D.) But suppose it were somehow possible to detect the atom in the superposition states $|S\pm\rangle = (|g\rangle \pm |f\rangle)/\sqrt{2}$. What field states are generated in these cases?

12. Reconsider the problem discussed in Section 4.10 but with the initial state of the atom taken to be the "balanced" superposition $|\psi(0)\rangle_{\text{atom}} = (|e\rangle + |g\rangle)/\sqrt{2}$. Obtain relevant plots of the components of the Bloch vector and of the von Neumann entropy. Contrast the results obtained in this case to the case where the atom is initially in the excited state.

13. The photon number parity operator is given by $\hat{\Pi} = (-1)^{\hat{a}^\dagger \hat{a}} = \exp(i\pi\hat{a}^\dagger\hat{a})$. Calculate the expectation value of this operator as a function of time for the Jaynes–Cummings model, assuming the atom is initially in the excited state and the field is initially in a coherent state of average photon number $\bar{n} = |\alpha|^2 = 15$. Compare your result with the evolution of the atomic inversion. Explain. See Ref. [22].

References

[1] J. J. Sakurai, *Modern Quantum Mechanics*, revised edition (Reading, MA: Addison-Wesley, 1994), p. 316.

[2] Two standard references on nuclear magnetic resonance, wherein the rotating wave approximation is discussed, are A. Abragam, *Principles of Nuclear Magnetism* (Oxford: Oxford University Press, 1961) and C. P. Slichter, *Principles of Magnetic Resonance*, 3rd edition (Berlin: Springer, 1992).

[3] See Ref. [1], p. 332.

[4] A. Einstein, *Verh. Deutsch. Phys. Ges.* **18**, 318 (1916); *Mitt. Phys. Ges. Zürich* **16**, 47 (1916); *Phys. Zeitschr.* **18**, 121 (1917). This last article appears in an English translation from the German in D. ter Haar, *The Old Quantum Theory* (Oxford: Pergamon, 1967) and is reprinted in P. L. Knight and L. Allen, *Concepts in Quantum Optics* (Oxford: Pergamon, 1983). See also A. Pais, *'Subtle is the Lord...' The Science and Life of Albert Einstein* (Oxford: Oxford University Press, 1982), pp. 402–415.

[5] P. W. Milonni, *Am. J. Phys.* **52**, 340 (1984).

[6] V. F. Weisskopf and E. Wigner, *Z. Phys.* **63**, 54 (1930). See also M. Sargent III, M. O. Scully, and W. E. Lamb Jr., *Laser Physics* (Reading, MA: Addison-Wesley, 1974), p. 236.

[7] I. I. Rabi, *Phys. Rev.* **51**, 652 (1937).

[8] See Refs. [2] and F. Bloch, *Phys. Rev.* **70**, 460 (1946); E. L. Hahn, *Phys. Rev.* **80**, 580 (1950).

[9] See, for example, B. W. Shore, *The Theory of Coherent Atomic Excitation* (New York: Wiley, 1990) and references cited therein.

[10] E. T. Jaynes and F. W. Cummings, *Proc. IEEE* **51**, 89 (1963).

[11] J. J. Sanchez-Mondragon, N. B. Narozhny, and J. H. Eberly, *Phys. Rev. Lett.* **51**, 550 (1983); G. S. Agarwal, *J. Opt. Soc. Am. B* **2**, 480 (1985).

[12] F. W. Cummings, *Phys. Rev.* **149**, A1051 (1965); S. Stenholm, *Phys. Rep.* **6**, 1 (1973); P. Meystre, E. Geneux, A. Quattropani, and A. Faust, *Nuovo Cim. B* **25**, 521 (1975); *J. Phys. A* **8**, 95 (1975).

[13] J. H. Eberly, N. B. Narozhny, and J. J. Sanchez-Mondragon, *Phys. Rev. Lett.* **44**, 1323 (1980); N. B. Narozhny, J. J. Sanchez-Mondragon, and J. H. Eberly, *Phys. Rev. A* **23**, 236 (1981); P. L. Knight and P. M. Radmore, *Phys. Rev. A* **26**, 676 (1982).

[14] J. Gea-Banacloche, *Phys. Rev. Lett.* **65**, 3385 (1990).

[15] H. J. Yoo and J. H. Eberly, *Phys. Rep.* **118**, 239 (1985); B. W. Shore and P. L. Knight, *J. Mod. Opt.* **40**, 1195 (1993).

[16] P. L. Knight and P. W. Milonni, *Phys. Rep.* **66**, 21 (1980).

[17] P. L. Knight and P. M. Radmore, *Phys. Lett. A* **90**, 342 (1982). Note that in this paper, the atom is assumed to initially be in the ground state.

[18] C. M. Savage, S. L. Braunstein, and D. F. Walls, *Opt. Lett.* **15**, 628 (1990).

[19] E. Schrödinger, *Naturwissenshaften* **23**, 807, 823, 844 (1935). An English translation can be found in J. Wheeler and W. H. Zurek (Eds.), *Quantum Theory of Measurement* (Princeton, NJ: Princeton University Press, 1983).

[20] F. Le Kien and A. S. Shumovsky, *Intl. J. Mod. Phys. B* **5**, 2287 (1991); A. Messina, S. Maniscalco, and A. Napoli, *J. Mod. Opt.* **50**, 1 (2003).

[21] S. J. D. Phoenix and P. L. Knight, *Phys. Rev. A* **44**, 6023 (1991). See also S. J. D. Phoenix and P. L. Knight, *Ann. Phys. (N.Y.)* **186**, 381 (1988).

[22] R. Birrittella, K. Cheng, and C. C. Gerry, *Opt. Commun.* **354**, 286 (2015).

Bibliography

The following, in addition to those references already cited, is a selection of useful books on atom–field interactions, discussing both classically prescribed fields and quantum fields, with an orientation toward an audience of quantum opticians.

W. H. Louisell, *Quantum Statistical Properties of Radiation* (New York: Wiley, 1973).

L. Allen and J. H. Eberly, *Optical Resonance and the Two-Level Atom* (New York: Wiley, 1975; Mineola, MN: Dover, 1987).

M. Weissbluth, *Photon–Atom Interactions* (New York: Academic Press, 1989).

C. Cohen-Tannoudji, J. Dupont-Roc, and G. Grynberg, *Atom–Photon Interactions* (New York: Wiley, 1992).

L. Mandel and E. Wolf, *Optical Coherence and Quantum Optics* (Cambridge: Cambridge University Press, 1995).

P. Meystre and M. Sargent III, *Elements of Quantum Optics*, 3rd edition (Berlin: Springer, 1999).

R. Loudon, *The Quantum Theory of Radiation*, 3rd edition (Oxford: Oxford University Press, 2000).

An exhaustive book covering numerous applications and extensions of the Jaynes–Cummings model in many areas of physics beyond its original setting in quantum optics is

J. Larson and T. Mavrogordatos, *The Jaynes–Cummings Model and Its Descendants* (Bristol: Institute of Physics Publishing, 2021). This can also be found at quant-phy 2202.00330.

5

Quantum Coherence Functions

In this chapter, we discuss the classical and quantum theories of coherence. Beginning with the example of Young's interference, we review the notions of classical first-order coherence theory and the classical first-order coherence functions. We then proceed to introduce the quantum mechanical first-order coherence functions and present a fully quantum mechanical formulation of Young's interference. We next introduce the second-order coherence functions known also as the intensity–intensity correlation functions, and their connection with the Hanbury-Brown and Twiss experiment, one of the pioneering developments of quantum optics. We finish the chapter by introducing all the higher-order coherence functions, using them to provide a criterion for full coherence of a quantized radiation field.

5.1 Classical Coherence Functions

We begin with a brief review of classical coherence [1] and motivate this with Young's two-slit experiment as sketched in Fig. 5.1. Under certain conditions, interference fringes will appear on the screen. If the source has a bandwidth of $\Delta\omega$ and if $\Delta s = |s_1 - s_2|$ is the path difference, then interference will occur if $\Delta s \leq c/\Delta\omega$. The quantity $\Delta s_{\mathrm{coh}} = c/\Delta\omega$ is called the coherence length. The quantity $\Delta t_{\mathrm{coh}} = \Delta s_{\mathrm{coh}}/c = 1/\Delta\omega$ is called the coherence time. Interference fringes will be visible for $\Delta t_{\mathrm{coh}}\Delta\omega \simeq 1$.

The field on the screen, or at the detector, at time t can be written as a linear superposition of the fields at the earlier times $t_1 = t - s_1/c$ and $t_2 = t - s_2/c$, that is

$$E(\mathbf{r}, t) = K_1 E(\mathbf{r}_1, t_1) + K_2 E(\mathbf{r}_2, t_2), \qquad (5.1)$$

where $E(\mathbf{r}_i, t_i)$ is the complex field on the screen arriving from the ith slit and the quantities K_1 and K_2 are complex geometric factors that depend on the distances s_1 and s_2, respectively. For simplicity, we are treating the fields as scalars here, equivalent to assuming that the fields have the same polarization. Diffraction effects associated with the slits are ignored. Optical light detectors have long response times and are capable only of measuring the average light intensity. Thus, the intensity of light on the detector is given by

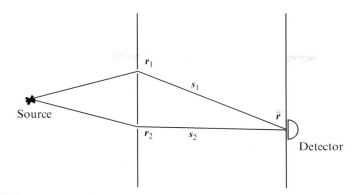

Fig. 5.1 Sketch of the standard setup for Young's double-slit interference experiment.

$$I(\mathbf{r}) = \langle |E(\mathbf{r}, t)|^2 \rangle, \tag{5.2}$$

where here the angular bracket means time average:

$$\langle f(t) \rangle = \lim_{T \to \infty} \frac{1}{T} \int_0^T f(t) dt. \tag{5.3}$$

We assume that this average is *stationary* (i.e. independent of the origin of the time axis). By the ergodic hypothesis, the time average will then be equivalent to an ensemble average. Using Eq. (5.1), we have

$$\begin{aligned} I(\mathbf{r}) = &|K_1|^2 \langle |E(\mathbf{r}_1, t_1)|^2 \rangle \\ &+ |K_2|^2 \langle |E(\mathbf{r}_2, t_2)|^2 \rangle \\ &+ 2\mathrm{Re}[K_1^* K_2 \langle E^*(\mathbf{r}_1, t_1) E(\mathbf{r}_2, t_2) \rangle]. \end{aligned} \tag{5.4}$$

The first two terms are just the intensities associated with the fields from each of the slits, while the third term gives rise to the interference. We set

$$\begin{aligned} I_1 &= |K_1|^2 \langle |E(\mathbf{r}_1, t_1)|^2 \rangle, \\ I_2 &= |K_2|^2 \langle |E(\mathbf{r}_2, t_2)|^2 \rangle. \end{aligned} \tag{5.5}$$

We now introduce the first-order normalized mutual coherence function

$$\gamma^{(1)}(x_1, x_2) = \frac{\langle E^*(x_1) E(x_2) \rangle}{\sqrt{\langle |E(x_1)|^2 \rangle \ \langle |E(x_2)|^2 \rangle}}, \tag{5.6}$$

where we have followed the standard convention of setting $x_i = \mathbf{r}_i, t_i$. Thus, we can write the intensity measured at the screen in terms of component intensities plus a coherence term

$$I(\mathbf{r}) = I_1 + I_2 + 2\sqrt{I_1 I_2} \ \mathrm{Re}[K_1 K_2 \gamma^{(1)}(x_1, x_2)]. \tag{5.7}$$

If we now set $K_i = |K_i| \exp(i\psi_i)$ and write

$$\gamma^{(1)}(x_1, x_2) = |\gamma^{(1)}(x_1, x_2)| \exp(i\Phi_{12}), \tag{5.8}$$

then we have

$$I(\mathbf{r}) = I_1 + I_2 + 2\sqrt{I_1 I_2}| \gamma^{(1)}(x_1, x_2)| \cos(\Phi_{12} - \psi), \tag{5.9}$$

where $\psi = \psi_1 - \psi_2$ is a phase difference arising from the difference in path length (assumed smaller than the coherence length). Interference will occur for $|\gamma^{(1)}(x_1, x_2)| \neq 0$. We may discern three types of coherence:

$$
\begin{aligned}
|\gamma^{(1)}(x_1, x_2)| &= 1 && \text{(complete coherence)}, \\
0 < |\gamma^{(1)}(x_1, x_2)| &< 1 && \text{(partial coherence)}, \\
|\gamma^{(1)}(x_1, x_2)| &= 0 && \text{(complete incoherence)}.
\end{aligned}
\tag{5.10}
$$

It is convenient to introduce Rayleigh's definition of fringe visibility:

$$\mathcal{V} = (I_{\text{max}} - I_{\text{min}})/(I_{\text{max}} + I_{\text{min}}), \tag{5.11}$$

where

$$I_{\substack{\text{max} \\ \text{min}}} = I_1 + I_2 \pm 2\sqrt{I_1 I_2} \, |\gamma^{(1)}(x_1, x_2)|, \tag{5.12}$$

so that

$$\mathcal{V} = \frac{2\sqrt{I_1 I_2} \, |\gamma^{(1)}(x_1, x_2)|}{I_1 + I_2}. \tag{5.13}$$

Clearly, for complete coherence, the visibility (or contrast) for the fringes is a maximum:

$$\mathcal{V} = \frac{2\sqrt{I_1 I_2}}{I_1 + I_2}, \tag{5.14}$$

whereas for complete incoherence, there is zero visibility, $V = 0$. We should further note an important property: going back to Eq. (5.6), it will be evident that if the numerator factorizes according to

$$\langle E^*(x_1) E(x_2) \rangle = \sqrt{\langle |E(x_1)|^2 \rangle} \sqrt{\langle |E(x_2)|^2 \rangle}, \tag{5.15}$$

then we have $|\gamma^{(1)}(x_1, x_2)| = 1$. Thus, the factorization property may be used as the criterion for complete optical coherence. The quantum analog of this criterion, the factorization of the expectation value of a product of field operators, will give rise to a criterion for the complete coherence in a quantum mechanical light field, as we shall see.

We now consider some examples of the classical first-order coherence functions. We consider first the *temporal* coherence of stationary light fields at a fixed position. For example, a monochromatic light field propagating in the z-direction is given at the position z at times t and $t + \tau$ as

$$E(z,t) = E_0 e^{i(kz - \omega t)},$$

$$E(z, t + \tau) = E_0 e^{i[kz - \omega(t + \tau)]}.$$

(5.16)

These are the quantities $E(x_1)$ and $E(x_2)$, respectively, from which we obtain

$$\langle E^*(z,t)E(z, t + \tau)\rangle = E_0^2 e^{-i\omega\tau}.$$

(5.17)

This is called the autocorrelation function and, dropping the position variable, is often written as $\langle E^*(t)E(t + \tau)\rangle$. It is clear in this case that

$$\gamma^{(1)}(x_1, x_2) = \gamma^{(1)}(\tau) = \gamma^{(1)*}(-\tau) = e^{-i\omega\tau},$$

(5.18)

and thus $|\gamma^{(1)}(\tau)| = 1$, so we have complete temporal coherence. But perfectly monochromatic sources of light do not exist. A more realistic model of a "monochromatic" source should take into account the random processes by which light is emitted from the decay of excited atoms in the source. These "monochromatic" sources will emit light as wave trains of finite length, where the trains are separated by a discontinuous change of phase. The average wave train time, τ_0, of a given source is called the coherence time of the source. This coherence time is inversely related to the natural line width of the spectral lines of the radiating atoms in the source. We can then define the coherence length ℓ_{coh} of a wave train as simply $\ell_{coh} = c\tau_0$. So we now consider the field

$$E(z,t) = E_0 e^{i(kz - \omega t)} e^{i\psi(t)},$$

(5.19)

where $\psi(t)$ is a random step function in the range $(0, 2\pi)$ with period τ_0, as shown in Fig. 5.2a.

The autocorrelation function in this case is

$$\langle E^*(t)E(t + \tau)\rangle = E_0^2 e^{-i\omega\tau} \langle e^{i[\psi(t + \tau) - \psi(t)]}\rangle,$$

(5.20)

so that

$$\gamma^{(1)}(\tau) = e^{-i\omega\tau} \lim_{T \to \infty} \frac{1}{T} \int_0^T e^{i[\psi(t + \tau) - \psi(t)]} dt,$$

(5.21)

where the difference in the phases is shown in Fig. 5.2b.

Now for $0 < t < \tau_0$, $\psi(t + \tau) - \psi(t) = 0$ for $0 < t < \tau_0 - \tau$. But for $\tau_0 - \tau < t < \tau_0$, $\psi(t + \tau) - \psi(t)$ is a random number in the range $(0, 2\pi)$. The same holds for subsequent coherence time intervals. Thus, the integral can be evaluated (for details see Refs. [1]) to give

$$\gamma^{(1)}(\tau) = \left(1 - \frac{\tau}{\tau_0}\right) e^{-i\omega\tau} \quad (\tau < \tau_0),$$

$$= 0 \quad (\tau \geq \tau_0),$$

(5.22)

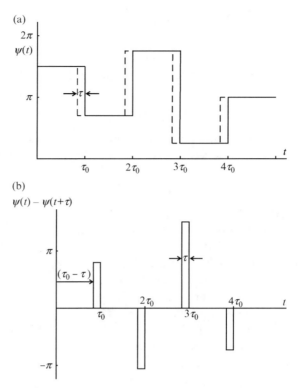

Fig. 5.2 (a) Random phase fluctuations of period τ_0 with $\psi(t)$ (solid line) and $\psi(t+\tau)$ (dashed line); (b) the difference $\psi(t) - \psi(t+\tau)$.

and so we have

$$|\gamma^{(1)}(\tau)| = 1 - \frac{\tau}{\tau_0} \quad (\tau < \tau_0),$$

$$= 0 \qquad (\tau \geq \tau_0). \tag{5.23}$$

Thus, if the time delay τ is longer than the coherence time τ_0, there is no coherence.

A more realistic model of collision-broadened light is one for which the radiation has a Lorentzian power spectrum centered at frequency ω_0. In such a case (see Appendix B) the autocorrelation function is (assuming again propagation along the z-direction)

$$\langle E^*(t)E(t+\tau)\rangle = E_0^2 e^{-i\omega_0\tau - |\tau|/\tau_0}, \tag{5.24}$$

where τ_0 here may be interpreted as the average time between collisions. This results in

$$\gamma^{(1)}(\tau) = e^{-i\omega_0\tau - |\tau|/\tau_0},$$

$$0 \leq |\gamma^{(1)}(\tau)| \leq 1. \tag{5.25}$$

Evidently, $|\gamma^{(1)}(\tau)| \to 0$ as $\tau \to \infty$, so the beam becomes increasingly incoherent with increased time delay. If $|\tau| \ll \tau_0$, complete coherence is approached. Further, if the average time between collisions is shortened, then the spectrum is broadened and $|\gamma^{(1)}(\tau)| \to 0$ more rapidly with increasing τ. Light beams with such an autocorrelation function are known as *chaotic*, although other light beams with different power spectra, Gaussian, for example, are also termed chaotic.

5.2 Quantum Coherence Functions

Glauber and others, in a series of papers in the 1960s [2], showed how a quantum theory of coherence can be constructed upon observables in a manner that closely parallels the classical theory. The intensity of a light beam is measured by devices that actually attenuate the beam by absorbing photons. The actual determination of the intensity of a beam is determined by measuring in some way the response of the absorbing system. We shall consider an ideal detector consisting of a single atom of small dimension compared to the wavelength of the light. The absorption of light over a broad band of wavelengths results in the ionization of the atom, and the subsequent detection of the photoelectron constitutes the detection of the photon. From the counting of the photoelectrons, the statistical properties of the light field may be determined.

The single-atom detector couples to the quantized field through the dipole interaction (as discussed in Chapter 4)

$$\hat{H}^{(I)} = -\hat{\mathbf{d}}.\hat{\mathbf{E}}(\mathbf{r}, t), \tag{5.26}$$

where we are using the Heisenberg picture in which the electric field operator can be written as

$$\hat{\mathbf{E}}(\mathbf{r}, t) = i \sum_{\mathbf{k}, s} \left(\frac{\hbar \omega_k}{2 \varepsilon_0 V} \right)^{1/2} \mathbf{e}_{\mathbf{k}s} [\hat{a}_{\mathbf{k}s}(t) e^{i \mathbf{k} \cdot \mathbf{r}} - \hat{a}_{\mathbf{k}s}^\dagger(t) e^{-i \mathbf{k} \cdot \mathbf{r}}]. \tag{5.27}$$

Because the wavelength of the light is long compared to the dimensions of the atom, then $|\mathbf{k} \cdot \mathbf{r}| \ll 1$ (the dipole approximation) and thus at the atomic detector the field is

$$\hat{\mathbf{E}}(\mathbf{r}, t) \approx i \sum_{\mathbf{k}, s} \sqrt{\frac{\hbar \omega_k}{2 \varepsilon_0 V}} \, \mathbf{e}_{\mathbf{k}s} [\hat{a}_{\mathbf{k}s}(t) - \hat{a}_{\mathbf{k}s}^\dagger(t)]. \tag{5.28}$$

Furthermore, the component of the field describing absorption is the positive frequency part

$$\hat{\mathbf{E}}^{(+)}(\mathbf{r}, t) = i \sum_{\mathbf{k}, s} \left(\frac{\hbar \omega_k}{2 \varepsilon_0 V} \right)^{1/2} \mathbf{e}_{\mathbf{k}s} \hat{a}_{\mathbf{k}s}(t), \tag{5.29}$$

as it contains the annihilation operator. We assume that the atom is initially in some state $|g\rangle$ and the field in some state $|i\rangle$. Upon the absorption of radiation, the atom makes a transition to state $|e\rangle$ and the field to state $|f\rangle$. We assume that $|e\rangle$ is an ionized atomic state and approximate it as a free-electron state (a highly idealized model of photodetection) with wave function:

$$\langle \mathbf{r}|e\rangle = \frac{1}{\sqrt{V}}\, e^{i\mathbf{q}\cdot\mathbf{r}}, \tag{5.30}$$

where $\hbar\mathbf{q}$ is the momentum of the ionized electron. Thus, with initial state $|I\rangle = |g\rangle|i\rangle$ and final state $|F\rangle = |e\rangle|f\rangle$, the matrix element for the transition is

$$\langle F|\hat{H}^{(I)}|I\rangle = -\langle e|\hat{\mathbf{d}}|g\rangle\langle f|\hat{\mathbf{E}}^{(+)}(\mathbf{r},t)|i\rangle. \tag{5.31}$$

The transition probability for the atom–field system is proportional to

$$|\langle F|\hat{H}^{(I)}|I\rangle|^2. \tag{5.32}$$

The probability that the *field* undergoes a transition from $|i\rangle$ to $|f\rangle$ is proportional to

$$|\langle f|\hat{\mathbf{E}}^{(+)}(\mathbf{r},t)|i\rangle|^2. \tag{5.33}$$

We are really only interested in the final state of the detector, not the field, so we must sum over all the possible final states. For all practical purposes, this set of final states may be regarded as complete (and can always *be* completed with the addition of states for unallowed transactions), so that

$$\sum_f |\langle f|\hat{\mathbf{E}}^{(+)}(\mathbf{r},t)|i\rangle|^2$$
$$= \sum_f \langle i|\hat{\mathbf{E}}^{(-)}(\mathbf{r},t)|f\rangle\langle f|\hat{\mathbf{E}}^{(+)}(\mathbf{r},t)|i\rangle \tag{5.34}$$
$$= \langle i|\hat{\mathbf{E}}^{(-)}(\mathbf{r},t)\cdot\hat{\mathbf{E}}^{(+)}(\mathbf{r},t)|i\rangle,$$

where, of course, $\hat{\mathbf{E}}^{(-)}(\mathbf{r},t) = [\hat{\mathbf{E}}^{(+)}(\mathbf{r},t)]^\dagger$.

In the preceding, we have assumed the field to be initially in a pure state. More likely, it will initially be in a mixed state described by a density operator of the form

$$\hat{\rho}_F = \sum_i P_i|i\rangle\langle i|. \tag{5.35}$$

In this case, the expectation value of Eq. (5.34) is replaced by the ensemble average

$$\mathrm{Tr}\{\hat{\rho}_F\hat{\mathbf{E}}^{(-)}(\mathbf{r},t)\cdot\hat{\mathbf{E}}^{(+)}(\mathbf{r},t)\}$$
$$=\sum_i P_i\langle i|\hat{\mathbf{E}}^{(-)}(\mathbf{r},t)\cdot\hat{\mathbf{E}}^{(+)}(\mathbf{r},t)|i\rangle. \tag{5.36}$$

Note the normal ordering of the operators, which is a consequence of our use of the absorbing detector. Henceforth, we shall resort to our earlier convention that the fields are of the same polarization (i.e. we can treat them as scalars). We again use the notation $x = (\mathbf{r}, t)$ and define the function

$$G^{(1)}(x,x) = \mathrm{Tr}\{\rho\hat{E}^{(-)}(x)\hat{E}^{(+)}(x)\}, \tag{5.37}$$

which is the intensity of light at the space-time point $x = \mathbf{r}, t$, that is $I(\mathbf{r}, t) = G^{(1)}(x, x)$ is the quantum analog of the classical expression for the time or ensemble averages of the Section 5.1. For Young's interference experiment as pictured in Fig. 5.1, the positive frequency part of the field at a photodetector located at position $\mathbf{r}$ at time t is of the superposition of the fields from the two slits:

$$\hat{E}^{(+)}(\vec{r},t) = K_1\hat{E}^{(+)}(\mathbf{r}_1,t_1) + K_2\hat{E}^{(+)}(\mathbf{r}_2,t_2), \tag{5.38}$$

which is just the positive frequency part of the quantum analog of the classical relation in Eq. (5.1). The intensity of light on the screen (actually the photodetector) is

$$I(\mathbf{r},t) = \mathrm{Tr}\{\hat{\rho}\hat{E}^{(-)}(\mathbf{r},t)\hat{E}^{(+)}(\mathbf{r},t)\}$$
$$= |K_1|^2 G^{(1)}(x_1x_1) + |K_2|^2 G^{(2)}(x_2x_2) \tag{5.39}$$
$$+ 2\mathrm{Re}[K_1^* K_2 G^{(1)}(x_1,x_2)],$$

where

$$G^{(1)}(x_i,x_i) = \mathrm{Tr}\{\hat{\rho}\hat{E}^{(-)}(x_i)\hat{E}^{(+)}(x_i)\}, \quad i = 1, 2, \ldots \tag{5.40}$$

and where

$$G^{(1)}(x_1,x_2) = \mathrm{Tr}\{\hat{\rho}\hat{E}^{(-)}(x_1)\hat{E}^{(+)}(x_2)\}, \tag{5.41}$$

for $x_1 = \mathbf{r}_1, t_1$ and $x_2 = \mathbf{r}_2, t_2$. This last quantity is a general first-order correlation function, Eq. (5.40) being special cases. Note that $G^{(1)}(x_i, x_i)$ is just the intensity at the detector of light arriving from x_i, whereas $G^{(1)}(x_1, x_2)$ is a measure of the correlation of light arriving from both x_1 and x_2, thus a measure of interference.

We may further define, in analogy with the classical coherence function $\gamma^{(1)}(x_1, x_2)$ of Eq. (5.6), the normalized first-order quantum coherence function

$$g^{(1)}(x_1,x_2) = \frac{G^{(1)}(x_1,x_2)}{[G^{(1)}(x_1,x_1)G^{(1)}(x_2,x_2)]^{1/2}}, \tag{5.42}$$

for which

$$0 \le |g^{(1)}(x_1, x_2)| \le 1. \tag{5.43}$$

As in the classical case, we may discern three types of coherence:

$$
\begin{aligned}
|g^{(1)}(x_1, x_2)| &= 1 &&\text{(complete coherence)}, \\
0 < |g^{(1)}(x_1, x_2)| &< 1 &&\text{(partial coherence)}, \\
|g^{(1)}(x_1, x_2)| &= 0 &&\text{(incoherent)}.
\end{aligned}
\tag{5.44}
$$

From the definition of the correlation function, it is possible to show that

$$
\begin{aligned}
G^{(1)}(x_1, x_2) &= [G^{(1)}(x_2, x_1)]^*, \\
G^{(1)}(x_i, x_i) &\ge 0, \\
G^{(1)}(x_1, x_1)G^{(1)}(x_2, x_2) &\ge |G^{(1)}(x_1, x_2)|^2.
\end{aligned}
\tag{5.45}
$$

Equality in the last expression implies maximum fringe visibility, that is $|g^{(1)}(x_1, x_2)| = 1$.

For a single-mode plane-wave quantized field propagating with wave vector $\mathbf{k}$, the positive frequency part of the field operator is

$$\hat{E}^{(+)}(x) = iK \, \hat{a}e^{i(\mathbf{k}\cdot\mathbf{r} - \omega t)}, \tag{5.46}$$

where $K = (\hbar\omega/2\varepsilon_0 V)^{1/2}$ is the electric field per photon. If the field is in a number state $|n\rangle$, then

$$
\begin{aligned}
G^{(1)}(x, x) &= \langle n|\hat{E}^{(-)}(x)\hat{E}^{(+)}(x)|n\rangle \\
&= K^2 n,
\end{aligned}
\tag{5.47}
$$

$$
\begin{aligned}
G^{(1)}(x_1, x_2) &= \langle n|\hat{E}^{(-)}(x_1)\hat{E}^{(+)}(x_2)|n\rangle, \\
&= K^2 n \exp\{i\mathbf{k}(\mathbf{r}_1 - \mathbf{r}_2) - i\omega(t_2 - t_1)\},
\end{aligned}
\tag{5.48}
$$

and thus

$$|g^{(1)}(x_1, x_2)| = 1. \tag{5.49}$$

For a coherent state $|\alpha\rangle$:

$$
\begin{aligned}
G^{(1)}(x, x) &= \langle \alpha|\hat{E}^{(-)}(x)\hat{E}^{(+)}(x)|\alpha\rangle \\
&= K^2 |\alpha|^2,
\end{aligned}
\tag{5.50}
$$

$$
\begin{aligned}
G^{(1)}(x_1, x_2) &= \langle \alpha|\hat{E}^{(-)}(x_1)\hat{E}^{(+)}(x_2)|\alpha\rangle, \\
&= K^2 |\alpha|^2 \exp\{i\mathbf{k}(\mathbf{r}_1 - \mathbf{r}_2) - i\omega(t_2 - t_1)\},
\end{aligned}
\tag{5.51}
$$

So, we obtain $|g^{(1)}(x_1, x_2)| = 1$ as before. As in the classical case, the key to first-order quantum coherence is the factorization of the expectation value of the correlation function in the numerator of Eq. (5.42), that is

$$G^{(1)}(x_1, x_2) = \langle \hat{E}^{(-)}(x_1)\hat{E}^{(+)}(x_2) \rangle = \langle \hat{E}^{(-)}(x_1) \rangle \langle \hat{E}^{(+)}(x_2) \rangle. \qquad (5.52)$$

This condition is satisfied for the number and coherent states. The case for a thermal state will be left as an exercise for the reader.

5.3 Young's Interference

As a last example of first-order quantum coherence, we examine in some detail Young's interference experiment. We follow the presentation of Walls [3]. Again referring to Fig. 5.1, we assume that the source of light is monochromatic and assume that the pinholes have dimensions on the order of the wavelength of the light. The latter assumption allows us to ignore diffraction effects and assume that the pinholes are point sources of spherical waves. The field at the screen (or at a detector) at position $\mathbf{r}$ at time t is the sum of the spherical waves from each pinhole:

$$\hat{E}^{(+)}(\mathbf{r}, t) = f(r)[\hat{a}_1 e^{iks_1} + \hat{a}_2 e^{iks_2}] \, e^{-i\omega t}, \qquad (5.53)$$

where

$$f(r) = i \left[\frac{\hbar\omega}{2\varepsilon_0(4\pi R)} \right]^{1/2} \frac{1}{r}. \qquad (5.54)$$

R is the radius of the normalization volume, s_1 and s_2 are the distances from the pinholes to the screen, and the $r = |\mathbf{r}|$ appearing in Eq. (5.54) is the result of the approximation $s_1 \approx s_2 = r$. Also, $k = |\mathbf{k}_1| = |\mathbf{k}_2|$ is the magnitude of the wave numbers of the beams from each of the pinholes. The field operators $\hat{a}_1$ and $\hat{a}_2$ are associated with the *radial* modes of the field for photons emitted from pinholes 1 and 2, respectively. The intensity is given by

$$\begin{aligned} I(\mathbf{r}, t) &= \text{Tr}\{\hat{\rho}\hat{E}^{(-)}(\mathbf{r}, t)\hat{E}^{(+)}(\mathbf{r}, t)\} \\ &= |f(r)|^2 \{\text{Tr}(\hat{\rho}\hat{a}_1^\dagger \hat{a}_1) + \text{Tr}(\hat{\rho}\hat{a}_2^\dagger \hat{a}_2) \\ &\quad + 2|\text{Tr}(\hat{\rho}\hat{a}_1^\dagger \hat{a}_2)|\cos\Phi\}, \end{aligned} \qquad (5.55)$$

where

$$\text{Tr}(\hat{\rho}\hat{a}_1^\dagger \hat{a}_2) = |\text{Tr}(\hat{\rho}\hat{a}_1^\dagger \hat{a}_2)|e^{i\psi} \qquad (5.56)$$

and

$$\Phi = k(s_2 - s_1) + \psi. \qquad (5.57)$$

Maximum visibility of the interference fringes occurs for $\Phi = 2\pi m$, m an integer, and falls off as $1/r^2$ with increasing distance of the detector from the central fringe.

But we note that the beam falling onto the pinholes can be approximated as a plane-wave mode, and take the corresponding field operator as $\hat{a}$. If the pinholes are of equal size and if we were to put detectors right behind each of them, then incident photons would be equally likely to go through either one. That is, the two pinholes act to split a single beam into two beams. Thus, we can write

$$\hat{a} = \frac{1}{\sqrt{2}}(\hat{a}_1 + \hat{a}_2), \qquad (5.58)$$

where

$$[\hat{a}_i, \hat{a}_j^\dagger] = \delta_{ij}, \quad [\hat{a}_i, \hat{a}_j^\dagger] = 0, \quad [\hat{a}, \hat{a}^\dagger] = 1. \qquad (5.59)$$

Unfortunately, the relation in Eq. (5.58) does not by itself constitute a unitary transformation. There needs to be another mode introduced, what we shall call a "fictitious" mode, to make it unitary. (The reasoning behind this is discussed in detail in Chapter 6.) We shall call this fictitious mode $\hat{b}$ and let $\hat{b} = (\hat{a}_1 - \hat{a}_2)/\sqrt{2}$, where $[\hat{b}, \hat{b}^\dagger] = 1$. This mode will always be in the vacuum state so may be ignored in what follows, but we shall keep it for expository purposes.

Assuming the photons incident on the pinholes are in the a mode, in keeping with our description above, an incident state containing n photons is the product state $|n\rangle_a|0\rangle_b$. It is transformed by the pinholes to a state occupying the a_1 and a_2 mode states according to

$$
\begin{aligned}
|n\rangle_a|0\rangle_b &= \frac{1}{\sqrt{n!}}\hat{a}^{\dagger n}|0\rangle_a|0\rangle_b \\
&\rightarrow \frac{1}{\sqrt{n!}}\left(\frac{1}{\sqrt{2}}\right)^n (\hat{a}_1^\dagger + \hat{a}_2^\dagger)^n|0\rangle_1|0\rangle_2,
\end{aligned}
\qquad (5.60)
$$

where $|0\rangle_1|0\rangle_2$ represents the product of the a_1 and a_2 mode vacuum states. (The vacuum states on the a and b modes transform to the vacuum states of the 1 and 2 modes.) For the case of only one photon, we have

$$
\begin{aligned}
|1\rangle_a|0\rangle_b &\rightarrow \frac{1}{\sqrt{2}}(|1\rangle_1|0\rangle_2 + |0\rangle_1|1\rangle_2) \\
&= \frac{1}{\sqrt{2}}(|1,0\rangle + |0,1\rangle),
\end{aligned}
\qquad (5.61)
$$

where in the second line we have introduced compact notation with $|1\rangle_1|0\rangle_2 = |1,0\rangle$, $|0\rangle_1|1\rangle_2 = |0,1\rangle$, and so on. Note that $\hat{b} = (\hat{a}_1 - \hat{a}_2)/\sqrt{2}$ operating on the state of Eq. (5.61) yields zero identically, consistent with the idea that photons are not present in the fictitious mode. If the incident state is $|1\rangle_a|0\rangle_b$, then the intensity in Young's experiment is given by

$$I(\mathbf{r}, t) = |f(\mathbf{r})|^2 \left\{ \frac{1}{2} \langle 1, 0 | \hat{a}_1^\dagger \hat{a}_1 | 1, 0 \rangle \right.$$
$$+ \frac{1}{2} \langle 0, 1 | \hat{a}_2^\dagger \hat{a}_2 | 0, 1 \rangle$$
$$+ \left. \langle 1, 0 | \hat{a}_1^\dagger \hat{a}_2 | 0, 1 \rangle \cos\Phi \right\}, \tag{5.62}$$

which reduces to

$$I(\mathbf{r}, t) = |f(\mathbf{r})|^2 [1 + \cos\Phi]. \tag{5.63}$$

In the case of two photons, we have from Eq. (5.60) that

$$|2\rangle_a |0\rangle_b \rightarrow \frac{1}{2} [|2, 0\rangle + \sqrt{2} \, |1, 1\rangle + |0, 2\rangle], \tag{5.64}$$

where the meanings of the state labels on the right-hand side should be clear. [Note that the application of $\hat{b} = (\hat{a}_1 - \hat{a}_2)/\sqrt{2}$ to Eq. (5.64) again yields zero.] This state results in the intensity

$$I(\mathbf{r}, t) = 2|f(\mathbf{r})|^2 [1 + \cos\Phi]. \tag{5.65}$$

The general n-photon state gives

$$I(\mathbf{r}, t) = n|f(\mathbf{r})|^2 [1 + \cos\Phi]. \tag{5.66}$$

For a coherent state incident on the pinholes, we have, using the displacement operator and Eq. (5.58),

$$|\alpha\rangle_a |0\rangle_b = \hat{D}_a(\alpha)|0\rangle_a |0\rangle_b = \hat{D}_1\left(\frac{\alpha}{\sqrt{2}}\right) \hat{D}_2\left(\frac{\alpha}{\sqrt{2}}\right)|0, 0\rangle$$
$$= \left| \frac{\alpha}{\sqrt{2}} \right\rangle_1 \left| \frac{\alpha}{\sqrt{2}} \right\rangle_2, \tag{5.67}$$

where

$$\hat{D}_i(\alpha/\sqrt{2}) = \exp\left(\frac{\alpha^*}{\sqrt{2}} \hat{a}_i - \frac{\alpha}{\sqrt{2}} \hat{a}_i^\dagger \right), \quad i = 1, 2. \tag{5.68}$$

The intensity in this case is

$$I(\mathbf{r}, t) = |\alpha|^2 |f(\mathbf{r})|^2 [1 + \cos\Phi]. \tag{5.69}$$

In all of these cases, the two interfering modes are derived from the same single-mode beam incident on the two holes. They are all first-order coherent, giving rise to the same interference pattern; only the overall intensity is affected by the precise number, or average number, of photons. Note that throughout, as promised, the fictitious b mode remains in the vacuum and as such we could have ignored it, as in fact did Walls [3], but we think it important to recognize that unitarity must be preserved in all transformations.

Equation (5.62) seems to confirm Dirac's [4] famous remark that "each photon interferes only with itself, interference between different photons does not occur." Unfortunately, this remark, probably intended as a metaphor to delineate the sharp contrast between the classical and quantum pictures of the interference of light, has been taken out of context and interpreted by some to imply that photons from independent light beams cannot interfere. Furthermore, as we discuss in Section 6.6, it is incorrect to speak of photon interference. In quantum mechanics, it is quantum amplitudes that interfere, whether one is talking about material particles or photons. If the modes a_1 and a_2 are truly independent, then the product number state $|n_1\rangle_1 |n_2\rangle_2$ in fact does not give rise to interference fringes, as is easy to see from the last term of Eq. (5.55). But if the modes are in coherent states $|\alpha_1\rangle|\alpha_2\rangle$ arising, say, from *independent* lasers, then

$$I(\mathbf{r}, t) = |f(\mathbf{r})|^2 \{|\alpha_1|^2 + |\alpha_2|^2 + 2|\alpha_1^* \alpha_2| \cos \Phi\}, \qquad (5.70)$$

which clearly exhibits interference fringes. Interference in light emitted by independent lasers was demonstrated experimentally by Magyar and Mandel [5] just a few years after the invention of the laser.

5.4 Higher-Order Coherence Functions

First-order coherence in Young's experiment may be understood mathematically as the result of the factorization of the expectation values in the correlation function of the fields in both the classical and quantum cases. Such an experiment is able to determine the degree to which a light source is monochromatic, or to determine the coherence length of the light, but it says nothing about the statistical properties of the light. That is, first-order coherence experiments are unable to distinguish between states of light with identical spectral distributions but with quite different photon number distributions. We have seen that a single-mode field in either a number state or a coherent state is first-order quantum coherent, yet the photon distributions of these states are strikingly different.

In the 1950s, Hanbury Brown and Twiss in Manchester [6] developed a new kind of correlation experiment that involved the correlation of intensities rather than of fields. A sketch of the experiment is shown in Fig. 5.3. Detectors $D1$ and $D2$ are the same distance from the beam splitter. This setup measures a delayed coincidence rate where one of the detectors registers a count at time t and the other a count at time $t + \tau$. If τ, the time delay, is smaller than the coherence time τ_0, then information on the statistics of the light beam striking the beam splitter can be determined.

The rate of coincident counts is proportional to the time, or ensemble, average

$$C(t, t + \tau) = \langle I(t) I(t + \tau) \rangle, \qquad (5.71)$$

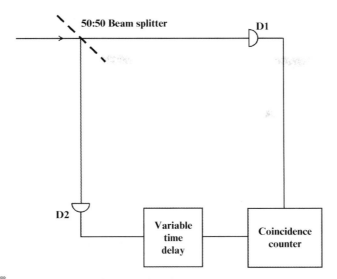

50:50 Beam splitter   D1

D2

Variable
time
delay

Coincidence
counter

Fig. 5.3 Sketch of the setup for the Hanbury Brown and Twiss experiment.

where $I(t)$ and $I(t + \tau)$ are the instantaneous intensities at the two detectors (these are classical quantities here). If we assume that the fields are stationary, the average is a function only of t. If the average of the intensity at each detector is $\langle I(t) \rangle$, then the probability of obtaining a coincidence count with time delay τ is

$$
\begin{aligned}
\gamma^{(2)}(\tau) &= \frac{\langle I(t)I(t + \tau) \rangle}{\langle I(t) \rangle^2} \\
&= \frac{\langle E^*(t)E^*(t + \tau)E(t + \tau)E(t) \rangle}{\langle E^*(t)E(t) \rangle^2}.
\end{aligned}
\tag{5.72}
$$

This is the classical second-order coherence function. If the detectors are at different distances from the beam splitter, then the second-order coherence function can be generalized to

$$
\begin{aligned}
\gamma^{(2)}(x_1, x_2; x_2, x_1) &= \frac{\langle I(x_1)I(x_2) \rangle}{\langle I(x_1) \rangle \langle I(x_2) \rangle} \\
&= \frac{\langle E^*(x_1)E^*(x_2)E(x_2)E(x_1) \rangle}{\langle |E(x_1)|^2 \rangle \langle |E(x_2)|^2 \rangle}.
\end{aligned}
\tag{5.73}
$$

By analogy to first-order coherence, there is said to be classical coherence to second order if $|\gamma^{(1)}(x_1, x_2)| = 1$ and $\gamma^{(2)}(x_1, x_2; x_2, x_1) = 1$. This second condition requires the factorization

$$
\langle E^*(x_1)E^*(x_2)E(x_2)E(x_1) \rangle = \langle |E(x_1)|^2 \rangle \, \langle |E(x_2)|^2 \rangle.
\tag{5.74}
$$

For a plane wave propagating in the z-direction, as given by Eq. (5.16), it is easy to show that

$$\langle E^*(t)E^*(t+\tau)E(t+\tau)E(t)\rangle = E_0^4, \tag{5.75}$$

and thus $\gamma^{(2)}(\tau) = 1$. For any light beam of constant (nonfluctuating) intensity, we have $I(t) = I(t+\tau) = I_0$, $\gamma^{(2)}(\tau) = 1$.

However, unlike the first-order coherence function, the second-order coherence function is not restricted to be unity or less. To see this, we first consider the zero time-delay coherence function

$$\gamma^{(2)}(0) = \frac{\langle I(t)^2\rangle}{\langle I(t)\rangle^2}. \tag{5.76}$$

For a sequence of N measurements taken at times $t_1, t_2, \ldots, t_N$, the required averages are given by

$$\langle I(t)\rangle = \frac{I(t_1) + I(t_2) + \cdots + I(t_N)}{N},$$

$$\langle I(t)^2\rangle = \frac{I(t_1)^2 + I(t_2)^2 + \cdots + I(t_N)^2}{N}. \tag{5.77}$$

Now according to Cauchy's inequality applied to a pair of measurements at times t_1 and t_2, we have

$$2I(t_1)I(t_2) \le I(t_1)^2 + I(t_2)^2. \tag{5.78}$$

Applying this to all the cross terms in $\langle I(t)\rangle^2$, it follows that

$$\langle I(t)^2\rangle \ge \langle I(t)\rangle^2, \tag{5.79}$$

and thus

$$1 \le \gamma^{(2)}(0) < \infty, \tag{5.80}$$

there being no way to establish an upper limit.

For nonzero time delays, the positivity of the intensity ensures that $0 \le \gamma^{(2)}(\tau) < \infty$ $(\tau \neq 0)$. But from the inequality of Eq. (5.80), it can be established that

$$[I(t_1)I(t_1+\tau) + \cdots + I(t_N)I(t_N+\tau)]^2$$

$$\le [I(t_1)^2 + \cdots + I(t_N)^2][I(t_1+\tau)^2 + \cdots + I(t_N+\tau)^2]. \tag{5.81}$$

For a long series of many measurements, the two series on the right-hand side are equivalent, so that

$$\langle I(t)I(t+\tau)\rangle \le \langle I(t)\rangle^2, \tag{5.82}$$

and thus we arrive at

$$\gamma^{(2)}(\tau) \leq \gamma^{(2)}(0). \tag{5.83}$$

The results reported in Eqs. (5.80) and (5.83) establish limits for classical light fields. Later we shall show that some quantum states of light violate the quantum mechanical version of the inequality of Eq. (5.83).

For a light source containing a large number of independently radiating atoms undergoing collisional broadening, it can be shown [7] that the first and second-order coherence functions are related according to

$$\gamma^{(2)}(\tau) = 1 + |\gamma^{(1)}(\tau)|^2, \tag{5.84}$$

a relation that holds for all kinds of chaotic light. Evidently, since $0 \leq |\gamma^{(1)}(\tau)| \leq 1$, it follows that $1 \leq |\gamma^{(2)}(\tau)| \leq 2$. Using the result in Eq. (5.25) we have, for sources with Lorentzian spectra,

$$\gamma^{(2)}(\tau) = 1 + e^{-2|\tau|/\tau_0}. \tag{5.85}$$

Although for $\tau \to \infty$, $\gamma^{(2)}(\tau) \to 1$, it is evident that for zero time delay, $\tau \to 0$, $\gamma^{(2)}(0) = 2$. In fact, for any kind of chaotic light, $\gamma^{(2)}(0) = 2$. The implication of this result is as follows: if light incident on one of the detectors is independent of the light incident on the other, there should be a uniform coincidence rate independent of t. This is what Hanbury Brown and Twiss [6] expected. Using an elementary (but wrong!) picture in which the photons are emitted independently by the source, and assuming that the beam splitter did not split photons themselves but merely reflected or transmitted them, Hanbury Brown and Twiss expected to be able to demonstrate the existence of photons. To their astonishment they found, for zero time delay, twice the detection rate compared to the rate at long time delays. If photons exist, they evidently arrive in pairs at zero time delay but independently at long time delays. That photons arrive in "bunched" pairs is now known as the *photon bunching effect* (also known as the Hanbury Brown and Twiss effect). Note that by measuring the coincidence counts at increasing delay times, it is possible to measure the coherence time τ_0 of the source.

In the last paragraph we spoke about the bunching of photons, even though the important result of Eq. (5.85) was not derived on the basis of the quantum theory of light. We now introduce the second-order quantum coherence function and show that light in a coherent state, as obtained (to a reasonable approximation) from a well-stabilized laser, is coherent to second order and that thermal light sources exhibit the photon bunching effect. In these cases, the quantum and classical pictures agree, but it will become clear that there are instances where the quantum theory predicts situations for which there is no classical counterpart.

We extend the argument used in the first-order case to the detection of two photons by absorption. The transition probability for the absorption of two photons is proportional to

$$|\langle f|\hat{E}^{(+)}(\mathbf{r}_2, t_2)\hat{E}^{(+)}(\mathbf{r}_1, t_1)|i\rangle|^2, \tag{5.86}$$

which, after summation over all final states, becomes

$$\langle i|\hat{E}^{(-)}(\mathbf{r}_1, t_1)\hat{E}^{(-)}(\mathbf{r}_2, t_2)\hat{E}^{(+)}(\mathbf{r}_2, t_2)\hat{E}^{(+)}(\mathbf{r}_1, t_1)|i\rangle. \tag{5.87}$$

Generalizing to cases of non-pure field states, we introduce the second-order quantum correlation function

$$G^{(2)}(x_1, x_2; x_2, x_1) = \text{Tr}\{\hat{\rho}\hat{E}^{(-)}(x_1)\hat{E}^{(-)}(x_2)\hat{E}^{(+)}(x_2)\hat{E}^{(+)}(x_1)\}, \tag{5.88}$$

which is to be interpreted as the ensemble average of $I(x_1)I(x_2)$. As in the first-order case, the normal ordering of the field operators for absorptive detection is important and must be preserved. We define the second-order quantum coherence function as

$$g^{(2)}(x_1, x_2; x_2, x_1) = \frac{G^{(2)}(x_1, x_2; x_2, x_1)}{G^{(1)}(x_1, x_1)G^{(1)}(x_2 x_2)}, \tag{5.89}$$

where $g^{(2)}(x_1, x_2; x_2, x_1)$ is the joint probability of detecting one photon at $\mathbf{r}_1$ at time t_1 and a second at $\mathbf{r}_2$ at time t_2. A quantum field is said to be second-order coherent if $|g^{(1)}(x_1, x_2)| = 1$ *and* $g^{(2)}(x_1, x_2; x_2, x_1) = 1$. This requires that $G^{(2)}(x_1, x_2; x_2, x_1)$ factorize according to

$$G^{(2)}(x_1, x_2; x_2, x_1) = G^{(1)}(x_1, x_1)G^{(1)}(x_2, x_2). \tag{5.90}$$

At a fixed position, $g^{(2)}$ depends only on the time difference $\tau = t_2 - t_1$:

$$g^{(2)}(\tau) = \frac{\langle\hat{E}^{(-)}(t)\hat{E}^{(-)}(t+\tau)\hat{E}^{(+)}(t+\tau)\hat{E}^{(+)}(t)\rangle}{\langle\hat{E}^{(-)}(t)\hat{E}^{(+)}(t)\rangle\langle\hat{E}^{(-)}(t+\tau)\hat{E}^{(+)}(t+\tau)\rangle}, \tag{5.91}$$

which is interpreted as the conditional probability that if a photon is detected at time t, then one is also detected at time $t + \tau$.

For a single-mode field of the form given by Eq. (5.46), $g^{(2)}(\tau)$ reduces to

$$g^{(2)}(\tau) = \frac{\langle\hat{a}^\dagger\hat{a}^\dagger\hat{a}\hat{a}\rangle}{\langle\hat{a}^\dagger\hat{a}\rangle^2} = \frac{\langle\hat{n}(\hat{n}-1)\rangle}{\langle\hat{n}\rangle^2}$$
$$= 1 + \frac{\langle(\Delta\hat{n})^2\rangle - \langle\hat{n}\rangle}{\langle\hat{n}\rangle^2}, \tag{5.92}$$

which is independent of τ. That is, for a single-mode field, $g^{(2)}(\tau) = g^{(2)}(0)$.

For the field in a coherent state $|\alpha\rangle$, it follows that

$$g^{(2)}(\tau) = g^{(2)}(0) = 1, \tag{5.93}$$

meaning that the probability of a delayed coincidence is independent of time. This state is second-order coherent. For a field in a single-mode thermal state (all other modes filtered out) given by Eq. (2.139), it can be shown that

$$g^{(2)}(\tau) = g^{(2)}(0) = 2, \qquad (5.94)$$

indicating a higher probability of detecting coincident photons. For a *multimode* (unfiltered) thermal state, it can be shown [7] that, just as in the classical case,

$$g^{(2)}(\tau) = 1 + |g^{(1)}(\tau)|^2, \qquad (5.95)$$

which lies in the range $1 \le g^{(2)}(\tau) \le 2$. For collision-broadened light with a Lorentzian spectrum and a first-order coherence function

$$g^{(1)}(\tau) = e^{-i\omega_0 \tau - |\tau|/\tau_0} \qquad (5.96)$$

(see Appendix B), we have

$$g^{(2)}(\tau) = 1 + e^{-2|\tau|/\tau_0}, \qquad (5.97)$$

which is just as in the classical case. But here it is legitimate to interpret the result in terms of the arrival of photons. For $|\tau| \ll \tau_0$, the probability of getting two photon counts within time $|\tau|$ is large compared to the random case. For zero time delay, $g^{(2)}(0) = 2$, and $g^{(2)}(\tau) < g^{(2)}(0)$. This inequality characterizes photon bunching. For a multimode coherent state, using the definition of Eq. (5.93), it can be shown that

$$g^{(2)}(\tau) = 1, \qquad (5.98)$$

and thus the photons arrive randomly as per the Poisson distribution, $g^{(2)}(\tau)$ being independent of the delay time. But there is another possibility, the case where $g^{(2)}(0) < g^{(2)}(\tau)$. This is the opposite of photon bunching: *photon antibunching* [8]. For this, the photons tend to arrive in a regular fashion; the probability of obtaining coincident photons in a time interval τ is less than for a coherent state (the random case). As we will show in Chapter 7, this situation is quite nonclassical in the sense that apparent negative probabilities are involved, meaningless for classical fields. But for now, let us consider the single-mode field in a number state $|n\rangle$, from which it follows that

$$g^{(2)}(\tau) = g^{(2)}(0) = \begin{cases} 0 & n = 0, 1, \\ 1 - \dfrac{1}{n} & n \ge 2. \end{cases} \qquad (5.99)$$

Evidently $g^{(2)}(0) < 1$, and this is outside the allowed range for its classical counterpart $\gamma^{(2)}(0)$, as discussed earlier. The fact that $g^{(2)}(0)$ takes on classically forbidden values may be interpreted as a quantum mechanical violation of the Cauchy inequality. Note that $g^{(2)}(0)$ will be less than unity whenever $\langle(\Delta\hat{n})^2\rangle < \langle\hat{n}\rangle$, according to Eq. (5.93). (For a number state $\langle(\Delta\hat{n})^2\rangle = 0$.) States for which this condition holds are *sub-Poissonian*. (States that possess sub-Poissonian statistics are also nonclassical for reasons we shall discuss in Chapter 7.) Since $g^{(2)}(\tau)$ is constant for the single-mode field, photon antibunching does not occur; the requirement *for* it to occur being $g^{(2)}(0) < g^{(2)}(\tau)$. The point is that photon antibunching and sub-Poissonian statistics are different effects, although they have often been confused as being essentially the same thing. They are not [9].

An alternative approach, widely used in the literature, to determining whether photon statistics are Poissonian, super-Poissonian, or sub-Poissonian is through calculation of Mandel's Q-parameter [10], given by

$$Q = \frac{\langle(\Delta\hat{n})^2\rangle - \langle\hat{n}\rangle}{\langle\hat{n}\rangle} = \langle\hat{n}\rangle[g^{(2)}(0) - 1]. \qquad (5.100)$$

The statistics are Poissonian, super-Poissonian, or sub-Poissonian for $Q = 0$, $Q > 0$, or $-1 \leq Q < 0$, respectively. For the coherent states $|\alpha\rangle$, as we showed in Chapter 3, $\langle\hat{n}\rangle = |\alpha|^2$ and $\langle(\Delta\hat{n})^2\rangle = |\alpha|^2$, so that $Q = 0$, which, of course, makes sense as these states have a Poissonian photon number distribution. States displaying sub-Poissonian statistics are inherently nonclassical and will be discussed in Chapter 7. Nonclassical states displaying super-Poissonian statistics will also be discussed in Chapter 7, but these states are nonclassical for reasons not directly associated with their photon number statistics.

The extension of the concept of quantum coherence to the nth order is straightforward. The nth-order quantum correlation function is given by

$$\begin{aligned} G^{(n)}&(x_1, \ldots, x_n; x_n, \ldots, x_1) \\ &= \mathrm{Tr}\{\hat{\rho}\hat{E}^{(-)}(x_1)\cdots\hat{E}^{(-)}(x_n)\hat{E}^{(+)}(x_n)\cdots\hat{E}^{(+)}(x_1)\}, \end{aligned} \qquad (5.101)$$

and the nth-order coherence function is then defined as

$$g^{(n)}(x_1, \ldots, x_n; x_n, \ldots, x_1) = \frac{G^{(n)}(x_1, \ldots, x_n; x_n, \ldots, x_1)}{G^{(1)}(x_1, x_1)\cdots G^{(1)}(x_n, x_n)}. \qquad (5.102)$$

Because $G^{(n)}$ contains counting rates (intensities and coincidence rates) which are always positive, it follows that

$$G^{(n)}(x_1, \ldots, x_n; x_n, \ldots, x_1) \geq 0. \qquad (5.103)$$

Generalizing on our definition of second-order coherence, a field is said to be nth-order coherent if

$$|g^{(n)}(x_1, \ldots, x_n; x_n, \ldots, x_1)| = 1, \qquad (5.104)$$

for all $n \geq 1$. If Eq. (5.104) holds for $n\to\infty$, the state is said to be fully coherent. The necessary and sufficient condition for Eq. (5.104) to hold is that the correlation function be factorable, that is

$$G^{(n)}(x_1, \ldots, x_n; x_n, \ldots, x_1) = G^{(1)}(x_1, x_1) \cdots G^{(1)}(x_n, x_n), \qquad (5.105)$$

a condition which automatically holds for coherent states.

Problems

1. Derive an expression for the interference pattern in a Young's double-slit experiment for an incident-field n-photon state; that is, verify Eq. (5.66).
2. Derive an expression for the interference pattern in a Young's double-slit experiment for an incident thermal light beam.
3. Show that thermal light is first-order coherent but not second or higher-order coherent.
4. Consider the superposition state of the vacuum and one-photon states,
    $$|\psi\rangle = C_0|0\rangle + C_1|1\rangle,$$
    where $|C_0|^2 + |C_1|^2 = 1$, and investigate its coherence properties. Note that quantum mechanically it is "coherent" because it is a pure state. Compare your result with that obtained for the mixture,
    $$\rho = |C_0|^2|0\rangle\langle 0| + |C_1|^2|1\rangle\langle 1|.$$

5. Assuming $|\alpha|$ large, discuss the coherence properties of the superposition of two coherent states (sometimes called the Schrödinger cat states)
    $$|\psi\rangle = \frac{1}{\sqrt{2}}(|\alpha\rangle + |-\alpha\rangle)$$
    and compare with those of the mixture,
    $$\rho = \frac{1}{2}(|\alpha\rangle\langle\alpha| + |-\alpha\rangle\langle-\alpha|).$$

6. Show that a single-mode field in a thermal state has super-Poissonian photon statistics.

References

[1] G. R. Fowles, *Introduction to Modern Optics*, 2nd edition (Mineola, NY: Dover, 1989), p. 58; F. L. Pedrotti and L. S. Pedrotti, *Introduction to Optics*, 3rd edition (Cambridge: Cambridge University Press, 2018), p. 224.
[2] R. J. Glauber, *Phys. Rev.* **130**, 2529 (1963); *Phys. Rev.* **131**, 2766 (1963); U. M. Titulaer and R. J. Glauber, *Phys. Rev.* **140**, B676 (165). See also C. L. Mehta and E. C. G. Sudarshan, *Phys. Rev.* **138**, B274 (1965).
[3] D. F. Walls, *Am. J. Phys.* **45**, 952 (1977).

[4] P. A. M. Dirac, *The Principles of Quantum Mechanics*, 4th edition (Oxford: Oxford University Press, 1958), p. 9.

[5] G. Magyar and L. Mandel, *Nature* **198**, 255 (1963).

[6] R. Hanbury Brown and R. Q. Twiss, *Nature* **177**, 27 (1956). For a complete and historical account of intensity interferometry, see R. Hanbury-Brown, *The Intensity Interferometer* (London: Taylor & Francis, 1974).

[7] See R. Loudon, *The Quantum Theory of Light*, 3rd edition (Oxford: Oxford University Press, 2000), Chap. 3.

[8] See R. Loudon, *Phys. Bull.* **27**, 21 (1976), reprinted in L. Allen and P. L. Knight, *Concepts of Quantum Optics* (Oxford: Pergamon, 1983), p. 174, and the review article D. F. Walls, *Nature* **280**, 451 (1979).

[9] X. T. Zou and L. Mandel, *Phys. Rev. A* **41**, 475 (1990).

[10] L. Mandel, *Opt. Lett.* **4**, 205 (1979).

Bibliography

A standard source of information on classical coherence theory, and classical optics in general, is

M. Born and E. Wolf, *Principles of Optics*, 7th edition (Cambridge: Cambridge University Press, 1999).

Another useful book, with a good chapter on classical coherence theory, is

S. G. Lipson, H. Lipson, and D. S. Tannhauser, *Optical Physics*, 3rd edition (Cambridge: Cambridge University Press, 1995).

A book discussing both classical and quantum coherence is

L. Mandel and E. Wolf, *Optical Coherence and Quantum Optics* (Cambridge: Cambridge University Press, 1995).

Some other books with good discussions of quantum optical coherence theory are

M. Weissbluth, *Photon–Atom Interactions* (New York: Academic Press, 1989). Our discussion of quantum coherence is fairly close to the one presented in this book.

D. F. Walls and G. J. Milburn, *Quantum Optics*, 2nd edition (Berlin: Springer, 1994).

R. Loudon, *The Quantum Theory of Light*, 3rd edition (Oxford: Oxford University Press, 2000).

Though dated, the following classic review article is recommended:

L. Mandel and E. Wolf, *Rev. Mod. Phys.* **37**, 231 (1965).

Beam Splitters and Interferometers

6.1 Experiments with Single Photons

Central to the entire discipline of quantum optics, as should be evident from the preceding chapters, is the concept of the photon. Yet it is perhaps worthwhile to pause and ask: what is the evidence for the existence of photons? Most of us first encounter the photon concept in the context of the photoelectric effect. As we showed in Chapter 5, the photoelectric effect is, in fact, used to indirectly detect the presence of photons – the photoelectrons being the entities actually counted. But it turns out that some aspects of the photoelectric effect can be explained without introducing the concept of the photon. In fact, one can go quite far with a semiclassical theory in which only the atoms are quantized, with the field treated classically. But we hasten to say that for a satisfactory explanation of *all* aspects of the photoelectric effect, the field *must* be quantized. As it happens, the other venerable "proof" of the existence of photons, the Compton effect, can also be explained without quantized fields.

In an attempt to observe quantum effects with light, Taylor, in 1909 [1], obtained interference fringes in an experiment with an extremely weak source of light. His source was a gas flame and the emitted light was attenuated by means of screens made of smoked glass. The "double slit" in the experiment was, in fact, a needle whose shadow on a screen exhibited the fringes of a diffraction pattern when exposed to direct light from the source. But Taylor found that the fringes persisted upon attenuation of the source, even down to the lowest intensities where, one could naively conclude, simply on the basis of energy considerations, that there was at most only one photon at a time between the source and the screen. Apparently, photons passing by the needle one at a time give rise to interference. Presumably, this is the origin of Dirac's famous, but incorrect, remark [2] (see Section 6.8 below) that "each photon interferes only with itself, interference between two photons never occurs." But we now know, as discussed in Chapter 5, that a thermal source such as the gas flame used by Taylor does not produce photons one at a time, but rather produces them in bunches. Hence it is truly naïve and wrong to use energy considerations alone to determine the number of photons between the source and the screen at any given time; there is a strong likelihood that there are two photons present, this being the photon bunching effect. A laser source produces photons randomly, so even when attenuated there is at least some chance that there may be more than one photon between the source and the screen. To get as close as possible to having a single photon between source and screen requires a source of antibunched photons. This in turn requires a source consisting of only a very few atoms, ideally a single atom.

Such a source was developed by Grangier *et al.* [3] for the purpose of a fundamental test of quantum mechanics, namely a search for violations of Bell's inequality. This source consists of a beam of calcium atoms irradiated by laser light, exciting the atoms to a high-lying s-state. The s-state undergoes a rapid decay to a p-state, emitting a photon of frequency v_1. Subsequently, the atom quickly undergoes another rapid decay, this time to the ground s-state, by emitting a second photon, this one having frequency v_2 (see Fig. 6.1). The photons, to conserve momentum, are emitted in opposite directions. The first photon was used as a "trigger" to alert a set of photodetectors placed at the outputs of a 50:50 beam splitter upon which the second photon falls, as illustrated in Fig. 6.2. The trigger tells the photodetectors to expect a photon to emerge from the beam splitter by "gating" the detection electronics for a

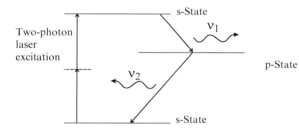

Fig. 6.1 Energy-level diagram of the single-photon source for the experiments of Grangier *et al.* A calcium atom is irradiated by a laser which excites the atom by two-photon absorption to a high-lying s-state. The atom then undergoes a cascade decay first to a p-state, emitting a photon of frequency v_1, used as the trigger photon, and then to the original s-state, emitting a photon of frequency v_2.

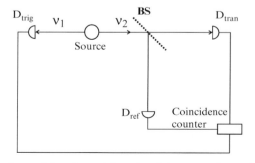

Fig. 6.2 Anti-correlation experiment of Grangier *et al.* Detection of the trigger photon alerts the coincidence counters to expect a photon to pass through the beam splitter. The beam splitter is 50:50.

brief time interval. This eliminates spurious photon counts due to photons entering the detectors from irrelevant sources. The experimental setup, as pictured in Fig. 6.2, is such that only the particle nature of the photons will be manifested. That is, a single photon falling on the beam splitter is expected to be either reflected into detector D_{ref} or transmitted into detector D_{tran} – hence it is a "which path" experiment and no interference effects are expected. There should be no simultaneous counts (the counts should be anti-correlated) of reflected and transmitted photons and, because the beam splitter is 50:50, repeated runs of the experiment should result in each of the two detectors clicking approximately 50% of the time. These expectations were confirmed by the investigators.

6.2 Quantum Mechanics of Beam Splitters

At this point we must pause and consider the beam splitter in a fully quantum mechanical context. In the Chapter 5, we used the notion of a beam splitter in a somewhat cavalier manner. We were able to get away with it because for "classical"-like light beams, coherent and thermal beams, the quantum and classical treatments of beam splitters agree. But at the level of a single or few photons, the classical approach to beam splitting produces erroneous and quite misleading results.

To see how classical reasoning over beam splitting goes wrong, let us consider first a classical light field of complex amplitude ε_1 incident upon a lossless beam splitter as indicated in Fig. 6.3. ε_2 and ε_3 are the amplitudes of the transmitted and reflected beams, respectively. If r and t are the (complex) reflectance and transmittance, respectively, of the beam splitter, then it follows that

$$\varepsilon_2 = t\varepsilon_1 \text{ and } \varepsilon_3 = r\varepsilon_1. \tag{6.1}$$

For a 50:50 beam splitter we would have $|r| = |t| = 1/\sqrt{2}$. However, for the sake of generality, we do not impose this condition here. Since the beam splitter is assumed lossless, the intensity of the input beam should equal the sum of the intensities of the two output beams:

$$|\varepsilon_1|^2 = |\varepsilon_2|^2 + |\varepsilon_3|^2, \tag{6.2}$$

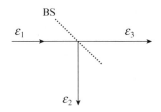

Fig. 6.3   Classical beam splitting. A classical field of amplitude ε_1 is split into fields of amplitudes ε_2 and ε_3.

Fig. 6.4 Naïve, and incorrect, quantum mechanical depiction of a beam splitter.

which requires that

$$|r|^2 + |t|^2 = 1. \tag{6.3}$$

To treat the beam splitter quantum mechanically, we might try replacing the classical complex field amplitudes ε_i by a set of annihilation operators $\hat{a}_i$ ($i = 1, 2, 3$), as indicated in Fig. 6.4. In analogy with the classical case, we might try to set

$$\hat{a}_2 = r\hat{a}_1 \text{ and } \hat{a}_3 = t\hat{a}_1. \tag{6.4}$$

However, the operators of each of the fields are supposed to satisfy the commutation relations

$$[\hat{a}_i, \hat{a}_j^\dagger] = \delta_{ij}, \ [\hat{a}_i, \hat{a}_j] = 0 = [\hat{a}_i^\dagger, \hat{a}_j^\dagger] \quad (i, j = 1, 2, 3), \tag{6.5}$$

but it is easy to see that for the operators of Eq. (6.4), we obtain

$$\begin{aligned}
[\hat{a}_2, \hat{a}_2^\dagger] &= |r|^2 [\hat{a}_1, \hat{a}_1^\dagger] = |r|^2, \\
[\hat{a}_3, \hat{a}_3^\dagger] &= |t|^2 [\hat{a}_1, \hat{a}_1^\dagger] = |t|^2, \\
[\hat{a}_2, \hat{a}_3^\dagger] &= rt^* \neq 0, \text{etc.}
\end{aligned} \tag{6.6}$$

Thus, the transformations in Eq. (6.4) do not preserve the commutation relations and therefore cannot provide the correct quantum description of a beam splitter. This conundrum is resolved as follows. In the classical picture of the beam splitter there is an unused "port" which, being empty of an input field, has no effect on the output beams. However, in the quantum mechanical picture, the "unused" port still contains a quantized field mode, albeit in the vacuum state, and (as we have repeatedly seen), the fluctuations of the vacuum lead to important physical effects. The situation with the beam splitter is no exception.

In Fig. 6.5 we indicate all the modes required for a proper quantum description of the beam splitter, $\hat{a}_0$, representing the field operator of the classically vacant input mode. Also indicated are two sets of transmittances and reflectances, allowing for the possibility of an asymmetric beam splitter. The beam splitter transformations for the field operators we now write as

$$\hat{a}_2 = t'\hat{a}_0 + r\hat{a}_1, \quad \hat{a}_3 = r'\hat{a}_0 + t\hat{a}_1, \tag{6.7}$$

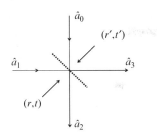

Fig. 6.5 The correct quantum mechanical depiction of a beam splitter.

or collectively as

$$\begin{pmatrix} \hat{a}_2 \\ \hat{a}_3 \end{pmatrix} = \mathbf{B} \begin{pmatrix} \hat{a}_0 \\ \hat{a}_1 \end{pmatrix}, \tag{6.8}$$

where $\mathbf{B}$ is the scattering matrix given by

$$\mathbf{B} = \begin{pmatrix} t' & r \\ r' & t \end{pmatrix}. \tag{6.9}$$

The parameters $t, t', r,$ and r' we assume satisfy the reciprocity relations

$$|r'| = |r|, \ |t'| = |t|, \ |r|^2 + |t|^2 = 1, \ r^*t' + r't^* = 0, \ \text{and} \ r^*t + r't'^* = 0. \tag{6.10}$$

The quantity $T = |t|^2$ is called the transmissivity of the beam splitter and $R = |r|^2$ is known as the reflectivity. Of course, $T + R = 1$, which follows from the reciprocity relations. The reciprocity relations can also be derived on the basis of energy conservation. Here we show that these relations must be satisfied in order to preserve the commutation relations across the beam splitter. We find that

$$\begin{aligned} [\hat{a}_2, \hat{a}_2^\dagger] &= [t'\hat{a}_0 + r\hat{a}_1, t'^*\hat{a}_0^\dagger + r^*\hat{a}_1^\dagger] \\ &= |t'|^2 [\hat{a}_0, \hat{a}_0^\dagger] + |r|^2 [\hat{a}_1, \hat{a}_1^\dagger] \\ &= |t|^2 + |r|^2 = 1, \end{aligned} \tag{6.11}$$

with a similar result for

$$[\hat{a}_3, \hat{a}_3^\dagger] = |r|^2 + |t|^2 = 1. \tag{6.12}$$

And we find that

$$\begin{aligned} [\hat{a}_2, \hat{a}_3^\dagger] &= [t'\hat{a}_0 + r\hat{a}_1, r'^*\hat{a}_0^\dagger + t^*\hat{a}_1^\dagger] \\ &= t'r'^* [\hat{a}_0, \hat{a}_0^\dagger] + rt^* [\hat{a}_1, \hat{a}_1^\dagger] \\ &= t'r'^* + rt^* = 0, \end{aligned} \tag{6.13}$$

and similarly that

$$[\hat{a}_3, \hat{a}_2^\dagger] = r't'^* + rt^* = 0. \tag{6.14}$$

Thus the commutation relations are preserved by the reciprocity relations, as must be the case. The transformation given by Eq. (6.8) is a unitary transformation. A beam splitter preserves unitarity.

Note that from Eqs. (6.7), we have

$$\hat{a}_2^\dagger a_2 = |t|^2 \hat{a}_0^\dagger a_0 + |r|^2 \hat{a}_1^\dagger a_1 + t'^* r \hat{a}_0^\dagger a_1 + r^* t' \hat{a}_1^\dagger a_0,$$
$$\hat{a}_3^\dagger a_3 = |r|^2 \hat{a}_0^\dagger a_0 + |t|^2 \hat{a}_1^\dagger a_1 + r'^* t \hat{a}_0^\dagger a_1 + r' t^* \hat{a}_1^\dagger a_0. \tag{6.15}$$

Adding these equations and using the reciprocity relations, we find that

$$\hat{n}_2 + \hat{n}_3 = \hat{n}_0 + \hat{n}_1, \tag{6.16}$$

where $\hat{n}_i = \hat{a}_i^\dagger a_i$, $i = 0, 1, 2, 3$. This result expresses the conservation of total photon number across the beam splitter (number of input photons = number of output photons) and it expresses the conservation of energy across the device as well.

An alternative way to express the beam splitter transformation is to introduce the angles θ and φ to write the scattering matrix $\mathbf{B}$ as

$$\mathbf{B} = \begin{pmatrix} \cos(\theta/2) & e^{i\varphi} \sin(\theta/2) \\ -e^{-i\varphi} \sin(\theta/2) & \cos(\theta/2) \end{pmatrix}, \tag{6.17}$$

where $0 \le \theta \le \pi$ and $0 \le \varphi \le 2\pi$. Comparison with Eq. (6.9) shows that the angle θ determines the transmissivity and reflectivity of the beam splitter, while the angle φ determines the phase shifts that occur on the reflected beam. We have set $t = t' = \cos(\theta/2)$ and $r = -r'^* = e^{i\varphi} \sin(\theta/2)$, so that $T = |t|^2 = \cos^2(\theta/2)$ and $R = |r|^2 = \sin^2(\theta/2)$ are the transmittance and reflectance, respectively. Thus, our beam splitter transformations can generally be written from Eq. (6.8) as

$$\hat{a}_2 = \hat{a}_0 \cos(\theta/2) + \hat{a}_1 e^{i\varphi} \sin(\theta/2),$$
$$\hat{a}_3 = -\hat{a}_0 e^{-i\varphi} \sin(\theta/2) + \hat{a}_1 \cos(\theta/2). \tag{6.18}$$

An alternative and rather flexible approach to beam splitter transformations, especially when working in the Schrödinger picture, is to write the above transformation in the form

$$\begin{pmatrix} \hat{a}_2 \\ \hat{a}_3 \end{pmatrix} = \hat{U}^\dagger(\theta, \varphi) \begin{pmatrix} \hat{a}_0 \\ \hat{a}_1 \end{pmatrix} \hat{U}(\theta, \varphi), \tag{6.19}$$

where $\hat{U}(\theta, \varphi)$ is the unitary operator given as

$$\hat{U}(\theta, \varphi) = \exp\left[\frac{\theta}{2}(\hat{a}_0^\dagger \hat{a}_1 e^{i\varphi} - \hat{a}_0 \hat{a}_1^\dagger e^{-i\varphi})\right]. \tag{6.20}$$

Then, by using the Baker–Hausdorff lemma

$$e^{\xi\hat{A}}\hat{B}e^{-\xi\hat{A}} = \hat{B} + \xi[\hat{A},\hat{B}] + \frac{\xi^2}{2!}[A,[\hat{A},\hat{B}]] + \cdots \qquad (6.21)$$

in Eq. (6.19), one again arrives at the results given in Eq. (6.18).

For a beam splitter that imparts a 90° phase shift to the reflected beam, we set $\varphi = \pi/2$ such that

$$\mathbf{B} = \begin{pmatrix} \cos(\theta/2) & i\sin(\theta/2) \\ i\sin(\theta/2) & \cos(\theta/2) \end{pmatrix}. \qquad (6.22)$$

The unitary transformation operator corresponding to the matrix given by Eq. (6.22) is

$$\hat{U}(\theta, \pi/2) = \exp\left[i\frac{\theta}{2}(\hat{a}_0^\dagger\hat{a}_1 + \hat{a}_0\hat{a}_1^\dagger)\right]. \qquad (6.23)$$

Note that if we now set $\theta = \pi$, this being the limiting case wherein the beam splitter is a mirror, we obtain from Eq. (6.22)

$$\mathbf{B} = \begin{pmatrix} 0 & i \\ i & 0 \end{pmatrix}, \qquad (6.24)$$

such that $\hat{a}_2 = i\hat{a}_1, \hat{a}_3 = i\hat{a}_0$, indicating each beam picking up the expected $\pi/2$ phase shift.

In the literature, one often encounters the phase-shift choice $\varphi = 0$ such that

$$\mathbf{B} = \begin{pmatrix} \cos(\theta/2) & \sin(\theta/2) \\ -\sin(\theta/2) & \cos(\theta/2) \end{pmatrix}, \qquad (6.25)$$

where the corresponding unitary transformation operator is

$$\hat{U}(\theta, 0) = \exp\left[\frac{\theta}{2}(\hat{a}_0^\dagger\hat{a}_1 - \hat{a}_0\hat{a}_1^\dagger)\right]. \qquad (6.26)$$

Notice that in all cases we have

$$\det(\mathbf{B}) = 1, \qquad (6.27)$$

as is required for a unitary transformation.

For a 50:50 beam splitter where the reflected beam picks up a 90° phase shift, the angles are $\theta = \pi/2$ and $\varphi = \pi/2$, such that

$$\mathbf{B} = \frac{1}{\sqrt{2}}\begin{pmatrix} 1 & i \\ i & 1 \end{pmatrix} \qquad (6.28)$$

and the operator transformations are

$$\hat{a}_2 = \frac{1}{\sqrt{2}}(\hat{a}_0 + i\hat{a}_1), \quad \hat{a}_3 = \frac{1}{\sqrt{2}}(i\hat{a}_0 + \hat{a}_1). \qquad (6.29)$$

The inverses of these relations,

$$\hat{a}_0 = \frac{1}{\sqrt{2}}(\hat{a}_2 - i\hat{a}_3), \quad \hat{a}_1 = \frac{1}{\sqrt{2}}(-i\hat{a}_2 + \hat{a}_3), \tag{6.30}$$

will prove useful in what follows. For the case of a 50:50 beam splitter that does not impart a $\pi/2$ phase shift to the reflected beam ($\theta = \pi/2$, $\varphi = 0$), we have

$$\mathbf{B} = \frac{1}{\sqrt{2}} \begin{pmatrix} 1 & 1 \\ -1 & 1 \end{pmatrix} \tag{6.31}$$

and the operator transformations are

$$\hat{a}_2 = \frac{1}{\sqrt{2}}(\hat{a}_0 + \hat{a}_1), \quad \hat{a}_3 = \frac{1}{\sqrt{2}}(\hat{a}_1 - \hat{a}_0), \tag{6.32}$$

with inverses

$$\hat{a}_0 = \frac{1}{\sqrt{2}}(\hat{a}_2 - \hat{a}_3), \quad \hat{a}_1 = \frac{1}{\sqrt{2}}(\hat{a}_2 + \hat{a}_3). \tag{6.33}$$

In the balance of this book, unless otherwise stated, we shall take our beam splitters to always impart a $\pi/2$ phase shift to the reflected beam.

In all the foregoing, we have tacitly assumed that our beam splitters have no effect on the polarization of the light beams that fall on them (i.e. that the beam splitters preserve polarization). We shall maintain that assumption through the balance of this chapter.

In the above, we mainly took a Heisenberg picture point of view. On the other hand, we may adopt the Schrödinger picture and ask the following question: for a given input state to the beam splitter, what is the output state? Remembering that all photon number states $|n\rangle$, hence any super-position or any statistical mixture of such states, may be constructed by the action of n powers of the creation operator on the vacuum, we may use Eqs. (6.7) and (6.8) to construct the output states from the action of the transformed creation operators on the vacuum states of the output modes, it being obvious that an input vacuum transforms to an output vacuum: $|0\rangle_0|0\rangle_1 \rightarrow |0\rangle_2|0\rangle_3$.

As an example, consider the single-photon input state $|0\rangle_0|1\rangle_1$, which we may write as $\hat{a}_1^\dagger|0\rangle_0|0\rangle_1$. For the beam splitter described by Eqs. (6.30), we find that $\hat{a}_1^\dagger = (i\hat{a}_2^\dagger + \hat{a}_3^\dagger)/\sqrt{2}$. Thus we may write, using that $|0\rangle_0|0\rangle_1 \xrightarrow{BS} |0\rangle_2|0\rangle_3$,

$$|0\rangle_0|1\rangle_1 \xrightarrow{BS} \frac{1}{\sqrt{2}}(i\hat{a}_2^\dagger + \hat{a}_3^\dagger)|0\rangle_2|0\rangle_3,$$

$$\tag{6.34}$$

$$|\text{out}\rangle = \frac{1}{\sqrt{2}}(i|1\rangle_2|0\rangle_3 + |0\rangle_2|1\rangle_3).$$

This is an important result. It says that a single photon incident at one of the input ports of the beam splitter, the other port containing only the vacuum, will be either transmitted or reflected with equal probability. Of course, this is precisely as we earlier claimed and explains why no coincident counts are to be expected with photon counters placed at the outputs of the beam splitter, as confirmed by the experiment of Grangier et al. [3]. Actually, because the beam splitter is well understood, the lack of coincident counts in the above experiment may be taken as an indication that the source is truly producing single-photon states. (Obviously, the beam splitter is a *passive* device in that it neither creates nor destroys photons. There are, of course, *active* devices that convert one photon into two, for example, and we shall encounter these in Chapter 7.)

One other point needs to be made about the output state of Eq. (6.34). It is an *entangled* state: it cannot be written as a simple product of states of the individual modes 2 and 3. The density operator (see Appendix A) for the (pure) state of Eq. (6.34) is

$$\hat{\rho}_{23} = \frac{1}{2}\{|1\rangle_2|0\rangle_3\,_2\langle 1|_3\langle 0| + |0\rangle_2|1\rangle_3\,_2\langle 0|_3\langle 1|$$

$$+ i|1\rangle_2|0\rangle_3\,_2\langle 0|_3\langle 1| - i|0\rangle_2|1\rangle_3\,_2\langle 1|_3\langle 0|\}. \tag{6.35}$$

In placing detectors in the two output beams, we are measuring the full "coherence" as described by the state vector of Eq. (6.34), or equivalently the density operator of Eq. (6.14). Suppose, on the other hand, we make no measurement of, say, mode 3. Mode 2 is then described by the reduced density operator obtained by performing a partial trace, that is tracing over the states of the unmeasured mode (see Appendix A):

$$\hat{\rho}_2 = \text{Tr}_3 \hat{\rho}_{23} = \sum_{n=0}^{\infty} {}_3\langle n|\hat{\rho}_{23}|n\rangle_3$$

$$= \frac{1}{2}(|0\rangle_2\,_2\langle 0| + |1\rangle_2\,_2\langle 1|). \tag{6.36}$$

This represents merely a statistical mixture, there being no "off-diagonal" coherence terms of the form $|0\rangle\langle 1|$ or $|1\rangle\langle 0|$. Thus, placing a detector in only one of the output beams yields random results, 0 or 1, each 50% of the time, just as we would expect.

Let us now consider mixing the vacuum with n photons at the 50:50 beam splitter. This means that we now have

$$|0\rangle_0|n\rangle_1 \xrightarrow{\text{BS}} \left(\frac{1}{\sqrt{2}}\right)^n \frac{1}{\sqrt{n!}}(i\hat{a}_2^\dagger + \hat{a}_3^\dagger)^n|0\rangle_2|0\rangle_3, \tag{6.37}$$

where we have used the fact that $|N\rangle = \hat{a}^N|0\rangle/\sqrt{N!}$. Then, by using the binomial expansion, we have

$$|0\rangle_0|n\rangle_1 \xrightarrow{\text{BS}} \left(\frac{1}{\sqrt{2}}\right)^n \frac{1}{\sqrt{n!}} \sum_{k=0}^{n} \binom{n}{k}(i\hat{a}_2^\dagger)^{n-k}(\hat{a}_3^\dagger)^k|0\rangle_2|0\rangle_3, \tag{6.38}$$

where $\begin{pmatrix} a \\ b \end{pmatrix}$ is the binomial expansion coefficient given by

$$\begin{pmatrix} a \\ b \end{pmatrix} = \frac{a!}{b!(a-b)!}. \tag{6.39}$$

The output state here [the left side of Eq. (6.38)] expands to

$$|\text{out}\rangle = \left(\frac{1}{\sqrt{2}}\right)^n \frac{1}{\sqrt{n!}} \sum_{k=0}^{n} \begin{pmatrix} n \\ k \end{pmatrix} i^{n-k} \sqrt{(n-k)!k!} |n-k\rangle_2 |k\rangle_3$$

$$= \left(\frac{1}{\sqrt{2}}\right)^n \sum_{k=0}^{n} \left(\frac{n!}{k!(n-k)!}\right)^{1/2} i^{n-k} |n-k\rangle_2 |k\rangle_3. \tag{6.40}$$

The joint probability of find n_2 photons in mode 2 and n_3 photons in mode 3 is given by

$$P(n_2, n_3) = |_2\langle n_2|_3\langle n_3|\text{out}\rangle|^2, \tag{6.41}$$

where

$$_2\langle n_2|_3\langle n_3|\text{out}\rangle = \left(\frac{1}{\sqrt{2}}\right)^n \sum_{k=0}^{n} \left(\frac{n!}{k!(n-k)!}\right)^{1/2} i^{n-k} \delta_{n_2,n-k} \delta_{n_2,k}. \tag{6.42}$$

The nonzero joint photon number probabilities, those for finding $n-k$ photons in mode 2 and k photons in mode 3, clearly form the binomial distribution

$$P(n-k, k) = \frac{1}{2^n} \frac{n!}{k!(n-k!)}. \tag{6.43}$$

In Fig. 6.6 we present a representative graph of $P(n_2, n_3)$ versus n_2 and n_3 for the case where $n = 20$.

In the preceding, we considered beam splitting in the cases of definite photon numbers mixing with the vacuum state at a 50:50 beam splitter. Of course, the number states, other than the vacuum, are highly nonclassical. So let us now consider the mixing of a vacuum state and light prepared in a coherent state, a classical-like state at a 50:50 beam splitter of the type described about. That is, the input state is $|0\rangle_0|\alpha\rangle_1 = \hat{D}_1(\alpha)|0\rangle_0|0\rangle_1$, where $\hat{D}_1(\alpha) = \exp(\alpha^* \hat{a}_1^\dagger - \alpha \hat{a}_1)$ is the displacement operator for mode 1. We may then, following the procedure above using Eqs. (6.30), obtain the output state according to

$$|0\rangle_0|\alpha\rangle_1 \xrightarrow{BS} \exp\left[\frac{\alpha}{\sqrt{2}}(i\hat{a}_2^\dagger + \hat{a}_3^\dagger) - \frac{\alpha^*}{\sqrt{2}}(-i\hat{a}_2 + \hat{a}_3)\right]|0\rangle_2|0\rangle_3$$

$$= \exp\left[\left(\frac{i\alpha}{\sqrt{2}}\right)\hat{a}_2^\dagger + \left(\frac{i\alpha^*}{\sqrt{2}}\right)\hat{a}_2\right] \exp\left[\left(\frac{\alpha}{\sqrt{2}}\right)\hat{a}_3^\dagger - \left(\frac{\alpha^*}{\sqrt{2}}\right)\hat{a}_3\right]|0\rangle_2|0\rangle_3$$

$$= \left|\frac{i\alpha}{\sqrt{2}}\right\rangle_2 \left|\frac{\alpha}{\sqrt{2}}\right\rangle_3.$$

$$\tag{6.44}$$

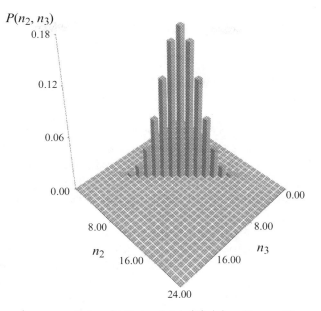
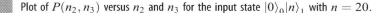

Fig. 6.6 Plot of $P(n_2, n_3)$ versus n_2 and n_3 for the input state $|0\rangle_0 |n\rangle_1$ with $n = 20$.

Evidently, we obtain the result expected for a classical light wave where the incident intensity is evenly divided between the two output beams (e.g. half the incident average photon number, $|\alpha|^2/2$, emerges in each beam). We also naturally obtain the phase shift $i = e^{i\pi/2}$ for the reflected wave. Finally, note that the output is not entangled.

Essentially everything about beam splitting with a coherent state is classical, and in that regard it is worth inserting here a note of caution. The case of a single photon incident on the beam splitter is *not* obtainable as the limiting case of an incident coherent state (i.e. for $|\alpha|$ small). It is easy to check that the single-photon result of Eq. (6.13) is not obtainable in any way as a limiting case of Eq. (6.16). This is quite obvious without doing any calculations, as the former is entangled while the latter is not. Entanglement cannot arise as a limiting case of a product state. The point is, as mentioned earlier in the chapter, that attempts at extrapolating low-field behavior from classical results are misleading and quite wrong.

We now return to the strictly quantum domain and consider the interesting situation where single photons are simultaneously injected into the two input ports of our 50:50 beam splitter, the incident state being $|1\rangle_0|1\rangle_1 = \hat{a}_0^\dagger \hat{a}_1^\dagger |0\rangle_0|0\rangle_1$. Again, following the above procedure, with $\hat{a}_0^\dagger = (\hat{a}_2^\dagger + i\hat{a}_3^\dagger)/\sqrt{2}$ and $\hat{a}_1^\dagger = (i\hat{a}_2^\dagger + \hat{a}_3^\dagger)/\sqrt{2}$, we have

$$|1\rangle_0|1\rangle_1 \xrightarrow{\text{BS}} \frac{1}{2}(\hat{a}_2^\dagger + i\hat{a}_3^\dagger)(i\hat{a}_2^\dagger + \hat{a}_3^\dagger)|0\rangle_2|0\rangle_3$$

$$= \frac{i}{2}(\hat{a}_2^\dagger \hat{a}_2^\dagger + \hat{a}_3^\dagger \hat{a}_3^\dagger)|0\rangle_2|0\rangle_3 \qquad (6.45)$$

$$= \frac{i}{\sqrt{2}}(|2\rangle_2|0\rangle_3 + |0\rangle_2|2\rangle_3).$$

Apparently, the two photons emerge together such that photodetectors placed in the output beams should not register simultaneous counts. But, unlike the case of a single incident photon, the physical basis for obtaining no simultaneous counts is not a result of the particle-like nature of photons. Rather, it is due to interference (a wave-like effect) between two possible ways of obtaining the (absent) output state $|1\rangle_2|1\rangle_3$: the process where both photons are transmitted (Fig. 6.7a) and the process where they are both reflected (Fig. 6.7b). Note the indistinguishability of the two processes for the output state $|1\rangle_2|1\rangle_3$. There is a simple and rather intuitive way of understanding this result. Recall Feynman's rule [5] for obtaining the probability of an outcome that can occur by several indistinguishable processes: one simply adds the probability *amplitudes* of all the processes and then calculates the square of the modulus. Assuming that our beam splitter is described by Eqs. (6.29), the reflected photons each pick an $e^{i\pi/2} = i$ phase shift. The amplitude for transmission for each photon is $A_T = 1/\sqrt{2}$ and the amplitude for reflection for each is $A_R = i/\sqrt{2}$. The amplitude that *both* photons are transmitted is $A_T \cdot A_T$ and that both are reflected is $A_R \cdot A_R$. Thus, the probability of having photons emerge in both output beams is

$$P_{11} = |A_T \cdot A_T + A_R \cdot A_R|^2 = \left|\frac{1}{\sqrt{2}} \cdot \frac{1}{\sqrt{2}} + \frac{i}{\sqrt{2}}\frac{i}{\sqrt{2}}\right|^2 = 0. \qquad (6.46)$$

An experimental demonstration of this effect was first performed by Hong, Ou, and Mandel [6], and is discussed in Chapter 9. The effect was apparently first noticed theoretically by Fearn and Loudon [7].

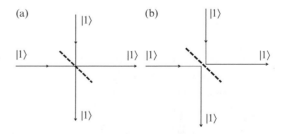

Fig. 6.7 Two indistinguishable processes with simultaneous single-photon inputs. (a) Both photons are transmitted. (b) Both photons are reflected. These processes interfere destructively with each other.

It may be tempting to interpret the result in Eq. (6.17) as due to the bosonic nature of photons, a kind of clustering in the sense of Bose–Einstein condensation (BEC). Indeed, for the case of fermions, such as in neutron interferometry, the output of a beam splitter for the corresponding input would find the fermions always in different beams, in accordance with the Pauli exclusion principle. Of course, this behavior and that of the photons is linked to the statistical properties of the particles, and hence to the nature of the operators that describe them. But for photons it would be misleading to push the clustering analogy á la BEC too far. For example, if n photons are injected into each port simultaneously (i.e. the input state being $|n\rangle_0|n\rangle_1$), the output state is *not* $\sim |2n\rangle_2|0\rangle_3 + e^{i\phi}|0\rangle_2|2n\rangle_3$, except for $n = 1$, as we will now show. Loudon [8] has discussed the beam splitter statistics for both fermion and boson inputs.

We can extend the above procedure to consider as inputs twin number states $|n\rangle_0|n\rangle_1$ as follows. Again using $|n\rangle = \hat{a}^{\dagger n}|0\rangle/\sqrt{n!}$, we have

$$|n\rangle_0|n\rangle_1 \xrightarrow{\text{BS}} \left(\frac{1}{\sqrt{2}}\right)^{2n} \frac{1}{n!} (\hat{a}_2^\dagger + i\hat{a}_3^\dagger)^n (i\hat{a}_2^\dagger + \hat{a}_3^\dagger)^n |0\rangle_2|0\rangle_3$$

$$= \frac{i^n}{2^n n!} (\hat{a}_2^\dagger \hat{a}_2^\dagger + \hat{a}_3^\dagger \hat{a}_3^\dagger)^n |0\rangle_2|0\rangle, \tag{6.47}$$

where we have used the fact that

$$(\hat{a}_2^\dagger + i\hat{a}_3^\dagger)(i\hat{a}_2^\dagger + \hat{a}_3^\dagger) = i(\hat{a}_2^\dagger \hat{a}_2^\dagger + \hat{a}_3^\dagger \hat{a}_3^\dagger). \tag{6.48}$$

Making a binomial expansion, we have

$$|n\rangle_0|n\rangle_1 \xrightarrow{\text{BS}} \frac{i^n}{2^n n!} \sum_{k=0}^{n} \binom{n}{k} (\hat{a}_2^\dagger)^{2n-2k}(\hat{a}_3^\dagger)^{2k}|0\rangle_2|0\rangle_3$$

$$= \frac{i^n}{2^n n!} \sum_{k=0}^{n} \binom{n}{k} \sqrt{(2n-2k)!(2k)!}|2n-2k\rangle_2|2k\rangle_3. \tag{6.49}$$

Using the relations

$$(2k)! = (k!)^2 \binom{2k}{k} \text{ and } (2n-2k)! = [(2n-2k)!!]^2 \binom{2n-2k}{n-k}, \tag{6.50}$$

the output state can be expressed as

$$|\text{out}\rangle = \frac{i^n}{2^n} \sum_{k=0}^{n} \left[\binom{2n-2k}{n-k}\binom{2k}{k}\right]^{1/2}|2n-2k\rangle_2|2k\rangle_3. \tag{6.51}$$

Note that only even photon number states appear in the output upon mixing identical (even or odd) number states at a 50:50 beam splitter.

The joint probability of finding n_2 photons in mode 2 and n_3 photons in mode 3 is given by

$$P(n_2, n_3) = |_2\langle n_2|_3\langle n_3|\text{out}\rangle|^2, \tag{6.52}$$

where

$$_2\langle n_2|_3\langle n_3|\text{out}\rangle = \frac{i^n}{2^n} \sum_{k=0}^{n} \left[\binom{2n-2k}{n-k} \binom{2k}{k} \right]^{1/2} \delta_{n_2,2n-2k}\delta_{n_3,2k}. \tag{6.53}$$

The nonzero probabilities are

$$P(2n-2k, 2k) = \binom{2n-2k}{n-k}\binom{2k}{k}\left(\frac{1}{2}\right)^{2n}, \quad k \in [0, n]. \tag{6.54}$$

In Fig. 6.8 we present a plot of $P(n_2, n_3)$ versus n_2 and n_3 for the case with $n = 10$, that is with 20 photons in total. We see the characteristic result of beam-mixing twin number states at a beam splitter, this being that the probabilities $P(2n, 0)$ and $P(0, 2n)$ are relatively high, with the probabilities for $k \neq 0, n$ forming a "bathtub"-shaped cross-section. The probability distribution of Eq. (6.54) is known in probability theory as the fixed-multiplicative discrete arcsine law of order n [9]. For this reason, the states given by Eq. (6.51) are sometime known as the arcsine states. They were first discussed in the quantum optics literature by Campos, Saleh, and Teich [10].

For completeness, we now consider the general case of there being m and n photons presented at the input of a beam splitter. That is, we

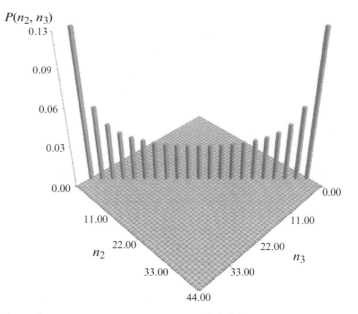

Fig. 6.8   Plot of $P(n_2, n_3)$ versus n_2 and n_3 for the input state $|n\rangle_0|n\rangle_1$ with $n = 20$.

take our input state as $|m\rangle_0|n\rangle_1 = (m!n!)^{-1/2}(\hat{a}_0^\dagger)^m(\hat{a}_1^\dagger)^n|0\rangle_0|0\rangle_1$ such that the beam splitter transformation is now

$$|m\rangle_0|n\rangle_1 \xrightarrow{\text{BS}} \left(\frac{1}{\sqrt{2}}\right)^{m+n} \frac{1}{\sqrt{m!n!}}(\hat{a}_2^\dagger + i\hat{a}_3^\dagger)^m(i\hat{a}_2^\dagger + \hat{a}_3^\dagger)^n|0\rangle_2|0\rangle_3. \quad (6.55)$$

Using the binomial expansion theorem twice as follows, we write

$$(\hat{a}_2^\dagger + i\hat{a}_3^\dagger)^m = \sum_{k=0}^{m}\binom{m}{k}(\hat{a}_2^\dagger)^{m-k}(i\hat{a}_3^\dagger)^k, \quad (6.56)$$

$$(i\hat{a}_2^\dagger + \hat{a}_3^\dagger)^n = \sum_{l=0}^{n}\binom{n}{l}(i\hat{a}_2^\dagger)^l(\hat{a}_3^\dagger)^{n-l} \quad (6.57)$$

(notice the slightly different orderings of the powers of the operators in these expressions, done for convenience) and thus obtain

$$|m\rangle_0|n\rangle_1 \xrightarrow{\text{BS}} \left(\frac{1}{\sqrt{2}}\right)^{m+n} \frac{1}{\sqrt{m!n!}}\sum_{k=0}^{m}\sum_{l=0}^{n} i^{k+l}\binom{m}{k}\binom{n}{l}(\hat{a}_2^\dagger)^{m-k+l}(\hat{a}_3^\dagger)^{n-l+k}|0\rangle_2|0\rangle_3.$$
$$(6.58)$$

The corresponding output state is then

$$|\text{out}(m|n)\rangle = \left(\frac{1}{\sqrt{2}}\right)^{m+n}\frac{1}{\sqrt{m!n!}}\sum_{k=0}^{m}\sum_{l=0}^{n} i^{k+l}\binom{m}{k}\binom{n}{l}$$

$$\times \sqrt{(m-k+l)!(n+k-l)!}\,|m-k+l\rangle_2|n+k-l\rangle_3. \quad (6.59)$$

It is not at all obvious that quantum interferences are lurking in formulas such as Eq. (6.59). But we know from our experience with the input state $|1\rangle_0|1\rangle_1$ that quantum interference effects will be present as a general rule. As another example, consider the input state $|1\rangle_0|n\rangle_1$ for n arbitrary, for which we have the output state

$$|\text{out}(1|n)\rangle = \left(\frac{1}{\sqrt{2}}\right)^{n+1}\frac{1}{\sqrt{n!}}\sum_{k=0}^{1}\sum_{l=0}^{n} i^{k+l}\binom{1}{k}\binom{n}{l}$$

$$\times \sqrt{(1-k+l)!(n+k-l)!}\,|1-k+l\rangle_2|n+k-l\rangle_3. \quad (6.60)$$

The case for $m=1$ and $n=1$ we know already. For $m=1$ and $n=2$, we have a total of three photons passing through the beam splitter, which means that the possible output states include $|1\rangle_2|2\rangle_3, |2\rangle_2|1\rangle_3,$ $|0\rangle_2|3\rangle_3,$ and $|3\rangle_2|0\rangle_3.$ Indeed, straightforward application of Eq. (6.60) results in

$$|\text{out}(1|2)\rangle = \frac{1}{4}\{\sqrt{2}|1\rangle_2|2\rangle_3 + i\sqrt{2}|2\rangle_2|1\rangle_3 + i\sqrt{3}|0\rangle_2|3\rangle_3 - \sqrt{3}|3\rangle_2|0\rangle_3\},$$

$$(6.61)$$

where clearly no complete destructive interference occurs. All the possible output states are present in the superposition.

On the other hand, for $n = 3$, the *possible* output states include the state $|2\rangle_2|2\rangle_3$. But is this state really present in the output? We can try to answer this question without the full machinery of Eq. (6.60) in the same manner we discussed the destructive interference removing the $|1\rangle_2|1\rangle_3$ in the Hong–Ou–Mandel case. We must add the amplitudes for the processes that lead to this state. These are: (a) the photon in mode 0 is transmitted and one of the photons from mode 1 is reflected, the other two being transmitted into mode 3, resulting in the output state $i|2\rangle_2|2\rangle_3$; (b) the photon in mode 0 is reflected into mode 3 and two of the photons from mode 1 are reflected into mode 2, resulting in the output state $i^3|2\rangle_2|2\rangle_3 = -i|2\rangle_2|2\rangle_3$. Thus, these two processes cancel each other out, so that the $|2\rangle_2|2\rangle_3$ state should not be present in the total output state. That this is so can be shown directly from Eq. (6.60). Both outcome states have the same numerical factor aside from the i and $-i$ factors. The demonstration of this is left as an exercise.

It turns out that for the general case where $m + n$ is an even number, that is where $m + n = 2N$, $N = 1, 2, \ldots$, there will be no twin-Fock state of the form $|N\rangle_2|N\rangle_3$ in the total output state as the result of quantum interference. This is a generalized form of the Hong–Ou–Mandel effect. If $m + n$ is an odd number, it is clear from the combinatorics that no twin-Fock state can appear in the output. Thus, in both cases, states of the form $|N\rangle_2|N\rangle_3$ will not appear in the output, but for different reasons. The effects of mixing a single photon with an arbitrary number state at a beam splitter were first discussed by Ou [11].

Before leaving this topic, we consider the mixing of a single photon with a coherent state at a beam splitter. That is, we take our input state to be

$$|1\rangle_0|\alpha\rangle_1 = e^{-|\alpha|^2/2}\sum_{N=0}^{\infty}\frac{\alpha^n}{\sqrt{n!}}|1\rangle_0|n\rangle_1.$$

$$(6.62)$$

The state after beam splitting is

$$|\text{out}(1|\alpha)\rangle = e^{-|\alpha|^2/2}\sum_{N=0}^{\infty}\frac{\alpha^n}{\sqrt{n!}}|\text{out}(1|n)\rangle.$$

$$(6.63)$$

We can deduce from the previous paragraph that all twin-Fock states $|N\rangle_2|N\rangle_3$ will be missing from $|\text{out}(1|\alpha)\rangle$. To show this explicitly, we express our output state as

$$|\text{out}(1|\alpha)\rangle = e^{-|\alpha|^2/2} \sum_{n=0}^{\infty} \frac{\alpha^n}{n!} \left(\frac{1}{\sqrt{2}}\right)^{n+1} \sum_{k=0}^{1}\sum_{l=0}^{n} i^{k+l} \binom{1}{k}\binom{n}{l}$$

$$\times \sqrt{(1-k+l)!(n+k-l)!}\,|1-k+l\rangle_2|n+k-l\rangle_3, \qquad (6.64)$$

and proceed to calculate the joint probability of finding n_2 photons in mode 2 and n_3 in mode 1 as given by

$$P(n_2, n_3) = |_2\langle n_2|_3\langle n_3|\text{out}(1|\alpha)\rangle|^2, \qquad (6.65)$$

where

$$_2\langle n_2|_3\langle n_3|\text{out}(1|\alpha)\rangle = e^{-|\alpha|^2/2} \sum_{n=0}^{\infty} \frac{\alpha^n}{n!} \left(\frac{1}{\sqrt{2}}\right)^{n+1} \sum_{k=0}^{1}\sum_{l=0}^{n} i^{k+l} \binom{1}{k}\binom{n}{l}$$

$$\times \sqrt{(1-k+l)!(n+k-l)!}\,\delta_{n_2,1-k+l}\delta_{n_3,n+k-l}. \qquad (6.66)$$

Doing the sum over n first and then the sum over l, we arrive at

$$P(n_2, n_3) = e^{-|\alpha|^2} \frac{|\alpha|^{2(n_2+n_3-1)}n_2!n_3!}{2^{M+N}[(n_2+n_3-1)!]^2}$$

$$\times \left|\sum_{k=0}^{1}(-i)^{2k}\binom{1}{k}\binom{n_2+n_3-1}{n_3-1+k}\right|^2, \qquad (6.67)$$

which, after doing the remaining sum and after expanding the factorials, yields

$$P(n_2, n_3) = e^{-|\alpha|^2}\frac{|\alpha|^{2(n_2+n_3-1)}}{2^{n_2+N}n_2!n_3!}(n_2-n_3)^2. \qquad (6.68)$$

It is immediately apparent that $P(N, N) = 0, \forall N$. That is, along the "diagonal" in the n_2, n_3 plane where $n_2 = n_3$ there is a line of zeros in the joint photon number probability distribution as a result of quantum interference. In Fig. 6.9 we plot this distribution along with the corresponding joint photon number distribution in the case where the input state is $|0\rangle_0|\alpha\rangle_1$, that being

$$P(n_2, n_3) = e^{-|\alpha|^2}\frac{|\alpha|^{2(n_2+n_3)}}{2^{n_2+n_3}n_2!n_3!}, \qquad (6.69)$$

which follows from Eq. (6.44). The plots are for $\alpha = 3$, and one can see the dramatic difference between distributions for the two cases. Figure 6.9a is that for the input state $|0\rangle_0|\alpha\rangle_1$, while that for $|1\rangle_0|\alpha\rangle_1$ in Fig. 6.9b has a line of zeros along the diagonal. We should stress that this effect does not disappear with increasing $|\alpha|$. Rather, it persists even if $\bar{n} = |\alpha|^2$ is a macroscopic number of photons, and in this sense we have here a counterintuitive effect due to the mixing of coherent light with just one photon. How can one photon be responsible for such a large change in the joint photon number distribution? Quantum interference is the answer. And the quantum interference due to mixing with one photon again illustrates the inherent limitations of taking coherent light to be essentially classical light.

This result, along with similar results for mixing two or three photons with coherent light at a beam splitter, appears in Birrittella, Mimih, and Gerry [12]. These results have been interpreted as an extension of the Hong–Ou–Mandel effect by Alsing *et al.* [13].

Finally in this section, we now reconsider the case of the vacuum and coherent state incident to a beam splitter as given in Eq. (6.44). We expand the product state $|0\rangle_0|\alpha\rangle_1$ as

$$|\text{in}\rangle = |0\rangle_0|\alpha\rangle_1 = e^{-|\alpha|^2/2}\sum_{N=0}^{\infty}\frac{\alpha^N}{\sqrt{N!}}|0\rangle_0|N\rangle_1, \tag{6.70}$$

where it is evident that we have a superposition of the product states $|0\rangle_0|N\rangle_1$. Because the beam splitter conserves the total number of photons, we should be able to take the expansion for the product state in the last line of Eq. (6.44) and, by partitioning it into sectors of total photon numbers N, deduce the beam splitter transformation for the input states $|0\rangle_0|N\rangle_1$. We write

$$|\text{out}\rangle = \left|\frac{i\alpha}{\sqrt{2}}\right\rangle_2\left|\frac{\alpha}{\sqrt{2}}\right\rangle_3 = e^{-|\alpha|^2/2}\sum_{n=0}^{\infty}\sum_{m=0}^{\infty}\frac{(i\alpha/\sqrt{2})^n(\alpha/\sqrt{2})^m}{\sqrt{n!m!}}|n\rangle_2|m\rangle_3, \tag{6.71}$$

and setting $m = N - n$ we have, after some rearrangement,

$$|\text{out}\rangle = e^{-|\alpha|^2/2}\sum_{N=0}^{\infty}\frac{\alpha^N}{\sqrt{N!}}|\psi_N\rangle_{2,3}, \tag{6.72}$$

where

$$|\psi_N\rangle_{2,3} = \frac{1}{2^{N/2}}\sum_{n=0}^{N}i^n\left[\frac{N!}{n!(N-n)!}\right]^{1/2}|n\rangle_2|N-n\rangle_3. \tag{6.73}$$

Thus we recover the beam splitter transformation of the state $|0\rangle_0|N\rangle_1$ as given in Eq. (6.40).

There is yet one other way to obtain this result, and that is by using projection operators (see Appendix B). (We stress that only the mathematics of the projection operators is to be used here; no state-reductive measurements are implied.) If we let $\hat{P}_N^{(0,1)} = |0\rangle_0|N\rangle_1{}_0\langle 0|_1\langle N|$ be the projection operator onto the N-photon sector of the 0–1 modes, the corresponding projection operator onto the 2–3 modes is

$$\hat{P}_N^{(2,3)} = \sum_{k=0}^{N}|k\rangle_2|N-k\rangle_3{}_2\langle k|_3\langle N-k|. \tag{6.74}$$

(a)

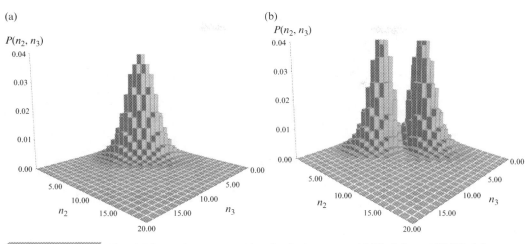

(b)

Fig. 6.9 Plot of $P(n_2, n_3)$ versus n_2 and n_3 for the input states (a) $|0\rangle_0|\alpha\rangle_1$ and (b) $|1\rangle_0|\alpha\rangle_1$ for $\alpha = 3$.

Evidently, we have

$$|0\rangle_0|N\rangle_1 = \frac{\hat{P}_N^{(0,1)}|\text{in}\rangle}{\langle\text{in}|\hat{P}_N^{(0,1)}|\text{in}\rangle^{1/2}},\tag{6.75}$$

and so

$$|\psi_N\rangle_{2,3} = \frac{\hat{P}_N^{(2,3)}|\text{out}\rangle}{\langle\text{out}|\hat{P}_N^{(2,3)}|\text{out}\rangle^{1/2}}.\tag{6.76}$$

6.3 Interferometry with a Single Photon

We return now to the single-photon experiments of Grangier *et al.* [3]. As we have said, those investigators used an atomic source to produce single photons, as verified by the lack of coincident photon counts with detectors placed at the output ports of a beam splitter. Of course, such a setup addresses only the particle nature of the photons and no interference is exhibited. To obtain interference effects with single photons, we must scramble the pathways to the detectors as interference occurs in the lack of "which path" information. A convenient way to accomplish this is to construct a Mach–Zehnder interferometer (MZI), which consists of two beam splitters and a set of mirrors as pictured in Fig. 6.10. Interference occurs because the detectors D_1 and D_2 cannot distinguish between the photon taking the clockwise or the counterclockwise paths. In the

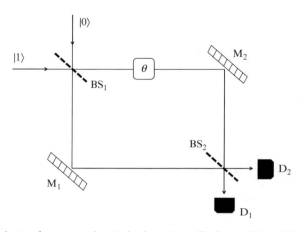

A Mach–Zehnder interferometer with a single-photon input. The beam splitters BS_1 and BS_2 are 50:50, M_1 and M_2 are mirrors, and the box labeled θ represents the relative phase shift between the two arms of the interferometer. The clockwise beam will be the a-mode and the counterclockwise beam will be the b-mode.

clockwise path is placed a phase shifter, which could simply be a length of optical fiber, so that the lengths of the paths about the MZI are different and adjustable. This phase shifting may be represented by the unitary operator $\exp(i\theta\hat{n})$, where $\hat{n}$ is to be understood as the number operator of the field in that section of the interferometer. The angle θ represents a relative phase shift between the two paths. In what follows, we change our mode labels to indicate that the first mode (the a-mode) is the counterclockwise beam and the second mode (the b-mode) is the clockwise beam all through the interferometer.

To see how single-photon interference arises in the MZI, we start with the input state $|0\rangle|1\rangle$, where we shall adopt the convention that the states propagating along the counterclockwise path precede those along the clockwise one, as indicated in Fig. 6.10. Assuming that the beam splitters are of the type that impart a $\pi/2$ phase shift to the reflected beam (which we assume to be the case throughout this book), the first beam splitter (BS_1) transforms the input according to

$$|0\rangle_a|1\rangle_b \xrightarrow{\;BS_1\;} \frac{1}{\sqrt{2}}\left(|0\rangle_a|1\rangle_b + i|1\rangle_a|0\rangle_b\right). \tag{6.77}$$

The mirrors contribute a factor of $e^{i\pi/2}$ to each term, amounting to an irrelevant overall phase, which we omit. The phase shifter in the clockwise path causes a phase change in the first component:

$$\frac{1}{\sqrt{2}}\left(|0\rangle_a|1\rangle_b + i|1\rangle_a|0\rangle_b\right) \xrightarrow{\;\theta\;} \frac{1}{\sqrt{2}}\left(e^{i\theta}|0\rangle_a|1\rangle_b + i|1\rangle_a|0\rangle_b\right). \tag{6.78}$$

At the second beam splitter (BS_2), we have the following transformations:

$$|0\rangle_a|1\rangle_b \xrightarrow{\text{BS}_2} \frac{1}{\sqrt{2}}(|0\rangle_a|1\rangle_b + i|1\rangle_a|0\rangle_b),$$

$$|1\rangle_a|0\rangle_b \xrightarrow{\text{BS}_2} \frac{1}{\sqrt{2}}(|1\rangle_a|0\rangle_b + i|0\rangle_a|1\rangle_b), \qquad (6.79)$$

where (mindful of the propagation direction convention) on the right-hand side the first state in each of the product states refers to the beam directed toward detector D_1, while the second is toward D_2. Thus the transformation on the total state due to the second beam splitter is

$$\frac{1}{\sqrt{2}}(|1\rangle_a|0\rangle_b + i|0\rangle_a|1\rangle_b) \xrightarrow{\text{BS}_1} \frac{1}{2}[(e^{i\theta} - 1)|0\rangle_a|1\rangle_b + i(e^{i\theta} + 1)|1\rangle_a|0\rangle_b]. \qquad (6.80)$$

The probability that the state $|0\rangle|1\rangle$ is detected (that only D_1 clicks) is

$$P_{01} = \frac{1}{2}(1 - \cos\theta), \qquad (6.81)$$

whereas the probability that the state $|1\rangle|0\rangle$ is detected (that only D_2 clicks) is

$$P_{10} = \frac{1}{2}(1 + \cos\theta). \qquad (6.82)$$

Obviously, as the path length (and hence θ) is changed, these probabilities exhibit oscillations which, of course, are interference fringes, these being a manifestation of "single-photon" interference. It is precisely these fringes that have been observed in the experiments of Grangier *et al.* [3].

Note that for $\theta = 0$, the output state becomes $i|1\rangle_a|0\rangle_b$. This is the output state of an interferometry wherein the interfering pathways are of the same length and where each beam splitter imparts a $\pi/2$ phase shift to the reflected beam. We shall return to this point at the end of Section 6.6.

6.4 Interaction-Free Measurement

We now wish to take advantage of the machinery developed so far to expose what surely must be one of the most peculiar features of quantum mechanics, namely the capability to detect the presence of an object without scattering any quanta (in this case photons) off it. Conventional wisdom would have it that in order to detect the presence of an object, one would minimally be required to detect one scattered photon from it. But Elitzur and Vaidman [14] showed theoretically, and Kwiat *et al.* [15] showed experimentally, that the conventional wisdom is not quite right.

To set the stage for our presentation of this counterintuitive effect, let us reconsider the MZI of Fig. 6.10 and choose the path lengths in both arms to be the same; that is, $\theta = 0$. Under this condition, only detector D_1 clicks.

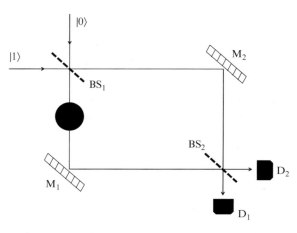

Fig. 6.11 Experimental setup for interaction-free measurement. A Mach–Zehnder interferometer contains an object, the shaded ellipse, in one arm.

D_2 does not click due to destructive interference between the two paths of the interferometer. Suppose now an object is placed in one of the arms of the MZI, as shown in Fig. 6.11. The presence of the object, in this case in the counterclockwise arm, changes everything. Imagine now that we place a detector, or for that matter a number of detectors, in the vicinity of the object such that we would be able to detect a photon scattered from it. Upon detection of such a photon, we know two things: that the photon took the counterclockwise path after the first beam splitter (i.e. we gain "which path" information) and that there is an object in that path. But it turns out that it is not even necessary to detect a scattered photon to determine the presence of the object. Suppose there are no extra detectors near the object and we confine our attention to the detectors at the outputs of the MZI. If, for any given run of the experiment, neither of these detectors clicks, then we again know the path of the photon and that it had to be scattered by the object. Recall that after the first beam splitter, there is a probability of $1/2$ that the photon is in the counterclockwise path and thus a probability of $1/2$ that the photon scatters from the object, hence in 50% of runs of the experiment neither detector clicks. But there is a probability of $1/2$ that the photon takes the clockwise path. Because there is no longer an open path to cause interference, at the second beam splitter the photon has a 50:50 chance of being transmitted to detector D_1 or reflected to detector D_2. So, in *those* cases where detectors do click, D_1 clicks half the time and D_2 the other half. And then, for the interferometer as a *whole*, there is a probability of $1/2$ that neither detector clicks, a probability of $1/4$ each that D_1 or D_2 clicks. Remember, the MZI was initially configured such that with no object in either path, *only* D_1 would click and would do so 100% of the time. But whenever D_2 clicks, we *know* that there must be an object present in one arm of the MZI. The fact that

D_2 clicks at all (and it will click 25% of the time) may be taken as a detection of the object in one of the arms of the interferometer, even though no photon has been scattered by it.

This bizarre result is ultimately a manifestation of a rather general feature of quantum mechanics known as nonlocality – the apparently instantaneous effects of certain kinds of influences due to interactions and measurements. Another example of nonlocal effects includes the Aharonov–Bohm effect, in which an electron wave function accumulates phase due to a distant and confined magnetic flux that never directly interacts with the electron [16], and the violation of various forms of Bell's inequalities. The violation of Bell's inequalities in quantum optical experiments is discussed in Section 9.6 of this book.

6.5 Interferometry with Coherent States of Light

Having examined interferometry at the level of a single photon where quantum effects are expected to be pre-eminent, we now examine interferometry at the other extreme, where we now assume an input coherent state. The result should be essentially the classical one.

To this end, we envision the MZI of Fig. 6.10 but where the mode previously containing a single photon now contains a coherent state $|\alpha\rangle$. Following the previous convention, and using Eq. (6.16), the first beam splitter performs the transformation

$$|0\rangle_a |\alpha\rangle_b \xrightarrow{\ BS_1\ } \left|\frac{i\alpha}{\sqrt{2}}\right\rangle_a \left|\frac{\alpha}{\sqrt{2}}\right\rangle_b . \tag{6.83}$$

Ignoring the common phase shifts from the mirrors, the effect of the phase shifter in the clockwise arm is

$$\left|\frac{i\alpha}{\sqrt{2}}\right\rangle_a \left|\frac{\alpha}{\sqrt{2}}\right\rangle_b \xrightarrow{\ \theta\ } \left|\frac{i\alpha}{\sqrt{2}}\right\rangle_a \left|\frac{\alpha e^{i\theta}}{\sqrt{2}}\right\rangle_b . \tag{6.84}$$

Lastly, BS_2 performs the transformation

$$\left|\frac{i\alpha}{\sqrt{2}}\right\rangle_a \left|\frac{\alpha e^{i\theta}}{\sqrt{2}}\right\rangle_b \xrightarrow{\ BS_2\ } \left|\frac{i(1 + e^{i\theta})\alpha}{2}\right\rangle_a \left|\frac{(e^{i\theta} - 1)\alpha}{2}\right\rangle_b . \tag{6.85}$$

In a typical phase shift measurement with an MZI, the shift θ is determined by subtracting the intensities obtained from the detectors D_1 and D_2. This difference is proportional to the number difference operator of the two modes. Let the operators $(\hat{a}, \hat{a}^\dagger)$ represent the beam directed to detector D_1 and $(\hat{b}, \hat{b}^\dagger)$ that toward D_2. The number difference operator we then designate as $\hat{O} = \hat{a}^\dagger \hat{a} - \hat{b}^\dagger \hat{b}$. Using the right-hand side of Eq. (6.85), we obtain $\langle \hat{O} \rangle = |\alpha|^2 \cos \theta$. This is essentially the result obtained using coherent classical light waves. But it is not the end of the story.

We must never forget that coherent states, even though they have many classical-like properties, are fundamentally quantum mechanical and thus carry with them quantum fluctuations, as we have seen earlier. These fluctuations impose limits on the accuracy of the phase measurements. Using the calculus of error propagation, the uncertainty of the phase measurements (also known as the sensitivity of the measurements) is given by

$$\Delta\theta = \Delta O / \left| \frac{\partial \langle \hat{O} \rangle}{\partial \varphi} \right|, \qquad (6.86)$$

where $\Delta O = \sqrt{\langle \hat{O}^2 \rangle - \langle \hat{O} \rangle^2}$. With the coherent state as input, the phase uncertainty is found to be

$$\Delta\theta = \frac{1}{\sqrt{\bar{n}}|\sin \varphi|}, \qquad (6.87)$$

where $\bar{n} = |\alpha|^2$. Clearly, the phase uncertainty depends on the relative phase φ and evidently we obtain the optimal, or minimum, phase uncertainty $\Delta\varphi_{\min} = 1/\sqrt{\bar{n}}$ for φ equal to odd multiples of $\pi/2$. This is the best we can do with classical-like states of light.

However, on fundamental grounds, it should be possible to exceed this limitation, often known as the standard quantum limit, through the use of manifestly nonclassical states of the field to do interferometry. In fact, the fundamental limit on the phase uncertainty, known as the Heisenberg limit, appears to be $\Delta\varphi_H = 1/\bar{n}$. Reaching this limit of sensitivity has been an important goal over the past four decades, largely driven by attempts to detect interferometrically astrophysically generated gravitational waves traveling through the Earth. The signals were and are expected to be very weak. However, in 2016, two groups, the LIGO (Laser Interferometer Gravitational-Wave Observatory) collaboration and the Virgo collaboration together announced the observation of gravitational waves from a binary black hole merger [17] using laser light (i.e. classical light) alone. This was made possible by the clever reduction of noise from within the interferometers themselves and from without. These are, by the way, very large-scale Michelson interferometers with arm lengths of about 4 km, and multiple interferometers separated by thousands of kilometers are employed to rule out spurious local disturbances. A description of LIGO can be found in Ref. [18].

Upgraded versions of these detectors are already employing nonclassical light sources in order to increase the sensitivities even further, which, in turn, will increase the astrophysical reach of such detectors. In particular, the so-called squeezed states of the light are the basis for the upgraded versions of LIGO [19] and Virgo [20]. The same is true for the British–German detector GEO 600 [21]. Again, the point of increasing the sensitivity of these detectors is to increase the volume of space over which sources of gravitational waves can be found.

The increase in sensitivity of interferometric measurements is made possible using squeezed states of light ultimately because the injection of squeezed light in an interferometer results in entanglement. Entanglement is the key. Nonclassical states of light, including squeezed states of light, are discussed in Chapter 7. The connection between improved sensitivity using *entangled* states in interferometers is discussed in Chapter 11.

6.6 The SU(2) Formulation of Beam Splitting and Interferometry

For more complicated input states to a beam splitter than have been considered so far, the techniques described above for obtaining the corresponding outputs may not be all that efficient or general. There is a well-known connection between the beam splitter transformations and the angular momentum algebra of quantum mechanics that can be brought to light here. To get at this in a rather direct way, let $\hat{J}_+ = \hat{a}_0^\dagger \hat{a}_1$ and $\hat{J}_- = \hat{a}_0 \hat{a}_1^\dagger = (\hat{J}_+)^\dagger$, which satify the commutation relation

$$[\hat{J}_+, \hat{J}_-] = \hat{a}_0^\dagger \hat{a}_0 - \hat{a}_1^\dagger \hat{a}_1 = 2\hat{J}_3, \quad \hat{J}_3 = \frac{1}{2}(\hat{a}_0^\dagger \hat{a}_0 - \hat{a}_1^\dagger \hat{a}_1). \qquad (6.88)$$

This is exactly the angular momentum algebra, or at least part of it. We can complete it by defining the additional operators

$$\hat{J}_1 = \frac{1}{2}(\hat{J}_+ + \hat{J}_-) = \frac{1}{2}(\hat{a}_0^\dagger \hat{a}_1 + \hat{a}_0 \hat{a}_1^\dagger),$$
$$\hat{J}_2 = \frac{1}{2i}(\hat{J}_+ - \hat{J}_-) = \frac{1}{2i}(\hat{a}_0^\dagger \hat{a}_1 - \hat{a}_0 \hat{a}_1^\dagger), \qquad (6.89)$$

which together with $\hat{J}_3$ satisfy the familiar angular momentum commutation relations $[\hat{J}_i, \hat{J}_j] = i\varepsilon_{ijk}\hat{J}_k$, also known as the SU(2) Lie algebra. The square of the angular momentum is given by

$$\hat{J}^2 = \hat{J}_1^2 + \hat{J}_2^2 + \hat{J}_3^2 = \hat{J}_0(\hat{J}_0 + 1), \text{ where } \hat{J}_0 = \frac{1}{2}(\hat{a}_0^\dagger \hat{a}_0 + \hat{a}_1^\dagger \hat{a}_1). \qquad (6.90)$$

The operators $\hat{J}^2$ and $\hat{J}_0$ commute with the operators $\hat{J}_i$, $i = 1, 2, 3$. Operators that commute with all the elements of a Lie algebra are known as Casimir operators. $\hat{J}^2$ and $\hat{J}_0$ are Casimir operators for the SU(2) algebra, though they are not independent. There is really only one Casimir operator for SU(2). The set of operators $\hat{J}_+, \hat{J}_1, \hat{J}_2, \hat{J}_3$, and $\hat{J}_0$ expressed in terms of two sets of Bose operators constitutes the Schwinger realization of the SU(2) Lie algebra [22].

With this background, the general beam splitter transformation given by Eq. (6.20) can be written as

$$\hat{U}(\theta, \varphi) = \exp\left[\frac{\theta}{2}(\hat{J}_+ e^{i\varphi} - \hat{J}_- e^{-i\varphi})\right], \quad (6.91)$$

and for the case where $\varphi = \pi/2$ we find that Eq. (6.20) now reads

$$\hat{U}(\theta, \pi/2) = \exp(i\theta\hat{J}_1), \quad (6.92)$$

whereas for the case $\varphi = 0$ we find that Eq. (6.20) now becomes

$$\hat{U}(\theta, 0) = \exp(i\theta\hat{J}_2). \quad (6.93)$$

Thus, these beam splitter transformations are rotations by angle θ about the fictitious "1" and "2" axes, respectively, and as such are elements of the Lie group SU(2). For 50:50 beam splitters we have $\theta = \pi/2$, so we define transformations for two beam splitter types as

$$\hat{U}_1 \equiv \hat{U}(\pi/2, \pi/2) = \exp\left(i\frac{\pi}{2}\hat{J}_1\right) \quad (6.94)$$

and

$$\hat{U}_2 \equiv \hat{U}(\pi/2, 0) = \exp\left(i\frac{\pi}{2}\hat{J}_2\right). \quad (6.95)$$

We refer to the beam splitters described by Eqs. (6.94) and (6.95) as J1 and J2 beam splitter types, respectively. The two beam splitter types are related by a $\pi/2$ rotation about the "3" axis according to

$$\exp(i\theta\hat{J}_1) = \exp\left(i\frac{\pi}{2}\hat{J}_3\right)\exp(i\theta\hat{J}_2)\exp\left(-i\frac{\pi}{2}\hat{J}_3\right), \quad (6.96)$$

where we have used the result

$$\exp\left(i\frac{\pi}{2}\hat{J}_3\right)\hat{J}_2\exp\left(-i\frac{\pi}{2}\hat{J}_3\right) = \hat{J}_1, \quad (6.97)$$

obtained from the Baker–Hausdorff lemma.

The J1 beam splitter imparts a $\pi/2$ phase shift to the reflected beam while the J2 beam splitter imparts no phase shift. To see this, we consider first the transformation

$$\begin{pmatrix} \hat{a}_2 \\ \hat{a}_3 \end{pmatrix} = \hat{U}_1^\dagger \begin{pmatrix} \hat{a}_0 \\ \hat{a}_1 \end{pmatrix} \hat{U}_1. \quad (6.98)$$

From the Baker–Hausdorff lemma we obtain

$$\begin{pmatrix} \hat{a}_2 \\ \hat{a}_3 \end{pmatrix} = \frac{1}{\sqrt{2}} \begin{pmatrix} \hat{a}_0 + i\hat{a}_1 \\ \hat{a}_1 + i\hat{a}_0 \end{pmatrix}, \quad (6.99)$$

in agreement with Eqs. (6.29). We know from Eq. (6.34) that this transformation gives a $\pi/2$ phase shift to the reflected beam. As for the J2 beam splitter, we have

$$\begin{pmatrix} \hat{a}_2 \\ \hat{a}_3 \end{pmatrix} = \hat{U}_2^\dagger \begin{pmatrix} \hat{a}_0 \\ \hat{a}_1 \end{pmatrix} \hat{U}_2, \tag{6.100}$$

where now we find that

$$\begin{pmatrix} \hat{a}_2 \\ \hat{a}_3 \end{pmatrix} = \frac{1}{\sqrt{2}} \begin{pmatrix} \hat{a}_0 + a_1 \\ \hat{a}_1 - \hat{a}_0 \end{pmatrix}, \tag{6.101}$$

in agreement with Eqs. (6.32).

One can now bring to bear on the problem of beam splitting the full machinery of the angular momentum formalism, provided we can map the corresponding angular momentum basis $|j, m\rangle$, satisfying the eigenvalue relations $\hat{J}_0| j, m\rangle = j| j, m\rangle$, $\hat{J}_3| j, m\rangle = m| j, m\rangle$, onto a product of number states $| p\rangle_0 |q\rangle_1$. The mapping is straightforward, namely that $| j, m\rangle = | j + m\rangle_0 | j - m\rangle_1 = |p\rangle_0 |q\rangle_1$ with $p = j + m$, $q = j - m$, or that $j = (p + q)/2$, $m = (p - q)/2$. Any product of number states can be mapped onto an angular moment state in this way, or vice versa.

Now consider an input state of the general form

$$|\text{in}\rangle = \sum_{p=0}^{\infty} \sum_{q=0}^{\infty} C_{pq}| p\rangle_0 |q\rangle_1, \tag{6.102}$$

which in terms of angular momentum states is

$$|\text{in}\rangle = \sum_{j=0,\frac{1}{2},1,\dots}^{\infty} \sum_{m=-j}^{j} C_{j+m, j-m}| j, m\rangle, \tag{6.103}$$

where the j and m values are given in the previous paragraph. Suppose we have a J1 beam splitter of arbitrary angle θ, as given by Eq. (6.93), acting on our input state. The output state is given as

$$|\text{out}\rangle = \exp(i\theta\hat{J}_1)|\text{in}\rangle = \sum_{j=0,\frac{1}{2},1,\dots}^{\infty} \sum_{m=-j}^{j} C_{j+m, j-m}\exp(i\theta\hat{J}_1)| j, m\rangle$$

$$= \sum_{j=0,\frac{1}{2},1,\dots}^{\infty} \sum_{m=-j}^{j} \sum_{m'=-j'}^{j'} C_{j+m, j-m}\langle j, m'|\exp(i\theta\hat{J}_1)| j, m\rangle| j, m'\rangle, \tag{6.104}$$

where a complete set of states was inserted by resolving unity according to

$$\hat{\mathbf{I}}_j = \sum_{m'=-j}^{j} | j, m'\rangle\langle j, m'|. \tag{6.105}$$

Using the result of Eq. (6.96), we have

$$|\text{out}\rangle = \sum_{j=0,\frac{1}{2},1,\dots}^{\infty} \sum_{m=-j}^{j} \sum_{m'=-j'}^{j'} C_{j+m,j-m}i^{m'-m}d_{m',m}^{j}(-\theta)| j, m'\rangle, \tag{6.106}$$

where [23]

$$d^j_{m',m}(\beta) = \langle j, m'|e^{-i\beta \hat{J}_2}|j, m\rangle$$

$$= \left[\frac{(j-m)!(j+m')!}{(j+m)!(j-m')!}\right]^{1/2} \frac{\left(\cos\frac{\beta}{2}\right)^{2j+m-m'}\left(-\sin\frac{\beta}{2}\right)^{m'-m}}{(m'-m)!}$$

$$\times {}_2F_1\left(m'-j, -m-j; m'-m+1; -\tan^2\frac{\beta}{2}\right), \quad \text{for } m' \geq m,$$

(6.107)

and where ${}_2F_1(a, b; c; z)$ is the hypergeometric function given by

$$_2F_1(a,b;c;z) = 1 + \frac{ab}{c}z + \frac{1}{2!}\frac{a(a+1)b(b+1)}{c(c+1)}z^2 + \cdots.$$ (6.108)

For the case where $m > m'$, one has $d^j_{m',m}(\beta) = d^j_{m,m'}(-\beta)$. This is a general result for a J1 beam splitter of arbitrary transmissivity. For a J2 beam splitter one need only remove the factors $i^{m'-m}$ from Eq. (6.106). Ultimately, it is the factor $d^j_{m',m}(-\theta)$ that determines the essential statistical properties of the output states of a beam splitter.

The result of Eq. (6.106) can be translated into two-mode photonic states as follows. We set

$$|\text{out}\rangle = \sum_{k=0}^{\infty}\sum_{l=0}^{\infty} F_{kl}|k\rangle_2|l\rangle_3,$$ (6.109)

from which it follows that

$$F_{kl} = \sum_{j=0,\frac{1}{2},1,\ldots}^{\infty}\sum_{m=-j}^{j}\sum_{m'=-j'}^{j'} C_{j+m,j-m}i^{m'-m}d^j_{m',m}(-\theta)_{23}\langle k, l|j, m'\rangle,$$ (6.110)

where we have set ${}_2\langle k|_3\langle l| = {}_{23}\langle k, l|$ for convenience. Now we write ${}_{23}\langle k, l| \equiv \langle J, M|$, where $J = (k+l)/2$ and $M = (k-l)/2$. Then Eq. (6.110) becomes

$$F_{kl} = \sum_{j=0,\frac{1}{2},1,\ldots}^{\infty}\sum_{m=-j}^{j}\sum_{m'=-j'}^{j'} C_{j+m,j-m}i^{m'-m}d^j_{m',m}(-\theta)\langle J, M|j, m'\rangle$$

$$= \sum_{j=0,\frac{1}{2},1,\ldots}^{\infty}\sum_{m=-j}^{j}\sum_{m'=-j'}^{j'} C_{j+m,j-m}i^{m'-m}d^j_{m',m}(-\theta)\delta_{J,j}\delta_{M,m'}$$

$$= \sum_{m=-J}^{J} C_{J+m, J-m}i^{M-m}d^J_{M,m}(-\theta), \quad \text{provided } J = \frac{k+l}{2} \text{ and } M = \frac{k-l}{2}.$$

(6.111)

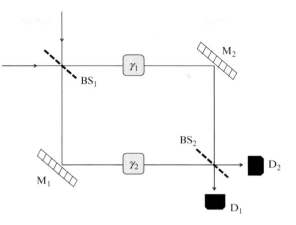

Fig. 6.12   Sketch of a Mach–Zehnder interferometer indicating the phase shifts γ_1 and γ_2 incurred along the counterclockwise and clockwise paths, respectively.

With these definitions of J and M, the joint photon number probability for finding k photons in mode 2 and l in mode 3 is

$$P_{23}(k, l) = \left| \sum_{m=-J}^{J} C_{J+m,J-m} i^{M-m} d_{M,m}^{J}(-\theta) \right|^{2}. \tag{6.112}$$

Finally in this section, we point out that not only can beam splitters be discussed in the context of the SU(2) formalism, but the entire Mach–Zehnder interferometer can be as well. We take our interferometer to be as sketched in Fig. 6.12, where for simplicity we again label all parts of the counterclockwise beam a and all parts of the clockwise beam b.

This means that our SU(2) operators always have the form

$$\hat{J}_0 = \frac{1}{2}(\hat{a}^\dagger\hat{a} + \hat{b}^\dagger\hat{b}), \ \hat{J}_1 = \frac{1}{2}(\hat{a}^\dagger\hat{b} + \hat{a}\hat{b}^\dagger),$$

$$\hat{J}_2 = \frac{1}{2i}(\hat{a}^\dagger\hat{b} - \hat{a}\hat{b}^\dagger), \ \hat{J}_3 = \frac{1}{2}(\hat{a}^\dagger\hat{a} - \hat{b}^\dagger\hat{b}). \tag{6.113}$$

We take both beam splitters to be 50:50 and of the J1 type. That is, both will be described by the operator $\exp(i\pi\hat{J}_1/2)$. Each path around the interferometer will incur phase shifts, which we take to be γ_1 and γ_2 for the counterclockwise and clockwise paths, respectively. Quantum mechanically, these phase shifts are generated by the operator $e^{i\gamma_1\hat{a}^\dagger\hat{a}}e^{i\gamma_2\hat{b}^\dagger\hat{b}}$, which can be rewritten in terms of the operators of Eqs. (6.113) according to

$$e^{i\gamma_1\hat{a}^\dagger\hat{a}}e^{i\gamma_2\hat{b}^\dagger\hat{b}} = e^{i(\gamma_1+\gamma_2)\hat{J}_0}e^{i(\gamma_1-\gamma_2)\hat{J}_3}. \tag{6.114}$$

Therefore, the entire interferometer is described by the unitary operator

$$\hat{U}'_{\text{MZI}}(\varphi) = e^{i\frac{\pi}{2}\hat{J}_1} e^{i(\gamma_1+\gamma_2)\hat{J}_0} e^{i(\gamma_1-\gamma_2)\hat{J}_3} e^{i\frac{\pi}{2}\hat{J}_1} = e^{i(\gamma_1+\gamma_2)\hat{J}_0} e^{i\frac{\pi}{2}\hat{J}_1} e^{i\varphi\hat{J}_3} e^{i\frac{\pi}{2}\hat{J}_1}, \quad (6.115)$$

where we have set $\varphi = \gamma_1 - \gamma_2$, this being the relevant phase shift to be measured, and where we have moved to the left the factor $e^{i(\gamma_1+\gamma_2)\hat{J}_0}$ as this factor commutes with the others. In fact, one can ignore the factor $e^{i(\gamma_1+\gamma_2)\hat{J}_0}$ altogether as it will contribute nothing, which can be seen as follows. The usual measurement scheme for the measurement of the relative phase shift involves measuring the expectation value of the difference in the photon numbers of the output beams, as discussed in Section 6.5. This amounts to taking the expectation value of the operator $\hat{J}_3$ with respective to the output state given by

$$|\text{out}\rangle = \hat{U}'_{\text{MZI}}(\varphi)|\text{in}\rangle, \quad (6.116)$$

such that

$$\begin{aligned} \langle \hat{J}_3(\varphi) \rangle &= \langle \text{out}|\hat{J}_3|\text{out}\rangle = \langle \text{in}|\hat{U}'^{\dagger}_{\text{MZI}}\hat{J}_3\hat{U}'_{\text{MZI}}|\text{in}\rangle \\ &= \langle \text{in}|\hat{U}^{\dagger}_{\text{MZI}}(\varphi)e^{-i(\gamma_1+\gamma_2)\hat{J}_0}\hat{J}_3 e^{i(\gamma_1+\gamma_2)\hat{J}_0}\hat{U}_{\text{MZI}}(\varphi)|\text{in}\rangle, \end{aligned} \quad (6.117)$$

where

$$\hat{U}_{\text{MZI}}(\varphi) = e^{i\frac{\pi}{2}\hat{J}_1} e^{i\varphi\hat{J}_3} e^{i\frac{\pi}{2}\hat{J}_1}. \quad (6.118)$$

But as $[\hat{J}_0, \hat{J}_3] = 0$, the factors $e^{\pm i(\gamma_1+\gamma_2)\hat{J}_0}$ cancel out and

$$\langle \hat{J}_3(\varphi) \rangle = \langle \text{in}|\hat{U}^{\dagger}_{\text{MZI}}(\varphi)\hat{J}_3\hat{U}_{\text{MZI}}(\varphi)|\text{in}\rangle. \quad (6.119)$$

Thus, for all practical purposes, the factor $e^{i(\gamma_1+\gamma_2)\hat{J}_0}$ can be ignored and the Mach–Zehnder interferometer is quantum mechanically described by the operator given in Eq. (6.118).

The error propagation formula for the measurement of the relative phase shift is now

$$\Delta\varphi = \Delta J_3 \Big/ \left|\frac{\partial \langle \hat{J}_3(\varphi) \rangle}{\partial \varphi}\right|, \quad (6.120)$$

where $\Delta J_3 = \sqrt{\langle \hat{J}_3^2(\varphi) \rangle - \langle \hat{J}_3(\varphi) \rangle^2}$.

For the case where the phase shift in the interferometer vanishes (i.e. for $\varphi = 0$), the MZI transformation operator becomes $\hat{U}_{\text{MZI}} = \exp(i\pi\hat{J}_1)$. The output mode operators of the MZI, which we label as $\hat{a}'$ and $\hat{b}'$, are related to the input mode operators $\hat{a}$ and $\hat{b}$ according to

$$\begin{pmatrix} \hat{a}' \\ \hat{b}' \end{pmatrix} = e^{-i\pi\hat{J}_1} \begin{pmatrix} \hat{a} \\ \hat{b} \end{pmatrix} e^{i\pi\hat{J}_1} = \begin{pmatrix} i\hat{b} \\ i\hat{a} \end{pmatrix}, \quad (6.121)$$

where we have again used the Baker–Hausdorff lemma. It follows from this that $\hat{a} = -i\hat{b}'$ and $\hat{b} = -i\hat{a}'$. Now consider the input state we used for single-photon interferometry in Section 6.3: $|0\rangle_a|1\rangle_b = \hat{b}^\dagger|0\rangle_a|0\rangle_b$, which, under the transformation just described, becomes $\hat{b}^\dagger|0\rangle_a|0\rangle_b \to i\hat{a}^\dagger|0\rangle_{a'}|0\rangle_{b'} = i|1\rangle_{a'}|0\rangle_{b'}$. This result is in agreement with that obtained at the very end of Section 6.3. This is the expected result for back-to-back J1 beam splitters, and it verifies that our SU(2) formulation of an MZI is in agreement with what we worked out in the earlier parts of this chapter.

A complete discussion of the SU(2) approach to interferometry can be found in the paper by Yurke, McCall, and Klauder [24]. Also see the paper by Campos, Saleh, and Teich [10], which heavily uses the SU(2) formalism.

6.7 The Beam Splitter as a Displacer

Consider a beam splitter described by the transformation

$$\hat{U}(\theta, \pi/2) = \exp\left[i\frac{\theta}{2}(\hat{a}_0^\dagger \hat{a}_1 + \hat{a}_0 \hat{a}_1^\dagger)\right]. \tag{6.122}$$

We consider the limit $\theta \to 0$ and assume that the 1-mode is populated by a very strong coherent state $|\beta\rangle_1(|\beta| \to \infty)$ such that the product $\theta|\beta| \to$ constant, and make the gross approximation that

$$_1\langle\beta|\hat{U}(\theta, \pi/2)|\beta\rangle_1 \simeq \exp\left[i\frac{\theta}{2}(\hat{a}_0^\dagger \beta + \hat{a}_0 \beta^*)\right]. \tag{6.123}$$

The right-hand side of this is effectively a displacement operator for the 0-mode. That is,

$$\exp\left[i\frac{\theta}{2}(\hat{a}_0^\dagger \beta + \hat{a}_0 \beta^*)\right] = \hat{D}_0(\alpha), \tag{6.124}$$

where $\alpha = i\theta\beta/2$. From Eqs. (6.18) we can see that the limit $\theta \to 0$ is the case of high transmissivity, that is $\hat{a}_2 \to \hat{a}_0$ and $\hat{a}_3 \to \hat{a}_1$, which means that the input modes do not become mixed in this limit. Having $|\beta| \to \infty$ and $\theta \to 0$ such that $\theta|\beta| \to$ constant ensures that we have a finite displacement parameter α as given above.

The approximation just described amounts to what is called a *parametric approximation*. A more rigorous and detailed justification for this approximation has been given by Paris [25] and D'Ariano *et al.* [26].

A displacement operator of this realization could be used, among other things, for the displacement of states other than the vacuum. For example, it could be used for the displacement of arbitrary number states. Displaced number states of light are known to have strong nonclassical properties [27].

6.8 Photons Do Not Interfere

It is common practice to speak of "single-photon interference" and of "multiphoton interference." Indeed, an important book by Ou has the title *Multi-Photon Quantum Interference* [28]. But as we have illustrated repeatedly in this chapter, only quantum probability amplitudes interfere. Photons do not interfere. And this is not merely an issue of semantics. As was pointed out in a letter by Roy Glauber to the *American Journal of Physics* in 1995 [29], the source of the trouble with the phrase "multiphoton interference" stems from Dirac's dictum, already alluded to in Ref. [1], that "[e]ach photon interferes only with itself. Interference between two photons never occurs." As Glauber says, Dirac's "statement was meant to dismiss absurd images of photons eating or reinforcing each other." Unfortunately, Dirac's statement has come to be "regarded as Scripture," says Glauber. As we have seen, even with just one photon in an interferometer, it is the addition of the complex quantum amplitudes associated with each path in the device that gives rise to "single-photon interference." The notion that "each photon interferes only with itself" is a rather meaningless statement and should be laid to rest. By "multiphoton interference" we really mean interference with states of light containing multiple photons.

6.9 Are Photons Entangled?

The prototype of an entangled system, as discussed by Bohm [30], is the spin-singlet state of two spin-½ particles

$$|SS\rangle = \frac{1}{\sqrt{2}}(|\uparrow\rangle_1|\downarrow\rangle_2 - |\downarrow\rangle_1|\uparrow\rangle_2), \qquad (6.125)$$

where, due to rotational symmetry of the state, the quantization axes are arbitrary (though they must be the same for each particle). The spin-singlet state cannot be factored into a product state. That is, $|SS\rangle \neq |\psi\rangle_1|\varphi\rangle_2$. If we do a partial trace, say, of particle 1 over the density operator $\rho_{SS} = |SS\rangle\langle SS|$, we obtain the reduced density operator for particle 2 as

$$\rho_{SS}^{(2)} = \frac{1}{2}(|\uparrow\rangle_{2\,2}\langle\uparrow| + |\downarrow\rangle_{2\,2}\langle\downarrow|), \qquad (6.126)$$

which is a statistical mixture.

On the other hand, we claimed above that the state obtained by beam splitting on just one photon, this being

$$|\psi\rangle_{23} \equiv \frac{1}{\sqrt{2}}(i|1\rangle_2|0\rangle_3 + |0\rangle_2|1\rangle_3), \qquad (6.127)$$

is an entangled state. Indeed, we showed above that by tracing over one of the modes we obtain a statistical mixture, as given in Eq. (6.36). Thus, there must be entanglement in the one-photon output state of the beam splitter. Yet there have been occasional claims in the literature that there is no entanglement in a state of the form of Eq. (6.127). The usual objection is that one needs at least two particles to have entanglement, as is the case in the spin-singlet state, whereas here there is only one particle, the photon. But a photon is an excitation of a field mode. All photons are excitations in some field mode or another, as explained in Chapter 2. Two field modes appear in Eq. (6.127), and the entanglement is between these two field modes. By analogy, the two "particles" in this case are the two field modes. And these modes are spatially separated and have different propagation directions associated with them. The number of photons excited, and their corresponding statistics, will affect the degree of entanglement between the modes, but it is still the modes that are entangled, not the photons *per se*. So, to the question posed in the title of this section: "Are photons entangled?" The answer is "no," photons are not entangled. The field modes *can* be entangled.

The issue of single-photon entanglement has been discussed by van Enk [31]. In Section 7.10 we discuss the fact that the beam splitter creates entanglement if at least one the input states is nonclassical. We've seen this already with input number states.

Final note: As mentioned above, we have assumed our beam splitters to be polarization preserving. The quantum mechanics of polarizing beam splitters is discussed in chapter 5 of the book by Agarwal [32].

Problems

1. Show that the beam splitter transformations in the form of Eq. (6.18) follow from Eqs. (6.19) and (6.20).
2. Consider as input to a beam splitter the product of coherent states $|\alpha\rangle_0 |\beta\rangle_1$. Assume the beam splitter to be 50:50 and that the reflected beam picks a $\pi/2$ phase shift. Determine the output state. Are the output beams entangled?
3. Consider the transformation of the state $|0\rangle_0 |\alpha\rangle_1$ across a 50:50 beam splitter for the case of a beam splitter that does not impart a $\pi/2$ phase shift to the reflected beam. Determine the output state of the beam splitter for this input state.
4. Let the input state to a beam splitter be $|\psi\rangle_0 |\beta\rangle_1$, where $|\beta\rangle_1$ is a coherent state and where $|\psi\rangle_0$ is a superposition of coherent states as given by $|\psi\rangle_0 = (|\alpha\rangle_0 + |-\alpha\rangle_0)/\sqrt{2}$, with $|\alpha|^2$ taken large enough such that $\langle -\alpha|\alpha\rangle_0 0 \simeq 0$. This is a kind of state known as a Schrödinger cat state, which is discussed in Chapter 7. Assume the beam splitter is 50:50 and imparts a $\pi/2$ phase shift to the reflected beam. Determine the output state and exam the joint photon number distribution. Exam the special case where $\beta = -i\alpha$. Will there be entanglement between the modes?

5.   Let the input state to a beam splitter of arbitrary θ and φ be $|4\rangle_0|0\rangle_1$. Determine the corresponding output state.

6.   For an arbitrary beam spitter, determine the output state for the input state $|2\rangle_0|2\rangle_1$.

7.   Let the input state of the 50:50 beam splitter described in the text be $|1\rangle_0|3\rangle_1$. Use Eq. (6.60) to work out in detail the output state. Especially, show that the $|2\rangle_2|2\rangle_3$ state does not occur in the output.

8.   Consider a beam splitter that is not 50:50. Find expressions for the output states if the input states are (a) $|n\rangle_0|n\rangle_1$ and (b) $|m\rangle_0|n\rangle_1$.

9.   In the text of this chapter, only pure states of light passing through a beam splitter were discussed. Suppose instead that light in a mixed state is passed through a beam splitter. To be definite, assume that mode 0 is in a vacuum state but that mode 1 is in a mixed state described by the density operator $\hat{\rho}_1$, such that the input state density operator is $\hat{\rho}_{in} = |0\rangle_0{}_0\langle 0|\otimes\hat{\rho}_1$. Further assume that the mixed state itself is a thermal state (see Chapter 2), and then determine the density operator of the output state.

10. Consider the case of the single photon mixed with a coherent state $|\alpha\rangle$ at a 50:50 beam splitter as discussed in the text. Calculate the linear entropy for the output state as a function of the amplitude $|\alpha|$. Compare the degree of entanglement of this state with the degree of entanglement found in the one-photon entangled state given by Eq. (6.34).

11. In Section 6.5 we studied interferometry with input vacuum and coherent states. Consider now interferometry with input states of the form $|0\rangle(|\alpha\rangle + |-\alpha\rangle)/\sqrt{2}$, with $|\alpha|^2$ large, such that $\langle -\alpha|\alpha\rangle \simeq 0$. Find the phase uncertainty as a function of the phase shift φ and compare with that for the case of interferometry with coherent light mixed with a vacuum.

12. The description of interaction-free measurements given in Section 6.4 assumed that the beam splitters involved are 50:50. The presence of an object in one arm is determined when one of the detectors, the one not clicking (because of interference, with no object in one of the arms) begins to click. It will click in a quarter of the runs. Examine the question: can the presence of an object in one arm be detected more efficiently (i.e. in a greater number of experimental runs) by using beam splitters that are not 50:50?

13. Consider the case of beam splitter input states $|3\rangle_0|n\rangle_1$ for n odd. Use the techniques developed in Section 6.6 to show that the joint probability for detecting $(n + 3)/2$ photons in each mode of the output vanishes in the limit of the beam splitter being 50:50.

14. Using the exact unitary beam splitter transformation as given in Eq. (6.122), perform numerical calculations to show that with the initial state of the 0-mode being vacuum state, to a good approximation, the coherent state $|\alpha\rangle_2$ is generated where $\alpha = i\theta\beta/2$ under the condition that $|\beta|\rightarrow\infty$ and $\theta\rightarrow 0$, such that $\theta|\beta|\rightarrow$constant.

15. Consider a 50:50 beam splitter that does *not* impart a $\pi/2$ phase shift to the reflected beam, as described by Eqs. (6.32) and (6.33). (a) For the input state $|1\rangle_0|1\rangle_1$, does the output of this beam splitter exhibit the Hong–Ou–Mandel effect? (b) Find an expression for the output state of this beam splitter for the general input state $|m\rangle_0|n\rangle_1$.

References

[1] G. I. Taylor, *Proc. Camb. Philos. Soc.* **15**, 114 (1909).

[2] P. A. M. Dirac, *The Principles of Quantum Mechanics*, 4th edition (London: Oxford University Press, 1958), p. 9.

[3] P. Grangier, G. Roger, and A. Aspect, *Europhys. Lett.* **1**, 173 (1986).

[4] See M. W. Hamilton, *Am. J. Phys.* **68**, 186 (2000).

[5] R. P. Feynman, R. B. Leighton, and M. Sands, *The Feynman Lectures on Physics*, Vol. III (Reading, MA: Addison-Wesley, 1965), chap. 3.

[6] C. K. Hong, Z. Y. Ou, and L. Mandel, *Phys. Rev. Lett.* **59**, 2044 (1987).

[7] H. Fearn and R. Loudon, *Opt. Commun.* **64**, 485 (1987).

[8] R. Loudon, *Phys. Rev. A* **58**, 4904 (1998). See also chapter 4 of Ref. [5].

[9] See W. Feller, *An Introduction to Probability Theory and its Applications*, Vol. I, 3rd edition (New York: Wiley, 1968), p. 79.

[10] R. A. Campos, B. E. A. Saleh, and M. C. Teich, *Phys. Rev.* **40**, 1371 (1989).

[11] Z. Y. Ou, *Quant. Semiclass. Opt.* **8**, 315 (1996).

[12] R. Birrittella, J. Mimih, and C. C. Gerry, *Phys. Rev. A* **86**, 063828 (2012).

[13] P. M. Alsing, R. J. Birrittella, C. C. Gerry, J. Mimih, and P. L. Knight, *Phys. Rev. A* **105**, 013712 (2022).

[14] A. C. Elitzur and L. Vaidman, *Found. Phys.* **23**, 987 (1993). See also R. H. Dicke, *Am. J. Phys.* **49**, 925 (1981).

[15] P. Kwiat, H. Weinfurter, T. Herzog, and A. Zielinger, *Phys. Rev. Lett.* **74**, 4763 (1995).

[16] Y. Aharonov and D. Bohm, *Phys. Rev.* **115**, 484 (1959).

[17] B. P. Abbott *et al.*, *Phys. Rev. Lett.* **116**, 061102 (2016).

[18] B. P. Abbott *et al.*, *Rep. Prog. Phys.* **72**, 076901 (2009).

[19] M. Tse *et al.*, *Phys. Rev. Lett.* **123**, 231107 (2019).

[20] F. Acernese *et al.*, *Phys. Rev. Lett.* **123**, 231108 (2019).

[21] C. Affeldt *et al.*, *Class. Quant. Grav.* **31**, 224002 (2014).

[22] J. Schwinger, *On Angular Momentum* (Mineola, NY: Dover, 2015).

[23] M. E. Rose, *Elementary Theory of Angular Momentum* (Mineola, NY: Dover, 1995), p. 53.

[24] B. Yurke, S. L. McCall, and J. R. Klauder, *Phys. Rev. A* **33**, 4033 (1986).

[25] M. G. A. Paris, *Phys. Lett. A* **217**, 78 (1996).

[26] G. M. D'Ariano, M. G. A. Paris, and M. F. Sacchi, *Nuovo Cimento* **114B**, 339 (1999).

[27] See F. A. M de Oliveira, M. S. Kim, P. L. Knight, and V. Bužek, *Phys. Rev. A* **41**, 2645 (1990) and references cited therein.

[28] Z. Y. J. Ou, *Multi-Photon Quantum Interference* (New York: Springer, 2007).

[29] R. Glauber, *Am. J. Phys.* **63**, 12 (1995).

[30] D. Bohm, *Quantum Theory* (New York: Prentice-Hall, 1951), p. 615.

[31] S. van Enk, *Phys. Rev. A* **72**, 064306 (2005).

[32] G. S. Agarwal, *Quantum Optics* (Cambridge: Cambridge University Press, 2013).

Bibliography

Additional papers on the quantum description of beam splitters:

A. Zeilinger, *Am. J. Phys.* **49**, 882 (1981).

Z. Y. Ou, C. K. Hong, and L. Mandel, *Opt. Commun.* **63**, 118 (1987).

S. Prasad, M. O. Scully, and W. Martienssen, *Opt. Commun.* **62**, 139 (1987).

Z. Y. Ou and L. Mandel, *Am. J. Phys.* **57**, 66 (1989).

Nonclassical Light

The word 'classical' means only one thing in science: it's wrong! [1]

We have previously emphasized the fact that *all* states of light are quantum mechanical and are thus nonclassical, deriving some quantum features from the discreteness of the photons. Of course, in practice, the nonclassical features of light are difficult to observe. (We shall use "quantum mechanical" and "nonclassical" more or less interchangeably here.) Already we have discussed what must certainly be the most nonclassical of all nonclassical states of light – the single-photon state. Yet, as we shall see, it is possible to have nonclassical states involving a very large number of photons. But we need a criterion for nonclassicality. Recall that in Chapter 5 we discussed such a criterion in terms of the quasi-probability distribution known as the P function, $P(\alpha)$. States for which $P(\alpha)$ is positive everywhere – or no more singular than a delta function – are classical, whereas those for which $P(\alpha)$ is negative – or more singular than a delta function – are nonclassical. We have shown, in fact, that $P(\alpha)$ for a coherent state *is* a delta function and Hillery [2] has shown that *all* other pure states of the field will have functions $P(\alpha)$ that are negative in some regions of phase space and are more singular than a delta function. It is evident that the variety of possible nonclassical states of the field is quite large.

In this chapter, we shall discuss some of the most important examples of nonclassical states. We begin with the squeezed states (i.e. the quadrature and the number squeezed states, the latter also known as states with sub-Poissonian statistics), then discuss again the nonclassical nature of photon antibunching, introduce the Schrödinger cat states, the states of correlated double beams, broadband squeezed light, and quantum state engineering by adding and subtraction photons from readily available states of light.

7.1 Quadrature Squeezing

If two operators $\hat{A}$ and $\hat{B}$ satisfy the commutation relation $[\hat{A}, \hat{B}] = i\hat{C}$, it follows that

$$\langle (\Delta \hat{A})^2 \rangle \langle (\Delta \hat{B})^2 \rangle \geq \frac{1}{4} |\langle \hat{C} \rangle|^2. \tag{7.1}$$

A state of the system is said to be squeezed if either

$$\langle(\Delta\hat{A})^2\rangle < \frac{1}{2}|\langle\hat{C}\rangle| \quad \text{or} \quad \langle(\Delta\hat{B})^2\rangle < \frac{1}{2}|\langle\hat{C}\rangle|. \tag{7.2}$$

[Because of Eq. (7.1), we obviously cannot have both variances less than $|\langle\hat{C}\rangle|/2$ simultaneously.] Squeezed states for which the $[\hat{X}_1,\hat{X}_2] = i/2$ equality holds in Eq. (7.1) are sometimes known as ideal squeezed states and are an example of the "intelligent" states we discussed in Chapter 3.

In the case of quadrature squeezing, we take $\hat{A} = \hat{X}_1$ and $\hat{B} = \hat{X}_2$, $\hat{X}_1$ and $\hat{X}_2$ being the quadrature operators of Eqs. (2.53) and (2.54), satisfying Eq. (2.56) and thus $\hat{C} = 1/2$. From Eq. (2.57) it follows that quadrature squeezing exists whenever [3]

$$\langle(\Delta\hat{X}_1)^2\rangle < \frac{1}{4} \quad \text{or} \quad \langle(\Delta\hat{X}_2)^2\rangle < \frac{1}{4}. \tag{7.3}$$

We have already established that for a coherent state $|\alpha\rangle$, equality holds in Eq. (2.57) and that the variances of two quadratures are equal: $\langle(\Delta\hat{X}_1)^2\rangle = \langle(\Delta\hat{X}_2)^2\rangle = 1/4$. Not only that, but the result for the coherent state is exactly the same as for the vacuum [see Eq. (2.60)]. States for which one of the conditions in Eq. (7.3) holds will have less "noise" in one of the quadratures than for a coherent state or a vacuum state – the fluctuations in that quadrature are squeezed. Of course, the fluctuations in the other quadrature must be enhanced so as not to violate the uncertainty relation. There are squeezed states for which the uncertainty relation is equalized, but this need not be the case in general. See Fig. 7.1 for a graphical representation of the range of squeezing.

Before presenting specific examples of squeezed states, we wish to demonstrate why quadrature squeezing must be considered a nonclassical effect.

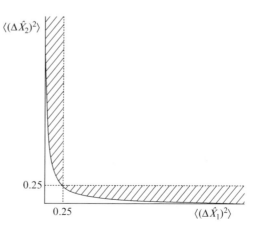

Fig. 7.1 A graphical characterization of quadrature squeezing. Squeezing exists for the points inside the shaded areas. The solid-line border is a hyperbola determined from the equalized uncertainty relation $\langle(\Delta\hat{X}_1)^2\rangle\langle(\Delta\hat{X}_2)^2\rangle = 1/16$.

To this end, we express the relevant expectation values in terms of the P function. Using Eq. (3.108) it is easy to show that

$$\langle(\Delta \hat{X}_1)^2\rangle = \frac{1}{4}\left\{1 + \int P(\alpha)[(\alpha + \alpha^*) - (\langle\hat{a}\rangle + \langle\hat{a}^\dagger\rangle)]^2 d^2\alpha\right\} \qquad (7.4)$$

and

$$\langle(\Delta \hat{X}_2)^2\rangle = \frac{1}{4}\left\{1 + \int P(\alpha)[(\alpha - \alpha^*)/i - (\langle\hat{a}\rangle - \langle\hat{a}^\dagger\rangle)/i]^2 d^2\alpha\right\}, \qquad (7.5)$$

where

$$\langle\hat{a}\rangle = \int P(\alpha)\alpha d^2\alpha \quad \text{and} \quad \langle\hat{a}^\dagger\rangle = \int P(\alpha)\alpha^* d^2\alpha. \qquad (7.6)$$

Since the squared bracket term is, of course, always positive, it is evident that the condition $\langle(\Delta \hat{X}_{1,2})^2\rangle < 1/4$ requires $P(\alpha)$ to be nonpositive at least in some regions of phase space.

It is sometimes convenient to introduce a generic quadrature operator

$$\hat{X}(\vartheta) = \frac{1}{2}(\hat{a}e^{i\vartheta} + \hat{a}^\dagger e^{-i\vartheta}), \qquad (7.7)$$

where obviously $\hat{X}(0) = \hat{X}_1$ and $\hat{X}(\pi) = \hat{X}_2$. To characterize squeezing, we introduce the parameter

$$s(\vartheta) = \frac{\left\langle\left(\Delta\hat{X}(\vartheta)\right)^2\right\rangle - 1/4}{1/4} = 4\left\langle\left(\Delta\hat{X}(\vartheta)\right)^2\right\rangle - 1. \qquad (7.8)$$

Squeezing exists for some value of the angle ϑ whenever $-1 \leq s(\vartheta) < 0$. Alternatively, we can introduce the normally ordered variance $\langle: (\Delta\hat{X}(\vartheta))^2 :\rangle$ such that squeezing exists whenever $-\frac{1}{4} \leq \langle: (\Delta\hat{X}(\vartheta))^2 :\rangle < 0$. In terms of the normally ordered variance, we have

$$s(\vartheta) = 4\left\langle: \left(\Delta\hat{X}(\vartheta)\right)^2 :\right\rangle. \qquad (7.9)$$

So, how can squeezed states be generated? One way of generating a squeezed state mathematically, in a manner corresponding to a physical process that we discuss below, is through the action of a "squeeze" operator, defined as

$$\hat{S}(\xi) = \exp\left[\frac{1}{2}(\xi^*\hat{a}^2 - \xi\hat{a}^{\dagger 2})\right], \qquad (7.10)$$

where $\xi = re^{i\theta}$, with r known as the squeeze parameter, and $0 \leq r < \infty$, $0 \leq \theta \leq 2\pi$. This operator $\hat{S}(\xi)$ is a kind of two-photon generalization of the displacement operator used to define the usual coherent states of a single-mode field, as discussed in Chapter 3. Evidently, the operator

$\hat{S}(\xi)$ acting on the vacuum would create some sort of "two-photon coherent state," as it is clear that photons will be created or destroyed in pairs by the action of this operator. To get some sense of what happens as a result of acting with this operator, let us consider the state

$$|\psi_s\rangle = \hat{S}(\xi)|\psi\rangle, \tag{7.11}$$

where $|\psi\rangle$ is for the moment arbitrary and $|\psi_s\rangle$ denotes the state generated by the action of $\hat{S}(\xi)$ on $|\psi\rangle$. To obtain the variances of $\hat{X}_1$ and $\hat{X}_2$, we need the expectation values of $\hat{a}$, $\hat{a}^2$, and so on. To this end we need the following results obtained from application of the Baker–Hausdorff lemma:

$$\hat{S}^{\dagger}(\xi)\hat{a}\hat{S}(\xi) = \hat{a}\cosh r - \hat{a}^{\dagger}e^{i\theta}\sinh r,$$

$$\hat{S}^{\dagger}(\xi)\hat{a}^{\dagger}\hat{S}(\xi) = \hat{a}^{\dagger}\cosh r - \hat{a}e^{-i\theta}\sinh r, \tag{7.12}$$

where $\hat{S}^{\dagger}(\xi) = \hat{S}(-\xi)$. Thus we have

$$\langle\psi_s|\hat{a}|\psi_s\rangle = \langle\psi|\hat{S}^{\dagger}(\xi)\hat{a}S(\xi)|\psi\rangle, \tag{7.13}$$

$$\langle\psi_s|\hat{a}^2|\psi_s\rangle = \langle\psi|\hat{S}^{\dagger}(\xi)\hat{a}S(\xi)\hat{S}^{\dagger}(\xi)\hat{a}S(\xi)|\psi\rangle, \tag{7.14}$$

and so on. For the special case where $|\psi\rangle$ is the vacuum state $|0\rangle$, $|\psi_s\rangle$ is the squeezed vacuum state, which we denote as $|\xi\rangle$:

$$|\xi\rangle = \hat{S}(\xi)|0\rangle. \tag{7.15}$$

Using Eqs. (7.12)–(7.15) we find that for the squeezed vacuum state

$$\langle(\Delta\hat{X}_1)^2\rangle = \frac{1}{4}[\cosh^2 r + \sinh^2 r - 2\sinh r \cosh r \cos\theta], \tag{7.16}$$

$$\langle(\Delta\hat{X}_2)^2\rangle = \frac{1}{4}[\cosh^2 r + \sinh^2 r + 2\sinh r \cosh r \cos\theta]. \tag{7.17}$$

For $\theta = 0$, these reduce to

$$\langle(\Delta\hat{X}_1)^2\rangle = \frac{1}{4}e^{-2r},$$

$$\langle(\Delta\hat{X}_2)^2\rangle = \frac{1}{4}e^{2r}, \tag{7.18}$$

and evidently squeezing exists in the $\hat{X}_1$ quadrature. For $\theta = \pi$, squeezing will appear in the $\hat{X}_2$ quadrature. Note that the product of the uncertainties yields $1/16$ and thus for $\theta = 0$ or π, the squeezed vacuum states equalize the uncertainty relation of Eq. (2.57). Squeezed states need not, and generally do not, equalize the uncertainty relation.

There is a simple way to represent squeezing graphically. Recall that in Figs. 3.1 and 3.2 we present phase-space representations of the noise associated with a coherent state (Fig. 3.1) and a vacuum state (Fig. 3.2), where in both cases the fluctuations of the quadrature operators were

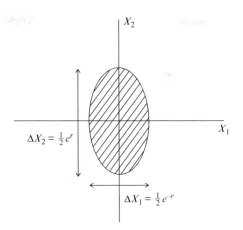

 Fig. 7.2   Error ellipse for a squeezed vacuum state where the squeezing is in the X_1 quadrature.

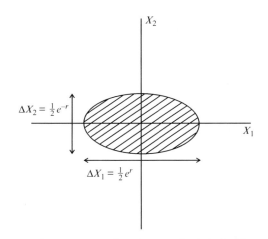

Fig. 7.3   Error ellipse for a squeezed vacuum state where the squeezing is in the X_2 quadrature.

equal: $\Delta X_1 = \Delta X_2 = 1/2$. A representation of the squeezed vacuum state for $\theta = 0$, where the fluctuations in X_1 are reduced at the expense of those of X_2, can be seen in Fig. 7.2, whereas for $\theta = \pi$ the situation is reversed, as can be seen in Fig. 7.3.

In the cases with $\theta = 0$ or π, the squeezing has been along either $\hat{X}_1$ or $\hat{X}_2$. For other values of θ, let us define the rotated quadrature operators $\hat{Y}_1$ and $\hat{Y}_2$ as

$$\begin{pmatrix} \hat{Y}_1 \\ \hat{Y}_2 \end{pmatrix} = \begin{pmatrix} \cos\theta/2 & \sin\theta/2 \\ -\sin\theta/2 & \cos\theta/2 \end{pmatrix} \begin{pmatrix} \hat{X}_1 \\ \hat{X}_2 \end{pmatrix} \tag{7.19}$$

or equivalently

$$\hat{Y}_1 + i\hat{Y}_2 = (\hat{X}_1 + i\hat{X}_2)e^{-i\theta/2}. \tag{7.20}$$

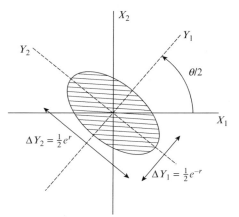

Fig. 7.4 Rotated error ellipse for the squeezed vacuum. The squeezing is along the $\theta/2$ direction.

For the squeezed vacuum state of Eq. (7.12) it can be shown that

$$\langle (\Delta \hat{Y}_1)^2 \rangle = \frac{1}{4} e^{-2r},$$
$$\langle (\Delta \hat{Y}_2)^2 \rangle = \frac{1}{4} e^{2r}, \tag{7.21}$$

which is represented graphically in Fig. 7.4.

The important point here is that the squeezing need not be along X_1 or X_2.

A more general squeezed state may be obtained by applying the displacement operator to Eq. (7.12):

$$|\alpha, \xi\rangle = \hat{D}(\alpha)\hat{S}(\xi)|0\rangle. \tag{7.22}$$

Obviously for $\xi = 0$ we obtain just a coherent state. Since the displacement operator effects the transformation

$$\hat{D}^\dagger(\alpha)\hat{a}\hat{D}(\alpha) = \hat{a} + \alpha,$$
$$\hat{D}^\dagger(\alpha)\hat{a}^+\hat{D}(\alpha) = \hat{a}^\dagger + \alpha^*, \tag{7.23}$$

the action of the product of the displacement and squeeze operators on $\hat{a}$ and $\hat{a}^\dagger$ can be obtained from Eqs. (7.12) and (7.23). Leaving the steps as an exercise, it can be shown that

$$\langle \hat{a} \rangle = \alpha, \tag{7.24}$$

which is independent of the squeeze parameter r, and that

$$\langle \hat{a}^2 \rangle = \alpha^2 - e^{i\theta}\sinh r \cosh r, \tag{7.25}$$

$$\langle \hat{a}^\dagger \hat{a} \rangle = |\alpha|^2 + \sinh^2 r. \tag{7.26}$$

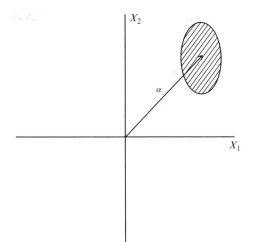

Fig. 7.5 Error ellipse of a displaced squeezed vacuum state with the squeezing in the X_1 quadrature.

Of course, the corresponding quantities for the coherent states are obtained for $r \to 0$ and for the squeezed vacuum as $\alpha \to 0$. Furthermore

$$\langle \hat{Y}_1 + i\hat{Y}_2 \rangle = \alpha e^{-i\theta/2} \tag{7.27}$$

and once again it follows that

$$\langle (\Delta \hat{Y}_1)^2 \rangle = \frac{1}{4} e^{-2r},$$
$$\langle (\Delta \hat{Y}_2)^2 \rangle = \frac{1}{4} e^{2r}. \tag{7.28}$$

For the case that $\theta = 0$, the squeezed state is represented in phase space as in Fig. 7.5, where the "error ellipse" is essentially that of the squeezed vacuum displaced by α.

Of course, the more general case of θ will simply be a displaced rotated error ellipse as in Fig. 7.6.

Before going on, let's look at the electric field for a squeezed state. The single-mode electric field operator of Eq. (2.15) may be written as

$$\hat{E}(\chi) = \mathcal{E}_0 \sin(kz)(\hat{a} e^{-i\chi} + \hat{a}^\dagger e^{i\chi}), \tag{7.29}$$

where the operators $\hat{a}$ and $\hat{a}^\dagger$ are at $t = 0$ and $\chi = \omega t$, the phase of the field, so that $\hat{E}(\chi)$ is the phase-dependent electric field. In terms of the quadrature operators, Eq. (7.29) can be written as

$$\hat{E}(\chi) = 2\mathcal{E}_0 \sin(kz)[\hat{X}_1 \cos\chi + \hat{X}_2 \sin\chi]. \tag{7.30}$$

From the commutator $[\hat{X}_1, \hat{X}_2] = i/2$ we have

$$[\hat{E}(0), \hat{E}(\chi)] = i\mathcal{E}_0^2 \sin^2(kz) \sin\chi, \tag{7.31}$$

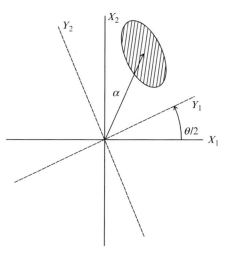

Rotated error ellipse of a displaced squeezed vacuum state.

and thus the electric field operators at different phases (or times) do not commute. From Eq. (7.28) it follows that

$$\left\langle \left(\Delta \hat{E}(0) \right)^2 \right\rangle \left\langle \left(\Delta \hat{E}(\chi) \right)^2 \right\rangle \ge \frac{1}{4} [\mathcal{E}_0^2 \sin^2(kz)\sin\chi]. \qquad (7.32)$$

For many states of the field, the field uncertainty is independent of the phase χ (e.g. coherent states, thermal states). But if the uncertainty in the field for $\chi = 0$ is such that

$$\Delta E(0) < \frac{1}{\sqrt{2}} \mathcal{E}_0 |\sin(kz)\sin\chi|, \qquad (7.33)$$

then at some other phase χ we must have

$$\Delta E(\chi) > \frac{1}{\sqrt{2}} \mathcal{E}_0 |\sin(kz)\sin\chi|. \qquad (7.34)$$

In this sense, squeezed light contains phase-dependent noise, reduced below that of the vacuum for some phases and enhanced above that of the vacuum for others. Quadrature squeezing, because it is phase sensitive, is related to the wave-like nature of light.

Using phasor diagrams, in Fig. 7.7 we illustrate the distribution of noise in the electric field for squeezed states. The case of the coherent state is illustrated in Fig. 3.8.

The fact that some parts of a light wave in a squeezed state are less noisy than a field in a vacuum state has possible technological applications, particularly in the detection of weak signals. An important example is

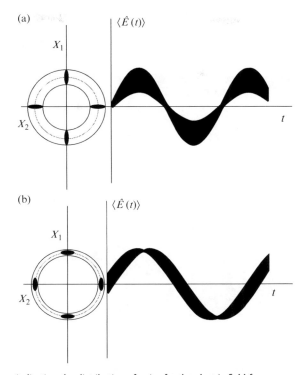

Phasor diagram indicating the distribution of noise for the electric field for a squeezed state for (a) squeezing initially in the X_1 quadrature and (b) squeezing initially in the X_2 quadrature.

the detection of gravity waves with large-scale interferometers [4], as in the LIGO [5] and Virgo projects [6]. Another possible application is in the area of optical communications and quantum information processing [7]. We shall not pursue these applications here.

To get a bit more insight into the squeezed states and ultimately find an expression for them in terms of the photon-number state, let us start with the vacuum state $|0\rangle$ satisfying

$$\hat{a}|0\rangle = 0. \tag{7.35}$$

Multiplying by $\hat{S}(\xi)$ from the left and using the fact that this operator is unitary, we may write

$$\hat{S}(\xi)\hat{a}\hat{S}^{\dagger}(\xi)\hat{S}(\xi)|0\rangle = 0, \tag{7.36}$$

or

$$\hat{S}(\xi)\hat{a}\hat{S}^{\dagger}(\xi)|\xi\rangle = 0. \tag{7.37}$$

Since

$$\hat{S}(\xi)\hat{a}\hat{S}^{\dagger}(\xi) = \hat{a}\cosh r + e^{i\theta}\hat{a}^{\dagger}\sinh r, \tag{7.38}$$

we may rewrite Eq. (7.37) as

$$(\hat{a}\mu + \hat{a}^\dagger v)|\xi\rangle = 0, \tag{7.39}$$

where $\mu = \cosh r$ and $v = e^{i\theta}\sinh r$. Thus the squeezed vacuum state is an eigenstate of the operator $\hat{a}\mu + \hat{a}^\dagger v$ with eigenvalue zero. For the more general state of Eq. (7.22) we can similarly write

$$\hat{D}(\alpha)\hat{S}(\xi)\hat{a}\hat{S}^\dagger(\xi)\hat{D}^\dagger(\alpha)\hat{D}(\alpha)\hat{S}(\xi)|0\rangle = 0, \tag{7.40}$$

which, using the relation

$$\hat{D}(\alpha)\hat{a}\hat{D}^\dagger(\alpha) = \hat{a} - \alpha, \tag{7.41}$$

and Eq. (7.38) can be rewritten as the eigenvalue problem

$$(\hat{a}\mu + \hat{a}^\dagger v)|\alpha, \xi\rangle = \gamma|\alpha, \xi\rangle, \tag{7.42}$$

where

$$\gamma = \alpha \cosh r + \alpha^* e^{i\theta}\sinh r. \tag{7.43}$$

Obviously, for $r = 0$ this is just the eigenvalue problem for the ordinary coherent state and for $\alpha = 0$ it is for the squeezed vacuum state. Now writing

$$\hat{a} = \hat{X}_1 + i\hat{X}_2 = (\hat{Y}_1 + i\hat{Y}_2)\, e^{i\theta/2}, \tag{7.44}$$

and substituting into Eq. (7.39), we obtain after some rearrangement

$$(\hat{Y}_1 + i\hat{Y}_2 e^{-2r})|\alpha, \xi\rangle = \beta|\alpha, \xi\rangle, \tag{7.45}$$

where

$$\beta = \alpha e^{-r}e^{-i\theta/2} = \langle\hat{Y}_1\rangle + i\langle\hat{Y}_2\rangle e^{-2r}. \tag{7.46}$$

Equation (7.45) is precisely of the form of an eigenvalue equation for the eigenstate that equalizes the uncertainty product for the operators $\hat{Y}_1$ and $\hat{Y}_2$, which, since $[\hat{Y}_1, \hat{Y}_2] = i/2$, is just

$$\langle(\Delta\hat{Y}_1)^2\rangle\langle(\Delta\hat{Y}_2)^2\rangle = \frac{1}{16}. \tag{7.47}$$

Comparing Eq. (7.45) with Eq. (3.19), setting $\hat{A} = \hat{Y}_1$, $\hat{B} = \hat{Y}_2$, and $\hat{C} = 1/2$, we deduce that

$$\langle(\Delta\hat{Y}_2)^2\rangle = \frac{1}{4}e^{2r} \tag{7.48}$$

and then from Eq. (7.43) that

$$\langle(\Delta\hat{Y}_1)^2\rangle = \frac{1}{4}e^{-2r}. \tag{7.49}$$

Thus the state $|\alpha, \xi\rangle$ is an "intelligent" state with respect to the rotated quadrature operators $\hat{Y}_1$ and $\hat{Y}_2$.

In terms of the original quadrature operators $\hat{X}_1$ and $\hat{X}_2$, Eq. (7.39) reads

$$(\hat{X}_1 + i\lambda \hat{X}_2)|\alpha, \xi\rangle = \gamma|\alpha, \xi\rangle, \tag{7.50}$$

where

$$\lambda = \left(\frac{\mu - v}{\mu + v}\right) \quad \text{and} \quad \gamma = \frac{\alpha}{\mu + v}. \tag{7.51}$$

For $\theta = 0$, Eq. (7.46) takes the form

$$(\hat{X}_1 + i \hat{X}_2 e^{-2r})|\alpha, \xi\rangle = \alpha\, e^{-r}|\alpha, \xi\rangle \tag{7.52}$$

and we obtain the results of Eqs. (7.15). But what about the case when $\theta \neq 0$? To understand the meaning of such a situation, we go back to the commutation relation $[\hat{A}, \hat{B}] = i\hat{C}$ and point out the most general statement of the uncertainty relation is actually not as given in Eq. (7.1) but rather

$$\langle(\Delta\hat{A})^2\rangle\langle(\Delta\hat{B})^2\rangle \geq \frac{1}{4}\left[\langle\hat{F}\rangle^2 + \langle\hat{C}\rangle^2\right], \tag{7.53}$$

where

$$\langle\hat{F}\rangle = \langle\hat{A}\hat{B} + \hat{B}\hat{A}\rangle - 2\langle\hat{A}\rangle\langle\hat{B}\rangle \tag{7.54}$$

is the covariance, essentially the measure of the *correlations* between the observables $\hat{A}$ and $\hat{B}$. Equality in Eq. (7.53) is obeyed for a state $|\psi\rangle$ satisfying

$$(\hat{A} + i\lambda\hat{B})|\psi\rangle = (\langle\hat{A}\rangle + i\lambda\langle\hat{B}\rangle)|\psi\rangle, \tag{7.55}$$

where λ may, in general, be complex. From Eq. (7.51) it is easy to show that

$$\langle(\Delta\hat{A})^2\rangle - \lambda^2\langle(\Delta\hat{B})^2\rangle = i\lambda\langle\hat{F}\rangle, \tag{7.56}$$

$$\langle(\Delta\hat{A})^2\rangle + \lambda^2\langle(\Delta\hat{B})^2\rangle = \lambda\langle\hat{C}\rangle. \tag{7.57}$$

If λ is real, then from Eq. (7.56) we must have $\langle\hat{F}\rangle = 0$ and thus there are no correlations between $\hat{A}$ and $\hat{B}$. If λ is pure imaginary, from Eq. (7.57) we must have $\langle\hat{C}\rangle = 0$. But for $\hat{A} = \hat{X}_1, \hat{B} = \hat{X}_2$, it is not possible to have $\lambda = (\mu - v)/(\mu + v)$ pure imaginary. When it is real ($\theta = 0$) then

$$\langle\hat{F}\rangle = \langle\hat{X}_1\hat{X}_2 + \hat{X}_2\hat{X}_1\rangle - 2\langle\hat{X}_1\rangle\langle\hat{X}_2\rangle = 0, \tag{7.58}$$

and there are no correlations between $\hat{X}_1$ and $\hat{X}_2$. But for $\theta \neq 0$ or 2π, λ is complex and it is easy to show that for this case $\langle\hat{F}\rangle \neq 0$. Thus, the general squeezed state of the form $|\alpha, \xi\rangle = \hat{D}(\alpha)\hat{S}(\xi)|0\rangle$ may be characterized by the presence of correlations between the observables $\hat{X}_1$ and $\hat{X}_2$. In the limit $\xi \to 0$ ($r \to 0$) we recover the coherent state $|\alpha\rangle$ for which there are no correlations between $\hat{X}_1$ and $\hat{X}_2$. We shall encounter other nonclassical states exhibiting correlations between the quadratures. Again, we emphasize that the squeezed state $|\alpha, \xi\rangle$ is explicitly constructed to have the property of squeezing, but many other pure states of the field may also exhibit squeezing, at least for some ranges of the relevant parameters.

We now decompose the squeezed states into the photon-number states in order to examine the photon statistics. We consider the squeezed vacuum state first. We write

$$|\xi\rangle = \sum_{n=0}^{\infty} C_n |n\rangle, \tag{7.59}$$

which, upon substituting into Eq. (7.36), leads to the recursion relation

$$C_{n+1} = -\frac{v}{\mu} \left(\frac{n}{n+1}\right)^{1/2} C_{n-1}. \tag{7.60}$$

Note that this relation connects only every other photon state. In fact, there are two distinct solutions, one involving only even photon states and a second involving only odd. Obviously, only the even solution contains the vacuum, so we address only this case here. The solution of the recursion relation is

$$C_{2m} = (-1)^m (e^{i\theta} \tanh r)^m \left[\frac{(2m-1)!!}{(2m)!!}\right]^{1/2} C_0. \tag{7.61}$$

C_0 is determined from the normalization

$$\sum_{m=0}^{\infty} |C_{2m}|^2 = 1, \tag{7.62}$$

which leads to

$$|C_0|^2 \left[\sum_{m=0}^{\infty} \frac{[\tanh r]^{2m}(2m-1)!!}{(2m)!!}\right] = 1. \tag{7.63}$$

Fortunately, there exists the mathematical identity

$$\sum_{m=0}^{\infty} z^m \left(\frac{(2m-1)!!}{(2m)!!}\right) = (1-z)^{-1/2}, \tag{7.64}$$

from which we easily obtain $C_0 = 1/\sqrt{\cosh r}$. Finally, we use the identities

$$(2m)!! = 2^m m!, \tag{7.65}$$

$$(2m-1)!! = \frac{1}{2^m} \frac{(2m)!}{m!}, \tag{7.66}$$

to obtain the most common form of the expansion coefficients for the squeezed vacuum state:

$$C_{2m} = (-1)^m \frac{\sqrt{(2m)!}}{2^m m!} \frac{(e^{i\theta} \tanh r)^m}{\sqrt{\cosh r}}. \tag{7.67}$$

Thus, the squeezed vacuum state is

$$|\xi\rangle = \frac{1}{\sqrt{\cosh r}} \sum_{m=0}^{\infty} (-1)^m \frac{\sqrt{(2m)!}}{2^m m!} e^{im\theta} (\tanh r)^m |2m\rangle. \tag{7.68}$$

This expression for the squeezed vacuum state is obviously a superposition of only the even photon-number states. For computational purposes it may be more convenient to express the state in terms of the full range set of photon numbers. By inserting the factor $\cos^2(n\pi/2)$, which has value 1 for n even and 0 for n odd, we can rewrite Eq. (7.68) according to

$$|\xi\rangle = \frac{1}{\sqrt{\cosh r}} \sum_{n=0}^{\infty} (-1)^{n/2} \frac{\sqrt{n!}}{2^{n/2}(n/2)!} e^{in\theta/2} (\tanh r)^{n/2} \cos^2\left(\frac{n\pi}{2}\right) |n\rangle. \tag{7.69}$$

The probability of detecting n photons in the field is

$$P_n = |\langle n|\xi\rangle|^2 = \begin{cases} \dfrac{n!}{2^n[(n/2)!]^2} \dfrac{(\tanh r)^n}{\cosh r} & n \text{ even,} \\[4mm] 0 & n \text{ odd.} \end{cases} \tag{7.70}$$

Thus, the photon probability distribution for a squeezed vacuum state is oscillatory, vanishing for all odd photon numbers. A typical distribution for the squeezed vacuum is given in Fig. 7.8.

Notice that, aside from the oscillatory nature of the distribution, the shape of it resembles that of thermal radiation. It must be remembered that the squeezed vacuum is a pure state whereas the thermal state is a mixed one.

We now seek solutions of Eq. (7.42) for the more general case of $\alpha \neq 0$. We again assume a solution of the form of Eq. (7.59), taking advantage of what we have learned in the case of $\alpha = 0$ to make the ansatz

$$C_n = \mathcal{N}(\cosh r)^{-1/2} \left[\frac{1}{2} e^{i\theta} \tanh r\right]^{n/2} f_n(x), \tag{7.71}$$

where $f_n(x)$ is as yet an unknown function and $\mathcal{N}$ is a normalization factor. The resulting recursion relation for $f_n(x)$ is

$$(n+1)^{1/2} f_{n+1}(x) - 2\alpha \left(e^{i\theta} \sinh(2r)\right)^{-1/2} f_n(x) + 2n^{1/2} f_{n-1}(x) = 0. \tag{7.72}$$

This is identical to the recursion relation for the Hermite polynomials $H_n(x)$:

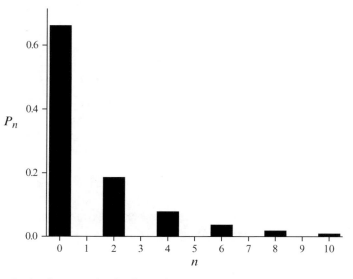

 Fig. 7.8   Histogram for the photon number distribution for squeezed vacuum state.

$$H_{n+1}(x) - 2xH_n(x) + 2nH_{n-1}(x) = 0, \qquad (7.73)$$

provided we make the identifications

$$f_n(x) = H_n(x)/\sqrt{n!} \text{ and } x = \gamma\left(e^{i\theta}\sinh(2r)\right)^{-1/2}. \qquad (7.74)$$

Thus we have

$$C_n = \mathcal{N}[n!\cosh r]^{-1/2}\left[\frac{1}{2}e^{i\theta}\tanh r\right]^{n/2}H_n\left[\gamma\left(e^{i\theta}\sinh(2r)\right)^{-1/2}\right]. \qquad (7.75)$$

To find $\mathcal{N}$, we set $n = 0$ in Eq. (7.75) to obtain

$$C_0 = \mathcal{N}[\cosh r]^{-1/2}, \qquad (7.76)$$

and note that

$$
\begin{aligned}
C_0 = \langle 0|\alpha, \xi\rangle &= \langle 0|\hat{D}(\alpha)\hat{S}(\xi)|0\rangle \\
&= \langle 0|\hat{D}^\dagger(-\alpha)\hat{S}(\xi)|0\rangle \qquad (7.77) \\
&= \langle -\alpha|\xi\rangle,
\end{aligned}
$$

where the last line is the inner product of the coherent state $|-\alpha\rangle$ and the squeezed vacuum state $|\xi\rangle$. This product can be expressed as

$$\langle -\alpha|\xi\rangle = \exp\left(-\frac{1}{2}|\alpha|^2\right)\sum_{m=0}^{\infty}(\alpha^*)^{2m}[(2m)!]^{-1/2}C_{2m}, \qquad (7.78)$$

where the C_{2m} is given in Eq. (7.67). Thus from Eqs. (7.76)–(7.78) and using Eq. (7.67), we obtain

$$\mathcal{N} = (\cosh r)^{1/2} \langle -\alpha | \xi \rangle = \exp\left[-\frac{1}{2}|\alpha|^2 - \frac{1}{2}\alpha^{*2}e^{i\theta}\tanh r\right]$$

$$= \exp\left[-\frac{1}{2}|\gamma|^2 - \frac{1}{2}\gamma^2 e^{-i\theta}\tanh r\right]. \tag{7.79}$$

Thus, our squeezed state $|\alpha, \xi\rangle$ has the number state decomposition

$$|\alpha, \xi\rangle = \frac{1}{\sqrt{\cosh r}}\exp\left[-\frac{1}{2}|\gamma|^2 + \frac{1}{2}\gamma^2 e^{-i\theta}\tanh r\right]$$

$$\times \sum_{n=0}^{\infty}\frac{\left[\frac{1}{2}e^{i\theta}\tanh r\right]^{n/2}}{\sqrt{n!}}H_n\left[\gamma\left(e^{i\theta}\sinh(2r)\right)^{-1/2}\right]|n\rangle, \tag{7.80}$$

where we remind the reader that $\gamma = \alpha\cosh r + \alpha^* e^{i\theta}\sinh r$, which can be inverted to give $\alpha = \gamma\cosh r - \gamma^* e^{i\theta}\sinh r$. The probability of finding n photons in the field is given by

$$P_n = |\langle n|\alpha, \xi\rangle|^2$$

$$= \frac{(\tanh r)^n}{2^n n!\cosh r}\exp\left[-|\gamma|^2 - \frac{1}{2}(\gamma^2 e^{-i\theta} + \gamma^{*2}e^{i\theta})\tanh r\right] \tag{7.81}$$

$$\times \left|H_n\left[\gamma\left(e^{i\theta}\sinh(2r)\right)^{-1/2}\right]\right|^2.$$

Recall that the average photon number for this state given by Eq. (7.26). If $|\alpha|^2 \gg \sinh^2 r$, we can say that the "coherent" part of the state dominates the squeezed part. But through Eq. (7.81) it is evident that the distribution will depend on the phase of α. In Fig. 7.9 we show distributions for two values of $\psi - \theta/2$, where is the phase of α (i.e. $\alpha = |\alpha|e^{i\psi}$) along with, for the sake of comparison, the distribution for a coherent state of average photon number $|\alpha|^2$.

Note that the distribution for $\psi - \theta/2 = 0$ is narrower than for the coherent state. This is in fact a manifestation of another form of squeezing, sometimes called number squeezing, in which the photon number distribution is sub-Poissonian (i.e. narrower than the Poisson distribution for a coherent state). Like quadrature squeezing, it is a nonclassical effect. We shall return to this matter a bit later. In the case where $\psi - \theta/2 = \pi/2$, the distribution is broader than for a coherent state and is therefore called super-Poissonian. This is *not* a nonclassical effect. The case for which the squeezed part dominates (i.e. $\sinh^2 r > |\alpha|^2$) is pictured in Fig. 7.10 for $\psi - \theta/2 = 0$. Oscillations, some rather large scale, once again appear in the distribution. Oscillating photon number distributions have been interpreted by Schleich and Wheeler [8] as resulting from the interference of error contours in phase space. We will not pursue this interesting idea here, but we refer the reader to the book by Schleich [9] which covers this and all other matters of interest in quantum optics from the point of view of phase space.

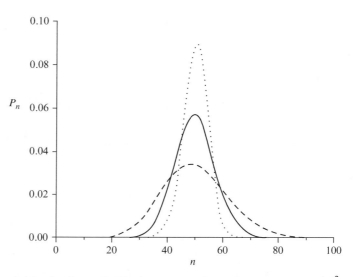

Fig. 7.9 Photon probability distributions (valid only at integers) for the coherent state with $|\alpha|^2 = 50$ (solid line) and the squeezed states for $|\alpha|^2 = 50$, $r = 0.5$, with $\psi - \theta/2 = 0$ (dotted line) and $\psi - \theta/2 = \pi/2$ (dashed line).

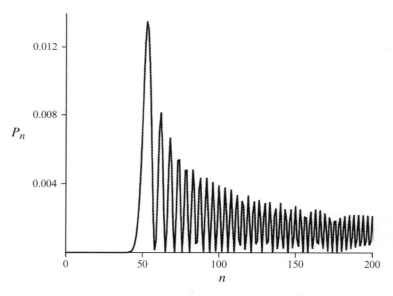

Fig. 7.10 Photon number probability distribution of a squeezed state with $\psi - \theta/2 = 0$ and with $|\alpha|^2 = 50$, $r = 4$.

Before leaving this section, let us calculate some quasi-probability distribution functions associated with the squeezed states. We already know, of course, that the P function for a squeezed state is problematic, at least if we wish to write down something as an ordinary function. Rather, it turns out that the P function for a squeezed state, and other nonclassical states, is highly singular – containing such things as derivatives of a delta function. But the Q and Wigner functions are always well behaved. Using the coherent state $|\beta\rangle$, the Q function for the squeezed state $|\alpha, \xi\rangle$ is given by

$$
\begin{aligned}
Q(\beta) &= \frac{1}{\pi} |\langle \beta | \alpha, \xi \rangle|^2 \\
&= \frac{1}{\pi \cosh r} \exp\{-(|\alpha|^2 + |\beta|^2) \\
&\quad -(\beta^* \alpha + \beta \alpha^*)/\cosh r \\
&\quad -\frac{1}{2}[e^{i\theta}(\beta^{*2} - \alpha^{*2}) + e^{-i\theta}(\beta^2 - \alpha^2)]\tanh r\}.
\end{aligned}
\tag{7.82}
$$

In Fig. 7.11 we plot Q as a function $x = \text{Re}(\beta)$ and $y = \text{Im}(\beta)$, for $\theta = 0$ and for (a) $\alpha = 0$ and (b) $\alpha = \sqrt{5}$. For this choice of θ, the squeezing is along the X_1 direction, the narrowing of the Gaussian profile of the Q function being along the X_1 direction with a corresponding expansion along the X_2 direction. For $\alpha = 0$, the peak of the Q function is centered at $\beta = 0$ apropos a squeezed vacuum state, whereas for $\alpha = \sqrt{5}$, the peak is simply translated to $\beta = \sqrt{5}$.

From the Q function we may obtain the anti-normally ordered characteristic function $\chi_A(\lambda)$ via the Fourier transform of Eq. (3.169). From Eq. (3.164) we obtain the Wigner characteristic function $\chi_A(\lambda)$ and, finally, from Eq. (3.173) we obtain the Wigner function

$$
W(\beta) = \frac{2}{\pi} \exp\left[-\frac{1}{2} X^2 e^{-2r} - \frac{1}{2} Y^2 e^{2r} \right],
\tag{7.83}
$$

specialized to the case with $\theta = 0$. Once again we have Gaussians, narrowed in the direction of squeezing and expanded in the orthogonal direction. Note that for the squeezed state the Wigner function is non-negative. In fact, it can be shown that squeezed states of the form $|\alpha, \xi\rangle$, including the special case of the coherent states ($\xi = 0$), are the only pure quantum states yielding non-negative Gaussian Wigner functions [10].

7.2 Generation of Quadrature Squeezed Light

Most schemes for the generation of quadrature squeezed light are based on some sort of parametric process utilizing various types of nonlinear optical devices. Generally, one desires an interaction Hamiltonian quadrature in the annihilation and creation operators of the field mode to be squeezed. We consider a device known as a degenerate parametric down-converter. A certain kind of nonlinear medium is pumped by a field of frequency ω_p

(a)

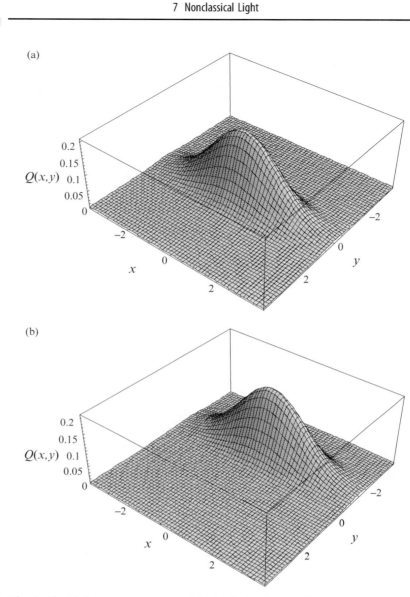

(b)

Fig. 7.11 Q function for (a) the squeezed vacuum and (b) the displaced squeezed vacuum.

and some photons of that field are converted into pairs of identical photons, of frequency $\omega = \omega_p/2$ each, into the "signal" field, the process being known as *degenerate parametric down conversion*. The Hamiltonian for this process is given by

$$\hat{H} = \hbar\omega\hat{a}^\dagger\hat{a} + \hbar\omega_P\hat{b}^\dagger\hat{b} + i\hbar\chi^{(2)}(\hat{a}^2\hat{b}^\dagger - \hat{a}^{\dagger 2}\hat{b}), \qquad (7.84)$$

where b is the pump mode and a is the signal mode. The object $\chi^{(2)}$ is a second-order nonlinear susceptibility. (For a discussion of nonlinear optics, the reader should consult, for example, the book by Boyd [11].) We now make the "parametric approximation," whereby we assume that the pump field is in a strong coherent classical field, which is strong enough to remain undepleted of photons over the relevant time scale. We assume that the field is in a coherent state $|\beta e^{-i\omega_p t}\rangle$ and approximate the operators $\hat{b}$ and $\hat{b}^\dagger$ by $\beta e^{-i\omega_p t}$ and $\beta^* e^{i\omega_p t}$, respectively. Dropping irrelevant constant terms, the parametric approximation to the Hamiltonian in Eq. (7.84) is

$$\hat{H}^{(PA)} = \hbar\omega\hat{a}^\dagger\hat{a} + i\hbar(\eta^*\hat{a}^2 e^{i\omega_p t} - \eta\hat{a}^{\dagger 2} e^{-i\omega_p t}), \tag{7.85}$$

where $\eta = \chi^{(2)}\beta$. Finally, transforming to the interaction picture, we obtain

$$\hat{H}_I(t) = i\hbar[\eta^*\hat{a}^2 e^{i(\omega_P - 2\omega)t} - \eta\hat{a}^{\dagger 2} e^{-i(\omega_P - 2\omega)t}], \tag{7.86}$$

which is generally time dependent. But if ω_P is chosen such that $\omega_p = 2\omega$, we arrive at the time-independent interaction Hamiltonian

$$\hat{H}_I = i\hbar(\eta^*\hat{a}^2 - \eta\hat{a}^{\dagger 2}). \tag{7.87}$$

The associated evolution operator

$$\hat{U}_I(t,0) = \exp(-i\hat{H}_I t/\hbar) = \exp(\eta^* t\hat{a}^2 - \eta t\hat{a}^{\dagger 2}) \tag{7.88}$$

obviously has the form of the squeeze operator of Eq. (7.7), $\hat{U}_I(t,0) = \hat{S}(\xi)$ for $\xi = 2\eta t$.

There is another nonlinear process that gives rise to squeezed light, namely *degenerate four-wave mixing*, in which two pump photons are converted into two signal photons of the same frequency. The fully quantized Hamiltonian for this process is

$$\hat{H} = \hbar\omega\hat{a}^\dagger\hat{a} + \hbar\omega\hat{b}^\dagger\hat{b} + i\hbar\chi^{(3)}(\hat{a}^2\hat{b}^{\dagger 2} - \hat{a}^{\dagger 2}\hat{b}^2), \tag{7.89}$$

where $\chi^{(3)}$ is a third-order nonlinear susceptibility. Going through similar arguments as above for the parametric down-converter, and once again assuming a strong classical pump field, we obtain the parametric approximation of Eq. (7.87), but this time with $\eta = \chi^{(3)}\beta^2$.

7.3 Detection of Quadrature Squeezed Light

It is, of course, not enough to generate squeezed light; one must be able to detect it. Several schemes for such detection have been proposed and implemented. The general idea behind all proposed methods is to mix the signal field, presumed to contain the squeezing, with a strong coherent field, called the "local oscillator." Here we shall consider only one method, known as *balanced homodyne detection*. A schematic of the method is shown in Fig. 7.12.

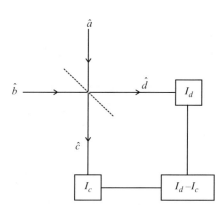

Schematic of the balanced homodyne method for the detection squeezing. The field to be detected enters along $\hat{a}$, while a strong coherent field is injected along $\hat{b}$. The boxes in the lower left and upper right represent photodetectors measuring the respective photocurrents. The box on the lower right represents a correlation device that subtracts the photocurrents.

Mode a contains the single field, which is possibly squeezed. Mode b contains a strong coherent classical field, which may be taken as a coherent state of amplitude β. The beam splitters are assumed to be 50:50 (hence the term *balanced* homodyne detection). Let us assume that the relation between the input $(\hat{a}, \hat{b})$ and output $(\hat{c}, \hat{d})$ operators is the same as in Eqs. (6.29):

$$\hat{c} = \frac{1}{\sqrt{2}}(\hat{a} + i\hat{b}),$$
$$\hat{d} = \frac{1}{\sqrt{2}}(\hat{b} + i\hat{a}). \tag{7.90}$$

The detectors placed in the output beams measure the intensities $I_c = \langle \hat{c}^\dagger \hat{c} \rangle$ and $I_d = \langle \hat{d}^\dagger \hat{d} \rangle$, and the difference in these intensities is

$$\begin{aligned} I_c - I_d &= \langle \hat{n}_{cd} \rangle = \langle \hat{c}^\dagger \hat{c} - \hat{d}^\dagger \hat{d} \rangle \\ &= i\langle \hat{a}^\dagger \hat{b} - \hat{a}\hat{b}^\dagger \rangle, \end{aligned} \tag{7.91}$$

where we have used Eq. (7.81) to obtain the last line and where we have set $\hat{n}_{cd} = \hat{c}^\dagger \hat{c} - \hat{d}^\dagger \hat{d}$. Assuming the b mode to be in the coherent state $|\beta e^{-i\omega t}\rangle$, where $\beta = |\beta|e^{-i\psi}$, we have

$$\langle \hat{n}_{cd} \rangle = |\beta|\{\hat{a}e^{i\omega t}e^{-i\theta} + \hat{a}^\dagger e^{-i\omega t}e^{i\theta}\}, \tag{7.92}$$

where $\theta = \psi + \pi/2$. Assuming that the a mode light is also of frequency ω (and in practice *both* the a and b mode light fields derive from the same laser), we can set $\hat{a} = \hat{a}_0 e^{-i\omega t}$, so that we may write

$$\langle \hat{n}_{cd} \rangle = 2|\beta|\langle \hat{X}(\theta) \rangle, \tag{7.93}$$

where

$$\hat{X}(\theta) = \frac{1}{2}\{\hat{a}_0 e^{-i\theta} + \hat{a}_0^\dagger e^{i\theta}\} \tag{7.94}$$

is the field quadrature operator at angle θ. By changing θ, which can be done by changing the phase ψ of the local oscillator, we can measure an arbitrary quadrature of the signal field. Of course, generally, θ will be chosen so as to achieve the maximum amount of quadrature squeezing. The variance of the output photon number difference operator, $\hat{n}_{cd}$, in the limit of a strong local oscillator is

$$\langle (\Delta \hat{n}_{cd})^2 \rangle = 4|\beta|^2 \langle (\Delta \hat{X}(\theta))^2 \rangle. \tag{7.95}$$

For the input squeezing condition $\langle (\Delta \hat{X}(\theta))^2 \rangle < \frac{1}{4}$, we have $\langle (\Delta \hat{n}_{cd})^2 \rangle < |\beta|^2$. In an actual experiment, the signal beam is first blocked in order to obtain the shot noise level. Furthermore, one has to account for the inefficiencies of the photodetectors. We ignore this concern here and refer the reader to the relevant literature.

The first experimental realization of squeezed light, by Slusher *et al.* [12], used a four-wave mixing interaction. A noise reduction of about 20% of the allowable below the shot noise level was achieved. A subsequent experiment by Wu *et al.* [13] achieved a reduction of 65% below the shot noise level. Their method used a parametric amplifier. Even greater noise reduction has been achieved [14].

7.4 Amplitude (or Number) Squeezed States

Recall from Chapter 2 the number-phase commutator $[\hat{n}, \hat{\varphi}] = i$ which, though technically unfounded, leads to the heuristically correct number-phase uncertainty relation $\Delta n \Delta \varphi \geq 1/2$ valid in the regime of large average photon number. For a coherent state $|\alpha\rangle$ for which $\Delta n = n^{-1/2}$ ($\bar{n} = |\alpha|^2$), it can be shown (see problem 2 in Chapter 3) that $\Delta \varphi = 1/(2\bar{n}^{1/2})$ if $\bar{n} \gg 1$, and thus the number-phase uncertainty relation is equalized. But it is possible to envision, in analogy to quadrature squeezing, states that are squeezed in number, where $\Delta n < \bar{n}^{1/2}$, or in phase, where $\Delta \varphi = 1/(2\bar{n}^{1/2})$. States that are squeezed in phase may be difficult to categorize as nonclassical, in part due to the lack of a Hermitian operator representing phase, as discussed in Chapter 2, and because it is not too clear what classical limit has to be overcome. After all, the phase fluctuations of a coherent state can be made arbitrarily small simply by increasing $\bar{n}$. But for the photon number it is a different story. We can write the variance as

$$\langle (\Delta \hat{n})^2 \rangle = \langle \bar{n} \rangle + (\langle \hat{a}^{\dagger 2} \hat{a}^2 \rangle - \langle \hat{a}^\dagger \hat{a} \rangle^2) \tag{7.96}$$

which, in terms of the P function, is

$$\langle(\Delta\hat{n})^2\rangle = \langle\hat{n}\rangle + \int d^2\alpha P(\alpha)[|\alpha|^2 - \langle\hat{a}^\dagger\hat{a}\rangle]^2, \qquad (7.97)$$

where

$$\langle\hat{a}^\dagger\hat{a}\rangle = \int d^2\alpha P(\alpha)|\alpha|^2. \qquad (7.98)$$

Obviously, the condition $\langle(\Delta\hat{n})^2\rangle < \langle\hat{n}\rangle$ for amplitude (or photon number) squeezing requires that $P(\alpha)$ take on negative values in some region of phase space. Hence amplitude squeezing is nonclassical. It is perhaps worth pointing out that the first term of Eq. (7.97) is sometimes referred to as the "particle-like" contribution, while the second term is the "wave-like" contribution. Indeed, for thermal light, where $\langle(\Delta\hat{n})^2\rangle = \langle\hat{n}\rangle + \langle\hat{n}\rangle^2$, or coherent light, where $\langle(\Delta\hat{n})^2\rangle = \langle\hat{n}\rangle$, such a separation holds up. But for amplitude squeezed light the interpretation of the second term as "wave-like" becomes a bit murky, as it too takes on properties of a distinctly quantum mechanical nature. A state exhibiting amplitude squeezing is said to possess sub-Poissonian statistics, the distribution being narrower than for a coherent state of the same average photon number. Not surprisingly, states with a photon number distribution broader than for a coherent state are said to have super-Poissonian statistics.

A simple way to gauge the nature of the photon statistics of any state is to calculate Mandel's Q-parameter

$$Q = \frac{\langle(\Delta\hat{n})^2\rangle - \langle\hat{n}\rangle}{\langle\hat{n}\rangle} \qquad (7.99)$$

(not to be confused with the Q function). For a state with Q in the range $-1 \le Q < 0$, the statistics are sub-Poissonian and if $Q > 0$, super-Poissonian. Obviously, $Q = 0$ for a coherent state.

In Fig. 7.13 we present a generic phase-space sketch of the amplitude squeezed state. For obvious reasons, such states are sometimes called "crescent" states. In the extreme case of a number state, as in Fig. 3.2, the "crescent" spreads out into a full circle and the Q-parameter goes to zero.

It is possible for some states of the field to exhibit simultaneously both quadrature and amplitude squeezing. Consider once again the states $|\alpha, \xi\rangle = \hat{D}(\alpha)\hat{S}(\xi)|0\rangle$. As we have already discussed, it can be shown that

$$\langle\hat{n}\rangle = |\alpha|^2 + \sinh^2 r. \qquad (7.100)$$

Also, it can be shown that

$$\langle(\Delta\hat{n})^2\rangle = |\alpha\cosh r - \alpha^* e^{i\theta}\sinh r|^2 + 2\sinh^2 r\cosh^2 r. \qquad (7.101)$$

Setting $\alpha = |\alpha|e^{i\psi}$ and choosing $\psi = \theta/2 - \pi/2$, we obtain

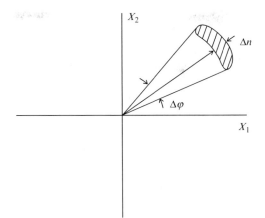

Fig. 7.13 The phase-space representation of a number squeezed state. Note that this particular state will also be quadrature squeezed.

$$\langle (\Delta \hat{n})^2 \rangle = |\alpha|^2 e^{2r} + 2\sinh^2 r \cosh^2 r, \tag{7.102}$$

which clearly does not exhibit amplitude squeezing. On the other hand, with the choice $\psi = \theta/2$, we obtain

$$\langle (\Delta \hat{n})^2 \rangle = |\alpha|^2 e^{-2r} + 2\sinh^2 r \cosh^2 r, \tag{7.103}$$

which for certain choices of $|\alpha|$ and r may exhibit number squeezing. For example, for $|\alpha|^2$ very large and r small, such that $\sinh r \approx 0$, $\langle \hat{n} \rangle \approx |\alpha|^2$ and $\langle (\Delta \hat{n})^2 \rangle \approx \langle \hat{n} \rangle e^{-2r}$, and thus $\langle (\hat{n})^2 \rangle < \langle \hat{n} \rangle$. The connection to quadrature squeezing can be made quite explicit by using Eq. (7.25) to write

$$\langle (\Delta \hat{n})^2 \rangle \approx 4 \langle \hat{n} \rangle \langle (\Delta \hat{Y}_1)^2 \rangle. \tag{7.104}$$

On the experimental side, Short and Mandel [15] were the first to observe, convincingly, sub-Poissonian photon statistics. This was in an experiment on single-atom resonance fluorescence.

7.5 Photon Antibunching

In Section 5.4 we discussed the second-order coherence function $g^{(2)}(\tau)$ of Eq. (5.42) characterizing the joint probability of detecting one photon followed by another within the delay time τ at some fixed position. If $g^{(2)}(\tau) = 1$ then the photons arrive independently, and, in fact, we should expect that $g^{(2)}(\tau) \to 1$ as $\tau \to \infty$ for any field state, the "memory" of the first detected photon being lost after a long enough time. For coherent states, $g^{(2)}(\tau) = 1$. For $g^{(2)}(\tau) < g^{(2)}(0)$, the probability of detecting a second photon after delay time τ *decreases* and this indicates a *bunching* of photons. For a thermal field, $g^{(2)}(0) = 2$. On the other hand, for $g^{(2)}(\tau) > g^{(2)}(0)$, the probability of detecting a second photon *increases* with the delay time. This is photon antibunching which, as we have previously stated, is a nonclassical effect. We can now explain why.

Let us first consider a single-mode field for which (as in Eq. (5.93))

$$
\begin{aligned}
g^{(2)}(\tau) = g^{(2)}(0) &= \frac{\langle \hat{a}^\dagger \hat{a}^\dagger \hat{a} \hat{a} \rangle}{\langle \hat{a}^\dagger \hat{a} \rangle^2} \\
&= 1 + \frac{\langle (\Delta \hat{n})^2 \rangle - \langle \hat{n} \rangle}{\langle \hat{n} \rangle^2}.
\end{aligned}
\tag{7.105}
$$

Strictly speaking, there cannot be any photon antibunching or bunching for a single-mode field, as $g^{(2)}(\tau)$ is independent of the delay time τ. Nevertheless, in terms of the P function we can write Eq. (7.105) as

$$
g^{(2)}(0) = 1 + \frac{\int P(\alpha) [|\alpha|^2 - \langle \hat{a}^\dagger \hat{a} \rangle]^2 d^2 \alpha}{\langle \hat{a}^\dagger \hat{a} \rangle^2},
\tag{7.106}
$$

where $\langle \hat{a}^\dagger \hat{a} \rangle$ is given by Eq. (7.98). For a classical field state where $P(\alpha) \geq 0$, we must have $g^{(2)}(0) \geq 1$. But for a nonclassical field state it is possible to have $g^{(2)}(0) < 1$, which, as previously stated, may be interpreted as a quantum mechanical violation of the Cauchy inequality. The alert reader will have noticed that the condition $g^{(2)}(0) < 1$ is the condition that the Q-parameter of Eq. (7.99) be negative, in other words, the condition for sub-Poissonian statistics. Indeed, Q and $g^{(2)}(0)$, for a single-mode field, are simply related:

$$
Q = \langle \hat{n} \rangle [g^{(2)}(0) - 1].
\tag{7.107}
$$

The fact that $Q < 0$ when $g^{(2)}(0) < 1$ has led to some confusion regarding the relationship of sub-Poissonian statistics and photon antibunching. Again, for a single-mode field, $g^{(2)}(\tau) = g^{(2)}(0) = \text{const.}$ and thus there can be no photon antibunching (or bunching for that matter). Bunching and antibunching occur only for multimode fields. For such fields in states with P function $P(\{\alpha_i\})$, where $\{\alpha_i\}$ denotes collectively the set of complex phase-space variables associated with each of the modes (the modes being distinguished by the labels i and j) it can easily be shown that

$$g^{(2)}(0) = 1 + \frac{\int P(\{\alpha_i\}) \left[\sum_j |\alpha_j|^2 - \langle \hat{a}_j^\dagger \hat{a}_j \rangle \right]^2 d^2\{\alpha_i\}}{\left(\sum_j \langle \hat{a}_j^\dagger \hat{a}_j \rangle \right)^2}, \qquad (7.108)$$

where $d^2\{\alpha_i\} = d^2\alpha_1 d^2\alpha_2 \dots$ and

$$\langle \hat{a}_j^\dagger \hat{a}_j \rangle = \int P(\{\alpha_i\}) |\alpha_j|^2 d^2\{\alpha_i\}. \qquad (7.109)$$

Again, for classical fields with $P(\{\alpha_i\}) \geq 0$, $g^{(2)}(0) \geq 1$. As shown in Appendix C, the Cauchy–Schwarz inequality applied to the corresponding classical coherence function $\gamma^{(2)}(\tau)$ implies that for classical fields it should always be the case that $g^{(2)}(\tau) \leq g^{(2)}(0)$, which does not allow for photon antibunching. The condition for antibunching, that $g^{(2)}(\tau) > g^{(2)}(0)$, is therefore an indication of nonclassical light, as is the condition $g^{(2)}(0) < 1$. Remembering that $g^{(2)}(\tau) \to 1$ for $\tau \to \infty$, it follows that the condition $g^{(2)}(0) < 1$ implies photon antibunching, *except* (and this is an important exception) in the case of a single-mode field, or if $g^{(2)}(\tau)$ is constant for some other reason. But the converse is not necessarily true. The condition for antibunching, $g^{(2)}(\tau) > g^{(2)}(0)$, does not imply sub-Poissonian statistics, $g^{(2)}(0) < 1$. Quite to the contrary, Zou and Mandel [16], in an article discussing the frequent confusion between the issues of sub-Poissonian statistics and photon antibunching, constructed a (somewhat artificial) two-mode state possessing simultaneously the properties of sub-Poissonian statistics *and* photon bunching.

In contrast to quadrature squeezing, sub-Poissonian statistics and photon antibunching are not phase dependent and are therefore related to the particle nature of light.

Photon antibunching was predicted in the resonance fluorescence from a two-level atom driven by a resonant laser field [17]. A full discussion of resonance fluorescence is beyond the scope of this book, but the results for the predicted second-order coherence function are easy to state and to interpret. With Ω the Rabi frequency of the driving field and γ the spontaneous emission decay rate, the stationary second-order correlation function $g^{(2)}(\tau)$ is given by

$$g^{(2)}(\tau) = [1 - \exp(-\gamma\tau/2)]^2, \quad \text{for} \quad \Omega \ll \gamma \qquad (7.110)$$

or

$$g^{(2)}(\tau) = 1 - \exp\left(-\frac{3\gamma}{4}\tau\right)\cos(\Omega\tau), \quad \text{for} \quad \Omega \gg \gamma. \qquad (7.111)$$

These functions, for which $g^{(2)}(0) = 0$ in both limits, are plotted in Fig. 7.14.

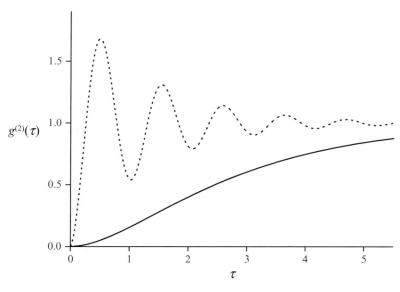

$g^{(2)}(\tau)$

Fig. 7.14 Second-order coherence function for resonance fluorescence for the cases where $\Omega \gg \gamma$ (dashed line) and $\Omega \gg \gamma$ (solid line).

The behaviors displayed in the figure may be interpreted as follows. Remember that $g^{(2)}(\tau)$ is essentially the joint probability of detecting a photon at time $t = 0$ and another photon at time $t = \tau$. But the source of the photons is a single atom driven by a coherent field. A detected fluorescent photon at $t = 0$ results, of course, from the electron "jump" from the excited to the ground state. But with the electron in the ground state, there is no possibility (i.e. the probability is zero) of emitting another photon at the time delay $\tau = 0$. There is a "dead time" before the laser field drives the electron back into the excited state with a finite probability, at which time another photon may be emitted. The correlation function $g^{(2)}(\tau)$ is, in fact, proportional to the probability that at time τ the atom will be in the excited state given that it was initially in the ground state.

Experiments by Mandel and collaborators [18] have confirmed these predictions. These experiments involved a low-intensity beam of sodium atoms irradiated at a right angle by a stabilized dye laser resonantly tuned to a particular hyperfine transition ($3P_{3/2}$, $F = 3$, $M_F = 2$ to $3^2S_{1/2}$, $F = 2$, $M_F = 2$). The beam intensity was so low that only one atom at a time occupied the region of observation, that region being observed at right angles to both the atomic and laser beams.

7.6 Schrödinger-Cat States

As another example of an important type of single-mode field state having strong nonclassical properties, we consider a class of states consisting of superpositions of two coherent states of equal amplitude but separated in phase by 180°, that is states of the form

$$|\psi\rangle = \mathcal{N}(|\alpha\rangle + e^{i\Phi}|-\alpha\rangle), \tag{7.112}$$

where the normalization factor $\mathcal{N}$ is given by

$$\mathcal{N} = [2 + 2\exp(-2\alpha^2)\cos\Phi]^{-1/2} \tag{7.113}$$

(we have taken α real). For large $|\alpha|$, the states $|\alpha\rangle$ and $|-\alpha\rangle$ are macroscopically distinguishable and superpositions of the form of Eq. (7.112) are frequently referred to as Schrödinger-cat states. Recall that the fate of Schrödinger's cat [19] was to end up in a superposition of macroscopically distinguishable states, the states of being alive and of being dead. It should be stressed that Schrödinger's intent was *not* to demonstrate how quantum weirdness could be elevated into the everyday classical world, but rather to satirize the Copenhagen interpretation of quantum mechanics, which he and many others (including Einstein) thought absurd. In spite of this, it has, in recent years, with the advancement of technology, become possible to legitimately consider laboratory realizations of superpositions of quantum states that are in some way macroscopically distinguishable. The driving force behind much of the activity is the attempt to address the question: where is the border between the quantum mechanical and classical worlds? Superposition states of the form of Eq. (7.112) are never seen in the macroscopic everyday world. Why not? Part of the answer seems to be as follows. No quantum system, particularly a macroscopic one, is truly an isolated system; it ultimately interacts with the rest of the universe, the "environment." The environment involves innumerable degrees of freedom which are not observed although, in some sense, the environment "observes" the system, effectively interacting with it in a dissipative fashion. In reality, of course, the entire universe is quantum mechanical and when a small part of it interacts with the rest, the two subsystems become entangled. As shown in Appendix A, tracing out the variables of the unobserved part of an entangled system leaves the system of interest in a mixed state. That's the general idea here. If a coherent macroscopic superposition state of the form of Eq. (7.112) can be created somehow, it should, upon interaction with the environment, quickly "decohere" into a statistical mixture of the form

$$\hat{\rho} = \frac{1}{2}(|\beta\rangle\langle\beta| + |-\beta\rangle\langle-\beta|) \tag{7.114}$$

where $\beta = \alpha e^{-\gamma t/2}$, and where here this γ is related to the rate of energy dissipation. Furthermore, it should be the case that the more macroscopic the components of initial superposition state (i.e. the greater $|\alpha|$), the more rapid should be the decoherence. We shall address these issues in more detail in Chapter 8. For the present, we concentrate on the properties of the cat states of Eq. (7.103), with a particular interest in those properties that are nonclassical. These are important as one must distinguish between superposition states and statistical mixtures, the latter having properties that are only classical.

There are three important cat states, depending on the choice of the relative phase ψ, that we shall consider. For $\Phi = 0$ we obtain the even coherent states

$$|\psi_e\rangle = \mathcal{N}_e(|\alpha\rangle + |-\alpha\rangle), \tag{7.115}$$

for $\Phi = \pi$ the odd coherent states

$$|\psi_o\rangle = \mathcal{N}_o(|\alpha\rangle - |-\alpha\rangle), \tag{7.116}$$

where

$$\mathcal{N}_e = \frac{1}{\sqrt{2}}[1 + \exp(-2\alpha^2)]^{-1/2}, \quad \mathcal{N}_o = \frac{1}{\sqrt{2}}[1 - \exp(-2\alpha^2)]^{-1/2} \tag{7.117}$$

are the respective normalization factors. These states were first introduced by Dodonov et al. [20]. For $\Phi = \pi/2$, we have the Yurke–Stoler states [21]

$$|\psi_{ys}\rangle \frac{1}{\sqrt{2}}(|\alpha\rangle + i|-\alpha\rangle). \tag{7.118}$$

All three states are eigenstates of the square of the annihilation operator, with α^2 as the eigenvalue:

$$\hat{a}^2|\psi\rangle = \alpha^2|\psi\rangle. \tag{7.119}$$

We must be able to distinguish between the cat states of Eqs. (7.115)–(7.119), which are coherent superpositions of $|\alpha\rangle$ and $|-\alpha\rangle$, and the statistical mixture

$$\hat{\rho}_{\text{mixture}} = \frac{1}{2}(|\alpha\rangle\langle\alpha| + |-\alpha\rangle\langle-\alpha|). \tag{7.120}$$

That is, as the corresponding density operator for a cat state (say the even coherent state) is

$$\hat{\rho}_e = |\psi_e\rangle\langle\psi_e| = |\mathcal{N}_e|^2(|\alpha\rangle\langle\alpha| + |-\alpha\rangle\langle-\alpha| + |-\alpha\rangle\langle\alpha| + |\alpha\rangle\langle-\alpha|), \tag{7.121}$$

we need to be able to detect the effects of the "coherence" terms $|\alpha\rangle\langle-\alpha|$ and $|-\alpha\rangle\langle\alpha|$ not present in the statistical mixture of Eq. (7.120)

Let's first consider the photon statistics. For $|\psi_e\rangle$ we obtain

$$P_n = \begin{cases} \dfrac{2\exp(-\alpha^2)}{1 + \exp(-2\alpha^2)} \dfrac{\alpha^{2n}}{n!} & n \text{ even}, \\ 0 & n \text{ odd} \end{cases} \tag{7.122}$$

and for the odd coherent state

$$P_n = \begin{cases} 0 & n \text{ even,} \\ \dfrac{2\exp(-\alpha^2)}{1+\exp(-2\alpha^2)}\dfrac{\alpha^{2n}}{n!} & n \text{ odd.} \end{cases} \tag{7.123}$$

For the Yurke–Stoler state, P_n is just a Poisson distribution, identical to that of a coherent state $|\alpha\rangle$ *and* of the statistical mixture of Eq. (7.110) – that is, $P_n = \langle n|\hat{\rho}_{\text{mixture}}|n\rangle$. At least for the even and odd coherent states, oscillating photon-number distributions distinguish those states from statistical mixtures. It is easy to see that the disappearance of the odd number states in $|\psi_e\rangle$, or the even in $|\psi_0\rangle$, is the result of quantum interference. Mandel's parameter for the even coherent states is

$$Q = \frac{4\alpha^2\exp(-\alpha^2)}{1+\exp(-4\alpha^2)} > 0, \tag{7.124}$$

indicating super-Poissonian statistics for all α and for the odd coherent states

$$Q = -\frac{4\alpha^2\exp(-\alpha^2)}{1+\exp(-4\alpha^2)} < 0, \tag{7.125}$$

indicating sub-Poissonian statistics, again for all α. As α becomes large, $Q \to 0$ for both the even and odd coherent states. For the Yurke–Stoler states, for all α, $Q = 0$.

Now we look at quadrature squeezing. With respect to the even coherent, we find that

$$\langle(\Delta\hat{X}_1)^2\rangle = \frac{1}{4} + \frac{\alpha^2}{1+\exp(-2\alpha^2)}, \tag{7.126}$$

$$\langle(\Delta\hat{X}_2)^2\rangle = \frac{1}{4} - \frac{\alpha^2\exp(-2\alpha^2)}{1+\exp(-2\alpha^2)}, \tag{7.127}$$

so reduced fluctuations appear in the $\hat{X}_2$ quadrature for α not too large. For the odd coherent states we have

$$\langle(\Delta\hat{X}_1)^2\rangle = \frac{1}{4} + \frac{\alpha^2}{1-\exp(-2\alpha^2)}, \tag{7.128}$$

$$\langle(\Delta\hat{X}_2)^2\rangle = \frac{1}{4} + \frac{\alpha^2\exp(-2\alpha^2)}{1-\exp(-2\alpha^2)}, \tag{7.129}$$

and thus no squeezing is evident. (Note the role reversal here between the even and odd coherent states regarding quadrature squeezing and sub-Poissonian statistics.) For the Yurke–Stoler state we find

$$\langle(\Delta\hat{X}_1)^2\rangle = \frac{1}{4} + \alpha^2, \tag{7.130}$$

$$\langle (\Delta \hat{X}_2)^2 \rangle = \frac{1}{4} - \alpha^2 \exp(-4\alpha^2), \tag{7.131}$$

so squeezing appears in $\hat{X}_2$. Finally, for the statistical mixture of Eq. (7.120) we find

$$\langle (\Delta \hat{X}_1)^2 \rangle = \frac{1}{4} + \alpha^2, \tag{7.132}$$

$$\langle (\Delta \hat{X}_2)^2 \rangle = \frac{1}{4}, \tag{7.133}$$

which, as expected, exhibits no squeezing.

Another possible way to distinguish between the coherent superpositions and the statistical mixture is to examine the phase distribution. Using the phase states $|\theta\rangle$ introduced in Chapter 2 we obtain, for the state of Eq. (7.112), in the limit of large α, the phase distribution (again with α real)

$$\mathcal{P}_{|\psi\rangle}(\theta) \approx \left(\frac{2\bar{n}}{\pi} \right)^{1/2} |\mathcal{N}|^2 \{ \exp[-2\bar{n}\theta^2]$$

$$+ \exp[-2\bar{n}(\theta - \pi)^2]$$

$$+ 2\cos(\bar{n}\pi - \Phi)\exp[-\bar{n}\theta^2 - \bar{n}(\theta - \pi)^2] \}, \tag{7.134}$$

where $\bar{n} = |\alpha|^2$. For the density operator of Eq. (7.120), we have

$$\mathcal{P}_\rho(\theta) \approx \frac{1}{2} \left(\frac{2\bar{n}}{\pi} \right)^{1/2} \{ \exp[-\bar{n}\theta^2] + \exp[-\bar{n}(\theta - \pi)^2] \}. \tag{7.135}$$

Both distributions have peaks at $\theta = 0$ and $\theta = \pi$, as expected, but only the first has an interference term. Unfortunately, that interference term is very small, the Gaussians in the product having very little overlap.

As for the quasi-probability distributions, the P function is highly singular, involving an infinite sum of higher-order derivatives of a delta function. The Q function is, of course, always positive and, as such, does not give a clear signal for nonclassicality. But the Wigner function, when it takes on negative values, does exhibit a clear signal for nonclassicality. Using some results from Chapter 3, we find the Wigner function for the statistical mixture to be

$$W_M(x, y) = \frac{1}{\pi} \{ \exp[-2(x - \alpha)^2 - 2y^2]$$

$$+ \exp[-2(x + \alpha)^2 - 2y^2] \}, \tag{7.136}$$

which is always positive and contains two Gaussian peaks centered at $x = \pm \alpha$, as shown in Fig. 7.15a. But the Wigner function for the even coherent state is

$$W_e(x,y) = \frac{1}{\pi[1 + \exp(-2\alpha^2)]}\{\exp[-2(x-\alpha)^2 - 2y^2]$$
$$+ \exp[-2(x+\alpha)^2 - 2y^2]$$
$$+ 2\exp[-2x^2 - 2y^2]\cos(4y\alpha)\}. \qquad (7.137)$$

The last term results from the interference between the two states $|\alpha\rangle$ and $|-\alpha\rangle$. It causes the Wigner function to become highly oscillatory, akin to exhibiting interference fringes, and to go negative, as is evident in Fig. 7.15b. The negativity does not allow the Wigner function in this case to be interpreted as a probability distribution. For the sake of completeness, we give the corresponding Wigner functions for the odd coherent state and the Yurke–Stoler state, respectively:

$$W_o(x,y) = \frac{1}{\pi[1 - \exp(-2\alpha^2)]}\{\exp[-2(x-\alpha)^2 - 2y^2]$$
$$+ \exp[-2(x+\alpha)^2 - 2y^2]$$
$$- 2\exp[-2x^2 - 2y^2]\cos(4y\alpha)\}, \qquad (7.138)$$

$$W_{ys}(x,y) = \frac{1}{\pi}\{\exp[-2(x-\alpha)^2 - 2y^2]$$
$$+ \exp[-2(x+\alpha)^2 - 2y^2]$$
$$- 2\exp[-2x^2 - 2y^2]\sin(4y\alpha)\}. \qquad (7.139)$$

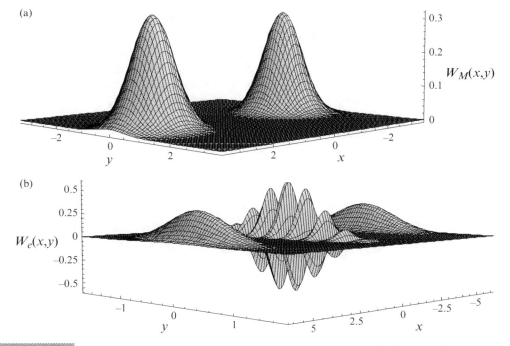

Fig. 7.15   Wigner function for (a) the statistical mixture of coherent states and (b) the even coherent state, both for $\alpha = \sqrt{5}$. It is evident that the former is always positive and that the latter becomes negative and is highly oscillatory (i.e. displays interference fringes) in some regions of phase space.

These functions are similar in character to that of the even coherent state.

Up until we discussed the cat states, most of the Wigner functions we have displayed have been Gaussian functions. The exception is that for the $n = 3$ photon-number state shown in Fig. 3.10. The number states are highly nonclassical, but so are the Schrödinger-cat states, so it should not be much of a surprise to see that the corresponding cat-state Wigner functions displayed in Eqs. (7.137)–(7.139) are non-Gaussian functions. States whose Wigner functions are non-Gaussian are referred to as non-Gaussian states. Operations to create non-Gaussian states are called non-Gaussian operations. Unitary operations such as squeezing with quadratic functions of annihilation and creation operators, as well as beam splitting, are called Gaussian operations. Unitary operators linear in the annihilation and creation operators are, of course, Gaussian operations as well.

How can the Schrödinger-cat states of Eqs. (7.115)–(7.118) be generated as a matter of principle (leaving aside for now the issue of decoherence)? We have earlier described how the squeezed states may be generated by parametric processes. The states are generated by unitary evolution, the evolution operator being a realization of the squeeze operator. If the initial state is a vacuum, the squeezed vacuum evolves, that state being populated by even photon numbers only. For the even coherent state of Eq. (7.115), again only even number states are populated but this time there does not seem to be a possible way to generate the states from the vacuum, or any other state, by unitary evolution. The same is true of the odd coherent state. Shortly, we shall describe a non-unitary method of generating such states, using projective state reduction. But the Yurke–Stoler state *can* be generated from the unitary evolution of a coherent state in a certain kind of nonlinear medium, namely a Kerr-like medium [21]. The interaction is a "self-Kerr" interaction (i.e. involving no parametric driving fields) and is modeled by the interaction Hamiltonian

$$\hat{H}_I = \hbar K (\hat{a}^\dagger \hat{a})^2 = \hbar K \hat{n}^2, \tag{7.140}$$

where K is proportional to a third-order nonlinear susceptibility $\chi^{(3)}$. An initial coherent state evolves into

$$
\begin{aligned}
|\psi(t)\rangle &= e^{-i\hat{H}_I t/\hbar}|\alpha\rangle \\
&= \exp(-|\alpha|^2/2) \sum_{n=0}^{\infty} \frac{\alpha^n}{\sqrt{n!}} \; e^{-iKn^2 t}|n\rangle.
\end{aligned}
\tag{7.141}
$$

As n^2 is an integer, the state $|\psi(t)\rangle$ is periodic with period $T = 2\pi/K$. If $t = \pi/K$, we have $e^{-iKn^2 t} = e^{-i\pi n^2} = (-1)^n$ and thus $|\psi(\pi/K)\rangle = |-\alpha\rangle$. But for $t = \pi/2K$,

$$
e^{-iKn^2 t} = e^{-i\pi n^2/2} = \begin{cases} 1 & n \text{ even}, \\ i & n \text{ odd}. \end{cases}
\tag{7.142}
$$

Thus we have

$$|\psi(\pi/2K)\rangle = \frac{1}{\sqrt{2}}e^{-i\pi/4}(|\alpha\rangle + i|-\alpha\rangle), \qquad (7.143)$$

which, apart from an overall irrelevant phase factor, is obviously the Yurke–Stoler state of Eq. (7.118). Though the operation involved here is unitary evolution, the Hamiltonian is quartic in the annihilation and creation operators.

A more realistic form of the Kerr interaction is given by the Hamiltonian [22]

$$\hat{H}_I = \hbar K \hat{a}^{\dagger 2}\hat{a}^2 = \hbar K(\hat{n}^2 - \hat{n}). \qquad (7.144)$$

Its action on a coherent state is as

$$|\psi(t)\rangle = e^{-i\,Kt(\hat{n}^2-\hat{n})}|\alpha\rangle = e^{-iKt\,\hat{n}^2}|\alpha e^{iKt}\rangle. \qquad (7.145)$$

For $t = \pi/2K$ we have

$$|\psi(\pi/2K)\rangle = \frac{1}{\sqrt{2}}e^{-i\,\pi/4}(|\beta\rangle + i|-\beta\rangle), \qquad (7.146)$$

where $\beta = i\alpha$, which is of the same form as Eq. (7.143), a Yurke–Stoler state apart from an irrelevant overall phase factor. The effect of the term linear in $\hat{n}$ is merely to produce a rotation of the amplitude of the coherent state.

We now present a non-unitary (non-Gaussian) method of generating the even and odd cat states, and the Yurke–Stoler states as well, for an optical field initially in a coherent state, as originally discussed by Gerry [23]. The central idea behind the approach we shall take is that of projective measurement where it is possible, in the case of entangled states, to project one of the subsystems of the entangled states into a *pure* state through a measurement on the other subsystem (see Appendix B).

In Fig. 7.16 we sketch the required experimental setup. Essentially, we have a Mach–Zehnder interferometer whose modes we label b and c as indicated, coupled to an external mode a through a nonlinear medium realizing a so-called "cross-Kerr" interaction given by the Hamiltonian

$$\hat{H}_{CK} = \hbar K \hat{a}^{\dagger}\hat{a}\hat{b}^{\dagger}\hat{b}, \qquad (7.147)$$

where K is proportional to a third-order nonlinear susceptibility $\chi^{(3)}$. The action of the medium in the interferometer will be given by the unitary evolution operator

$$\hat{U}_{CK} = \exp(-itK\hat{a}^{\dagger}\hat{a}\hat{b}^{\dagger}\hat{b}), \qquad (7.148)$$

where t is the interaction time $t = l/v$, with l the length of the medium and v the velocity of light in the medium. In the upper beam of the interferometer is placed a phase shift of θ, which is generated by the operator

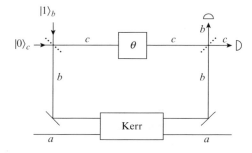

Fig. 7.16 Sketch of a proposed experiment for generating Schrödinger-cat states. A single photon is injected into the first beam splitter of the interferometer and an external mode, coupled to an internal mode through a cross-Kerr interaction, is initially in a coherent state. Projective state reduction occurs upon detection of a photon in either of the detectors.

$$\hat{U}_{\text{PS}}(\theta) = \exp(i\theta\hat{c}^{\dagger}\hat{c}). \tag{7.149}$$

The phase shift θ is to be adjusted. Beam splitters BS$_1$ and BS$_2$ we take to be of the type described by Eqs. (6.29). D$_1$ and D$_2$ are detectors placed in the output b and c beams. We assume that a coherent state $|\alpha\rangle$ is present in the a mode and that a single photon is injected into BS$_1$ as indicated in Fig. 7.16. That is, the input state at BS$_1$ is $|1\rangle_b|0\rangle_c \equiv |10\rangle_{bc}$. Just after BS$_1$ we have the state

$$|\text{out}\rangle_{\text{BS1}} = |\alpha\rangle_a \frac{1}{\sqrt{2}}(|10\rangle_{bc} + i|01\rangle_{bc}). \tag{7.150}$$

The actions of the phase shifter and the cross-Kerr medium yield the state

$$|Kt, \theta\rangle = \frac{1}{\sqrt{2}}(|e^{-iKt}\alpha\rangle_a|10\rangle_{bc} + ie^{i\theta}|\alpha\rangle_a|01\rangle_{bc}), \tag{7.151}$$

with the obvious notation $|0\rangle_b|1\rangle_c \equiv |01\rangle_{bc}$. The rotated coherent state of the first term results from the presence of the single photon in the b mode. In the second term, the photon is in the c mode and so picks up the phase shift θ. We now assume that $Kt = \pi$, so that we have

$$|\pi, \theta\rangle = \frac{1}{\sqrt{2}}(|-\alpha\rangle_a|10\rangle_{bc} + ie^{i\theta}|\alpha\rangle_a|01\rangle_{bc}). \tag{7.152}$$

Next, assuming a beam splitter of the type described by Eqs. (6.29), BS$_2$ causes the transformations

$$|10\rangle_{bc} \xrightarrow{\text{BS}_2} \frac{1}{\sqrt{2}}(|10\rangle_{bc} + i|01\rangle_{bc}),$$

$$|01\rangle_{bc} \xrightarrow{\text{BS}_2} \frac{1}{\sqrt{2}}(|01\rangle_{bc} + i|10\rangle_{bc}), \tag{7.153}$$

so that just after BS$_2$ we have the state

$$|\text{out}\rangle_{BS_2} = \frac{1}{2}[|-\alpha\rangle_a(|10\rangle_{bc} + i|01\rangle_{bc}) + ie^{i\theta}|\alpha\rangle_a(|01\rangle_{bc} + i|10\rangle_{bc})]$$

(7.154)

$$= \frac{1}{2}[(|-\alpha\rangle_a - e^{i\theta}|\alpha\rangle_a)|10\rangle_{bc} + i(|-\alpha\rangle_a + e^{i\theta}|\alpha\rangle_a)|01\rangle_{bc}].$$

We now choose the phase shift $\theta = \pi$ so that we have

$$|\text{out}\rangle_{BS_2} = \frac{1}{2}[(|\alpha\rangle_a + |-\alpha\rangle_a)|10\rangle_{bc} - i(|\alpha\rangle_a - |-\alpha\rangle_a)|01\rangle_{bc}]. \quad (7.155)$$

Now, if the detector D$_1$ clicks, indicating that a photon has emerged from the beam splitter in the b mode, the a mode is projected into the even coherent state $|\alpha\rangle_a + |-\alpha\rangle_a$ which upon normalization has the form of Eq. (7.115). But if detector D$_2$ clicks, indicating that a photon has emerged from the beam splitter in mode c, then the a mode is projected into the odd coherent state $|\alpha\rangle_a - |-\alpha\rangle_a$, given in normalized form by Eq. (7.116). These projective measurements bring about state reduction. State reduction is a discontinuous (and thus non-unitary) process, which we can use to great effect as a method for generating a variety of quantum states in entangled systems. Projective measurements are non-Gaussian operations. Note that had we chosen to perform state reduction *before* the second beam splitter [i.e. with the system in the state of Eq. (7.152)], we would not have been able to project onto the even and odd coherent states. The beam splitter effectively provides a means of scrambling the information on path, and thus provides the required interference to generate superpositions upon measurement. In contrast, a measurement on the state of Eq. (7.152) produces path information and hence there is no interference. Note that had we chosen $\theta = \pi/2$, the projective measurements would have yielded the states $|\alpha\rangle \pm i|-\alpha\rangle$, essentially Yurke–Stoler states.

For the scheme just discussed, cross-Kerr interactions of the form given by Eq. (7.147) are required to be large enough so that the condition $Kt = \pi$ holds. However, in readily available materials, the necessary $\chi^{(3)}$ are many orders of magnitude too small to maintain such a condition. For a long time it was hoped that large cross-Kerr interactions could be made possible by the techniques of electromagnetically induced transparency (EIT), but this has failed to be realized as far as we are aware.

In Section 7.10 we describe a method to produce approximations to traveling-wave low-amplitude cat-like states by the subtraction of photons from single-mode squeezed vacuum states. In Chapter 10 we describe a means of generating the cat states in the context of cavity QED.

7.7 Two-Mode Squeezed Vacuum States

So far in this chapter, we have considered nonclassical states of a single-mode field. However, for multimode fields, the prospects for nonclassical effects become even richer, as long as the field states are not merely product states of each of the modes. In other words, if we have *entangled* multimode field states we may expect, in general, strong nonclassical effects of greater richness than is possible for a single-mode field. In this section we shall consider the simple, and yet very important, example of the two-mode squeezed vacuum state.

We begin by introducing an analogy to the single-mode squeeze operator of Eq. (7.10), the two-mode squeeze operator

$$\hat{S}_2(\xi) = \exp(\xi^* \hat{a}\hat{b} - \xi \hat{a}^\dagger \hat{b}^\dagger), \tag{7.156}$$

where $\xi = re^{i\theta}$ as before and where $\hat{a}$ and $\hat{b}$ are the operators for the two modes. Of course, $[\hat{a}, \hat{b}^\dagger] = 0$, and so on. Note that $\hat{S}_2(\xi)$ does not factor into a product of the single-mode squeeze operators of the form of Eq. (7.10) for each mode. We now define the two-mode squeezed vacuum state by the action of $\hat{S}_2(\xi)$ on the two-mode vacuum $|0\rangle_a |0\rangle_b = |0, 0\rangle$:

$$\begin{aligned} |\xi\rangle_2 &= \hat{S}_2(\xi)|0, 0\rangle \\ &= \exp(\xi^* \hat{a}\hat{b} - \xi \hat{a}^\dagger \hat{b}^\dagger)|0, 0\rangle. \end{aligned} \tag{7.157}$$

Because $\hat{S}_2$ does not factor as a product of two single-mode squeeze operators, the two-mode squeezed vacuum state is *not* a product of two single-mode squeezed vacuum states, rather, as we shall later show, it is an entangled state containing strong correlations between the two modes. Notice that the two-mode squeeze operator contains terms that either create or annihilate photons pairwise from each mode.

It turns out that because of the correlations between the modes, the squeezing of quantum fluctuations is not in the individual modes but rather in a superposition of the two modes. We define the superposition quadrature operators as follows:

$$\begin{aligned} \hat{X}_1 &= \frac{1}{2^{3/2}}(\hat{a} + \hat{a}^\dagger + \hat{b} + \hat{b}^\dagger), \\ \hat{X}_2 &= \frac{1}{2^{3/2}} i(\hat{a} - \hat{a}^\dagger + \hat{b} - \hat{b}^\dagger). \end{aligned} \tag{7.158}$$

The operators satisfy the same commutation relations as for the single-mode case, $[\hat{X}_1, \hat{X}_2] = i/2$, and thus squeezing exists in the superposition quadratures if the conditions of Eq. (7.3) are satisfied. That such squeezing is nonclassical may be seen by writing the two-mode analogs to Eqs. (7.4) and (7.5) using a two-mode P function $P(\alpha, \beta) = P(\alpha)P(\beta)$. For squeezing to exist, $P(\alpha, \beta)$ must be nonpositive or singular in some regions of phase space. We omit the details here.

To investigate the squeezing properties of our state, we first use the Baker–Hausdorff lemma to obtain

$$\hat{S}_2^\dagger(\xi)\hat{a}\hat{S}_2(\xi) = \hat{a}\cosh r - e^{i\theta}\hat{b}^\dagger \sinh r,$$
$$\hat{S}_2^\dagger(\xi)\hat{b}\hat{S}_2(\xi) = \hat{b}\cosh r - e^{i\theta}\hat{a}^\dagger \sinh r. \tag{7.159}$$

Using these results we can show (we leave the details as an exercise) that for the two-photon squeezed vacuum state

$$\langle \hat{X}_1 \rangle = 0 = \langle \hat{X}_2 \rangle, \tag{7.160}$$

and thus the variances of the quadrature operators are

$$\langle (\Delta \hat{X}_1)^2 \rangle = \frac{1}{4}[\cosh^2 r + \sinh^2 r - 2\sinh r \cosh r \cos \theta],$$
$$\langle (\Delta \hat{X}_2)^2 \rangle = \frac{1}{4}[\cosh^2 r + \sinh^2 r + 2\sinh r \cosh r \cos \theta]. \tag{7.161}$$

For the choice $\theta = 0$ we obtain

$$\langle (\Delta \hat{X}_1)^2 \rangle = \frac{1}{4}e^{-2r}, \quad \langle (\Delta \hat{X}_2)^2 \rangle = \frac{1}{4}e^{2r}. \tag{7.162}$$

For the case $\theta = \pi$ the squeezing switches mode – that is we now have

$$\langle (\Delta \hat{X}_1)^2 \rangle = \frac{1}{4}e^{2r}, \quad \langle (\Delta \hat{X}_2)^2 \rangle = \frac{1}{4}e^{-2r}. \tag{7.163}$$

Note that these two choices of the phase θ minimize the uncertainty product – that is we have

$$\langle (\Delta \hat{X}_1)^2 \rangle \langle (\Delta \hat{X}_2)^2 \rangle = \frac{1}{16}. \tag{7.164}$$

On the other hand, note that for $\theta = \pi/2$ we have identical variances in both modes, these being

$$\langle (\Delta \hat{X}_1)^2 \rangle = \langle (\Delta \hat{X}_2)^2 \rangle = \frac{1}{4}(\cosh^2 r + \sinh^2 r), \tag{7.165}$$

where there is no squeezing in either mode, even though the corresponding state is colloquially referred to as a form of "two-mode squeezed vacuum state." Furthermore, the product of the uncertainties is not a minimum:

$$\langle (\Delta \hat{X}_1)^2 \rangle \langle (\Delta \hat{X}_2)^2 \rangle = \frac{1}{16}\cosh^2(2r). \tag{7.166}$$

We now seek a decomposition of our state $|\xi\rangle_2$ in terms of the two-mode number states $|n\rangle_a \otimes |m\rangle_b \equiv |n, m\rangle$. We follow our earlier procedure by starting with

$$\hat{a}|0,0\rangle = 0 \tag{7.167}$$

and using

$$\hat{S}_2(\xi)\hat{a}\hat{S}_2^\dagger(\xi) = \hat{a}\cosh r + e^{i\theta}\hat{b}^\dagger \sinh r, \tag{7.168}$$

to write

$$\hat{S}_2(\xi)\hat{a}\hat{S}_2^\dagger(\xi)\hat{S}_2(\xi)|0,0\rangle = (\mu\hat{a} + v\hat{b}^\dagger)|\xi\rangle_2 = 0, \tag{7.169}$$

where $\mu = \cosh r$ and $v = e^{i\theta}\sinh r$. Expanding $|\xi\rangle_2$ as

$$|\xi\rangle_2 = \sum_{n,m} C_{n,m}|n,m\rangle, \tag{7.170}$$

the eigenvalue problem becomes

$$\sum_{n,m} C_{n,m}\left[\mu\sqrt{n}|n-1,m\rangle + v\sqrt{m+1}|n,m+1\rangle\right] = 0. \tag{7.171}$$

Of the many *possible* solutions, we are only interested in the one containing the two-mode vacuum state $|0,0\rangle$. This solution has the form

$$C_{n,m} = C_{0,0}\left(-\frac{v}{\mu}\right)^n \delta_{n,m} = C_{0,0}(-1)^n e^{in\theta}\tanh^n r\,\delta_{n,m}, \tag{7.172}$$

where $C_{0,0}$ is determined from normalization to be $C_{0,0} = (\cosh r)^{-1}$. Thus the two-mode squeezed vacuum states, in terms of the number states, are

$$|\xi\rangle_2 = \frac{1}{\cosh r}\sum_{n=0}^{\infty}(-1)^n e^{in\theta}\tanh^n r|n,n\rangle. \tag{7.173}$$

Aside from the obvious fact that our states are entangled, perhaps their most striking feature is the evident strong correlations between the two modes: only the paired states $|n,n\rangle$ occur in the superposition of Eq. (7.173). This is one reason why such states are often called "twin beams." Another way to express this is to note that $|\xi\rangle_2$ is also an eigenstate of the number difference operator $\hat{n}_a - n_b$, where $\hat{n}_a = \hat{a}^\dagger\hat{a}$ and $\hat{n}_b = \hat{b}^\dagger\hat{b}$, with eigenvalue zero:

$$(\hat{n}_a - \hat{n}_b)|\xi\rangle_2 = 0. \tag{7.174}$$

Because of the correlations and the symmetry between the two modes, the average photon number in each mode is the same and can easily be shown to be

$$\langle\hat{n}_a\rangle = \langle\hat{n}_b\rangle = \sinh^2 r, \tag{7.175}$$

the same expression as for the single-mode squeezed vacuum [Eq. (7.91) with $\alpha = 0$]. The variances in the number operators are

$$\langle (\Delta \hat{n}_a)^2 \rangle = \langle (\Delta \hat{n}_b)^2 \rangle = \sinh^2 r \cosh^2 r = \frac{1}{4}\sinh^2(2r). \qquad (7.176)$$

Obviously $\langle (\Delta \hat{n}_i)^2 \rangle > \langle \hat{n}_i \rangle$, $i = a, b$, so both modes exhibit super-Poissonian photon statistics.

In order to quantify the quantum correlation between the two modes, we must examine operators acting on both systems. Here we take the combination photon number operators $\hat{n}_a \pm \hat{n}_b$ and examine the variances

$$\langle [\Delta(\hat{n}_a \pm \hat{n}_b)]^2 \rangle = \langle (\Delta \hat{n}_a)^2 \rangle + \langle (\Delta \hat{n}_b)^2 \rangle \pm 2\,\mathrm{cov}(\hat{n}_a, \hat{n}_b), \qquad (7.177)$$

where the covariance of the photon numbers is defined as

$$\mathrm{cov}(\hat{n}_a, \hat{n}_b) \equiv \langle \hat{n}_a \hat{n}_b \rangle - \langle \hat{n}_a \rangle \langle \hat{n}_b \rangle. \qquad (7.178)$$

For states containing *no* intermodal correlations, the expectation value of the product of the number operators factors as $\langle \hat{n}_a \hat{n}_b \rangle = \langle \hat{n}_a \rangle \langle \hat{n}_b \rangle$ and the covariance will be zero. But for the two-mode squeezed vacuum states, we know, by virtue of Eq. (7.174), that the variance of the number difference operator $\hat{n}_a - \hat{n}_b$ must be zero:

$$\langle [\Delta(\hat{n}_a - \hat{n}_b)]^2 \rangle = 0. \qquad (7.179)$$

It follows, from Eqs. (7.176) and (7.177), that

$$\mathrm{cov}(\hat{n}_a, \hat{n}_b) = \frac{1}{4}\sinh^2(2r). \qquad (7.180)$$

Thus, the linear correlation coefficient is defined as

$$J(\hat{n}_a, \hat{n}_b) = \frac{\mathrm{cov}(\hat{n}_a, \hat{n}_b)}{\langle (\Delta \hat{n}_a)^2 \rangle^{1/2} \langle (\Delta \hat{n}_b)^2 \rangle^{1/2}}, \qquad (7.181)$$

which, from Eqs. (7.176) and (7.180), takes the value

$$J(\hat{n}_a, \hat{n}_b) = 1, \qquad (7.182)$$

the maximal value indicating strong intermodal correlations.

From Eq. (7.173) we can obtain the joint probability of finding n_1 photons in mode a and n_2 photons in mode b, $P_{n_1 n_2} = |\langle n_1, n_2 | \xi \rangle_2|^2$, where

$$\langle n_1, n_2 | \xi \rangle_2 = \frac{1}{\cosh r} \sum_{n=0}^{\infty} (-1)^n \tanh^n r\, \delta_{n_1, n} \delta_{n_2, n}. \qquad (7.183)$$

If we plot $P_{n_1 n_2}$ versus n_1 and n_2, as in Fig. 7.17, it is clear that the joint probability is a monotonically decreasing function along the diagonal $n_1 = n_2$.

In order to examine properties of the individual modes, it is useful to introduce the density operator for the state $|\xi\rangle_2$:

$$\hat{\rho}_{ab} = |\xi\rangle_{2\,2}\langle \xi|. \qquad (7.184)$$

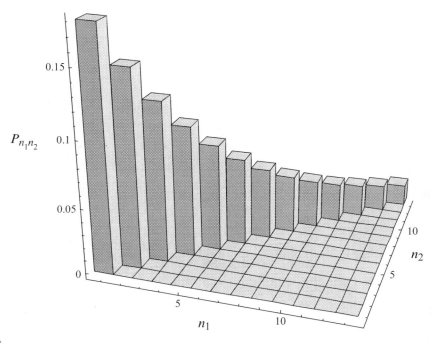

Fig. 7.17 $P_{n_1 n_2}$ versus n_1 and n_2 for the two-mode squeezed vacuum state for $r = 2$.

The selected density operators for the individual modes are easily found to be (see Appendix A)

$$
\hat{\rho}_a = \sum_{n=0}^{\infty} \frac{1}{(\cosh r)^2} (\tanh r)^{2n} |n\rangle_{a\, a}\langle n|,
$$

$$
\hat{\rho}_b = \sum_{n=0}^{\infty} \frac{1}{(\cosh r)^2} (\tanh r)^{2n} |n\rangle_{b\, b}\langle n|.
$$

(7.185)

In both cases, the probability of finding n photons in a single mode a or b is

$$
P_n^{(i)} =\, _i\langle n|\hat{\rho}_i|n\rangle_i = \frac{(\tanh r)^{2n}}{(\cosh r)^2} \quad (i = a, b).
$$

(7.186)

From Eq. (7.175) we can write

$$
P_n^{(i)} = \frac{\langle \hat{n}_i \rangle^n}{(1 + \langle \hat{n}_i \rangle^{n+1})},
$$

(7.187)

which is a thermal distribution with an average photon number $\langle \hat{n}_i \rangle = \sinh^2 r$. In other words, the radiation field in one of the modes obtained by disregarding (i.e. making no measurement on) the other mode of the two-mode squeezed vacuum is indistinguishable from the radiation from a thermal source [24]. The effective temperature of this "source" can be obtained from Eq. (2.142) by replacing $\bar{n}$ with $\langle \hat{n}_i \rangle$ to yield

$$T_{eff} = \frac{\hbar \omega_i}{2k_\beta \ln(\cosh r)}, \qquad (7.188)$$

where ω_i is the frequency of the mode.

Note that none of the quantities calculated in the previous three paragraphs depend on the phase angle θ. This means that the degree of entanglement between the two modes is independent of the existence or not of two-mode squeezing, as defined through the superposition quadratures given in Eq. (7.158). Also, because of the tight photon-number correlations between the modes, it is easy to see that there can be no squeezing in the quadratures of the individual modes.

The alert reader will have noted that the expansion in Eq. (7.173) is already of the form of a Schmidt decomposition, in this case containing an infinite number of terms. Furthermore, we already have the reduced density operators of both modes and they are identical in form. Thus, we can use the results in Appendix A to obtain the von Neumann entropy, a measure of the degree of entanglement, as

$$S(\hat{\rho}_a) = -\sum_{n=0}^{\infty} \left[\frac{\tanh^{2n} r}{\cosh^2 r} \right] \ln \left[\frac{\tanh^{2n} r}{\cosh^2 r} \right] = S(\hat{\rho}_b). \qquad (7.189)$$

In Fig. 7.18 we plot the von Neumann entropy against the squeeze parameter r. Evidently, the degree of entanglement increases with increasing r.

Our two-mode squeezed vacuum state, as for the single-mode squeezed vacuum state, may be generated by a parametrically driven nonlinear medium. The form of the Hamiltonian for this case, with all modes quantized, is

$$\hat{H} = \hbar \omega_a \hat{a}^\dagger \hat{a} + \hbar \omega_b \hat{b}^\dagger \hat{b} + \hbar \omega_p \hat{c}^\dagger \hat{c} + i\hbar \chi^{(2)} (\hat{a}\hat{b}\hat{c}^\dagger - \hat{a}^\dagger \hat{b}^\dagger \hat{c}), \qquad (7.190)$$

where the c mode contains the pump field. Again, we assume the pump field to be a strong, undepleted, coherent state $|\gamma e^{-i\omega_p t}\rangle$. The parametric approximation to Eq. (7.190) is then

$$\hat{H}^{(PA)} = \hbar \omega_a \hat{a}^\dagger \hat{a} + \hbar \omega_b \hat{b}^\dagger \hat{b} + i\hbar(\eta^* e^{i\omega_p t} \hat{a}\hat{b} - \eta e^{-i\omega_p t} \hat{a}^\dagger \hat{b}^\dagger), \qquad (7.191)$$

where $\eta = \chi^{(2)} \gamma$. Transforming to the interaction picture, we obtain the time-dependent interaction Hamiltonian

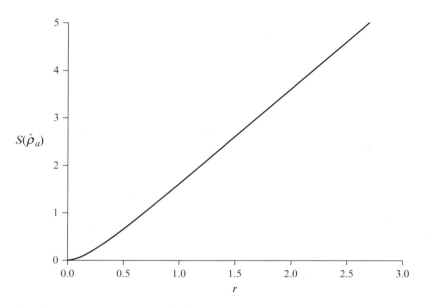

Fig. 7.18 A plot of the von Neumann entropy $S(\hat{\rho}_a)$ against the squeeze parameter r.

$$\hat{H}_I(t) = i\hbar\left[\eta^* e^{i(\omega_P - \omega_a - \omega_b)t}\hat{a}\hat{b} - \eta e^{-i\,(\omega_P - \omega_a - \omega_b)t}\hat{a}^\dagger\hat{b}^\dagger\right]. \qquad (7.192)$$

With the pump field frequency adjusted so that $\omega_P = \omega_a + \omega_b$, we obtain the time-independent interaction Hamiltonian

$$\hat{H}_I = i\hbar(\eta^*\hat{a}\hat{b} - \eta\hat{a}^\dagger\hat{b}^\dagger). \qquad (7.193)$$

This describes the process of *nondegenerate parametric down-conversion*, whereby a photon of the pump field is converted into two photons, one for each of the modes a and b. Obviously the associated evolution operator is

$$\hat{U}_I(t,0) = \exp[-i\hat{H}_I t/\hbar] = \hat{S}_2(\xi), \qquad (7.194)$$

where $\xi = \eta t$.

The states produced by the nondegenerate down-conversion process (i.e. the two-mode squeezed vacuum states) are entangled and, as such, they have been much discussed and have even been employed experimentally for fundamental testing of quantum mechanics, some aspects of which will be addressed in Chapter 9. They have also been studied in connection with various applications in quantum information processing, notably for purposes of quantum teleportation [25].

7.8 Broadband Squeezed Light

In our discussion of nonclassical light, we have, so far, confined our interest to a single-mode field or a double-mode field with widely separated frequencies. Of course, these are highly ideal situations, difficult to realize experimentally. In a more realistic approach, the presence of other modes of frequencies near the one of interest, or those in the bandwidth of the sensitivity of the detectors, must be taken into account. To give a sense of what is involved, we consider here (briefly) broadband squeezed light.

To keep things as uncluttered as possible, we assume that all the modes are identically polarized and thus treat the fields as scalars. We further assume that some frequency ω_0 is the mid-frequency of a quasi-monochromatic field. The positive frequency part of the field, from Eq. (2.126), is given by the collective annihilation operator

$$\hat{E}^{(+)}(\mathbf{r},t) = \sum_{[\mathbf{k}]} l(\omega_k)\hat{a}_\mathbf{k} e^{i(\mathbf{k}\cdot\mathbf{r}-\omega_k t)}, \tag{7.195}$$

where $l(\omega_k) = i(\hbar\omega_k/2\varepsilon_0 V)^{1/2}$ and where $[\mathbf{k}]$ represents the set of relevant plane-wave modes. Of course, the collective creation operator is just $\hat{E}^{(-)}(\mathbf{r},t) = [\hat{E}^{(+)}(\mathbf{r},t)]^\dagger$. These operators satisfy the equal-time commutation relation

$$\left[\hat{E}^{(+)}(t), \hat{E}^{(-)}(t)\right] = \sum_{[\mathbf{k}]} |l(\omega_k)|^2 \equiv C, \tag{7.196}$$

where we have suppressed the position vector for simplicity. We may now define two collective quadrature operators according to

$$\hat{X}_1^{(c)}(t) = \frac{1}{2}\left[\hat{E}^{(+)}(t) + \hat{E}^{(-)}(t)\right],$$
$$\hat{X}_2^{(c)}(t) = \frac{1}{2}\left[\hat{E}^{(+)}(t) - \hat{E}^{(-)}(t)\right], \tag{7.197}$$

satisfying

$$\left[\hat{X}_1^{(c)}, \hat{X}_2^{(c)}\right] = \frac{i}{2}C, \tag{7.198}$$

from which it follows that

$$\left\langle\left(\Delta\hat{X}_1^{(c)}\right)^2\right\rangle\left\langle\left(\Delta\hat{X}_2^{(c)}\right)^2\right\rangle \geq \frac{1}{16}|C|^2. \tag{7.199}$$

Broadband squeezing then exists if

$$\left\langle\left(\Delta\hat{X}_i^{(c)}\right)^2\right\rangle < \frac{1}{4}|C|, \; i = 1 \text{ or } 2. \tag{7.200}$$

7.9 Pair Coherent States

The two-mode squeezed vacuum state is a generalized coherent state in the sense that the displacement operator $\hat{D}(\alpha) = \exp(\alpha\hat{a}^\dagger - \alpha^*\hat{a})$ that acts on the vacuum state to produce the coherent state according to $|\alpha\rangle = \hat{D}(\alpha)|0\rangle$ is replaced by its two-mode analog, the squeeze operator $\hat{S}_2(\xi) = \exp(\xi^* \hat{a}\hat{b} - \xi \hat{a}^\dagger \hat{b}^\dagger)$ acting on the two-mode vacuum state $|0\rangle_a |0\rangle_b$, as in Eq. (7.157). Photons are created or destroyed pairwise in equal numbers in each mode. Recall that the two-mode squeezed vacuum state $|\xi\rangle_2$ automatically satisfies the condition $(\hat{a}^\dagger \hat{a} - \hat{b}^\dagger \hat{b})|\xi\rangle_2 = 0$. That is, the two-mode squeezing operator acting on the double-vacuum state yields states of the form $|\xi\rangle_2$, as given by Eq. (7.173) where there is strong pairwise correlation of photon numbers between the two modes.

On the other hand, let us consider the problem of finding right eigenstates $|\zeta\rangle_{ab}$ of the pair-annihilation operator $\hat{a}\hat{b}$ such that

$$\hat{a}\hat{b}|\zeta\rangle_{ab} = \zeta|\zeta\rangle_{ab}, \tag{7.201}$$

where ζ is a complex number, but where the supplementary condition

$$(\hat{a}^\dagger \hat{a} - \hat{b}^\dagger \hat{b})|\zeta\rangle_{ab} = q|\zeta\rangle_{ab} \tag{7.202}$$

must be satisfied and where, without loss of generality, q is taken to be a positive integer or zero. The most important case is that for which $q = 0$, so the states we seek here will satisfy

$$(\hat{a}^\dagger \hat{a} - \hat{b}^\dagger \hat{b})|\zeta\rangle_{ab} = 0. \tag{7.203}$$

One obvious solution to the eigenvalue problem of Eq. (7.201) is in terms of products of coherent states for each mode, that is $|\zeta\rangle_{ab} = |\alpha\rangle_a |\beta\rangle_b$ with $\zeta = \alpha\beta$. But this is a rather trivial result in which the states of the beam modes are not entangled. The supplementary condition of Eq. (7.202) ensures that there are correlations between the photon numbers in the two modes. The condition of Eq. (7.203) assures that $|\zeta\rangle_{ab}$ can be expanded in terms of twin number states according to

$$|\zeta\rangle_{ab} = \sum_{n=0}^{\infty} C_n |n\rangle_a |n\rangle_b. \tag{7.204}$$

With this, the eigenvalue problem in Eq. (7.201) can be solved to yield (when normalized)

$$|\zeta\rangle_{ab} = \frac{1}{\sqrt{I_0(2|\zeta|)}} \sum_{n=0}^{\infty} \frac{\zeta^n}{n!} |n\rangle_a |n\rangle_b, \tag{7.205}$$

where $I_0(2|\zeta|)$ is the modified Bessel function of order zero.

The pair coherent states were first studied in the context of quantum optics, and with arbitrary values of q, by Agarwal [26].

In addition to being eigenstates of the pair-annihilation operator subject to a constraint, the pair coherent states can also be understood to occur as a result of pairing projections from two single-mode coherent states where the pairing projection operator enforces the condition given by Eq. (7.203). This works as follows: we assume modes a and b contain coherent states $|\alpha\rangle_a$ and $|\beta\rangle_b$, respectively, where

$$|\alpha\rangle_a|\beta\rangle_b = \exp\left[-(|\alpha|^2 + |\beta|^2)/2\right]\sum_{n=0}^{\infty}\sum_{m=0}^{\infty}\frac{\alpha^n\beta^m}{\sqrt{n!m!}}|n\rangle_a|m\rangle_b. \qquad (7.206)$$

Our pairing projection operator, which is given by

$$\hat{P}_0 = \sum_{N=0}^{\infty}|N\rangle_a|N\rangle_b\,_a\langle N|_b\langle N|, \qquad (7.207)$$

has the action

$$\hat{P}_0|\alpha\rangle_a|\beta\rangle_b \propto \sum_{N=0}^{\infty}\frac{(\alpha\beta)^N}{N!}|N\rangle_a|N\rangle_b, \qquad (7.208)$$

which with $\zeta = \alpha\beta$ is the pair coherent state up to the normalization factor.

The pair coherent states have occasionally been referred to as "correlated" or "entangled" laser beams [27], though one must be careful not to confuse them with a state of the structure $|\alpha\rangle|\beta\rangle + |\delta\rangle|\gamma\rangle$, which is an entangled coherent state. The analysis of the above paragraph shows that the pair coherent states are correlated (or entangled) on the level of the photon-number states of each beam, as the result of a certain kind of projection on a product of coherent states.

The pair coherent states occasionally appear in the literature as "circle states" [28] as they may be written as

$$|\zeta\rangle = \mathcal{N}\int_0^{2\pi}|\alpha e^{i\theta}\rangle_a|\beta e^{-i\theta}\rangle_b d\theta, \qquad (7.209)$$

where again $\zeta = \alpha\beta$ and where $\mathcal{N}$ is a normalization factor. That the states defined by Eq. (7.209) are indeed pair coherent states is left as an exercise.

Of first interest is the joint photon-number statistics of the state. Because of the photon-number-state pairing in Eq. (7.205), we only need concern ourselves with the probabilities of finding the two modes to be in the states $|n_1\rangle_a|n_2\rangle_b$. This is given by $P_{n_1 n_2} = |\langle n_1, n_2|\zeta\rangle|^2$, where

$$\langle n_1, n_2|\zeta\rangle = \frac{1}{\sqrt{I_0(2|\zeta|)}}\sum_{n=0}^{\infty}\frac{\zeta^n}{n!}\delta_{n_1,n}\delta_{n_2,n}. \qquad (7.210)$$

In Fig. 7.19 we plot $P_{n_1 n_2}$ versus n_1 and n_2 for $\zeta = 3$, where we notice that the distribution has a very different character in comparison with the distribution for the two-mode squeezed vacuum states displayed in Fig. 7.17. Because the distribution in Eq. (7.210) goes as the inverse square of $n!$ rather

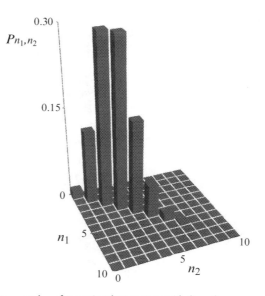

Fig. 7.19 Plot of $P_{n_1 n_2}$ versus n_1 and n_2 for a pair coherent state with $\zeta = 3$.

than as the inverse of it, one might suspect that the statistics revealed by the probabilities $P_{n_1 n_2}$ given here for the pair coherent states are sub-Poissonian. That turns out to be the case, as can be revealed by calculation of the Mandel Q-parameter for one of the modes, as we will now show.

The Mandel Q-parameter of the a mode is given by

$$Q_a = \frac{\langle \hat{a}^\dagger \hat{a} \hat{a}^\dagger \hat{a} \rangle - \langle \hat{a}^\dagger \hat{a} \rangle^2 - \langle \hat{a}^\dagger \hat{a} \rangle}{\langle \hat{a}^\dagger \hat{a} \rangle}. \tag{7.211}$$

Because of the symmetry between the modes of the pair coherent state, we have $\langle \hat{a}^\dagger \hat{a} \rangle = \langle \hat{b}^\dagger \hat{b} \rangle$ and so $Q_a = Q_b$. It is easily shown that

$$\begin{aligned} \langle \hat{a}^\dagger \hat{a} \rangle &= \mathcal{N}^2 \sum_{n=0}^{\infty} n \frac{|\zeta|^{2n}}{(n!)^2} = |\zeta|^2 \mathcal{N}^2 \sum_{n=0}^{\infty} \frac{|\zeta|^{2n}}{n!(n+1)!} \\ &= |\zeta| \frac{I_1(2|\zeta|)}{I_0(2|\zeta|)}, \end{aligned} \tag{7.212}$$

where $I_1(x)$ is a first-order modified Bessel function. To calculate $\langle \hat{a}^\dagger \hat{a} \hat{a}^\dagger \hat{a} \rangle$ we can write

$$\hat{a}^\dagger \hat{a} \hat{a}^\dagger \hat{a} = (\hat{a}^\dagger \hat{a} - \hat{b}^\dagger \hat{b}) \hat{a}^\dagger \hat{a} + \hat{a}^\dagger \hat{a} \hat{b}^\dagger \hat{b}, \tag{7.213}$$

so that we can take advantage of the fact that the pair coherent state is an eigenstate of $(\hat{a}^\dagger \hat{a} - \hat{b}^\dagger \hat{b})$ with eigenvalue zero, such that we obtain

$$\langle \hat{a}^\dagger \hat{a} \hat{a}^\dagger \hat{a} \rangle = \langle \hat{a}^\dagger \hat{a} \hat{b}^\dagger \hat{b} \rangle = \langle \hat{a}^\dagger \hat{b}^\dagger \hat{a} \hat{b} \rangle = |\zeta|^2. \tag{7.214}$$

Thus we have

$$Q_a = |\zeta| \left[\frac{I_0(2|\zeta|)}{I_1(2|\zeta|)} - \frac{I_1(2|\zeta|)}{I_0(2|\zeta|)} \right] - 1. \tag{7.215}$$

This quantity is in the range $-1 \le Q_a < 0$, which can be seen as follows. In the limit where $|\zeta|$ is small, we have the approximations $I_0(2|\zeta|) \sim 1$, $I_1(2|\zeta|) \sim |\zeta|$ such that

$$Q_a \sim |\zeta| \left[\frac{1}{|\zeta|} - \frac{|\zeta|}{1} \right] - 1 = -|\zeta|^2 \quad (\text{small } |\zeta|). \tag{7.216}$$

In the limit that $|\zeta|$ is large, we use the asymptotic forms of the Bessel functions given by [29]:

$$I_0(2|\zeta|) \sim \frac{e^{2|\zeta|}}{\sqrt{4\pi|\zeta|}} \sim I_1(2|\zeta|), \tag{7.217}$$

from which it follows that $Q_a \to -1$ in the limit of large $|\zeta|$. The sub-Poissonian statistics of the pair coherent states are in sharp contrast to the super-Poissonian statistics displayed by the two-mode squeezed vacuum states.

The pair coherent states are significantly more nonclassical than the two-mode squeezed vacuum states. Indeed, the pair coherent states are non-Gaussian states. They have other nonclassical properties as well, such as two-mode squeezing and violations of a Cauchy–Schwarz inequality. The exploration of these properties will be left to a problem at the end of this chapter.

As of this writing, the pair coherent states have yet to be generated in the laboratory. The principal difficulty is that being non-Gaussian states, they cannot be generated by unitary evolution. Agarwal [26] suggested they could be generated in the steady-state regime of a competing two-mode unitary process, such as the one that generates the two-mode squeezed vacuum state, and an incoherent non-unitary and dissipative process involving pair two-photon absorption. However, a discussion of the details of this generation scheme is beyond the scope of this book.

A method for generating the pair coherent states involving state reduction by projection out of two single-mode coherent states was proposed by Gerry et al. [30]. The method proposed is essentially a realization of the projection operation given in Eq. (7.208), where the pair coherent states are projected out of a product of ordinary coherent states. As mentioned below, states with properties similar to the pair coherent states have been generated out of a two-mode squeezed vacuum state by performing symmetric photon subtractions from each of the field modes.

7.10 Entanglement Generation via Beam Splitting

In Chapter 6 we noticed that a beam splitter could generate entangled states of light. For example, for a 50:50 beam splitter, the input state $|0\rangle_0|1\rangle_1$ results in the output state $(i|1\rangle_2|0\rangle_3 + |0\rangle_2|1\rangle_3)/\sqrt{2}$, and the input state $|1\rangle_0|1\rangle_1$ results in the output state $i(|2\rangle_2|0\rangle_3 + |0\rangle_2|2\rangle_3)/\sqrt{2}$. The input states in both cases are product states, but states containing a single photon are highly nonclassical. The vacuum state is, of course, a classical-like state of light, a coherent state of amplitude zero. The foregoing output states are entangled two-mode states of light. On the other hand, we know that coherent states mixed with the vacuum or another coherent state at a beam splitter lead to just products of coherent states in the outputs. Coherent states, as we have noted many times, are pure quantum states of a single-mode light beam that are as close as possible to classical coherent light. One can then perceive that it might be the case that nonclassicality for at least one of the input beams is a perquisite for the generation of entanglement by a beam splitter. This, in fact, is the case [31].

Several cases were studied in Ref. [31]. These include the input states $|0\rangle|N\rangle$, $|N\rangle|N\rangle$, and two squeezed vacuum states $|\xi_1\rangle|\xi_2\rangle$. Here, as a simple example involving a continuous variable non-Gaussian state, we consider an even Schrödinger-cat state mixing with the vacuum. That is, from Eq. (7.115) our input state is of the form $|0\rangle_0|\psi_e\rangle_1 = \mathcal{N}_e|0\rangle_0(|\alpha\rangle_1 + |-\alpha\rangle_1)$. Using Eq. (6.38) we easily find the output state to be

$$|\psi_{\text{out}}\rangle = \mathcal{N}_e(|i\beta\rangle_2|\beta\rangle_3 + |-i\beta\rangle_2|-\beta\rangle_3), \qquad (7.218)$$

where $\beta = \alpha/\sqrt{2}$. This is an entanglement of coherent states of the same amplitude but different phases. The reduced density operator for mode 2 after tracing out mode 3 is

$$\begin{aligned}
\hat{\rho}_2 &= \sum_{n=0}^{\infty} \langle n|\psi_{\text{out}}\rangle\langle\psi_{\text{out}}|n\rangle_3 \\
&= \mathcal{N}_e^2\{|i\beta\rangle_{22}\,\langle i\beta| + |-i\beta\rangle_{22}\,\langle -i\beta| \\
&\quad + |i\beta\rangle_{22}\,\langle -i\beta|_3\langle -\beta|\beta\rangle_3 + |i\beta\rangle_{22}\langle -i\beta|\langle\beta|-\beta\rangle_3\}.
\end{aligned} \qquad (7.219)$$

If $|\alpha|$ is large, and hence $|\beta|$ is large, then $\mathcal{N}_e^2 = 1/2$ and the inner products on the lower line of Eq. (7.219) go to zero. We are left with the balanced statistical mixture

$$\hat{\rho}_2 = \frac{1}{2}\{|i\beta\rangle_{22}\langle i\beta| + |-i\beta\rangle_{22}\langle -i\beta|\}. \qquad (7.220)$$

Upon taking the trace of the square of this operator, we obtain, under the same approximations, $\mathbf{Tr}\hat{\rho}_2^2 = 1/2$.

Another example is the mixing of one photon, a highly nonclassical state of light, with a coherent state, the most classical-like pure state of light, at a beam splitter – as discussed in Chapter 6 (see also problem 11 in Chapter 6).

That beam splitters are sources of entangled states of light for nonclassical input states is useful for directly generating entanglement, but as we will see below, when coupled with conditional measurements on one of the output modes of a beam splitter, the entanglement can be used to create non-Gaussian single-mode states of light by von Neumann projection.

7.11 Quantum State Engineering: Generation of Nonclassical States by Photon-Level Operations

Quantum state engineering is a collection of techniques by which one starts with readily available states of light, such as the coherent state, and performs one or another type of action on the state to modify it in some way. These actions are non-Gaussian as they modify the Gaussian states into non-Gaussian states which have strong nonclassical properties. An important class of non-Gaussian states are those generated by the action of annihilation or creation operators on Gaussian states, and an important subclass are the states generated by the action or multiple actions of the creation operator on coherent states. We start with photon-added coherent states.

Recall that the coherent state $|\alpha\rangle$ is a right eigenstate of the annihilation operator $\hat{a}$ such that $\hat{a}|\alpha\rangle = \alpha|\alpha\rangle$. On the other hand, there are no right eigenstates of the creation $\hat{a}^\dagger$ (see problem 1 in Chapter 3). But there is nothing to prevent us from operating on a coherent state with the creation operator – that is performing the operation $\hat{a}^\dagger|\alpha\rangle$, or performing multiple such operations $\hat{a}^{\dagger m}|\alpha\rangle$, where m is an integer and $m > 0$. The states obtained by these operations on coherent states were first studied by Agarwal and Tara [32].

The aforementioned operations constitute non-unitary "evolution" of the initial coherent state into photon-added coherent states which, when normalized, take the form

$$|\alpha, m\rangle = \frac{\hat{a}^{\dagger m}|\alpha\rangle}{\left(\langle\alpha|\hat{a}^m\hat{a}^{\dagger m}|\alpha\rangle\right)^{1/2}}, \tag{7.221}$$

where

$$\langle\alpha|\hat{a}^m\hat{a}^{\dagger m}|\alpha\rangle = \sum_{p=0}^{m}\frac{(m!)^2}{[(m-p)!]^2 p!}|\alpha|^{2(m-p)} = m!L_m(-|\alpha|^2), \tag{7.222}$$

and where $L_m(x)$ is the Laguerre polynomial of order m:

$$L_m(x) = m! \sum_{n=0}^{\infty} \frac{(-1)^n x^n}{(n!)^2 (m-n)!}. \qquad (7.223)$$

In terms of the photon-number states, the m-photon added coherent state can be written as

$$|a, m\rangle = \frac{\exp(-|\alpha|^2/2)}{[m! L_m(-|\alpha|^2)]^{1/2}} \sum_{n=0}^{\infty} \frac{\alpha^n \sqrt{(n+m)!}}{n!} |n+m\rangle. \qquad (7.224)$$

Note that the number states $|0\rangle, |1\rangle, \ldots, |m-1\rangle$ are absent from the superposition in this last equation. The general nonclassicality of the photon-added coherent states can be seen by a calculation of the corresponding P function, which, using the methods described in Chapter 3, is given by

$$P(\beta) = \frac{\exp(|\beta|^2 - |\alpha|^2)}{m! L_m(-|\alpha|^2)} \frac{\partial^{2m}}{\partial \beta^{*m} \partial \beta^m} \delta^{(2)}(\beta - \alpha). \qquad (7.225)$$

Note the similarity of this result to the P functions of number states given in Eq. (3.109). As in that case, the presence of derivatives of a delta function means we have a tempered distribution meaningful only under the integral sign, and an indication of the stronger nonclassical nature of the corresponding field states. These photon-added coherent states have been studied in detail by Agarwal and Tara [32], who showed explicitly that they possess strong nonclassical properties such as quadrature squeezing and sub-Poissonian photon statistics. In fact, the corresponding Wigner functions of these states take on negative values in phase space, and hence the states are non-Gaussian. Photon addition is a non-Gaussian operation. Note that adding even one photon to a coherent state $|\alpha\rangle$ causes a classical-to-quantum transition, no matter how large $|\alpha|$.

Agarwal and Tara [32] suggested that photon-added coherent states could be generated by nonlinear processes in atom–field interactions in cavities. However, for the case $m = 1$ they can also be generated in traveling-wave fields using parametric down-conversion and projective measurements. In fact, these one-photon-added coherent states in traveling-wave fields have been produced in the laboratory by the group of M. Bellini [33] using a scheme as sketched in Fig. 7.20. This procedure uses nondegenerate parametric down-conversion, with the state to which a photon is to be added seeding to the input signal mode, where the idler mode is initially in the vacuum state. That is, the signal and idler input state to the down-converter is $|\Psi(0)\rangle = |\psi\rangle_a |0\rangle_b$, where for the moment $|\psi\rangle_a$ is a pure but otherwise arbitrary state. We take the Hamiltonian for the parametric down-converter as given in Eq. (7.193) such that we have

$$\begin{aligned}|\Psi(t)\rangle &= \exp[-i\hat{H}_1 t/\hbar]|\Psi(0)\rangle \\ &= \exp[\eta(\hat{a}^\dagger \hat{b}^\dagger - \hat{a}\hat{b})]|\Psi(0)\rangle,\end{aligned} \tag{7.226}$$

where we have taken $\eta = \chi t$ to be real for convenience. For short times where $\eta \ll 1$, we have

$$\begin{aligned}|\Psi(t)\rangle &\simeq [1 + \eta(\hat{a}^\dagger \hat{b}^\dagger - \hat{a}\hat{b})]|\psi\rangle_a|0\rangle_b \\ &= |\psi\rangle_a|0\rangle_b + \eta(\hat{a}^\dagger|\psi\rangle_a)|1\rangle_b.\end{aligned} \tag{7.227}$$

Detection of one photon in mode b projects mode a into the one-photon-added state $\hat{a}^\dagger|\psi\rangle_a$, or $|\psi, 1\rangle_a = \hat{a}^\dagger|\psi\rangle_a/(_a\langle\psi|\hat{a}\hat{a}^\dagger|\psi\rangle_a)^{1/2}$ in normalized form. Obviously, if $|\psi\rangle_a = |\alpha\rangle_a$ this scheme will produce the one-photon-added coherent state $|\alpha, 1\rangle_a$. To verify that this state is generated, Bellini's group used the methods of quantum-state tomography (alluded to in Chapter 3) to reconstruct the corresponding Wigner function, which, unlike for the coherent state, has a region in phase space where it can take on negative values. They also detected the quadrature squeezing predicted for this state.

One can add more photons by repeating the above process, now with the input a-mode state $|\psi, 1\rangle_a$ to get $|\psi, 2\rangle_a$, and so on. One could arrange to have a sequence of down-converters and single-photon detections as sketched in Fig. 7.20 for the generation of $|\psi, 2\rangle_a$. Note that the production of this state is conditional on the detection of single photons in the idler beams of both down-converters. Generalization for the N-photon added coherent state

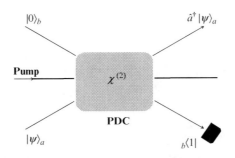

Fig. 7.20   Sketch of the method Zavatta *et al.* [33] used to generate a photon-added coherent state using nondegenerate down-conversion with a coherent state initially populating the signal a mode. The detection of a single photon in the idler b mode signals the projection of the state $|\alpha, 1\rangle_a$.

$$|\psi, N\rangle_a = \hat{a}^{\dagger N}|\psi\rangle_a / (_a\langle\psi|\hat{a}^N\hat{a}^{\dagger N}|\psi\rangle_a)^{1/2} \qquad (7.228)$$

is straightforward.

One can also subtract photons from a state. That is, one can apply the annihilation operator $\hat{a}$ a number of times N on an arbitrary single-mode state $|\psi\rangle$ to obtain the normalized state $|\psi, -N\rangle = \hat{a}^N|\psi\rangle / (\langle\psi|\hat{a}^{\dagger N}\hat{a}^N|\psi\rangle)^{1/2}$. Of course, in this case $|\psi\rangle$ must be other than a coherent state for a new state to be generated. A major difference between the states $|\psi, \pm N\rangle$ is that if the initial state $|\psi\rangle$ contains the low number states $|0\rangle, |1\rangle, \ldots$ so will $|\psi, -N\rangle$, whereas the photon-added state $|\psi, N\rangle$ will be missing the number states $|0\rangle, |1\rangle, \ldots, |N-1\rangle$.

Photon subtraction is much easier to perform in the laboratory than photon addition. Nonlinear devices are not needed to perform subtraction. The idea of subtracting photons from a single-mode field state using a beam splitter with a low reflectivity and photon detection in the reflected beam was due to Dakna *et al.* [34]. The beam splitter transformation for arbitrary transmissivity and reflectivity (assuming a 90° phase shift for the reflected beam) is described by the unitary operator

$$\hat{U}(\theta) = \exp\left[i\frac{\theta}{2}(\hat{a}^\dagger\hat{b} + \hat{b}^\dagger\hat{a})\right], \qquad (7.229)$$

where for simplicity we are labeling the beam modes a and b. For a 50:50 beam splitter, $\theta = \pi/2$ and we recover the operator given in Eq. (6.26). Here we are interested in the regime where the angle θ is small, which is the case for a highly transmitting beam splitter with low reflectance. If the input state is $|\psi\rangle_a|0\rangle_b$ then the output state is approximately

$$\begin{aligned}\hat{U}(\theta)|\psi\rangle_a|0\rangle_b &\approx \left[1 + i\frac{\theta}{2}(\hat{a}^\dagger\hat{b} + \hat{b}^\dagger\hat{a})\right]|\psi\rangle_a|0\rangle_b \\ &\approx |\psi\rangle_a|0\rangle_b + i\frac{\theta}{2}(\hat{a}|\psi\rangle_a)|1\rangle_b.\end{aligned} \qquad (7.230)$$

Placing a photon detector, such as an avalanche photodiode (APD) [35], in the reflected beam and detecting the reflected photon conditions the state in the transmitted beam to be in the one-photon subtracted state $\hat{a}|\psi\rangle_a$ (apart from normalization).

If one has available detectors that can resolve counts at the level of a single photon, such as the TES devices mentioned at the end of Chapter 6, one can use beam splitters of higher reflectivity to subtract a greater number of photons simultaneously. Expanding to second order in θ, we have

$$\hat{U}(\theta)|\psi\rangle_a|0\rangle_b \approx \left[1 + i\frac{\theta}{2}(\hat{a}^\dagger\hat{b} + \hat{b}^\dagger\hat{a}) - \frac{1}{2!}\left(\frac{\theta}{2}\right)^2(\hat{a}^\dagger\hat{b} + \hat{b}^\dagger\hat{a})^2 \right]|\psi\rangle_a|0\rangle_b$$

$$\approx |\psi\rangle_a|0\rangle_b + i\frac{\theta}{2}(\hat{a}|\psi\rangle_a)|1\rangle_b \qquad (7.231)$$

$$-\frac{1}{2!}\left(\frac{\theta}{2}\right)^2\left[(\hat{a}^\dagger\hat{a}|\psi\rangle_a)|0\rangle_b + \sqrt{2}(\hat{a}^2|\psi\rangle_a)|2\rangle_b\right],$$

where it is evident that the detection of two photons in mode b projects mode a into the two-photon subtracted state $\hat{a}^2|\psi\rangle_a$ (modulo normalization). Of course, the probability of detecting two photons is much lower than that of detecting one photon. One can continue in the above manner to subtract two, three, or more photons (probabilistically!) if one has available detectors with single-photon resolution. But it is important to keep in mind that neither photon addition nor photon subtraction is deterministic. Quantum state engineering via von Neumann state projection is non-unitary evolution and is probabilistic.

Obviously, photon subtraction has no effect on an input coherent state, such a state being a right eigenstate of the annihilation operator. On the other hand, suppose the input state is instead the single-mode squeezed vacuum state $|\xi\rangle$ given by Eq. (7.68). The state $|\xi\rangle$ is a superposition of only the even photon number states with a distribution that is otherwise thermal-like. Obviously, if we subtract one photon from it, the resulting state will contain only odd photon number states. But the nature of the distribution will change, too: it will be less thermal-like. If more photons are subtracted, the evenness or oddness of the resulting superpositions will depend on precisely the number of photons subtracted, but it also happens that the shape of the distribution becomes markedly altered, meaning that the statistical properties can be much altered from those of the initial state. This aspect of adding and subtracting photons has been discussed in detail by Barnett *et al.* [36].

Returning to the initial state at hand, the squeezed vacuum state: upon subtracting photons, the peak of the distribution shifts closer to the average photon number, which itself becomes higher as the number of photons subtracted is increased. That is, we have the seemingly paradoxical result that subtraction of photons from the squeezed vacuum state has the counter-intuitive effect that the state produced has an increased average photon number! The details we leave to the reader as a homework problem.

The increase in the average photon number via photon subtraction is, of course, not universal. If the photon-number distribution of the initial state is super-Poissonian, subtracting photons will increase the average photon number. The distribution for the squeezed vacuum is super-Poissonian. On the other hand, if the initial state has sub-Poissonian statistics, then subtraction of photons causes the average photon number to decrease. Of course, a coherent state is Poissonian and subtraction of photons leaves its average photon number unchanged. For a full explanation of these results, see the paper of Mizrahi and Dodonov [37] where the effects were first discussed.

Let us suppose that N photons are subtracted from a squeezed vacuum state $|\xi\rangle = \sum_{m=0}^{\infty} C_{2m}|2m\rangle$, where the coefficients C_{2m} are given by Eq. (7.67) and detected in the reflected beam, thus projecting the transmitted beam (approximately) into the state

$$|\xi, -N\rangle = \frac{\hat{a}^N|\xi\rangle}{\sqrt{\langle\xi|\hat{a}^{\dagger N}\hat{a}^N|\xi\rangle}}, \qquad (7.232)$$

where

$$\hat{a}^N|\xi\rangle = \sum_{m=0}^{\infty} C_{2m}\left[\frac{(2m)!}{(2m-N)!}\right]^{1/2}|2m-N\rangle, \qquad (7.233)$$

and we recall that $n! = \pm\infty$ for n a negative integer. The squeezed vacuum states are parity eigenstates, that is eigenstates of the photon-number parity operator $\hat{\Pi} = \exp(i\pi\hat{a}^{\dagger}\hat{a})$ with eigenvalue $+1$ for even parity: $\hat{\Pi}|\xi\rangle = |\xi\rangle$. It should be clear that the states $|\xi, -N\rangle$ will be even or odd-parity eigenstates according to $\hat{\Pi}|\xi, -N\rangle = (-1)^N|\xi, N\rangle$. It turns out that the states resulting from photon subtraction of the squeezed vacuum states are close to the odd and even Schrödinger-cat states $|\alpha\rangle \mp |-\alpha\rangle$, discussed earlier in this chapter, for N small and $|\alpha|$ small. The closeness of the obtained states to the even or odd cat states may be characterized by the *fidelity* which, for the comparison of two pure states $|\psi\rangle$ and $|\phi\rangle$, is given by $\mathcal{F} = |\langle\phi|\psi\rangle|^2$.

Experiments have been performed by Gerrits *et al.* [38] wherein one, two, and three photons have been subtracted from a squeezed vacuum state. In the case of one photon subtracted, that photon. was detected by an APD, whereas the subtracted two and three photons were detected by transition edge sensors (TES) [39], which are super-conductor devices that are capable of performing photon-number detection with a resolution at the level of a single photon. The range of photon numbers over which single-photon resolution is possible has increased to the point where, as of this writing, it is possible to count up to 100 photons with this level of resolution [40].

In the case of one-photon subtraction from the single-mode squeezed vacuum state, Biswas and Agarwal [41] showed that the resulting state is identical to a squeezed one-photon state [42]. Let the squeezed vacuum state now be denoted as $|\xi, 0\rangle = \hat{S}(\xi)|0\rangle = |\xi\rangle$, where again $\xi = re^{i\theta}$. The squeezed one-photon state is given by $|\xi, 1\rangle = \hat{S}(\xi)|1\rangle$. It follows from the unitarity of the squeezing operator that

$$\hat{a}|\xi, 0\rangle = \hat{S}(\xi)|0\rangle = \hat{S}(\xi)\hat{S}(\xi)^{\dagger}\hat{a}\hat{S}(\xi)|0\rangle, \qquad (7.234)$$

which, upon using the Baker–Hausdorff lemma, gives [repeating from Eqs. (7.12)]

$$\hat{S}(\xi)^{\dagger}\hat{a}\hat{S}(\xi) = \hat{a}\cosh r - \hat{a}^{\dagger}e^{-i\theta}\sinh r, \qquad (7.235)$$

resulting in

$$\hat{a}|\xi, 0\rangle = e^{i\theta}\sinh r\hat{S}(\xi)|1\rangle. \qquad (7.236)$$

It follows that

$$|\xi, 1\rangle = \hat{S}(\xi)|1\rangle = -\frac{e^{-i\theta}}{\sinh r}\hat{a}|\xi, 0\rangle, \qquad (7.237)$$

which apart from an overall factor is the one-photon subtracted squeezed vacuum.

We point out that there is extensive literature on squeezed number states [43], defined as states of the form $|\xi, n\rangle = \hat{S}(\xi)|n\rangle$, as well as displaced number states [44], defined as $|\alpha, n\rangle = \hat{D}(\alpha)|n\rangle$. Both types of states are non-Gaussian.

The technique of transforming a readily available Gaussian state into a non-Gaussian state by photon addition or subtraction is not restricted to single-mode fields. Consider, for example, the two-mode squeezed vacuum state of Eq. (7.173) which, for simplicity, we write as

$$|z\rangle = (1 - |z|^2)^{1/2}\sum_{n=0}^{\infty} z^n|n, n\rangle, \qquad (7.238)$$

where $z = -e^{i\theta}\tanh r$. Photons could be subtracted simultaneously (though not deterministically!) either asymmetrically or symmetrically from each mode. We consider here only symmetric subtraction. Subtracting l photons from each mode means performing the operation $\hat{a}^l\hat{b}^l|z\rangle$, which results in the normalized projected state [45]

$$|z, -l\rangle = \sum_{n=l}^{\infty} B_n^{(l)}|n - l, n - l\rangle, \qquad (7.239)$$

where

$$B_n^{(l)} = \mathcal{N}_{-l}\left(\frac{n!}{(n-l)!}\right)z^n, \quad \mathcal{N}_{-l} = \left[\sum_{n=0}^{\infty}\left(\frac{n!}{(n-l)!}\right)^2|z|^{2n}\right]^{-1/2}. \qquad (7.240)$$

We now set $N = n - l$ to write our state as

$$|z, -l\rangle = \sum_{N=0}^{\infty} D_N^{(l)} |N, N\rangle, \quad D_N^{(l)} = \mathcal{N}_{-l} \left(\frac{(N+l)!}{N!} \right) z^{N+l}. \tag{7.241}$$

The density operator of this two-mode field state is $\hat{\rho}_{ab} = |z, -l\rangle\langle z, -l|$ and the joint photon-number distribution is given by

$$P_{n,m} = \langle n, m | \hat{\rho}_{ab} | n, m \rangle = \left| \sum_{N=0}^{\infty} D_N^{(l)} \delta_{n,N} \delta_{m,N} \right|^2. \tag{7.242}$$

Because the pairing of the photon numbers of the two modes remains intact due to the symmetrical subtraction, to get a sense of the statistics of the fields it is sufficient to examine the marginal distribution of one of the modes. The reduced density of the a mode is given by

$$\hat{\rho}_a = \sum_{m=0}^{\infty} {}_b\langle m | \hat{\rho}_{ab} | m \rangle_b = \sum_{m=0}^{\infty} \left| D_m^{(l)} \right|^2 |m\rangle_{aa}\langle m|, \tag{7.243}$$

and the photon-number distribution for the a mode is given by

$$P_n = {}_a\langle n | \hat{\rho}_a | n \rangle_a = |D_n^{(l)}|^2. \tag{7.244}$$

Plots of this distribution are given in Fig. 7.21 for $l = 0, 1, 2, 3$ photons subtracted from each mode simultaneously and symmetrically, and for $|z| = 0.7$. Calculating the Mandel Q-parameter for one of the modes shows that the statistics can go from being super-Poissonian to Poissonian to sub-Poissonian as we go from $l = 0$ to $l = 3$ over a range of $|z|$ values, as can be seen in Fig. 7.22. As the two-mode squeezed vacuum state displays super-Poissonian photon statistics, we would expect the average photon number of the modified state to increase upon photon subtraction. This was predicted by Carranza and Gerry [45] and can be seen in Fig. 7.21 by noting the shift to the right of the peak of the marginal photon-number distribution with increasing numbers of photons subtracted.

The states described in the previous paragraph have been generated in the laboratory by Magaña-Loiza et al. [46], again using TES devices to perform photon counting with a resolution of one photon. They subtracted one, two, and three photons symmetrically from each beam of a two-mode squeezed vacuum state with a best-fit squeezing parameter of $r = 0.66$, corresponding to $z \approx 0.58$. They found the predicted increase in the average photon numbers, going from $\bar{n} = 0.7$ for $l = 0$ to $\bar{n} = 0.9$ for $l = 1$, $\bar{n} = 1.68$ for $l = 2$, and $\bar{n} = 2.05$ for $l = 3$. They also observed a change in the shape of the photon-number distribution for the individual modes (the marginal distributions) from thermal to essentially Poissonian after

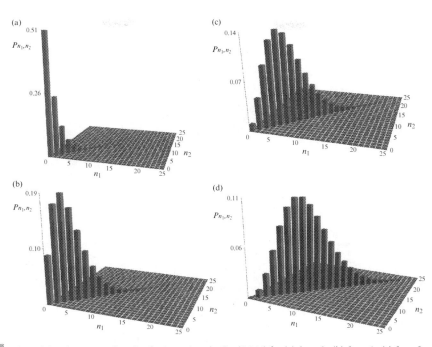

Fig. 7.21 Plots of the photon-number distributions given by Eq. (7.244) for (a) $l = 0$, (b) $l = 1$, (c) $l = 2$, (d) $l = 3$, with $|z| = 0.7$.

subtracting three photons. Symmetrically subtracting enough photons from the two-mode squeezed vacuum will render it a sub-Poissonian correlated state whose character begins to approach that of the pair coherent state discussed above.

The above discussion does not exhaust the possible approaches to quantum state engineering by manipulations at the level of a few photons. For example, there is the technique known as quantum-optical catalysis [47] or photon catalysis [48], wherein, in its original presentation, a single photon is mixed with a coherent state at a beam splitter, generally not a 50:50 beam splitter, and where a single photon is detected in one of the output beams, thus heralding the generation of a nonclassical single-mode state in the other output beam. More generally, the procedure can be extended to, say, k photons into the beam splitter, heralding the detection of, say, l photons in one of the output beams [49]. Photon catalysis is really an extension of photon subtraction, but where a Fock state other than the vacuum state is mixed with a continuous-variable state at a beam splitter which need not be of low reflectance. In fact, the beam splitter generally would not be 50:50, thus allowing for a greater variety of non-Gaussian states to be catalyzed. For details, we refer the reader to the literature cited [47–50].

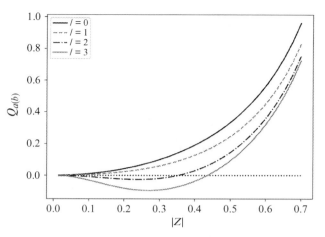

Fig. 7.22 Plots of the Mandel Q-parameter for (a) $l = 0$, (b) $l = 1$, (c) $l = 2$, (d) $l = 3$, over a range of $|z|$ values.

Problems

1. Investigate the time evolution of the wave packets associated with the squeezed vacuum state and of the more general squeezed state. Graphically display the wave packets (you should consider the probability distribution over the harmonic oscillator potential) at different times. Show that the squeezed vacuum state is a "breather," that is, its centroid is stationary but its width oscillates at twice the frequency of the harmonic oscillator potential.

2. Carry out the steps to obtain the squeezed vacuum Wigner function of Eq. (7.83).

3. Obtain the Wigner function for the displaced squeezed vacuum.

4. We defined a general squeezed state in Eq. (7.22) as a displaced squeezed vacuum state, that is the state generated by the operator product $\hat{D}(\alpha)\hat{S}(\xi)$ acting on the vacuum. Consider now the state generated by the reversed operator product, that is $\hat{S}(\xi)\hat{D}(\alpha)$, acting on the vacuum, the squeezed coherent state. Find the appropriate eigenvalue problem and solve it via a recursion relation. Investigate the properties of the state: photon statistics, quadrature and number squeezing. How is this state related to the displaced squeezed vacuum?

5. For the one-photon-added coherent state, obtain both the Q and Wigner functions and plot them as functions or the relevant phase-space variables.

6. Investigate (numerically!) the question of whether or not squeezing (both in quadrature and in number) occurs in the dynamics of the Jaynes–Cummings model. Assume the atom to be initially in the excited state and the field initially in a coherent state. Examine the effects of increasing the average photon number from as low as $\bar{n} = 5$ to as high as $\bar{n} = 100$.

7. Study the dynamics of the Jaynes–Cummings model where the initial coherent state of the field is replaced by the squeezed state of Eq. (7.80). Check the limiting cases where $r \rightarrow 0$ (the coherent state) and where $\alpha \rightarrow 0$ [the squeezed vacuum state of Eq. (7.68)].

8. Consider the Kerr interaction as described by Eq. (7.144).
    (a) Obtain the Heisenberg equation for the annihilation operator and solve it.
    (b) Show that for an initial coherent state, the photon-counting statistics remain Poissonian for all times. The state obtained for arbitrary times is often called a Kerr state.
    (c) Examine the possible occurrence of quadrature squeezing. We know that for times $t = \pi/2K$, the condition for generating Yurke–Stoler states, squeezing will be present in $\hat{X}_2$ as per Eq. (7.131), but what about at other, shorter, times? Be sure to examine both quadratures.
    (d) Obtain and graphically display the Wigner function of the Kerr state for times other than $t = \pi/2K$. Is the Kerr state a Gaussian state?

9. Consider the so-called "real" and "imaginary" cat states consisting of the superpositions of the forms $|\alpha\rangle \pm |\alpha^*\rangle$. Assuming α is complex, normalize the superposition and then investigate the states for quadrature and number squeezing. Obtain the Wigner function for these states.

10. Consider the set of operators defined as $\hat{K}_1 = \frac{1}{2}(\hat{a}^{\dagger 2} + \hat{a}^2)$, $\hat{K}_2 = \frac{1}{2i}(\hat{a}^{\dagger 2} - \hat{a}^2)$, and $\hat{K}_3 = \frac{1}{2}(\hat{a}^{\dagger}\hat{a} + \frac{1}{2})$. The first two are sometimes referred to as the quadratures of the square of the field [51]. (a) Show that these operators form a closed set under commutation; that is, the commutator of any two yields the third. This, by the way, is a realization of the SU(1, 1) Lie algebra [52]. (b) Obtain the uncertainty product for the operator $\hat{K}_1$ and $\hat{K}_2$. (c) Use the coherent states $|\alpha\rangle$ to see if they equalize this product. (d) Examine the question of squared-field squeezing as a form of higher-order squeezing: is it a nonclassical effect? (e) If this is a nonclassical effect, check to see if the Schrödinger-cat states, either the even, odd, or Yurke–Stoler types, are squeezed in this sense.

11. Consider the case of mixing of a Schrödinger-cat state of the form given by Eq. (7.112) at a 50:50 beam splitter with a coherent state $|\beta\rangle$,

where β is arbitrary. Show that by a proper choice of β, one can arrange the joint photon-number probability distribution after beam splitting, $P_{n,m}$, to have the property that $P_{n,m} = 0$ except along the lines $n = 0$ and $m = 0$.

12.　Consider the mixing of a one-photon-added coherent state $|\alpha\rangle$ with a vacuum state at a beam splitter. (a) Determine the output state of the beam splitter and graphically display the joint photon-number distribution for the case where $\alpha = 3$. (b) Determine the degree of entanglement obtained by calculating the linear entropy of one of the output field modes as a function of $|\alpha|$.

13.　The displaced number states are defined according to $|\alpha, n\rangle = \hat{D}(\alpha)|n\rangle$. Use the result given in Eq. (3.48) to obtain this state as an expansion in the number basis, and find the photon-number distribution for cases where $n = 1, 2, 3$, for $\alpha = 3$. Calculate the Wigner function for the state where $\alpha = 3$ and $n = 1$ and comment on the nonclassicality of this state.

14.　Consider the following operators composed of two sets of Bose operators $(\hat{a}, \hat{a}^\dagger)$ and $(\hat{b}, \hat{b}^\dagger)$: $\hat{V}_1 = (\hat{a}\hat{b} + \hat{a}^\dagger\hat{b}^\dagger)/2$, $\hat{V}_2 = i(\hat{a}\hat{b} - \hat{a}^\dagger\hat{b}^\dagger)/2$, and $\hat{V}_3 = (\hat{a}^\dagger\hat{a} + \hat{b}^\dagger\hat{b})/2$. (a) Show that these operators satisfy the same commutation relations as the operators introduced in problem 10. The corresponding reduction in the quantum noise of either $\hat{V}_1$ or $\hat{V}_2$ is called "sum squeezing" [52]. (b) Examine the two-mode squeezed vacuum states for sum squeezing; (c) do the same for the pair coherent states.

15.　Suppose that the signal and idler beams of a two-mode squeezed vacuum state are mixed at a 50:50 beam splitter. (a) Find the joint photon-number distribution of the output beams. (b) Show that there is no entanglement in the output beams. Show that the output consists of a product of two single-mode squeezed vacuum states (i.e. show that each output beam is in a single-mode squeezed vacuum state).

16.　Show that the "circle states" of Eq. (7.209) are, in fact, the pair coherent states.

17.　Consider the inequality $\langle \hat{a}^{\dagger 2}\hat{a}^2\rangle\langle \hat{b}^{\dagger 2}\hat{b}^2\rangle \geq \langle \hat{a}^\dagger\hat{a}\hat{b}^\dagger\hat{b}\rangle^2$. This is a form of the Cauchy–Schwarz inequality [53]. For a product of coherent states, it is easy to see that *equality* is satisfied. Violation of this inequality indicates a strong nonclassical correlation. Do the two-mode squeezed vacuum states violate inequality? What about the pair coherent states?

18.　Obtain the von Neumann entropy in the case of pair coherent states.

19.　Consider a Mach–Zehnder interferometer, as described in Chapter 6, and assume that one input mode is in a coherent state and the other is in a squeezed vacuum state. Show that the phase fluctuations at the output are reduced below the standard quantum limit ($\Delta\varphi_{SQL} = 1/\sqrt{\bar{n}}$) to $\Delta\varphi = e^{-r}/\sqrt{\bar{n}}$.

20. Reconsider the previous problem but with the squeezed vacuum replaced by even or odd coherent states. What effect, if any, do these states have on the reduction of noise at the output of the interferometer?

21. Consider the Hamiltonian

$$\hat{H} = \hbar\omega_P \hat{a}^\dagger \hat{a} + \hbar\omega_b \hat{b}^\dagger \hat{b} + \hbar\omega_c \hat{c}^\dagger \hat{c} + i\hbar\chi^{(2)}(\hat{a}\hat{b}\hat{c}^\dagger - \hat{a}^\dagger \hat{b}^\dagger \hat{c}),$$

which is similar to that of Eq. (7.190) but where we take the a mode as the pump. In this situation the nonlinear medium will perform *frequency conversion*.

(a) Make the parametric approximation and show that, under the condition $\omega_P = \omega_b - \omega_c$, the interaction picture Hamiltonian has the form

$$\hat{H}_I = -i\hbar(\eta\hat{b}^\dagger \hat{c} - \eta^*\hat{b}\hat{c}^\dagger).$$

(b) Using the interaction picture evolution operator $\hat{U}_I(t) = \exp(-i\hat{H}_I t/\hbar) = \exp[t(\eta\hat{b}^\dagger \hat{c} - \eta^*\hat{b}\hat{c}^\dagger)]$, obtain expressions for the $\hat{b}$ and $\hat{c}$ operators at times $t > 0$.

(c) It should be clear that under this interaction, a vacuum state (i.e. $|0\rangle_b|0\rangle_c$) remains a vacuum state. But what about a state of the form $|0\rangle_b|N\rangle_c$? (Hint: Start with the relation $\hat{b}|0\rangle_b|N\rangle_c = 0$ and obtain an eigenvalue problem as we did for nondegenerate down-conversion.) Show that the joint photon-number distribution is a binomial distribution of the N photons over the two modes. Is this an entangled state? Are there intermodal correlations for this state?

(d) Consider the initial state to be $|0\rangle_b|\alpha\rangle_c$. Are there intermodal correlations for the state at times $t > 0$?

22. Work out a general formula for photon-added squeezed vacuum states. For the case of one photon added, calculate the Fock-state expansion coefficients and determine the conditions under which the state approximates an odd Schrödinger-cat state.

23. Consider a one-photon-added thermal state. Show that such a state does not exhibit squeezing or sub-Poissonian statistics. Then show that it is, nevertheless, a nonclassical (in fact, non-Gaussian state) by calculating and displaying its Wigner function.

References

[1] J. R. Oppenheimer as paraphrased by B. R. Frieden, *Probability, Statistical Optics, and Data Testing* (Berlin: Springer, 1991), p. 363.

[2] M. Hillery, *Phys. Lett. A* **111**, 409 (1985).

[3] Early papers on squeezing in optical systems are B. R. Mollow and R. J. Glauber, *Phys. Rev.* **160**, 1076 (1967); **160**, 1097 (1967); D. Stoler, *Phys. Rev. D* **1**, 3217 (1970); E. C. Y. Lu, *Lett. Nuovo Cimento* **2**, 1241 (1971); H. P. Yuen, *Phys. Rev. A* **13**, 2226 (1976).

[4] See R. Schnabel, *Phys. Rep.* **684**, 1 (2017).

[5] B. Caron *et al.*, *Class. Quantum. Grav.* **14**, 1461 (1997).

[6] See Y. Yamamoto and H. A. Haus, *Rev. Mod. Phys.* **58**, 1001 (1986) and references cited therein.

[7] R. E. Slusher and B. Yurke, *IEEE J. Lightwave Tech.* **8**, 466 (1990). See also E. Giacobino, C. Fabre, and G. Leuchs, *Physics World* **2**, 31 (1989).

[8] W. Schleich and J. A. Wheeler, *Nature* **326**, 574 (1987); *J. Opt. Soc. Am. B* **4**, 1715 (1987).

[9] W. Schleich, *Quantum Optics in Phase Space* (Berlin: Wiley-VCH, 2001). See chapters 7 and 8.

[10] R. L. Hudson, *Rep. Math. Phys.* **6**, 249 (1974).

[11] R. W. Boyd, *Nonlinear Optics*, 2nd edition (New York: Academic Press, 2003).

[12] R. E. Slusher, L. W. Hollberg, B. Yurke, J. C. Mertz, and J. F. Valley, *Phys. Rev. Lett.* **55**, 2409 (1985).

[13] L.-A. Wu, H. J. Kimble, J. L. Hall, and H. Wu, *Phys. Rev. Lett.* **57**, 2520 (1987).

[14] G. Breitenbach, T. Mueller, S. F Pereira, J.-Ph. Poizat, S. Schiller, and J. Mlynek, *J. Opt. Soc. Am. B* **12**, 2304 (1995).

[15] R. Short and L. Mandel, *Phys. Rev. Lett.* **51**, 384 (1983).

[16] X. T. Zou and L. Mandel, *Phys. Rev. A* **41**, 475 (1990).

[17] H. J. Carmichael and D. F. Walls, *J. Phys. B* **9**, L43 (1976); C. Cohen-Tannoudji in R. Balian, S. Haroche, and S. Liberman (Eds.), *Frontiers in Laser Spectroscopy* (Amsterdam: North Holland, 1977); H. J. Kimble and L. Mandel, *Phys. Rev. A* **13**, 2123 (1976).

[18] H. J. Kimble, M. Dagenais, and L. Mandel, *Phys. Rev. Lett.* **39**, 691 (1977).

[19] E. Schrödinger's original articles on the so-called "cat paradox," originally published in *Naturwissenshaften* in 1935, can be found, in an English translation, in J. A. Wheeler and W. H. Zurek (Eds.), *Quantum Theory of Measurement* (Princeton, NJ: Princeton University Press, 1983), pp. 152–167.

[20] V. V. Dodonov, I. A. Malkin, and V. I. Man'ko, *Physica* **72**, 597 (1974). See also V. Bužek, A. Vidiella-Barranco, and P. L. Knight, *Phys. Rev. A* **45**, 6570 (1992); C. C. Gerry, *J. Mod. Opt.* **40**, 1053 (1993).

[21] B. Yurke and D. Stoler, *Phys. Rev. Lett.* **57**, 13 (1986).

[22] N. Imoto, H. A. Haus, and Y. Yamamoto, *Phys. Rev. A* **32**, 2287 (1985); M. Kitagawa and Y. Yamamoto, *Phys. Rev. A* **34**, 3974 (1986).

[23] C. C. Gerry, *Phys. Rev. A* **59**, 4095 (1999).

[24] S. M. Barnett and P. L. Knight, *J. Opt. Soc. Am. B* **2**, 467 (1985); B. Yurke and M. Potasek, *Phys. Rev. A* **36**, 3464 (1987); S. M. Barnett and P. L. Knight, *Phys. Rev. A* **38**, 1657 (1988).

[25] See, for example, P. T. Cochrane, G. J. Milburn, and W. J. Munro, *Phys. Rev. A* **62**, 062307 (2000).

[26] G. S. Agarwal, *J. Opt. Soc. B* **5**, 1940 (1988).

[27] See, for example, C. V. Usenko and V. C. Usenko, *J. Russ. Laser Res.* **25**, 361 (2004).

[28] See, for example, A. Gilchrist and W. J. Munro, *J. Opt. B* **2**, 47 (2000).

[29] M. Abramowitz and I. A. Stegun, *Handbook of Mathematical Functions* (Washington, D.C.: US Department of Commerce, 1964), p. 377.

[30] C. C. Gerry, J. Mimih, and R. Birrittella, *Phys. Rev. A* **84**, 023810 (2011).

[31] M. S. Kim, W. Son, V. Bużek, and P. L. Knight, *Phys. Rev. A* **65**, 032323 (2002).

[32] G. S. Agarwal and K. Tara, *Phys. Rev.* **43**, 492 (1991).

[33] A. Zavatta, S. Viciani, and M. Bellini, *Science* **306**, 660 (2004).

[34] M. Dakna, T. Anhut, T. Opatrný, L. Knöll, and D.-G. Welsch, *Phys. Rev. A* **55**, 3184 (1997).

[35] See, for example, C. Silberhorn, *Contemp. Phys.* **48**, 143 (2007).

[36] S. M. Barnett, G. Ferenczi, C. R. Gilson, and F. C. Speirits, *Phys. Rev. A* **98**, 013809 (2018).

[37] S. S. Mizrahi and V. V. Dodonov, *J. Phys. A: Math. Gen.* **35**, 8847 (2002).

[38] T. Gerrits, S. Clancy, T. S. Clements, B. Calkins, A. E. Lita, A. Miller, *et al.*, *Phys. Rev. A* **82**, 031802(R) (2010).

[39] A. E. Lita, A. J. Miller, and S. W. Nam, *Opt. Express* **16**, 3032 (2008).

[40] M. Eaton, A. Hossameldin, R. J. Birrittella, P. M. Alsing, C. C. Gerry, C. Cuevas, *et al.*, arXiv:2205.01221 [phys.ins-det].

[41] A. Biswas and G. S. Agarwal, *Phys. Rev. A* **75**, 032104 (2007).

[42] M. S. Kim, E. Park, P. L. Knight, and H. Jeong, *Phys. Rev. A* **71**, 043805 (2005).

[43] M. S. Kim, F. A. M. de Oliviera, and P. L. Knight, *Phys. Rev. A* **40**, 2492 (1989).

[44] See F. A. M. de Oliveira, M. S. Kim, and P. L. Knight, *Phys. Rev. A* **41**, 2645 (1990) and references cited therein.

[45] R. Carranza and C. C. Gerry, *J. Opt. Soc. Am. B* **29**, 2581 (2012).

[46] O. S. Magaña-Loiza, R. de J. León-Montiel, A. Perez-Leija, A. B. U'Ren, C. You, K. Busch, *et al.*, *NPJ Quant. Inf.* **5**, 80 (2019).

[47] A. I. Lvovsky and J. Mlynek, *Phys. Rev. Lett.* **88**, 250401 (2002).

[48] T. J. Bartley, G. Donati, J. B. Spring, X.-M. Jin, M. Barbieri, A. Datta, *et al.*, *Phys. Rev. A* **86**, 043820 (2012).

[49] R. J. Birrittella, M. El Baz, and C. C. Gerry, *J. Opt. Soc. Am. B* **35**, 1514 (2018).

[50] M. S. Kim, W. Son, V. Bužek, and P. L. Knight, *Phys. Rev. A* **65**, 032323 (2002).

[51] M. Hillery, *Phys. Rev. A* **36**, 3796 (1987).

[52] M. Hillery, *Phys. Rev. A* **40**, 3147 (1989).

[53] R. Loudon, *Rep. Prog. Phys.* **43**, 58 (1980).

Bibliography

Below is a selection of early to recent review articles and books on nonclassical states of light.

D. F. Walls, "Evidence for the quantum nature of light," *Nature* **280**, 451 (1979).

D. F. Walls, "Squeezed states of light," *Nature* **306**, 141 (1983).

L. Mandel, "Non-classical states of the electromagnetic field," *Physica Scripta* T12, 34 (1985).

R. W. Henry and S. C. Glotzer, "A squeezed-state primer," *Am. J. Phys.* **56**, 318 (1988).

M. C. Teich and B. E. A. Saleh, "Tutorial: Squeezed states of light," *Quantum Opt.* **1**, 153 (1989).

J. J. Gong and P. K. Aravind, "Expansion coefficients of a squeezed coherent state in a number state basis," *Am. J. Phys.* **58**, 1003 (1990).

V. Bužek and P. L. Knight, "Quantum interference, superposition states of light, and nonclassical effects," in E. Wolf (Ed.), *Progress in Optics XXXIV* (Amsterdam: Elsevier, 1995), p. 1.

L. Davidovich, "Sub-Poissonian processes in quantum optics," *Rev. Mod. Phys.* **68**, 127 (1996).

D. N. Klyshko, "The nonclassical light," *Physics-Uspeki* **39**, 573 (1996).

C. C. Gerry and P. L. Knight, "Quantum superpositions and Schrödinger cat states in quantum optics," *Am. J. Phys.* **65**, 964 (1997).

P. L. Knight, "Quantum fluctuations in optical systems," in S. Reynaud, E. Giacobini, and J. Zinn-Justin (Eds.), *Lectures in Les Houches, Session LXIII, 1995, Quantum Fluctuations* (Amsterdam: Elsevier, 1997).

V. V. Dodonov and V. I. Man'ko (Eds.), *Theory of Nonclassical States of Light* (London: Taylor & Francis, 2003).

J. Bauchowitz, T. Westphal, and R. Schnabel, "A graphical description of optical parametric generation of squeezed states of light," *Am. J. Phys.* **81**, 767 (2013).

A book dedicated to a discussion of experimental techniques in quantum optics, including the generation of squeezed light, is

H.-A. Bachor, *A Guide to Experiments in Quantum Optics*, 3rd edition (Weinheim: Wiley-VCH, 2019).

The application of squeezed light to interferometry was discussed early on by

C. M. Caves, "Quantum-mechanical noise in an interferometer," *Phys. Rev. D* **23**, 1693 (1981).

A recent review of the application of squeezed light to interferometry is

R. Schnabel, "Squeezed states of light and their application in laser interferometers," *Phys. Rep.* **684**, 1 (2017).

Another application of nonclassical light is in the area of quantum imaging. See

M. I. Kolobov, "The spatial behavior of nonclassical light," *Rev. Mod. Phys.* **71**, 1539 (1999).

More recent review articles on quantum imaging are

M. J. Padgett and R. W. Boyd, "An introduction to ghost imaging: Quantum and classical," *Phil. Trans. R. Soc. A* **375**, 20160233 (2016).

M. G. Basset, F. Setzpfandt, F. Steinlechner, E. Beckert, T. Pertsch, and M. Gräfe, "Perspectives for applications of quantum imaging," *Laser Photon. Rev.* **13**, 1900097 (2019).

R. Berchera and I. P. Degiovanni, "Quantum imaging with sub-Poissonian light: Challenges and perspectives in optical metrology," *Metrologia* **56**, 024001 (2019).

O. S. Magaña-Loaiza and R. W. Boyd, "Quantum imaging and information," *Rep. Prog. Phys.* **82**, 124401 (2019).

A book thoroughly covering, from a quantum optics point of view, the nonlinear processes that give rise to many of the nonclassical states discussed in this chapter is

P. D. Drummond and M. Hillery, *The Quantum Theory of Nonlinear Optics* (Cambridge: Cambridge University Press, 2014).

A book covering classical nonlinear optics is

R. W. Boyd, *Nonlinear Optics*, 4th edition (London: Academic Press, 2020).

For quantum state engineering, see

F. Dell'Anno, S. De Siena, and F. Illuminati, "Multiphoton quantum optics and quantum state engineering," *Phys. Rep.* **428**, 53 (2006).

M. S. Kim, "Recent developments in photon-level operations on travelling wave light field," *J. Phys. B: At. Mol. Opt. Phys.* **41**, 133001 (2008).

M. Bellini and A. Zavatta, "Manipulating states by single-photon addition and subtraction," *Progr. Opt.* **55**, 41 (2010).

8 Dissipative Interactions and Decoherence

8.1 Introduction

So far, we have discussed closed systems involving a single quantized mode of the field interacting with atoms, for example in the Jaynes–Cummings model of Chapter 4. As we saw in this model, the transition dynamics are coherent and reversible: the atom and field mode exchange excitation to and from without loss of energy. As we add more modes for the atom to interact with, the coherent dynamics becomes more complicated as the relevant atom–field states come in and out of phase and beat together to determine the total state occupation probabilities. As time goes on, these beats get out of phase, leading to an apparent decay of the initial-state occupation probability. But at later times, the beating eigenfrequencies get back in phase in a manner rather reminiscent of the Jaynes–Cummings revival discussed earlier in this book, and this leads to a partial recurrence or revival of the initial-state probability. The time scale for this partial revival depends on the number of participating electromagnetic field modes; as these increase to the level appropriate for an open system in free space, the recurrence disappears off to the remote future and the exponential decay law appropriate for decay is recovered as an excellent approximation [1].

We have already discussed the origin of spontaneous emission and the Einstein A coefficient using perturbation theory in Chapter 4. For many purposes, all that we need to do to include decay in our discussion of the dynamics of coherently excited atoms is to add a loss term to upper-state populations, a term which reflects the gain in population in lower states to which the excited atom can decay, and to allow for an appropriate decay for the coherences or dipole moments. The resultant equations are known as the optical Bloch equations [2]. The dynamics is often well approximated by an ansatz which makes the upper unstable state energy complex, with an imaginary part which reflects the lifetime of the upper state (but care needs to be exercised in the use of this simple ansatz applied to probability amplitudes as it cannot allow for the repopulation of the lower level in any convincing way).

More rigorous models of spontaneous emission decay have to take into account the quantized nature of the free-space vacuum with which the atom interacts. One can develop a multimode Schrödinger picture approach to this problem, which involves coupled equations of motion for the initial excited state of the atom with no photons present and the infinite number of states where there is one excitation in the field and the atom is in its ground state. These equations cannot be solved analytically, but when the coupling between the atom and the field is weak enough for only small changes in excitation probabilities to be generated within the coherence time of the relevant radiation (and for the vacuum this is a very short time indeed!), the exponential decay law is obtained. This is the essence of the standard theory of spontaneous emission due to Weisskopf and Wigner; if instead we work with the density matrix, the equivalent approach is called the Born–Markov approximation. An account of these models is beyond the scope of this book, but details can be found in the bibliography at the end of this chapter.

8.2 Single Realizations or Ensembles?

Quantum mechanics is usually introduced as a theory for ensembles, but the invention of ion traps, for example, offers the possibility to observe and manipulate single particles, where observability of quantum jumps – which are not seen directly in the ensemble – lead to conceptual problems of how to describe single realizations of these systems [3]. Usually, Bloch equations or Einstein rate equations are used to describe the time evolution of ensembles of atoms or ions driven by light. New approaches via conditional time evolution – say, when no photon has been emitted – have been developed to describe single experimental realizations of quantum systems. This leads to a description of the system via wave functions instead of density matrices. This conditional "quantum trajectory" approach [4] is still an ensemble description, but for a sub-ensemble where we know when photons have been emitted. The jumps that occur in this description can be considered as due to the increase in our knowledge about the system which is represented by the wave function (or density operator) describing the system. In the formalism to be presented, one usually imagines that Gedanken measurements are performed in rapid succession, for example, on the emitted radiation field. These will either have the result that a photon has been found in the environment, or that no photon has been found. A sudden change in our information about the radiation field (e.g. through detection of a photon emitted by the system into the environment) leads to a sudden change in the wave function of the system. However, not only does the detection of a photon lead to an increase in information; so does the failure to detect a photon (i.e. a null result). New insights have been obtained into atomic dynamics and dissipative processes, and new powerful theoretical approaches developed. Apart from the new insights into physics, these methods also allow the simulation of complicated problems (e.g. in laser cooling) that were completely intractable using the master equation approach.

Quantum mechanics is a statistical theory which makes probabilistic predictions of the behavior of ensembles (an ideally infinite number of identically prepared quantum systems) using density operators. This description was completely sufficient for the first 60 years of the existence of quantum mechanics, because it was generally regarded as completely impossible to observe and manipulate single quantum systems. For example, in 1952 Schrödinger wrote [5]

> ... we never experiment with just one electron or atom or (small) molecule. In thought experiments we sometimes assume that we do; this invariably entails ridiculous consequences... In the first place it is fair to state that we are not experimenting with single particles, any more than we can raise ichthyosauria in the zoo.

This (rather extreme) opinion was challenged by a remarkable idea of Dehmelt, which he first made public in 1975 [6]. He considered the problem of high-precision spectroscopy, where one wants to measure the transition frequency of an optical transition as accurately as possible [e.g. by observing the resonance fluorescence from that transition as part (say) of an optical frequency standard]. However, the accuracy of such a measurement is fundamentally limited by the spectral width of the observed transition. The spectral width is due to spontaneous emission from the upper level of the transition, which leads to a finite lifetime τ of the upper level. Basic Fourier considerations then imply a spectral width of the scattered photons on the order of $1/\tau$. To obtain a precise value of the transition frequency, it would therefore be advantageous to excite a metastable transition which scatters only a few photons within the measurement time. On the other hand, one then has the problem of detecting these few photons and this turns out to be practically impossible by direct observation. So, obviously, one has arrived at a major dilemma here.

Dehmelt's proposal, however, suggests a solution to these problems, provided one would be able to observe and manipulate single ions or atoms, which became possible with the invention of single-ion traps [7]. We illustrate Dehmelt's idea in its original simplified rate equation picture. It runs as follows. Instead of observing the photons emitted on the metastable two-level system directly, he proposed using an optical double-resonance scheme as depicted in Fig. 8.1.

One laser drives the metastable $0 \leftrightarrow 2$ transition, while a second strong laser saturates the strong $0 \leftrightarrow 1$; the lifetime of the upper level 1 is 10 ns, for example, while that of level 2 is of the order of 1s. If the initial state of the system is the lower state 0, then the strong laser will start to excite the system to the rapidly decaying level 1, which will then lead to the emission

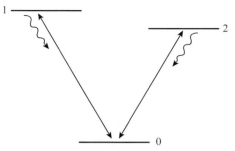

Fig. 8.1 The V system. Two upper levels 1 and 2 couple to a common ground state 0. The transition frequencies are assumed to be largely different, so that each of the two lasers driving the system couples to only one of the transitions. The $0 \leftrightarrow 1$ transition is assumed to be strong while the $0 \leftrightarrow 2$ transition is weak.

of a photon after a time which is usually very short (of the order of the lifetime of level 1). This emission restores the system back to the lower level 0; the strong laser can start to excite the system again to level 1, which will emit a photon on the strong transition again. This procedure repeats until at some random time the laser on the weak transition manages to excite the system into its metastable state 2, where it remains shelved for a long time, until it jumps back to the ground state either by spontaneous emission or by stimulated emission due to the laser on the $0 \leftrightarrow 2$ transition. During the time the electron remains in the metastable state 2, no photons will be scattered on the strong transition and only when the electron jumps back to state 0 can the fluorescence on the strong transition start again. Therefore, from the switching on and off of the resonance fluorescence on the strong transition (which is easily observable), we can infer the extremely rare transitions on the $0 \leftrightarrow 2$ transition. Therefore, we have a method to monitor rare quantum jumps (transitions) on the metastable $0 \leftrightarrow 2$ transition by observation of the fluorescence from the strong $0 \leftrightarrow 1$ transition. A typical experimental fluorescence signal is depicted in Fig. 8.2 [8], where the fluorescence intensity $I(t)$ is plotted.

However, this scheme only works if we observe a single quantum system, because if we observe a large number of systems simultaneously, then the random nature of the transitions between levels 0 and 2 implies that some systems will be able to scatter photons on the strong transition while others cannot, because they are in their metastable state at that moment. From a large collection of ions observed simultaneously, one would then obtain a more or less constant intensity of photons emitted on the strong transition. The calculation of single-system properties, such as the distribution of the lengths of the periods of strong fluorescence, required some effort that eventually led to the development of the quantum jump approach. Apart from the interesting theoretical implications of the study of individual quantum systems, Dehmelt's proposal obviously has

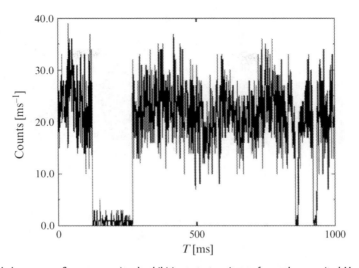

Fig. 8.2 Recorded resonance fluorescence signal exhibiting quantum jumps from a laser-excited Mg ion [8]. Periods of high photon count rate are interrupted by periods with negligible count rate (except for an unavoidable dark count rate).

important practical applications. An often-cited example is the realization of a new time standard using a single atom in a trap. The key idea here is to use either the instantaneous intensity or the photon statistics of the emitted radiation on the strong transition (the statistics of the bright and dark periods) to stabilize the frequency of the laser on the weak transition. This is possible because the photon statistics of the strong radiation depends on the detuning of the laser on the weak transition. Therefore, a change in the statistics of bright and dark periods indicates that the frequency of the weak laser has shifted and has to be adjusted. However, for continuously radiating lasers, this frequency shift will also depend on the intensity of the laser on the strong transition [9]. Therefore, in practice, pulsed schemes are preferable for frequency standards [10]. Due to the inability of experimentalists to store, manipulate, and observe single quantum systems (ions) at the time of Dehmelt's proposal, both the practical as well as the theoretical implications of his proposal were not immediately investigated. It was about 10 years later that this situation changed. At that time Cook and Kimble published a paper [11] in which they made the first attempt to analyze the situation described above theoretically. Their advance was stimulated by the fact that by that time, it had become possible to actually store *single* ions in an ion trap. In their simplified rate-equation approach, Cook and Kimble started with the rate equations for an incoherently driven three-level system as shown in Fig. 8.1 and assumed that the strong $0 \leftrightarrow 1$ transition is driven to saturation. (By saturation, we mean that the population of the states 0 and 1 is so strongly influenced by the stimulating radiation field that no further increase in the transition probabilities can be achieved by increasing the intensity.) They consequently simplified their rate equations, introducing the probabilities P_+ of being in the metastable state and P_- of being in the strongly fluorescing 0↔1 transition. This simplification now allows the description of the resonance fluorescence to be reduced to that of a two-state random telegraph process. Either the atomic population is in the levels 0 and 1 and therefore the ion is strongly radiating (on), or the population rests in the metastable level 2 and no fluorescence is observed (off). They then proceed to calculate the distributions for the lengths of bright and dark periods and find that their distribution is Poissonian. Their analysis, which we have outlined very briefly here, is of course very much simplified in many respects. The most important point is certainly the fact that Cook and Kimble assume incoherent driving and therefore adopt a rate-equation model. In real experiments, coherent radiation from lasers is used. The complications arising in coherent excitation finally led to the development of the quantum jump approach. Despite these problems, the analysis of Cook and Kimble showed the possibility of direct observation of quantum jumps in the fluorescence of single ions, a prediction that was confirmed shortly afterwards in a number of experiments [12]. The following effort of a great number of physicists eventually culminated in the development of the quantum jump approach [13].

8.3 Individual Realizations

The quantum jump method provides a simple description of decay processes which avoids the technicalities of more formal approaches. The procedure adopted in quantum jump simulations, a kind of Monte Carlo procedure, can be summarized as follows [3, 14] for the simple case of photons in a single-mode cavity field, where photons are lost from the field and absorbed by the walls of the cavity, which we assume to be at $T = 0$ K. When we speak of "emission" in what follows, we shall mean emission *from the field* (the loss of photons from the field), the emission being represented mathematically by the action of an annihilation operator acting on the field state. The rate at which photons are lost by the field (the emission rate) we denote as γ. Then the jump procedure is as follows.

1. Determine the current probability of an emission: this depends on the decay rate and the photon occupation number according to

$$\Delta P = \gamma \langle \Psi | \hat{a}^\dagger \hat{a} | \Psi \rangle \Delta t, \tag{8.1}$$

$|\Psi\rangle$ being a particular member of the ensemble.

2. Obtain a random number r between 0 and 1. Compare this with ΔP and decide on emission as follows.

3. Emit if $r < \Delta P$, so that the system jumps to the renormalized form

$$|\Psi\rangle \rightarrow |\Psi_{\text{emit}}\rangle = \frac{\hat{a}|\Psi\rangle}{\langle\Psi|\hat{a}^\dagger\hat{a}|\Psi\rangle^{1/2}}, \tag{8.2}$$

where the annihilation operator takes care of the loss of a cavity field photon emitted by the wall external to the cavity.

4. No emission if $r > \Delta P$, so the system evolves according to

$$
\begin{aligned}
|\Psi\rangle \rightarrow |\Psi_{\text{no emit}}\rangle &= \frac{e^{-i\Delta t\hat{H}_{\text{eff}}}|\Psi\rangle}{[\langle\Psi|e^{i\Delta t\hat{H}_{\text{eff}}^\dagger}e^{-i\Delta t\hat{H}_{\text{eff}}}|\Psi\rangle]^{1/2}} \\
&= \frac{e^{-i\Delta t\hat{H}_{\text{eff}}}|\Psi\rangle}{[\langle\Psi|e^{-\Delta t\hat{a}^\dagger\hat{a}}|\Psi\rangle]^{1/2}} \\
&\approx \frac{\{1-(i/\hbar)\hat{H}\Delta t-(\gamma/2)\Delta t\,\hat{a}^\dagger\hat{a}\}|\Psi\rangle}{(1-\Delta P)^{1/2}},
\end{aligned} \tag{8.3}
$$

valid to order Δt, where $\hat{H}_{\text{eff}} = \hat{H} - i\hbar(\gamma/2)\hat{a}^\dagger\hat{a}$ is the effective non-Hermitian Hamiltonian describing non-unitary evolution, and where the second term represents the decay in the energy of the cavity field. We refer the reader to the literature [8] for a more complete discussion of the no-jump evolution.

5. Repeat this whole process to obtain an individual trajectory, or history, which will describe the conditional evolution of one member of the ensemble of decaying systems, contingent on us recording jump events at times $t_1, t_2, \ldots$.

6. Average observables over many such trajectories to obtain a description of how the whole ensemble behaves on average, where all possible histories are included.

To reassure ourselves that this is all true, we note that the history of a particular trajectory splits into two alternatives in a time Δt short enough to allow at most one jump. Either a jump is recorded or it is not:

$$|\Psi\rangle \rightarrow \begin{cases} |\Psi_{\text{emit}}\rangle & \text{with probability } \Delta P, \\ |\Psi_{\text{no emit}}\rangle & \text{with probability } 1-\Delta P. \end{cases} \tag{8.4}$$

Then, in terms of the density operator of the associated pure state, $|\Psi\rangle\langle\Psi|$, the evolution for a short time step of length Δt, where $|\Psi(t)\rangle \equiv |\Psi\rangle$ and $|\Psi(\Delta t)\rangle \equiv |\Psi(t+\Delta t)\rangle$, becomes a sum of the two possible outcomes:

$$|\Psi\rangle\langle\Psi| \rightarrow |\Psi(\Delta t)\rangle\langle\Psi(\Delta t)|$$

$$= \Delta P|\Psi_{\text{emit}}\rangle\langle\Psi_{\text{emit}}| + (1 - \Delta P)|\Psi_{\text{no emit}}\rangle\langle\Psi_{\text{no emit}}|$$

$$= \gamma \, \Delta t \, \hat{a}|\Psi\rangle\langle\Psi|\hat{a}^\dagger$$

$$+ \{1 - (i/\hbar)\hat{H}\Delta t - (\gamma/2)\Delta t \hat{a}^\dagger \hat{a}\}|\Psi\rangle\langle\Psi|\{1 + (i/\hbar)\hat{H}\Delta t - (\gamma/2)\Delta t \hat{a}^\dagger \hat{a}\}$$

$$\approx |\Psi\rangle\langle\Psi| - \frac{i}{\hbar}\Delta t[\hat{H}, |\Psi\rangle\langle\Psi|]$$

$$+ \frac{\gamma}{2}\Delta t\{2\hat{a}|\Psi\rangle\langle\Psi|\hat{a}^\dagger - \hat{a}^\dagger \hat{a}|\Psi\rangle\langle\Psi| - |\Psi\rangle\langle\Psi|\hat{a}^\dagger \hat{a}\}.$$

$$(8.5)$$

Averaging over all members of the ensemble for this short time evolution, we obtain

$$\hat{\rho}(t + \Delta t) = \hat{\rho}(t) - \frac{i}{\hbar}\Delta t[\hat{H}, \hat{\rho}] + \frac{\gamma}{2}\Delta t\{2\hat{a}\hat{\rho}\hat{a}^\dagger - \hat{a}^\dagger \hat{a}\hat{\rho} - \hat{\rho}\hat{a}^\dagger \hat{a}\}, \qquad (8.6)$$

$\hat{\rho}(t)$ being the density operator of the ensemble. This result has the appearance of Euler's approximation for obtaining the solution of an ordinary differential equation. Indeed, in the limit $\Delta t \rightarrow 0$, we do obtain a first-order ordinary differential equation for the density operator:

$$\frac{d\hat{\rho}}{dt} = -\frac{i}{\hbar}[\hat{H}, \hat{\rho}] + \frac{\gamma}{2}\{2\hat{a}\hat{\rho}\hat{a}^\dagger - \hat{a}^\dagger \hat{a}\hat{\rho} - \hat{\rho}\hat{a}^\dagger \hat{a}\}. \qquad (8.7)$$

This we refer to as the master equation, and it is precisely of the form obtained from more complicated approaches for this zero-temperature ($T = 0$ K) field damping problem.

Before proceeding, let us look at the implications of the quantum jumps on some familiar quantum states of the field. We assume that no interactions occur other than the dissipative one, so that we have $\hat{H} = 0$. Suppose the cavity field is prepared in a coherent state $|\alpha\rangle$. When a quantum jump occurs and a photon is emitted by the field, then because the coherent state is a right eigenstate of the annihilation operator, $\hat{a}|\alpha\rangle = \alpha|\alpha\rangle$, the field is unchanged and thus the coherent state remains a coherent state. Then, the no-jump evolution results in

$$\frac{e^{-(\gamma/2)\Delta t \hat{a}^\dagger \hat{a}}|\alpha\rangle}{\langle\alpha|e^{-\gamma \, \Delta t \hat{a}^\dagger \hat{a}}|\alpha\rangle} = |\alpha e^{-\gamma\Delta t/2}\rangle \qquad (8.8)$$

for Δt small. Thus, it appears that the coherent state – the state we already know to be very much a classical state – remains a coherent state in spite of the dissipative interaction, though the amplitude of the state undergoes exponential decay.

On the other hand, suppose the cavity field has been prepared in, say, the even cat state

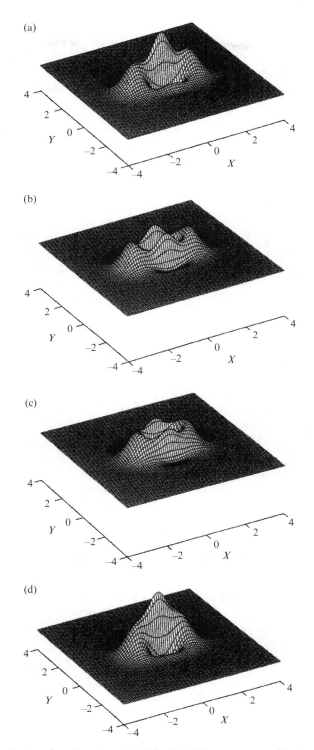

 The Wigner functions for a "jumping cat": (a) for the initial even cat state; (b) the odd cat state obtained after the first jump. (c) Wigner function shrinking. (d) The result of a second jump. Notice that the jumps reverse the sign of the interference fringes. See Ref. [13].

$$|\psi_e\rangle = \mathcal{N}_e(|\alpha\rangle + |-\alpha\rangle), \tag{8.9}$$

as discussed in Chapter 7. (A method for generating such a state in a microwave cavity is discussed in Chapter 10.) The occurrence of a quantum jump has the following effect:

$$\hat{a}(|\alpha\rangle + |-\alpha\rangle) = \alpha(|\alpha\rangle - |-\alpha\rangle).$$

The emission of a photon by the cavity field converts the even cat state into the odd cat state

$$|\psi_o\rangle = \mathcal{N}_o(|\alpha\rangle - |-\alpha\rangle). \tag{8.10}$$

The abrupt transition of an even cat state to an odd cat state is sometimes referred to as a "jumping cat" transition. For a no-jump transition, the field state undergoes decay:

$$\exp(-\gamma\Delta t\,\hat{a}^\dagger\hat{a}/2)(|\alpha\rangle \pm |-\alpha\rangle) = (|\alpha e^{-\gamma\Delta t/2}\rangle \pm |-\alpha e^{-\gamma\Delta t/2}\rangle), \tag{8.11}$$

thus showing that the cat states "shrink" during the intervals of no jumping. The jumping between even and odd cat states is particularly visible in the behavior of the Wigner function, because the interference fringes change sign. This transformation is seen in going from Fig. 8.3a to 8.3b. Subsequently, the cat state shrinks under "no-jump" evolution in going to Fig. 8.3c. After another jump, the interference fringes change sign again, as shown in Fig. 8.3b. Adding the Wigner functions of the even and odd cat states, as happens during averaging to describe the ensemble of cavity fields, causes the cancellation of the interference fringes altogether. The consequence of these uncontrollable jumps, when ensemble averaged, is the *decoherence* of a quantum superposition state into a classical statistical mixture. Decoherence will be addressed further in Section 8.5.

8.4 Shelving and Telegraph Dynamics in Three-Level Atoms

We now study in slightly more detail how the dynamics of the system determines the statistics of bright and dark periods. Again assume a three-level system as shown in Fig. 8.1. Provided the $0 \leftrightarrow 1$ and $0 \leftrightarrow 2$ Rabi frequencies are small compared with the decay rates, one finds for the population in the strongly fluorescing level 1 as a function of time something like the behavior shown in Fig. 8.3 (we derive this in detail later in this chapter).

Before we develop detailed theoretical models to describe individual quantum trajectories (i.e. state evolution conditioned on a particular sequence of observed events), it is useful to examine how the entire ensemble evolves. This is in line with the historical development, as initially it was tried to find quantum jump characteristics in the ensemble behavior of the system. We do this in detail for the particular three-level V configuration (shown in Fig. 8.1) appropriate for Dehmelt's quantum jump phenomena. For simplicity, we examine the case of *incoherent* excitation. Studies for coherent excitation using Bloch equations can be found, for example, in Ref. [15]. But for our purposes, the following Einstein rate equations for the elements of the atomic density operator demonstrate the key effects [16]:

$$\frac{d\rho_{11}}{dt} = -(A_1 + B_1 W_1)\rho_{11} + B_1 W_1 \rho_{00},$$

$$\frac{d\rho_{22}}{dt} = -(A_2 + B_2 W_2)\rho_{22} + B_2 W_2 \rho_{00}, \tag{8.12}$$

$$\frac{d\rho_{00}}{dt} = -(B_1 W_1 + B_2 W_2)\rho_{00} + (A_1 + B_1 W_1)\rho_{11} + (A_2 + B_2 W_2)\rho_{22},$$

where A_i, B_i are the Einstein A and B coefficients for the relevant spontaneous and induced transitions, W_i is the applied radiation field energy density at the relevant transition frequency, and ρ_{ii} is the relative population in state $|i\rangle$, where $\rho_{00} + \rho_{11} + \rho_{22} = 1$ for this *closed* system. In shelving, we assume that both $B_1 W_1$ and A_1 are much larger than $B_2 W_2$ and A_2, and furthermore that $B_2 W_2 \gg A_2$. The steady-state solutions of these rate equations are straightforward to obtain, and we find

$$\rho_{11}(t \to \infty) = \frac{B_1 W_1 (A_2 + B_2 W_2)}{A_1 (A_2 + 2B_2 W_2) + B_1 W_1 (2A_2 + 3B_2 W_2)},$$

$$\rho_{22}(t \to \infty) = \frac{B_2 W_2 (A_1 + B_1 W_1)}{A_1 (A_2 + 2B_2 W_2) + B_1 W_1 (2A_2 + 3B_2 W_2)}. \tag{8.13}$$

The expression for $\rho_{00}(t \to \infty)$ can be obtained from the relation $\rho_{00} + \rho_{11} + \rho_{22} = 1$, which must hold at all times. Now if the allowed $0 \leftrightarrow 1$ transition is saturated so that $B_2 W_2 \gg A_2$, we have

$$\rho_{11}(t \to \infty) \approx \rho_{00}(t \to \infty) \approx \frac{A_2 + B_2 W_2}{2A_2 + 3B_2 W_2} \tag{8.14}$$

and

$$\rho_{22}(t \to \infty) \approx \frac{B_2 W_2}{2A_2 + 3B_2 W}. \tag{8.15}$$

Now we see that a small $B_2 W_2$ transition rate to the shelf state has a major effect on the dynamics. Note that if the induced rates are much larger than the spontaneous rates, then the steady-state populations are $\rho_{00} = \rho_{11} = \rho_{22} = 1/3$: the populations are evenly distributed amongst the constituent states of the transition. However, the dynamics reveals a different story to that suggested by the steady-state populations. Again, if the allowed transition is saturated, then the time-dependent solutions of the excited state rate equations tell us that for $\rho_{00}(0) = 1$ we have

$$\rho_{11}(t) = \frac{B_2 W_2}{2(2A_2 + 3B_2 W_2)} e^{-(A_2 + 3B_2 W_2/2)t} - \frac{1}{2} e^{-(2B_1 W_1 + A_1 + B_2 W_2/2)t}$$

$$+ \frac{A_2 + B_2 W_2}{2A_2 + 3B_2 W_2}, \tag{8.16}$$

and

$$\rho_{22}(t) = \frac{B_2 W_2}{2A_2 + 3B_2 W_2} \{1 - e^{-(A_2 + 3B_2 W_2/2)t}\}. \tag{8.17}$$

Note that these expressions are good only for strong driving of the $0 \leftrightarrow 1$ transition. For a very long-lived shelf state 2, we see that for saturated transitions $(B_i W_i \gg A_i)$

$$\rho_{11}(t) \approx \frac{1}{3} \left\{ 1 + \frac{1}{2} (e^{-3B_2 W_2 t/2} - 3e^{-2B_1 W_1 t}) \right\} \tag{8.18}$$

and

$$\rho_{22}(t) \approx \frac{1}{3} \{1 - e^{-3B_2 W_2 t}\}. \tag{8.19}$$

These innocuous-looking expressions contain a lot of physics. We remember that state 1 is the strongly fluorescing state. On a time scale short compared with $(B_2 W_2)^{-1}$, we see that the populations attain a *quasi*-steady state appropriate for the *two-level* $0 \leftrightarrow 1$ dynamics, which for strong driving equalizes the populations in the states 0 and 1:

$$\rho_{11}(t, \text{ short}) \sim \frac{1}{2} \{1 - e^{-2B_1 W_1 t}\} \rightarrow \frac{1}{2}. \tag{8.20}$$

For truly long times, the third (shelving) state makes its effect and in the steady state for strong driving, the populations will equilibrate equally among the three levels so that

$$\rho_{11}(t, \text{long}) \sim \frac{1}{3}, \tag{8.21}$$

as we see qualitatively in Fig. 8.4.

As we saw earlier in the discussion of Eq. (8.3), the change from two-level to three-level dynamics already gives us a signature of quantum jumps and telegraphic fluorescence, provided we are wise enough to recognize the signs. Figure 8.5 illustrates the change from two to three-level dynamics.

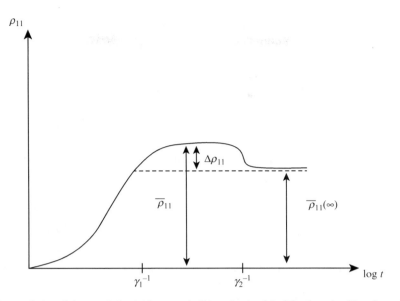

Fig. 8.4   Time evolution of the population in the strongly fluorescing level 1 of the three-level ion shown in Fig. 8.1. The lifetimes γ_1^{-1} and γ_2^{-1} are marked on the time axis. The short-time quasi-steady-state density matrix element is denoted $\overline{\rho}_{11}$, where the transition dynamics is three-level-like; $\overline{\rho}_{11}(\infty)$ is the true three-level steady-state matrix element; and $\Delta\rho_{11}$ is the difference between these two. What is crucial here is the "hump" $\Delta\rho_{11}$: this is a signature of the telegraphic nature of the fluorescence.

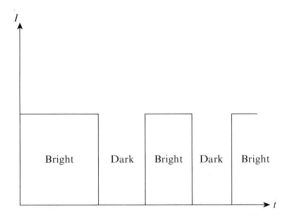

Fig. 8.5   A few periods of bright and dark sequences in the fluorescence intensity I from a three-level system. The bright periods last on average T_B and the dark periods T_D. In the mid-1980s, when studies of quantum jump dynamics of laser-driven three-level atoms began in earnest, a great deal of effort was expended in determining the relationship between joint probabilities of detection of a photon at time t and the *next* or *any* photon a time τ later. This was addressed in detail by many physicists (see e.g. Ref. [3] and references cited therein).

8.5 Modeling Losses with Fictitious Beam Splitters

In the case of the single-mode free field ($\hat{H} = 0$) at temperature $T = 0$ K, the master equation (8.7) reduces to

$$\frac{d\hat{\rho}}{dt} = \frac{\gamma}{2}\{2\hat{a}\hat{\rho}\hat{a}^{\dagger} - \hat{a}^{\dagger}\hat{a}\hat{\rho} - \hat{\rho}\hat{a}^{\dagger}\hat{a}\}. \tag{8.22}$$

If the field is prepared in a coherent state $|\alpha\rangle$ such that $\hat{\rho}(0) = |\alpha\rangle\langle\alpha|$, then it is easy to show that

$$\hat{\rho}(t) = |\alpha e^{-\gamma t/2}\rangle\langle\alpha e^{-\gamma t/2}| \tag{8.23}$$

is the correct solution to Eq. (8.22) by direct substitution. An initial coherent remains a coherent state under losses, though with amplitude undergoing exponential decay with time. The average photon number as a function of time is

$$\bar{n}(t) = |\alpha|^2 e^{-\gamma t}, \tag{8.24}$$

and thus the decay time of the field energy is given by $T_{\text{decay}} = 1/\gamma$.

There is an alternative way to describe losses that works in very general situations, and avoids the necessity of solving rather complicated master equations. The idea is to introduce for every field mode of the problem a fictitious field mode which we refer to as the "environment," whose states we label $|\psi\rangle_E$. We assume that this field is initially in the vacuum state $|0\rangle_E$. We further assume that the physical-field mode of interest is prepared in a coherent state such that the initial state of the combined systems is $|\alpha\rangle_F \otimes |0\rangle_E \equiv |\alpha\rangle_F |0\rangle_E$, where we have added the label F to the state of the field mode of interest. We now introduce a fictitious beam splitter of variable transmissivity η, where $0 \leq \eta \leq 1$, such that beam splitting the "input state" $|\alpha\rangle_F |0\rangle_E$, based on what we discussed in Chapter 6, results in the transformation

$$|\alpha\rangle_F |0\rangle_E \to |\alpha\sqrt{\eta}\rangle_F |\alpha\sqrt{1-\eta}\rangle_E, \tag{8.25}$$

where we have assumed, without loss of generality, that no phase shift occurs on the reflected beam. The phase shift is irrelevant because at the end of the calculation we shall trace out the fictitious E-mode. Note that for $\eta = 1$ we have complete transmissibility and there is no loss into the fictitious environment mode, that is we have $|\alpha\rangle_F |0\rangle_E \to |\alpha\rangle_F |0\rangle_E$, whereas for $\eta = 0$ we have $|\alpha\rangle_F |0\rangle_E \to |0\rangle_F |\alpha\rangle_E$, representing the complete loss of photons from the field of interest into the environment.

The density operator for the combined field and environment systems in the state given by the right-hand side of Eq. (8.25) is

$$\hat{\rho}_{FE} = |a\sqrt{\eta}\rangle_F\langle a\sqrt{\eta}| \otimes |a\sqrt{1-\eta}\rangle_E\langle a\sqrt{1-\eta}|. \tag{8.26}$$

Tracing out the environment leaves us with the density operator of the real-field mode as

$$\hat{\rho}_F = \text{Tr}_E\hat{\rho}_{FE} = |a\sqrt{\eta}\rangle_F\langle a\sqrt{\eta}|, \tag{8.27}$$

which has the same form as the right-hand side of Eq. (8.23), provided we make the identification $\eta = e^{-\gamma t}$. Note that for $t = 0$, $\eta = 1$, and that for $t \to \infty$, $\eta \to 0$, as we would expect. Thus, with this identification, our time-dependent field density operator is again

$$\hat{\rho}_F(t) = |ae^{-\gamma t/2}\rangle_F\langle ae^{-\gamma t/2}|. \tag{8.28}$$

The method of modeling losses via fictitious beam splitting works for fields initially prepared in states other than coherent states. Consider a field prepared in the number state $|N\rangle_F$. How can we describe losses from this state? To apply the fictitious beam splitter trick, we first go back to the right-hand side of Eq. (8.25), which we decompose in terms of number states of the two modes according to

$$|a\sqrt{\eta}\rangle_F|a\sqrt{1-\eta}\rangle_E = e^{-|a|^2/2}\sum_{n=0}^{\infty}\sum_{m=0}^{\infty}\frac{(a\sqrt{\eta})^n(a\sqrt{1-\eta})^m}{\sqrt{n!m!}}|n\rangle_F|m\rangle_E. \tag{8.29}$$

Using a technique shown in Chapter 6, we then apply the N-total photon sector projection operator

$$\begin{aligned}\hat{\Pi}_N &= \sum_{k=0}^{N}|N-k\rangle_F|k\rangle_{EF}\langle N-k|_E\langle k| \\ &= \sum_{k=0}^{N}|N-k\rangle_F\langle N-k| \otimes |k\rangle_E\langle k|\end{aligned} \tag{8.30}$$

to Eq. (8.29), so that we project out the state

$$\hat{\Pi}_N|a\sqrt{\eta}\rangle_F|a\sqrt{1-\eta}\rangle_E = e^{-|a|^2/2}\sum_{k=0}^{N}\frac{(a\sqrt{\eta})^{N-k}(a\sqrt{1-\eta})^k}{\sqrt{(N-k)!k!}}|N-k\rangle_F|k\rangle_E, \tag{8.31}$$

which, when normalized, may be written as

$$|\psi_N\rangle \equiv \sum_{k=0}^{N}\binom{N}{k}^{1/2}\eta^{(N-k)/2}(1-\eta)^{k/2}|N-k\rangle_F|k\rangle_E. \tag{8.32}$$

The corresponding density operator is $\hat{\rho}_{FE} = |\psi_N\rangle\langle\psi_N|$, from which we obtain, after tracing out the environment mode, the density operator for the field mode of interest:

$$\hat{\rho}_F = \sum_{k=0}^{N} \binom{N}{k} \eta^{(N-k)} (1-\eta)^k |N-k\rangle_F \langle N-k|. \tag{8.33}$$

Substituting for η, we obtain this density operator as a function of time:

$$\hat{\rho}_F(t) = \sum_{k=0}^{N} \binom{N}{k} e^{-\gamma t(N-k)} (1 - e^{-\gamma t})^k |N-k\rangle_F \langle N-k|. \tag{8.34}$$

The diagonal matrix elements of the operator are

$$_F\langle m|\hat{\rho}_F(t)|m\rangle_F = \binom{N}{m} e^{-\gamma t m} (1 - e^{-\gamma t})^{N-m}, \, m \leq N,$$

$$_F\langle m|\hat{\rho}_F(t)|m\rangle_F = 0, \, m > N. \tag{8.35}$$

For the case $m = N$, we have

$$_F\langle N|\hat{\rho}_F(t)|N\rangle_F = e^{-N\gamma t}, \tag{8.36}$$

from which we see that the lifetime of the Fock state with N photons is $\tau_N = 1/N\gamma$. For $N = 1$, the lifetime for a single photon is identical to the decay time for a coherent state, but for $N > 1$, the lifetimes are shorter for increasing N. In Chapter 3 we discussed the fact that the Wigner function for number states is not a probability distribution on account of it taking on negative values in some regions of phase space, this indicating a strong degree of nonclassicality. But another indication of the nonclassicality of number states has just been revealed, namely their extreme fragility in the face of losses (i.e. their increasingly short decay times for increasing photon number). Fragility in the face of losses is a feature of *all* nonclassical states of the quantized electromagnetic field.

The results just obtained algebraically are in agreement with those obtained by Lu [17] from solving the master equation (8.22) subject to the initial condition $\hat{\rho}_F(0) = |N\rangle_F \langle N|$. This means that our method of describing losses via the fictitious beam splitter approach should work for any superposition of number states or a statistical mixture of them. For multimode fields one can introduce a fictitious beam splitter and the corresponding environment states for every mode. In that way, one can algebraically incorporate the effects of losses, at least for the case temperature $T = 0$ K.

8.6 Decoherence

Decoherence, as already stated, is the transformation, over time, of a quantum mechanical superposition state into a classical statistical mixture as a result of the quantum systems interacting with the "environment." As a simple (perhaps even simple-minded) model demonstrating this effect, we consider an even cat state (it could be any cat state) and use the fictitious beam-splitter model of the environment as described in Section 8.5. That is, we take our initial system–environment state to be

$$|\psi_e\rangle_F |0\rangle_E = \mathcal{N}_e[|\alpha\rangle_F + |-\alpha\rangle_F]|0\rangle_E, \tag{8.37}$$

which transforms into

$$|\psi_e\rangle_F |0\rangle_E \rightarrow \mathcal{N}_e[|\alpha\sqrt{\eta}\rangle_F |\alpha\sqrt{1-\eta}\rangle_E + |-\alpha\sqrt{\eta}\rangle_F |-\alpha\sqrt{1-\eta}\rangle_E]. \tag{8.38}$$

The corresponding density operator is given by

$$\hat{\rho}_{FE} = \mathcal{N}_e^2 \{ |\alpha\sqrt{\eta}\rangle_F \langle\alpha\sqrt{\eta}| \otimes |\alpha\sqrt{1-\eta}\rangle_E \langle\alpha\sqrt{1-\eta}|$$

$$+ |-\alpha\sqrt{\eta}\rangle_F \langle-\alpha\sqrt{\eta}| \otimes |-\alpha\sqrt{1-\eta}\rangle_E \langle-\alpha\sqrt{1-\eta}|$$

$$+ |\alpha\sqrt{\eta}\rangle_F \langle-\alpha\sqrt{\eta}| \otimes |\alpha\sqrt{1-\eta}\rangle_E \langle-\alpha\sqrt{1-\eta}|$$

$$+ |-\alpha\sqrt{\eta}\rangle_F \langle\alpha\sqrt{\eta}| \otimes |-\alpha\sqrt{1-\eta}\rangle_E \langle\alpha\sqrt{1-\eta}| \}. \tag{8.39}$$

Tracing out the environment results in the reduced density operator of the field being

$$\hat{\rho}_F = \mathrm{Tr}_E \hat{\rho}_{FE}$$

$$= \mathcal{N}_e^2 \{ |\alpha\sqrt{\eta}\rangle_F \langle\alpha\sqrt{\eta}| + |-\alpha\sqrt{\eta}\rangle_F \langle-\alpha\sqrt{\eta}|$$

$$+ |\alpha\sqrt{\eta}\rangle_F \langle-\alpha\sqrt{\eta}|_E \langle-\alpha\sqrt{1-\eta}|\alpha\sqrt{1-\eta}\rangle_E$$

$$+ |-\alpha\sqrt{\eta}\rangle_F \langle\alpha\sqrt{\eta}|_E \langle\alpha\sqrt{1-\eta}|-\alpha\sqrt{1-\eta}\rangle_E \}, \tag{8.40}$$

which, upon using Eq. (3.66) and making the substitution $\eta = e^{-\gamma t}$, finally gives

$$\hat{\rho}_F(t) = \mathcal{N}_e^2 \{ |\alpha e^{-\gamma t/2}\rangle_F \langle\alpha e^{-\gamma t/2}| + |-\alpha e^{-\gamma t/2}\rangle_F \langle-\alpha e^{-\gamma t/2}|$$

$$+ e^{-2|\alpha|^2(1-e^{-\gamma t})}[|\alpha e^{-\gamma t/2}\rangle_F \langle-\alpha e^{-\gamma t/2}|$$

$$+ |-\alpha e^{-\gamma t/2}\rangle_F \langle\alpha e^{-\gamma t/2}|]\}. \tag{8.41}$$

That this result satisfies the master equation (8.22) is left as an exercise (see problem 10 at the end of this chapter).

Of special interest in this result is the exponential factor $\exp[-2|\alpha|^2(1-e^{-\gamma t})]$. For short times we may expand to yield $\exp[-2|\alpha|^2(1-e^{-\gamma t})] \approx \exp[-2\gamma t|\alpha|^2]$, from which we obtain the decoherence time $T_{\mathrm{decoh}} = 1/(2\gamma|\alpha|^2) = T_{\mathrm{decay}}/(2|\alpha|^2)$, this being the characteristic time over which the coherences tend to vanish. That is, for times $t \gtrsim T_{\mathrm{decoh}}$, the "off-diagonal" terms in Eq. (8.41) essentially disappear, such that

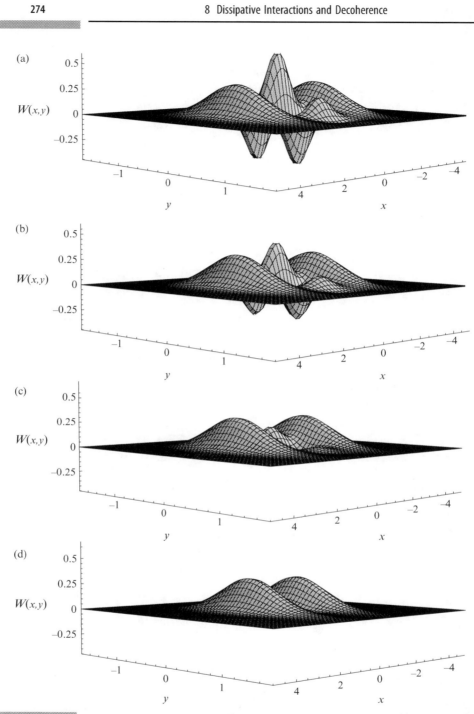

 A sequence showing the evolution of the Wigner function for an initial even cat state with $\alpha = 2$ for times (a) $\gamma t = 0$, (b) $\gamma t = 0.1$, (c) $\gamma t = 0.3$, and (d) $\gamma t = 1.0$.

$$\hat{\rho}_F(t \geq T_{\text{decoh}}) \rightarrow \mathcal{N}_e^2 \{|\alpha e^{-\gamma t/2}\rangle_F \langle \alpha e^{-\gamma t/2}| + |-\alpha e^{-\gamma t/2}\rangle_F \langle -\alpha e^{-\gamma t/2}|\}, \quad (8.42)$$

which is a statistical mixture of coherent states separated by 180° in phase space. The disappearance of the coherences is, of course, the essence of decoherence. Note that the coherences decay faster than the energy, as long as $|\alpha|^2$ is sufficiently high. If $|\alpha|^2$ is large enough to place the initial cat states in the mesoscopic or macroscopic domains, then decoherence will be virtually instantaneous, which explains why we never see cat-like states in the everyday world.

From Eqs. (8.41) and (3.173) we can obtain, for the even coherent state undergoing decoherence, the corresponding Wigner function

$$W(\beta) = W(x, y, t) = \frac{2\mathcal{N}_e^2}{\pi} \{\exp[-2(x - \alpha\sqrt{e^{-\gamma t}})^2 - 2y^2]$$
$$+ \exp[-2(x + \alpha\sqrt{e^{-\gamma t}})^2 - 2y^2]$$
$$+ 2\exp[-2\alpha^2(1 - e^{-\gamma t})]$$
$$\exp[-2(x^2 + y^2)] \cos(4y\alpha\sqrt{e^{-\gamma t}})\}, \quad (8.43)$$

where $\beta = x + iy$. The complete derivation of this result is left as an exercise.

In Fig. 8.6 we follow the evolution of this Wigner function for the even cat state. At sufficiently long time, the interference fringes disappear while the Gaussians associated with each of the coherent-state components of the original superposition remain distinct, though at very long times these two components merge to form the Wigner function of the vacuum.

For more details on the phenomenon of decoherence, the reader should consult the article by Zurek [18] and the many references cited therein.

8.7 Generation of Coherent States from Decoherence: Nonlinear Optical Balance

In this chapter we have shown how dissipative effects tend to destroy the quantum coherence inherent in quantum superposition effects. It may therefore come as a bit of a surprise to learn that, under some circumstances, quantum coherence can be *created* by dissipative effects. If we look at the master equation (8.7), we see that both unitary and non-unitary evolutionary terms are present, the former being the term with the commutator, and the latter being the dissipative terms, which are nonlinear in the annihilation and creation operators and associated with incoherent losses. These interactions are in competition with each other and thus it becomes possible that a system might evolve, as $t \rightarrow \infty$, into some state other than the vacuum. The condition just described is known as nonlinear optical balance and we demonstrate it by assuming that a pumped single-mode field, described by the interaction Hamiltonian

$$\hat{H}_{\mathrm{I}} = \hbar G(\hat{a} + \hat{a}^\dagger), \tag{8.44}$$

where G represents the strength of the classical pump field, is propagating in a photon-absorbing medium where the photon absorption is described by the dissipative terms in Eq. (8.7). Thus, our master equation becomes

$$\frac{d\hat{\rho}}{dt} = -iG[\hat{a} + \hat{a}^\dagger, \hat{\rho}] + \frac{\gamma}{2}\{2\hat{a}\hat{\rho}\hat{a}^\dagger - \hat{a}^\dagger\hat{a}\hat{\rho} - \hat{\rho}\hat{a}^\dagger\hat{a}\}. \tag{8.45}$$

We introduce the new operator $\hat{b} = \hat{a} + i\,G/\gamma$ so that we may rewrite Eq. (8.45) suggestively as

$$\frac{d\hat{\rho}}{dt} = -\frac{\gamma}{2}\{2\hat{b}\hat{\rho}\hat{b}^\dagger - \hat{b}^\dagger\hat{b}\hat{\rho} - \hat{\rho}\hat{b}^\dagger\hat{b}\}. \tag{8.46}$$

In the steady state we have $d\hat{\rho}/dt = 0$, and thus it follows that we must have $\hat{b}\hat{\rho} = 0 = \hat{\rho}\hat{b}^\dagger$, which in turn means that

$$\left(\hat{a} + \frac{iG}{\gamma}\right)\hat{\rho} = 0 = \hat{\rho}\left(\hat{a}^\dagger - \frac{iG}{\gamma}\right). \tag{8.47}$$

These last equations are solved by

$$\hat{\rho} = |\alpha\rangle\langle\alpha|, \quad \text{where } \alpha = -iG/\gamma. \tag{8.48}$$

Thus, competing coherent and incoherent processes lead to coherent states in the steady-state regime [19]. Note that if the initial state is a vacuum state, $\hat{\rho}(0) = |0\rangle\langle0|$, then at times short enough to ignore the effects of dissipation, the evolution will be unitary and will also bring about a coherent state:

$$\begin{aligned}
\hat{\rho}(t) &\approx e^{-i\,Gt(\hat{a}+\hat{a}^\dagger)}\hat{\rho}(0)e^{i\,Gt(\hat{a}+\hat{a}^\dagger)} \\
&= \hat{D}(-i\,Gt)|0\rangle\langle0|\hat{D}^\dagger(-i\,Gt) \\
&= |-i\,Gt\rangle\langle-i\,Gt|,
\end{aligned} \tag{8.49}$$

valid for very short times t.

8.8 Conclusions

In this chapter we have shown how quantum optics can address the interesting question of how one may understand the dynamics of single quantum systems. In current work on quantum computing, such questions are at the heart of how we can address, manipulate, and read out the state of quantum systems.

Problems

1. Consider a cavity field initially in a number state $|n\rangle$. Describe what happens to this state during the various steps of the quantum jump algorithm.

2. Suppose the cavity field is prepared in the superposition state $(|0\rangle + |10\rangle)/\sqrt{2}$. What is the average photon number for this state? Suppose now that a quantum jump occurs. What is the average number of photons in the field now that a single photon has been emitted? Does the result make any sense?

3. Obtain a set of coupled first-order differential equations for the elements of the density matrix for the density operator satisfying the master Eq. (8.22).

4. With reference to problem 3, solve the resulting equations numerically, if necessary, if the initial state of the system is the number state $|5\rangle$. Obtain the Wigner function, making three-dimensional plots, for the field state at various stages in the evolution.

5. Carry through the derivation of the Wigner function of Eq. (8.43).

6. Perform a numerical integration of the master equation for the case of an initial even cat state with $|\alpha|^2 = 2$. As the state evolves, monitor (i.e. plot $\mathrm{Tr}\hat{\rho}(t)$ vs. time) to check that it maintains a value of unity (a good way to check on the numerics), and monitor $\mathrm{Tr}\hat{\rho}^2(t)$ as a measure of entanglement.

7. Show that the solution to the master Eq. (8.22) can be given in iterative form as [20]

$$\rho_{mn}(t) = \exp\left(-\frac{\gamma t(m+n)}{2}\right)$$

$$\sum_l \left(\frac{(m+l)!(n+l)!}{m!n!}\right)^{1/2} \frac{\left(1 - \exp(-\gamma t)\right)^l}{l!} \rho_{m+l,n+l}(0), \quad (8.50)$$

where $\rho_{mn} = \langle m|\hat{\rho}|n\rangle$.

8. Use the iterative method to obtain the exact solutions of the master equation for initial coherent states and even cat states, in agreement with the above solutions.

9. Numerically integrate the master equation (8.45) assuming the field is initially in a vacuum state. Verify numerically in some way the existence of coherent states for both short time and steady-state regimes. Also address the question of whether or not the evolving state is always a coherent state. Is it a pure state at all times?

10. Show that the result given in Eq. (8.41), obtained by the use of fictitious beam splitters, satisfies the master equation (8.22).

References

[1] M. Ligare and R. Oliveri, *Am. J. Phys.* **70**, 58 (2002); V. Buzek, G. Drobny, M. G. Kim, M. Havukainen, and P. L. Knight, *Phys Rev A* **60**, 582 (1999).

[2] See L. Allen and J. H. Eberly, *Optical Resonance and the Two-Level Atom* (New York: Wiley, 1975; Mineola, NY: Dover, 1987).

[3] M. B. Plenio and P. L. Knight, *Rev. Mod. Phys.* **70**, 101(1998).

[4] T. A. Brun, *Am. J. Phys.* **70**, 719 (2002).

[5] E. Schrödinger, *Br. J. Philos. Sci.* **3**, 109; 233 (1952).

[6] H. Dehmelt, *Bull. Am. Phys. Soc.* **20**, 60 (1975).

[7] G. Z. K. Horvath, P. L. Knight and R. C. Thompson, *Contemp. Phys.* **38**, 25 (1997).

[8] R. C. Thompson, private communication (1996).

[9] This is merely the AC Stark effect: the strong radiation shifts the atomic energy levels in an intensity dependent fashion.

[10] H. A. Klein, G. P. Barwood, P. Gill, and G. Huang, *Phys. Scr.* **86**, 33 (2000).

[11] R. J. Cook and H. J. Kimble, *Phys. Rev. Lett.* **54**, 1023 (1985).

[12] J. C. Bergquist, R. B. Hulet, W. M. Itano, and D. J. Wineland, *Phys. Rev. Lett.* **57**, 1699 (1986); W. Nagourney, J. Sandberg, and H. G. Dehmelt, *Phys. Rev. Lett.* **56**, 2797 (1986); T. Sauter, R. Blatt, W. Neuhauser, and P. E. Toschek, *Opt. Comm.* **60**, 287 (1986).

[13] See e.g. P. L. Knight in S. Reynaud, E. Giacobino, and J. Zinn-Justin (Eds.), *Lectures in Les Houches, Session LXIII, 1995, Quantum Fluctuations* (Amsterdam: Elsevier, 1997).

[14] J. Dalibard, Y. Castin, and K. Mølmer, *Phys. Rev. Lett.* **68**, 580 (1992); G. C. Hegerfeldt and T. S. Wilser in H. D. Doebner, W. Scherer, and F. Schroek (Eds.), *Proceedings of the II International Wigner Symposium* (Singapore: World Scientific, 1991).

[15] M. S. Kim and P. L. Knight, *Phys. Rev. A* **36**, 5265 (1987); see also references in [2].

[16] D. T. Pegg, R. Loudon, and P. L. Knight, *Phys. Rev. A* **33**, 4085 (1986).

[17] N. Lu, *Phys. Rev. A* **40**, 1707 (1989).

[18] W. Zurek, *Rev. Mod. Phys.* **75**, 715 (2003).

[19] See G. S. Agarwal, *J. Opt. Soc. Am. B* **5**, 1940 (1988).

[20] J. Škvarček and M. Hillery, *Acta Phys. Slovaca* **49**, 756 (1999).

Bibliography

W. H. Louisell, *Quantum Statistical Properties of Radiation* (New York: Wiley, 1973).

H. J. Carmichael, *An Open Systems Approach to Quantum Optics* (Berlin: Springer Verlag, 1993).

S. Stenholm and M. Wilkens, "Jumps in quantum theory", *Contemp. Phys.* **38**, 257 (1997).

I. Percival, *Quantum State Diffusion* (Cambridge: Cambridge University Press, 1998).

P. Ghosh, *Testing Quantum Mechanics on New Ground* (Cambridge: Cambridge University Press, 1999).

D. Manzano, "A short introduction to the Lindblad master equation," *AIP Advances* **10**, 025106 (2020).

Over the last three decades or so, experiments of the type called Gedanken have become real. Recall the Schrödinger quote from Chapter 8: "… we never experiment with just one electron or atom or (small) molecule." This is no longer true. We can do experiments involving single atoms or molecules, and even on single photons, and thus it becomes possible to demonstrate that the "ridiculous consequences" alluded to by Schrödinger are, in fact, quite real. We have already discussed some examples of single-photon experiments in Chapter 6, and in Chapter 10 we shall discuss experiments performed with single atoms and single trapped ions. In the present chapter, we shall elaborate further on experimental tests of the fundamentals of quantum mechanics involving a small number of photons. By fundamental tests we generally mean tests of quantum mechanics against the predictions of local realistic theories (i.e. hidden variable theories). Specifically, we discuss optical experiments demonstrating violations of Bell's inequalities, violations originally discussed by Bell in the context of two spin one-half particles [1]. Such violations, if observed experimentally, falsify local realistic hidden variable theories. Locality refers to the notion, familiar in classical physics, that there cannot be a causal relationship between events with space-like separations. That is, the events cannot be connected by any signal moving at, or less than, the speed of light (i.e. the events are outside the light-cone). But in quantum mechanics, it appears that nonlocal effects – effects seemingly violating the classical notion of locality in a certain restricted sense – are possible. For example, the fact that a measurement on one part of an entangled system seems to project instantaneously the other part of the system into a particular state, even though the parts may be widely separated, is a nonlocal quantum effect. "Realism" in the context of hidden-variable theories means that a quantum system has objectively definite attributes (quantum numbers) for all observables at all times. For example, a particle with spin $+^1/_2$ along the z-direction is known to be in a superposition state with spin $\pm^1/_2$ along the x-direction. The standard Copenhagen interpretation of quantum mechanics holds that the particle's spin along the x-direction is objectively indefinite (i.e. has no particular value of spin as a matter of principle until a measurement reduces the state vector to one of the possible states in the superposition). In a hidden-variable theory, the spin along the x-axis is assumed to be definite and the experiment merely reveals the already existing state of the particle. But there are two types of hidden-variable theories, local and nonlocal. Nonlocal hidden-variable theories of the type considered by Bohm [2] reproduce the predictions of standard quantum mechanics. But they exhibit nonlocality, the feature of quantum mechanics that seems to have bothered Einstein even more than the apparent lack of realism. However, local hidden-variable theories do make predictions different from those of standard quantum theory, and Bell's theorem provides a way to perform comparative experimental tests of the two theories. In this chapter, we shall discuss optical tests of local hidden-variable theories using polarization-entangled photons.

In what follows we shall first discuss modern sources of paired and entangled single-photon states obtained from down-conversion processes (as opposed to the technique of cascaded emissions from atomic transitions as discussed in Chapter 6). We then review some one and two-photon interference experiments, introduce the notion of the "quantum eraser" in this context, and discuss an experiment on induced quantum coherence. Next, an experiment on photon tunneling exhibiting superluminal effects is discussed. Finally, we describe two experiments on tests of Bell's inequalities, one involving polarization-entangled photons and a second (Franson's experiment) based on the time–energy uncertainty relation.

9.1 Photon Sources: Spontaneous Parametric Down-Conversion

We have already discussed, in Chapter 7, the generation of nonclassical light via parametrically driven nonlinear media characterized by a second-order nonlinear susceptibility $\chi^{(2)}$. We focus on the nondegenerate case, and assuming the pump field to be quantized, the interaction Hamiltonian takes the form

$$\hat{H}_I \sim \chi^{(2)} \hat{a}_p \hat{a}_s^\dagger \hat{a}_i^\dagger + h.c., \tag{9.1}$$

where we have altered our notation from Chapter 7 so that now $\hat{a}_p$ is the annihilation operator of the pump beam and $\hat{a}_s^\dagger$ and $\hat{a}_i^\dagger$ are the creation operators of the "signal" and "idler" beams, respectively. The denotations "signal" and "idler" appear for historical reasons and have no special significance, the choice of beam labels being somewhat arbitrary. In the simplest case, with the signal and idler beams initially in vacuum states, a single pump beam photon, typically in the ultraviolet spectral range, is converted into two optical photons, one in the signal beam, the other in the idler:

$$|1\rangle_p |0\rangle_s |0\rangle_i \Rightarrow \hat{a}_p \hat{a}_s^\dagger \hat{a}_i^\dagger |1\rangle_p |0\rangle_s |0\rangle_i = |0\rangle_p |1\rangle_s |1\rangle_i. \tag{9.2}$$

As the signal and idler modes are initially in vacuum states, the process is "spontaneous." Note that the photons produced in the signal and idler modes are assumed to be generated almost simultaneously. That this is the case was demonstrated many years ago by Burnham and Weinberg [3], who used coincidence counting with detectors arranged to satisfy momentum and energy conservation and to have equal time delays. The simultaneous production of signal and idler photons is key to the applications of such parametric sources to the fundamental test of quantum mechanics to be described. In order for the down-conversion process to go forward, however, certain other conditions must be satisfied. Letting ω_p, ω_s, and ω_i represent the frequencies of the pump, signal, and idler, respectively, energy conservation requires that

$$\hbar\omega_p = \hbar\omega_s + \hbar\omega_i. \tag{9.3}$$

Further, if $\mathbf{k}_p$, $\mathbf{k}_s$, and $\mathbf{k}_i$ represent the respective wave numbers, then we must have, inside the crystal,

$$\hbar\mathbf{k}_p \approx \hbar\mathbf{k}_s + \hbar\mathbf{k}_i, \tag{9.4}$$

where the $\approx$ sign appears as the result of an uncertainty given by the reciprocal of the length of the nonlinear medium [4]. Equations (9.3) and (9.4) are known as the "phase-matching" conditions and they can be attained in certain types of nonlinear media, such as non-centrosymmetric crystals [5]. Only non-centrosymmetric crystals have a nonvanishing $\chi^{(2)}$. The most commonly used crystals are KDP (KD_2PO_4) and BBO (β-BaB_2O_4). The connection to nonlinear optics is given in Appendix D.

There are, in fact, two types of spontaneous parametric down-conversion (SPDC) processes. In type I, the signal and idler photons have the same polarization, but these are orthogonal to that of the pump. The interaction Hamiltonian for this process is given by

$$\hat{H}_I = \hbar\eta\hat{a}_s^\dagger\hat{a}_i^\dagger + h.c., \tag{9.5}$$

where the parametric approximation has been made and where $\eta \propto \chi^{(2)}\mathcal{E}_p$, where $\mathcal{E}_p$ is the amplitude of the classical coherent field. The phase-matching condition of Eq. (9.4) imposes a constraint such that the signal and idler photons (conjugate photons) must emerge from the crystal on opposite sides of concentric cones centered on the direction of the pump beam, as shown in Fig. 9.1. Evidently there are an infinite number of ways of selecting the signal and idler beams. Examples are shown in Fig. 9.2. The Hamiltonian of Eq. (9.5) in practice represents a particular "post-selection" of the momenta of the output beams.

In type II down-conversion, the signal and idler photons have orthogonal polarizations. Because of birefringence effects, the generated photons are emitted along two cones, one for the ordinary (o) wave and another for the extraordinary (e) wave, as indicated in Fig. 9.3. The intersection of the cones provides a means of generating polarization-entangled states. We use the notation $|V\rangle$ and $|H\rangle$ to represent vertically and horizontally polarized *single-photon* states. For photons that emerge along the

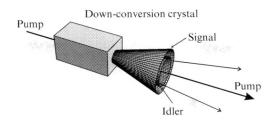

Fig. 9.1 Type I down-conversion. Photons from the pump beam are converted into signal and idler photons that emerge from the crystal along different directions. The signal and idler photons have identical polarization, but orthogonal to that of the pump. The possible directions form a set of concentric cones. The light from the different cones is of different colors, typically orange near the center to a deep red at wider angles. The pump beam is in the ultraviolet range.

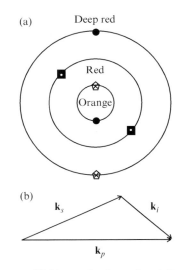

Fig. 9.2 (a) A cross-section of the cones of light emerging from a type I down-converter. Like symbols represent conjugate photons satisfying the phase-matching condition. Note that those on the middle circle are degenerate in frequency. (b). A graphical view of the phase-matching condition.

intersections of the cones, with photons from other parts of the cones being excluded by screens with pinholes in front of the intersection points, there will be an ambiguity as to whether the signal or idler photons will be vertically or horizontally polarized, as indicated in Fig. 9.4. The Hamiltonian describing this is given by

$$\hat{H}_{\mathrm{I}} = \hbar\eta(\hat{a}_{Vs}^{\dagger}\hat{a}_{Hi}^{\dagger} + \hat{a}_{Hs}^{\dagger}\hat{a}_{Vi}^{\dagger}) + h.c., \qquad (9.6)$$

where the operators $\hat{a}_{Vs}^{\dagger}, \hat{a}_{Hs}^{\dagger}, \hat{a}_{Vi}^{\dagger}$, and $\hat{a}_{Hi}^{\dagger}$ are the creation operators for photons with vertical and horizontal polarization for the signal and idler

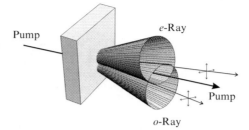

Fig. 9.3 Type II down-conversion. The signal and idler photons have orthogonal polarizations. Birefringence effects cause the photons to be emitted along two intersecting cones, one of the ordinary (o)-ray and the other of the extraordinary (e)-ray.

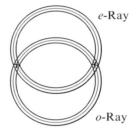

Fig. 9.4 The intersections of the cones for the o-ray and the e-ray are the sources for polarization-entangled light. From these points it is not possible to tell from which beam a photon is obtained. The Hamiltonian in Eq. (9.6) describes the light taken from both intersections.

beams, respectively. Again, this Hamiltonian represents the post-selection obtained by placing a screen in front of the source with pinholes located just in the regions of the overlapping beams.

Let us take the initial state of the signal and idler modes to be represented by $|\Psi_0\rangle = |\{0\}\rangle$, which is the collective vacuum state for either type I or type II down-conversion. In either case, the state vector evolves according to

$$|\Psi(t)\rangle = \exp(-i\hat{H}_I t/\hbar)|\Psi_0\rangle \tag{9.7}$$

which we expand, since $\hat{H}_I$ has no explicit time dependence, as

$$|\Psi(t)\rangle \approx [1 - i\hat{H}_I t/\hbar + (-i\hat{H}_I t/\hbar)^2]|\Psi_0\rangle \tag{9.8}$$

to second order in time. If we consider the type I SPDC, then $|\Psi_0\rangle = |0\rangle_s|0\rangle_i$ and we have

$$|\Psi(t)\rangle = (1 - \mu^2/2)|0\rangle_s|0\rangle_i - i\mu|1\rangle_s|1\rangle_i, \tag{9.9}$$

where $\mu = \eta t$. This state vector is normalized to first order in μ and we have dropped the term of order μ^2 containing the state $|2\rangle_s|2\rangle_i$. In the case of type II SPDC with the initial state $|\Psi_0\rangle = |0\rangle_{Vs}|0\rangle_{Hs}|0\rangle_{Vi}|0\rangle_{Hi}$, we have

$$|\Psi(t)\rangle = (1 - \mu^2/2)|0\rangle_{Vs}|0\rangle_{Hs}|0\rangle_{Vi}|0\rangle_{Hi}$$
$$- i\mu \frac{1}{\sqrt{2}}(|1\rangle_{Vs}|0\rangle_{Hs}|0\rangle_{Vi}|1\rangle_{Hi} + |0\rangle_{Vs}|1\rangle_{Hs}|1\rangle_{Vi}|0\rangle_{Hi}). \qquad (9.10)$$

We now define the vertically and horizontally polarized vacuum and single-photon states as $|0\rangle := |0\rangle_V|0\rangle_H$, $|V\rangle := |1\rangle_V|0\rangle_H$, and $|H\rangle := |0\rangle_V|1\rangle_H$, so that we may write

$$|\Psi(t)\rangle = (1 - \mu^2/2)|0\rangle_s|0\rangle_i$$
$$-i\mu(|V\rangle_s|H\rangle_i + |H\rangle_s|V\rangle_i). \qquad (9.11)$$

The state in the second term, which when normalized reads

$$|\Psi^+\rangle = \frac{1}{\sqrt{2}}(|V\rangle_s|H\rangle_i + |H\rangle_s|V\rangle_i), \qquad (9.12)$$

is one member out of a set of four states known as Bell states. The full set of Bell states is

$$|\Psi^\pm\rangle = \frac{1}{\sqrt{2}}(|H\rangle_1|V\rangle_2 \pm |V\rangle_1|H\rangle_2), \qquad (9.13)$$

$$|\Phi^\pm\rangle = \frac{1}{\sqrt{2}}(|H\rangle_1|H\rangle_2 \pm |V\rangle_1|V\rangle_2). \qquad (9.14)$$

We shall discuss these states, and their implications, in Section 9.6.

9.2 The Hong–Ou–Mandel Interferometer

In Chapter 6 we discussed what happens when twin single-photon states are simultaneously incident at each input port of a 50:50 beam splitter: the photons emerge together in one output beam or the other; no single photons ever emerge in both beams. Recall that for the input state $|1\rangle_s|1\rangle_i$, one has after the beam splitter the state

$$|\psi_{BS}\rangle = \frac{1}{\sqrt{2}}(|2\rangle_1|0\rangle_2 + |0\rangle_1|2\rangle_2), \qquad (9.15)$$

where we have labeled the output modes 1 and 2 in accordance with Fig. 9.5. Detectors placed at the outputs should never register simultaneous counts. In fact, one could take the lack of simultaneous counts as an indication that the photons were incident on the beam splitter simultaneously. The first experimental demonstration of this effect was by Hong, Ou, and Mandel (HOM) in a now classic experiment performed in 1987 [6]. In fact, the experiment was designed to measure the time separation

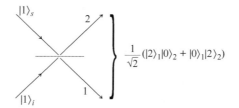

Fig. 9.5 Labeling of the output modes of a 50:50 beam splitter with inputs from the signal and idler beams of a type I down-converter. If single-photon states fall onto the beam splitter simultaneously, the output does not contain the term $|1\rangle_1|1\rangle_2$ for reasons discussed in Chapter 6.

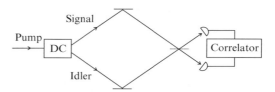

Fig. 9.6 The Hong–Ou–Mandel experiment. When the path lengths are equal, no coincident counts are detected.

between the two photons striking a beam splitter. A sketch of their experiment is given in Fig. 9.6. The nonlinear crystal is pumped to produce, assuming type I down-conversion, twin single-photon states whose beams are then directed to the input ports of a 50:50 beam splitter. Photon detectors are placed at the outputs of the beam splitter and the count signals are fed into a correlator. Changing the position of the beam splitter causes a slight time delay between the times at which the photons fall onto the beam splitter. With a slight nonzero time delay, the photons independently reflect or transmit through the beam splitter, causing both detectors to sometimes click within a short time of each other. It can be shown [6] that the rate of coincident detections, R_{coin}, has the form

$$R_{\text{coin}} \sim \left[1 - e^{-(\Delta\omega)^2 (\tau_s - \tau_i)^2} \right], \tag{9.16}$$

where $\Delta\omega$ is the bandwidth of the light and $c\tau_s$ and $c\tau_i$ are the distances that the signal and idler photons, respectively, travel from the down-converter to the beam splitter. The bandwidth $\Delta\omega$ incorporates the reality that the signal and idler beams are not monochromatic, and its appearance in Eq. (9.16) results from the assumption that the spectral distribution is Gaussian. Obviously, for $\tau_s - \tau_i = 0$, we have $R_{\text{coin}} = 0$. The rate of coincidence counts rises to a maximum for $|\tau_s - \tau_i| \gg \tau_{\text{corr}}$, where $\tau_{\text{corr}} = 1/\Delta\omega$ is the correlation time of the photons. The correlation time is of the order of a few nanoseconds, which is hard to measure using conventional techniques

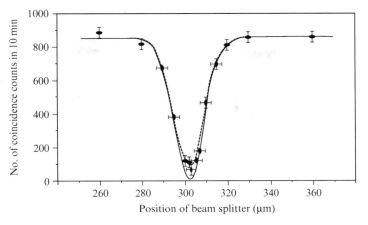

Fig. 9.7 From the Hong–Ou–Mandel paper, the number of coincident counts in a 10-minute time interval as a function of the position of the beam splitter. The count rate does not go to zero exactly, due to the fact that the beams do not perfectly overlap at the beam splitter. Reprinted with permission from Ref. [6]. Copyright 1987 by the American Physical Society.

as the detectors commonly used do not often have such short resolving times. But this correlation time can be measured with the HOM interferometer. The experimental results are plotted in Fig. 9.7, taken from Ref. [6]. Plotted here is the number of counts over a 10-minute interval against the position of the beam splitter (essentially the time separation), with the solid line representing the theoretical prediction. The experimental data do not go exactly to the predicted minimum because it is not possible for the beams to precisely overlap at the beam splitter. From the distribution of the counts, the correlation time of the two photons can be determined to be ~ 100 fs.

9.3 The Quantum Eraser

In the HOM experiment just described, the fact that the photons are indistinguishable is the key to understanding the results. Because type I SPDC is used as the source, both photons have the same polarization. They may have slightly different energies, but the photon detectors are not really sensitive to the energy difference. We have not needed to specify the polarization direction of the photons; it was only important that they be the same. But now suppose, for the sake of definiteness, that we take them to be horizontally polarized, denoting the twin single-photon states as $|H\rangle_s|H\rangle_i$. With such an input state to the beam splitter (assumed to not affect the polarization of the photons), the output state can be written as $i(|2H\rangle_1|0\rangle_2 + |0\rangle_1|2H\rangle_2)/\sqrt{2}$, where $|2H\rangle$ is a state with two horizontally polarized photons.

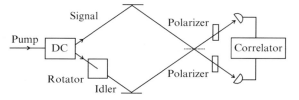

Fig. 9.8 A quantum erasure experiment. The rotator in the idler beam rotated the polarization of idler photons, thus marking them and yielding path information. This destroys the quantum interference exhibited in the HOM experiment. Inserting polarizers in the beams after the interferometer restores quantum interference.

Suppose we now place a rotator in, say, the idler beam, such that the polarization is rotated by angle θ to the horizontal, as indicated in Fig. 9.8. This transforms the idler polarization state to

$$|\theta\rangle_i = |H\rangle_i \cos\theta + |V\rangle_i \sin\theta. \tag{9.17}$$

The polarization state vector for a photon with polarization orthogonal to this is given by

$$|\theta^{\perp}\rangle_i = -|H\rangle_i \sin\theta + |V\rangle_i \cos\theta. \tag{9.18}$$

The input state to the beam splitter is now

$$|H\rangle_s|\theta\rangle_i = \cos\theta|H\rangle_s|H\rangle_i + \sin\theta|H\rangle_s|V\rangle_i. \tag{9.19}$$

The output state will be

$$\begin{aligned}|\psi_{\text{out}}(\theta)\rangle = &\frac{i}{\sqrt{2}} \cos\theta(|2H\rangle_1|0\rangle_2 + |0\rangle_1|2H\rangle_2) \\ &+ \frac{1}{2} \sin\theta(|H\rangle_1|V\rangle_2 - |V\rangle_1|H\rangle_2) \\ &+ \frac{i}{2} \sin\theta(|H,V\rangle_1|0\rangle_2 + |0\rangle_1|H,V\rangle_2),\end{aligned} \tag{9.20}$$

where the symbol $|H,V\rangle$ means that there is one horizontally polarized photon and one vertically polarized one in the same beam. If the polarization of the idler beam is rotated all the way to the vertical, i.e. to $\theta = \pi/2$, then the output state will be

$$\begin{aligned}|\psi_{\text{out}}(\pi/2)\rangle = |\Psi^-\rangle = &\frac{1}{2}(|H\rangle_1|V\rangle_2 - |V\rangle_1|H\rangle_2) \\ &+ \frac{i}{2}(|H,V\rangle_1|0\rangle_2 + |0\rangle_1|H,V\rangle_2),\end{aligned} \tag{9.21}$$

There will now be *only* coincident counts in the detectors; neither detector, assuming 100% efficiency, will fire alone. The "marking" of one of the beams by the rotation of its polarization by 90° has the effect of removing photon indistinguishability and thus it is *possible* to determine the path taken by each of the photons before the beam splitter. Interestingly, in the scenario discussed, photon polarization is never measured. The experimenter need not know the polarization of the photons; only the counts are measured. Evidently, the interference can be destroyed by the mere potential of obtaining which-path information, even if that information is not known to the experimenter. Now, in the case of one photon input to a beam splitter, one can observe the particle-like nature of the photon by placing detectors at the beam splitter outputs, or one can observe the wave-like nature, with interference effects, by using a Mach–Zehnder interferometer, as we have discussed in Chapter 6, providing an example of Bohr complementarity. Usually, it is the loss of coherence engendered by the availability of which-path information that is ascribed to the disappearance of interference. This "decoherence" supposedly renders the state vector into a statistical mixture where only probabilities, and not probability amplitudes, appear. But in the case of the two-photon interferometry experiment discussed here, there is no time, up to the point when the photons are detected, where we do not have a pure state.

Suppose now that a linear polarizer is placed at an angle θ_1 to the horizontal in front of the detector in output beam 1. The placing of a polarizer followed by photon detection constitutes a von Neumann projection (see Appendix B) onto the state vector

$$|\theta_1\rangle_1 = |H\rangle_1 \cos\theta_1 + |V\rangle_1 \sin\theta_1. \tag{9.22}$$

That is, only photons with polarization state $|\theta_1\rangle_1$ will be registered by the detector. The state of Eq. (9.21) is reduced to the pure state

$$
\begin{aligned}
|\psi, \theta_1\rangle &= \frac{|\theta_1\rangle\langle\theta_1|\psi_{\text{out}}(\pi/2)\rangle}{\langle\psi_{\text{out}}(\pi/2)|\theta_1\rangle\langle\theta_1|\psi_{\text{out}}(\pi/2)\rangle^{1/2}} \\
&= \frac{1}{2}|\theta_1\rangle(|V\rangle_2 \cos\theta_1 - |H\rangle_2 \sin\theta_1),
\end{aligned}
\tag{9.23}
$$

where the photon of mode 2 has a polarization orthogonal to that of the detected photon [see Eq. (9.18)]. Suppose we similarly place a polarizer at angle θ_2 to the horizontal in front of the detector for output beam 2. The probability that there will be coincident detections of a photon polarized at angle θ_1 and another at θ_2 is then

$$
\begin{aligned}
P_{\text{coin}} &= |\langle\theta_1|\langle\theta_2|\psi_{\text{out}}(\pi/2)\rangle|^2 \\
&= \frac{1}{4}\sin^2(\theta_2 - \theta_1).
\end{aligned}
\tag{9.24}
$$

From this result we see that the dip in the coincident count rate can be revived, depending on the relative angle of the polarizers (see Fig. 9.9). The effect of the polarizers placed just before the detectors is to erase the information encoded onto one of the beams by the rotator. The rotator creates which-path (*Welcher-Weg*, as it is often called) information and the polarizers erase it. Notice that the erasure takes place *outside* the HOM interferometer, that is

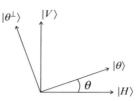

Fig. 9.9 The relationship between the $|\theta\rangle, |\theta^{\perp}\rangle$ and $|H\rangle, |V\rangle$ polarization vectors.

after the beam splitter. The optical quantum erasure effect has been demonstrated experimentally by Kwiat, Steinberg, and Chiao [7].

9.4 Induced Coherence

We now describe and explain another experiment, one that has been described, quite properly, as "mind boggling" [8], where again which-path information destroys interference but now where the mechanism of obtaining such information is not even in the interfering pathways. The experiment was performed by Zou, Wang, and Mandel [9]. A sketch of the experiment is presented in Fig. 9.10. Notice that without beam stop B blocking the idler of down-converter DC1, the idler modes of the two down-converters are aligned.

Qualitatively, here's what happens. A pump photon for an argon ion laser enters beam splitter A and goes to either DC1 or DC2, both assumed to be of type I. The down-converters, at a very low conversion rate of about $\gamma = 10^{-6}$, produce pairs of photons. With the path lengths adjusted appropriately, and with beam-stop B removed, the detector D_i monitoring the aligned idler beams cannot determine in which crystal down-conversion has occurred (the crystals are transparent). As a result, interference can be seen in the detector D_s monitoring one of the outputs of beam splitter C upon which the two signal beams fall. The interference can be observed by adjusting the path length, or equivalently the phase φ as indicated. The interference occurs whether the aligned idler beams are monitored or not. However, if the idler beam of DC1 is blocked, detection of a photon by D_i would indicate that that photon had to have been generated in DC2, which in turn means that the initial input photon had taken the path to DC2. Thus, which-path information is available and the interference at the other detector disappears. The distinctive feature of this experiment is that idler beams are not even in the paths leading to interference.

We now try to understand this experiment quantitatively. Using the labeling of the modes as given in Fig. 9.10, the initial state of the system is $|1\rangle_a|0\rangle_v$ with all other modes in vacuum states. The first beam splitter causes the transformation

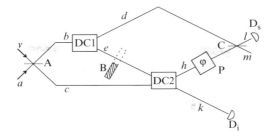

Fig. 9.10   Schematic of the Zou, Wang, and Mandel experiment. The mode labeled v contains a vacuum state. B is a beam stop that can be moved in and out of the beam e.

$$|1\rangle_a|0\rangle_v \rightarrow \frac{1}{\sqrt{2}}(|1\rangle_b|0\rangle_c + i|0\rangle_b|1\rangle_c). \qquad (9.25)$$

Now at DC1 we have $|1\rangle_b|0\rangle_d|0\rangle_e \rightarrow \gamma|0\rangle_b|1\rangle_d|1\rangle_e$ and similarly at DC2 we have $|1\rangle_c|0\rangle_h|0\rangle_k \rightarrow \gamma|0\rangle_c|1\rangle_h|1\rangle_k$, so that just after the down-converters we have

$$|\psi_{DC1,2}\rangle \sim \frac{\gamma}{\sqrt{2}}|0\rangle_b|0\rangle_c[|1\rangle_d|1\rangle_e|0\rangle_h|0\rangle_k + i|0\rangle_d|0\rangle_e|1\rangle_h|1\rangle_k]. \qquad (9.26)$$

The phase shift in the h beam gives us

$$|\psi_{PS}\rangle \sim \frac{\gamma}{\sqrt{2}}|0\rangle_b|0\rangle_c[|1\rangle_d|1\rangle_e|0\rangle_h|0\rangle_k + ie^{i\varphi}|0\rangle_d|0\rangle_e|1\rangle_h|1\rangle_k]. \qquad (9.27)$$

The beam splitter at C performs the transformations

$$|1\rangle_d|0\rangle_h \rightarrow \frac{1}{\sqrt{2}}(|1\rangle_m|0\rangle_l + i|0\rangle_m|1\rangle_l),$$
$$|0\rangle_d|1\rangle_h \rightarrow \frac{1}{\sqrt{2}}(|0\rangle_m|1\rangle_l + i|1\rangle_m|0\rangle_l), \qquad (9.28)$$

so finally we have

$$|\psi_{out}\rangle \sim \frac{\gamma}{2}|0\rangle_b|0\rangle_c[|1\rangle_e|0\rangle_k(|1\rangle_m|0\rangle_l + i|0\rangle_m|1\rangle_l)$$
$$+ ie^{i\varphi}|0\rangle_e|1\rangle_k(|0\rangle_m|1\rangle_l + i|1\rangle_m|0\rangle_l)]. \qquad (9.29)$$

Now, if the idler beams are perfectly aligned, modes e and k are identical and that mode contains one photon so that e and k are essentially the same mode. That is, we can write $(|0\rangle_e|1\rangle_k, |1\rangle_e|0\rangle_k) \rightarrow |1\rangle_k$, reflecting the fact that the detector can't tell which down-converter produced the photon. Thus, Eq. (9.29) can be written as

$$|\psi_{\text{out}}\rangle \sim \frac{\gamma}{2}|0\rangle_b|0\rangle_c|1\rangle_k[(1 - e^{i\varphi})|1\rangle_m|0\rangle_l + i(1 + e^{i\varphi})|0\rangle_m|1\rangle_l]. \qquad (9.30)$$

The factorization of the state $|1\rangle_k$ containing the photon from the aligned idler beams is crucial to predicting interference in the single-photon counts. The probability amplitude for D_s and D_i detecting single photons jointly in modes l and k, respectively, is then $i\gamma(1 + e^{i\varphi})/2$ and the corresponding probability is

$$P(1_l, 1_k) \sim \frac{\gamma^2}{2}(1 + \cos\varphi), \qquad (9.31)$$

which clearly contains information on the phase and thus exhibits interference effects. Actually, because the state $|1\rangle_k$ is factored out in Eq. (9.30), again the result of the alignment of the e and k beams, we really need not do any detection at all on the idler beams; all we need is to perform detection with D_s, where the detection probability for a single photon in D_s is

$$P(1_l) = P(1_l, 1_k) \sim \frac{\gamma^2}{2}(1 + \cos\varphi). \qquad (9.32)$$

But now suppose we insert the beam-stop B, which, in fact, could be just a mirror reflecting the photon in the e mode away from any detector. In any case, the idler beams e and k are no longer aligned and thus it is no longer possible to factor out the states of these modes in Eq. (9.29). The probability of obtaining coincident counts by D_s and D_i is now

$$P(1_l, 1_k) \sim \left|\frac{ie^{i\varphi}\gamma}{2}\right|^2 = \frac{\gamma^2}{4}, \qquad (9.33)$$

which contains no interference term. The detection of a photon by D_i necessarily means that the photon was generated by DC2 and thus yields path information. Other detection combinations are possible, but all the probabilities are the same:

$$P(1_l, 1_k) = P(0_l, 0_k) = P(1_l, 0_k) = P(0_l, 1_k) \sim \frac{\gamma^2}{4}. \qquad (9.34)$$

Note that non-detection of a photon by D_i also yields which-path information.

Suppose we don't do *any* measurement on the k beam: we just remove detector D_i. We shall now obtain no which-path information. Will the quantum interference revive in the single-photon counts by D_s? Detection of a single-photon state in the l mode can be represented by the projection operator $|1\rangle_{ll}\langle1|$. Using Eq. (9.29), the probability that a single photon is detected in mode l is

$$P(1_I) = \langle \psi_{\text{out}} | 1 \rangle_{II} \langle 1 | \psi_{\text{out}} \rangle \sim \frac{\gamma^2}{2}, \qquad (9.35)$$

as can easily be checked. But this contains no phase information and hence there is *still no interference* in the single-photon counting at detector D_s. Again, conventional wisdom says that lack of which-path information is a condition for quantum interference. It must not be sufficient. Here we have no which-path information, but we have no interference either. This is another example, perhaps more dramatic than the previous one, demonstrating that it is not necessarily the *possession* of which-path information that destroys quantum interference, but rather it is the mere *potential* of possessing such information.

9.5 Superluminal Tunneling of Photons

An interesting application of the HOM interferometer is to the measurement of tunneling times of photons through a barrier. Tunneling, one of the most striking predictions of quantum mechanics, is, of course, a strictly quantum mechanical phenomenon, having no classical analog. A perennial question has been: what is the time, the tunneling time, for a particle to tunnel through a barrier? The answers to this question have been various over the years, and we do not wish to review the literature on that topic here (but see the reviews in Ref. [10]). However, we shall point out that the HOM interferometer can indeed be used to measure the time for tunneling photons in a manner that would simply not be possible with ordinary electronic laboratory instruments, these devices being too slow. The key, again, is that the signal and idler photons are produced simultaneously.

So consider once again the HOM interferometer but with a tunneling barrier placed in the signal arm, as pictured in Fig. 9.11. In the experiment of Steinberg, Kwiat, and Chiao [11], the barrier was a one-dimensional photonic band-gap material consisting of a multilayered dielectric mirror with a thickness of $d = 1.1$ μm. For a particle traveling at the speed of light

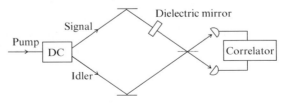

Fig. 9.11 Experiment on superluminal tunneling of photons. With the barrier placed in the signal beam, the path length of that beam must be increased in order to bring back the cancellation of coincident photon counts after the beam splitter.

c, the traversal time across the barrier would be $d/c = 3.6$ fs. Taking into account the index of refraction, of course, leads to longer traversal time.

Suppose that the HOM interferometer, without the barrier, is balanced such that there are no coincident counts by the detectors: both signal and idler photons reach the beam splitter simultaneously. Now we insert the barrier. Naively, we might expect that the photon from the idler mode will arrive first at the beam splitter. But this is another example where naïve intuition fails. If the idler *were* to arrive at the beam splitter first, we would expect that to regain balance coincidence by lengthening its path. However, experiments show that it is the path of the *signal* that needs to be lengthened, indicating that the photon passing through that barrier arrives at the beam splitter first. In fact, tunneling photons seem to be traveling at a speed greater than the speed of light, at about $1.7c$!

Needless to say, the interpretation of this result has led to some controversy. The conventional explanation of this effect is as follows. The emission time of a photon (or the photon pair) is not precisely known. Thus, its location is not precisely known either, though one can picture its position as being described by a Gaussian probability distribution – a Gaussian wave packet. Because of the correlations between the photons, both will have identical Gaussian wave packets. But the packet striking the barrier becomes split, the bulk of it being reflected with a very small part transmitted. Both of these packets describe the same photon. Most of the time the photon is reflected. The part representing the tunneling of the photon through the barrier is thought to be, as a result of pulse reshaping, composed mostly of the front part of the original wave packet. Thus, that section of the wave packet, because of the uncertainty in the production of the photons, reaches the beam splitter before the centroid of the packet in the other beam. At no time is any photon moving faster than the speed of light, and one certainly cannot use this phenomenon for faster-than-light signaling. However, this explanation has not satisfied everyone and controversy remains [12]. In Section 9.7 we examine another consequence of the uncertainty in the time of generation of a pair of photons.

9.6 Optical Test of Local Realistic Theories and Bell's Theorem

We have already shown how the Bell state $|\Psi^-\rangle$ can be produced from type I down-conversion by placing a $\pi/2$ rotator in one of the beams. The other Bell states can be generated as well via down-conversion in some form. For example, one can stack two type I down-converters whose optical axes are $90°$ to each other and pump with photons polarized at $45°$ to the optical axes of the crystals [13]. This source turns out to be rather a bright source of entangled light. Such an arrangement can produce the Bell states $|\Phi^\pm\rangle$.

Type II down-conversion, as described by the Hamiltonian of Eq. (9.6), can produce any of the four Bell states [14].

Let us now suppose, to make a definite choice, that we are able to routinely produce the Bell state

$$|\Psi^-\rangle = \frac{1}{\sqrt{2}}(|H\rangle_1|V\rangle_2 - |V\rangle_1|H\rangle_2), \qquad (9.36)$$

through a down-conversion process. Further, we define, in each of the modes, polarization states along and orthogonal to the directions θ (for mode 1) and ϕ (for mode 2):

$$
\begin{aligned}
|\theta\rangle_1 &= \cos\theta|H\rangle_1 + \sin\theta|V\rangle_1, \\
|\theta^\perp\rangle_1 &= -\sin\theta|H\rangle_1 + \cos\theta|V\rangle_1, \\
|\phi\rangle_2 &= \cos\phi|H\rangle_2 + \sin\phi|V\rangle_2, \\
|\phi^\perp\rangle_2 &= -\sin\phi|H\rangle_2 + \cos\phi|V\rangle_2.
\end{aligned}
\qquad (9.37)
$$

In terms of these states, we can write

$$
\begin{aligned}
|\Psi^-\rangle = \frac{1}{\sqrt{2}}\{&(\cos\theta\sin\phi - \sin\theta\cos\phi)|\theta\rangle_1|\phi\rangle_2 \\
&+ (\cos\theta\cos\phi + \sin\theta\sin\phi)|\theta\rangle_1|\phi^\perp\rangle_2 \\
&- (\sin\theta\sin\phi + \cos\theta\cos\phi)|\theta^\perp\rangle_1|\phi\rangle_2 \\
&- (\sin\theta\cos\phi - \cos\theta\sin\phi)|\theta^\perp\rangle_1|\phi^\perp\rangle_2\}. \quad (9.38)
\end{aligned}
$$

For the choice $\phi = \theta$, this result reduces to

$$|\Psi^-\rangle = \frac{1}{\sqrt{2}}(|\theta\rangle_1|\theta^\perp\rangle_2 - |\theta^\perp\rangle_1|\theta\rangle_2), \qquad (9.39)$$

demonstrating the rotational invariance of $|\Psi^-\rangle$. The state $|\Psi^-\rangle$ is mathematically equivalent to a singlet spin state, a spin-zero state composed of two spin-$1/2$ particles. The other Bell states are *not* rotationally invariant. This property, or the lack of it, has no bearing on what follows.

We are now in a position to discuss tests of quantum mechanics against the predictions of local realistic theories through a theorem enunciated 40 years ago by John Bell [1]. But first we should describe how photon polarization states can be measured. It is really very simple and requires only a piece of birefringent (double-refracting) material such as calcite, familiar in classical optics, and photon detectors. The action of a calcite crystal on an unpolarized light beam is to separate the beams into horizontally and vertically polarized beams where the former, called the ordinary (o) beam (or ray), is along the path of the input beam and the latter, called the extraordinary (e) beam (or ray) emerges along a different path, as illustrated in Fig. 9.12a. Operationally, the polarization state of a single photon is determined by observing which beam emits the photon: $|H\rangle$ if in the o-ray and $|V\rangle$ if in the e-ray. If we rotate the calcite crystal about the incident beam, the o-ray

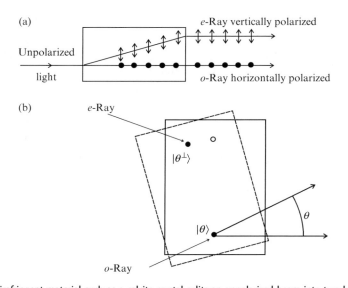

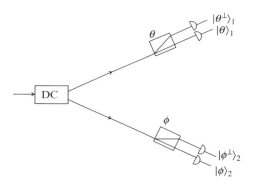

Fig. 9.12   (a) A birefringent material such as a calcite crystal splits an unpolarized beam into two beams of orthogonal polarizations. The o-ray is undeflected while the e-ray, whose polarization is orthogonal to that of the o-ray, is refracted. (b) In rotating the crystal about an axis through the o-ray, the polarizations of the two beams are still orthogonal.

Fig. 9.13   Schematic of a test of Bell's theorem using down-converted photons. The down-converter could be of type II or of type I with a polarization rotator placed in one of the beams.

remains unchanged while the e-ray follows the rotation as indicated in Fig. 9.12b. For a crystal rotated by an angle θ, single photons emerging in the o-ray are in the state $|\theta\rangle$ and those in the e-ray are in the state $|\theta^\perp\rangle$.

Suppose we have prepared a state of the form Eq. (9.39) and that we now place calcite crystals in each of the beams in order to do polarization measurements, as indicated in Fig. 9.13. We have assumed that the beams are propagating in directions such that the calcite measuring devices are widely separated. Ideally, we need to have the separations great enough so that the measurement events have a space-light, space-time interval.

Assuming this to be the case, and assuming both calcite crystals are set at an angle θ to the horizontal, here is what will happen. If the photon in mode 1 is found to be in the polarization state $|\theta\rangle_1$, then the other photon will be detected in the state $|\theta^\perp\rangle_2$, and vice versa. This is, of course, due to the strong correlations that exist in the state of Eq. (9.39). But there is a mystery that goes to the heart of the interpretation of the quantum mechanical state vector. Does the state described by Eq. (9.39) mean that, for a given pair of photons, the polarization of each within the pair is determined at the time of formation as either $|\theta\rangle_1|\theta^\perp\rangle_2$ or $|\theta^\perp\rangle_1|\theta\rangle_2$ and the measurements merely reveal which; or is it the measurement itself that randomly collapses one of the photons onto a particular polarization state and somehow forces the other, very distant, photon into the state with orthogonal polarization? If the latter, it would seem that some sort of violation of locality must be involved. If the measurements are performed with a space-like interval, how do the distant photons "know" what states the detectors must see in order to maintain the correlations apparent in their two-particle entangled state $|\Psi^-\rangle$? But there is more mystery. To illustrate that photon polarization is isomorphic to spin-$^1/_2$, we may construct the following operators:

$$\hat{\Sigma}_1 = |\theta\rangle\langle\theta^\perp| + |\theta^\perp\rangle\langle\theta|,$$
$$\hat{\Sigma}_2 = -i(|\theta\rangle\langle\theta^\perp| - |\theta^\perp\rangle\langle\theta|), \qquad (9.40)$$
$$\hat{\Sigma}_3 = |\theta\rangle\langle\theta| - |\theta^\perp\rangle\langle\theta^\perp|,$$

which satisfy the algebra $\left[\hat{\Sigma}_i, \hat{\Sigma}_j\right] = 2i\varepsilon_{i,j,k}\hat{\Sigma}_k$, the same algebra as satisfied by the Pauli spin operators. Thus, the Heisenberg uncertainty principle comes into play with regard to the components of the polarization. If, for example, the photon is known to be in the state $|\theta\rangle$ giving $\hat{\Sigma}_3$ the value of 1, then the values of $\hat{\Sigma}_1$ and $\hat{\Sigma}_2$ are undetermined as these operators do not commute with each other or with $\hat{\Sigma}_3$. But in the case of the Bell state of Eq. (9.39), we have the following situation. If mode 1 is detected in, say, state $|\theta\rangle_1$, then it follows that we can say with certainty that *if* a measurement *were* performed on beam 2 with the calcite again along θ, the result would be $|\theta^\perp\rangle_2$. But if we do no such measurement, can we conclude that the photon nevertheless must be in the state $|\theta^\perp\rangle_2$? If we draw such a conclusion (i.e. that the photon in mode 2 is in the eigenstate $|\theta^\perp\rangle_2$ of that mode's "spin" operator $\hat{\Sigma}_3^{(2)}$), then we are free to perform a measurement of, say, the operator $\hat{\Sigma}_1^{(2)}$, and find the photon in one of the eigenstates of that operator, which in turn implies a violation of the uncertainty relation as these operators do not commute.

The conundrum presented by the above discussion was first addressed, in a slightly different form, by Einstein, Podolsky, and Rosen (EPR) in

1935 [15], and later by Bohm [16] in the context of a spin-singlet state mathematically identical to polarization states above. We are faced with two problems: the locality problem and the problem of assigning definite values to observables at all times, the problem of "realism" (what EPR referred to as the "elements of reality"). The lack of determinate elements of reality led EPR to the conclusion that quantum mechanics must be an incomplete theory. Regarding the locality issue, we would expect a reasonable physical theory to be local. That is, it should not predict effects having a physical consequence, such as the possibility of superluminal signaling. As for the latter, classical-like reasoning might suggest the existence of "hidden variables" which supplement, and complete, quantum mechanics. In 1952, Bohm [2] proposed a hidden-variable theory that reproduced all the results of quantum mechanics. But it was nonlocal. Bell, in 1964 [1], was able to show that *local* hidden-variable theories make predictions that differ from those of quantum mechanics and thus provide a way to falsify either quantum mechanics or local hidden-variable theories.

To see how this comes about, we follow Bell and define the correlation function

$$C(\theta, \phi) = \text{Average}[A(\theta)B(\phi)] \tag{9.41}$$

where the average is taken over a large number of experimental runs. The quantity $A(\theta) = +1$ if photon 1 comes out in the o-beam (i.e. it is in the state $|\theta\rangle_1$) and -1 if it comes out in the e-beam (the state $|\theta^\perp\rangle_1$). A similar definition holds for $B(\phi)$ with regard to photon 2. As for the product $A(\theta)B(\phi)$, it has the value $+1$ for the outcome pairs $|\theta\rangle_1|\phi\rangle_2$ and $|\theta^\perp\rangle_1|\phi^\perp\rangle_2$ and -1 for the pairs $|\theta\rangle_1|\phi^\perp\rangle_2$ and $|\theta^\perp\rangle_1|\phi\rangle_2$. The average must then be

$$\begin{aligned} C(\theta, \phi) = &\Pr(|\theta\rangle_1|\phi\rangle_2) + \Pr(|\theta^\perp\rangle_1|\phi^\perp\rangle_2) \\ &- \Pr(|\theta\rangle_1|\phi^\perp\rangle_2) - \Pr(|\theta^\perp\rangle_1|\phi\rangle_2), \end{aligned} \tag{9.42}$$

where $\Pr(|\theta\rangle_1|\phi\rangle_2)$ is the probability of getting outcomes $|\theta\rangle_1|\phi\rangle_2$, and so on. For the state $|\Psi^-\rangle$ of Eq. (9.38), we can read off the probability amplitudes and then square to obtain, after a bit of manipulation,

$$C(\theta, \phi) = -\cos[2(\theta - \phi)]. \tag{9.43}$$

The same result is obtained from the expectation value

$$C(\theta, \phi) = \langle \Psi^- | \hat{\Sigma}_3^{(1)}(\theta) \hat{\Sigma}_3^{(2)}(\phi) | \Psi^- \rangle = -\cos[2(\theta - \phi)], \tag{9.44}$$

where

$$\begin{aligned} \hat{\Sigma}_3^{(1)}(\theta) &= |\theta\rangle_{11}\langle\theta| - |\theta^\perp\rangle_{11}\langle\theta^\perp|, \\ \hat{\Sigma}_3^{(2)}(\phi) &= |\phi\rangle_{22}\langle\phi| - |\phi^\perp\rangle_{22}\langle\phi^\perp| \end{aligned} \tag{9.45}$$

are the "spin" projections along a fictitious "3" axis. Clearly, these operators have as eigenvalues the numbers specified above for the values of $A(\theta)$ and $B(\phi)$. Thus, the correlation function of Eq. (9.43) is a straightforward quantum mechanical prediction.

To incorporate locality, we assume the existence of hidden variables, denoted λ, whose role is to determine the correlation between the photons in a pair when they are created. The variables can vary randomly from one run of the experiment to the next, but the values of $A(\theta)$ and $B(\phi)$ must depend on the hidden variables for a given run. Thus we should write these now as $A(\theta, \lambda)$ and $B(\phi, \lambda)$, where we still have $A(\theta, \lambda) = \pm 1$ and $B(\phi, \lambda) = \pm 1$. According to the principle of locality, the result of a measurement on one photon cannot depend on the angle setting of the calcite for the measurement of the other, and vice versa. For this reason we exclude functions of the form $A(\theta, \phi, \lambda)$ and $B(\theta, \phi, \lambda)$. We assume that for a large number of runs of the experiment, $\rho(\lambda)$ represents the probability distribution of the hidden variables λ such that

$$\int \rho(\lambda)d\lambda = 1. \tag{9.46}$$

Thus, for the local hidden-variable theory, the correlation function will be

$$C_{\mathrm{HV}}(\theta, \phi) = \int A(\theta, \lambda)B(\phi, \lambda)\rho(\lambda)d\lambda. \tag{9.47}$$

We must determine whether or not this expression is consistent with the quantum mechanical expression in Eq. (9.43). To this end, we need the following simple mathematical result. If $X_1, X_1', X_2, X_2' = \pm 1$, then we have

$$\begin{aligned} S &= X_1 X_2 + X_1 X_2' + X_1' X_2 - X_1' X_2' \\ &= X_1(X_2 + X_2') + X_1'(X_2 - X_2') = \pm 2, \end{aligned} \tag{9.48}$$

as can easily be checked. If we set $X_1 = A(\theta, \lambda)$, $X_1' = A(\theta', \lambda)$, $X_2 = B(\phi, \lambda)$, and $X_2' = B(\phi', \lambda)$, multiply by $\rho(\lambda)$ and integrate over λ, we obtain

$$-2 \le C_{\mathrm{HV}}(\theta, \phi) + C_{\mathrm{HV}}(\theta', \phi) + C_{\mathrm{HV}}(\theta', \phi) - C_{\mathrm{HV}}(\theta', \phi') \le +2. \tag{9.49}$$

This is a Bell inequality in the form given by Clauser, Horne, Shimony, and Holt [17] (the CHSH inequality for short). If we set $\theta = 0$, $\theta' = \pi/4$, $\phi = \pi/8$, and $\phi' = -\pi/8$, we find from the quantum mechanical correlation function of Eq. (9.43) that

$$S = C(\theta, \phi) + C(\theta', \phi) + C(\theta', \phi) - C(\theta', \phi') = 2\sqrt{2}, \tag{9.50}$$

a result clearly outside the bounds placed by the inequality of Eq. (9.49) obtained from local hidden-variable theory. This violation of the CHSH inequality means that the quantum mechanical result cannot be explained by any local hidden-variable theory. Of course, experiment must be the final arbiter in this matter.

Experimental tests of inequalities of the Bell type using polarized light generated from atomic $J = 0 \rightarrow 1 \rightarrow 0$ cascade transitions in calcium began in the early 1970s with the experiment of Freedman and Clauser [18], culminating in the experiments of Aspect *et al.* in 1982 [19, 20]. In the second of the latter experiments, the polarization analyzers were rotated while the photons were in flight. Subsequently, and independently, Ou and Mandel [21] and Shih and Alley [22] performed the first test using down-conversion as a source of polarization-entangled photons.

Most of the experiments exhibit violations of the inequalities being tested, but not without certain supplementary assumptions being used in the data analysis. In any real experiment done with photons, one must take into account the fact that the detector efficiency, which we represent by η_{det}, is never at the ideal, $\eta_{\text{det}} = 1$, although we assumed the contrary in deriving our results. For the non-ideal case, Eq. (9.43) should be replaced by

$$C(\theta, \phi) = -\eta_{\text{det}}^2 \cos[2(\theta - \phi)]. \tag{9.51}$$

This comes about because each term in Eq. (9.42) gets multiplied by η_{det}^2, which in turn is the result of the fact that each term is a joint probability of two measurements where a factor of η_{det} appears in the measurements. Thus, from Eq. (9.50), we must now have $S = \eta_{\text{det}}^2 2\sqrt{2}$. Recalling that Bell's inequality is violated whenever $S > 2$, it is evident that we must place the constraint $\eta_{\text{det}} > 1/\sqrt[4]{2} \approx 0.84$ in order to obtain a violation.

With inefficient detectors, not all photon pairs will register counts. In a practical experiment it is necessary to introduce an untestable supplementary assumption. The assumption is that the detectors perform a fair sampling of the ensemble of all events. That is, those events for which both photons are detected are representative of the entire ensemble. We thus redefine the correlation function according to

$$C(\theta, \phi) = \frac{\text{Average}[A(\theta)B(\phi)]}{\text{Average}[N_1 N_2]}, \tag{9.52}$$

where N_1 is the total number of photons detected in beam 1 (in any given run, $N_1 = 0$ or 1) with a similar definition for N_2, where

$$\begin{aligned} \text{Average}[N_1 N_2] = {} & \text{Pr}(\theta, \phi) + \text{Pr}(\theta^\perp, \phi^\perp) \\ & + \text{Pr}(\theta, \phi^\perp) + \text{Pr}(\theta^\perp, \phi). \end{aligned} \tag{9.53}$$

Because both numerator and denominator are proportional to the factor μ^2, the correlation function will be independent of it. With this fair sampling assumption, the CHSH of Bell's inequality will be violated according to quantum mechanics for detectors of any efficiency.

Our description of these experiments is, of course, considerably oversimplified. Nevertheless, the experiments, or experiments like them, are not too complicated to be done, at modest expense, in an undergraduate laboratory setting. Dehlinger and Mitchell [23] have described apparatus for the generation of polarization-entangled states that may be used to demonstrate the violation of Bell's inequalities in the undergraduate laboratory. Their approach is to use two face-to-face type I BBO down-converters, where the second is rotated by 90° around the normal to the first. The crystals are pumped by a beam of 45° polarized light obtained from a diode laser. This arrangement produces an entangled state of the form

$$|\Phi^+\rangle = \frac{1}{\sqrt{2}}(|H\rangle_1|H\rangle_2 + |V\rangle_1|V\rangle_2). \qquad (9.54)$$

The two-crystal scheme described is an efficient method of generating polarization states and was first used by Kwiat *et al.* [24]. Their results violated Bell's inequality by 21 standard deviations.

Not all experiments demonstrating a violation of a Bell-type inequality using down-converted light have involved polarization entanglement. Rarity and Tapster [25] have done experiments using momentum entanglement of the beams. Another, proposed by Franson [26], involves time–energy uncertainty. We now discuss this experiment.

9.7 Franson's Experiment

Consider the experimental setup pictured in Fig. 9.14. A pair of photons, signal and idler, is simultaneously produced by the down-converter and directed toward the interferometers. There is no single-photon interference inside a single interferometer as the photon coherence lengths are much shorter than the differences between path lengths of the arms of the interferometers. However, there is two-photon interference in the coincidence detection between the detectors D_1 and D_2. To see how this comes about, consider the following. To reach the detectors, the photons can both take the short paths (S, S), both take the long paths (L, L), the short–long (S, L) or the long–short (L, S) paths. The first two cases are indistinguishable as we do not know when the photons are created. Both detectors will tend to fire simultaneously, assuming identical path lengths in the two interferometers. The last two cases are distinguishable because of the delay between the "clicking" of the two detectors, the detector of the photon taking the short path will "click" first. So, the last two cases are distinguishable from each other *and* from the first two cases in which the detectors "click" simultaneously. Under experimental conditions, the fast electronics of the correlator can be set for a sufficiently narrow timing window such that counts from the two distinguishable processes are rejected. This has the effect of post-selecting (reducing) the two-photon output state to be

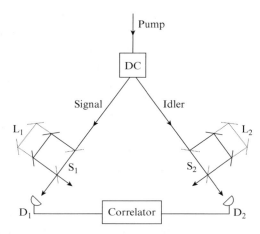

Fig. 9.14 Schematic for Franson's experiment. The path lengths of the interferometer are adjustable, as indicated.

$$|\psi\rangle = \frac{1}{2}(|S\rangle_1|S\rangle_2 + e^{i\Phi}|L_1\rangle_1|L_2\rangle_2), \tag{9.55}$$

evidently another form of Bell state, where we have assumed that $S_1 = S_2 = S$ but that L_1 and L_2 can be adjusted, as indicated in Fig. 9.14. The phase Φ, the relative phase between the (S, S) and (L, L) processes, is the sum of the relative phases acquired by the individual photons:

$$\begin{aligned}
\Phi &= \omega_s \Delta L_1/c + \omega_i \Delta L_2/c \\
&= \frac{\omega_s + \omega_i}{2c}(\Delta L_1 + \Delta L_2) + \frac{\omega_s - \omega_i}{2c}(\Delta L_1 - \Delta L_2) \\
&\approx \frac{\omega_p}{2c}(\Delta L_1 + \Delta L_2),
\end{aligned} \tag{9.56}$$

valid as $\Delta L_1 - \Delta L_2$ is taken to be small compared to the inverse bandwidth of the signal and idler frequencies ω_s and ω_i. The frequency $\omega_p = \omega_s + \omega_i$ is, of course, the frequency of the pump field. If we now use the Feynman dictum of adding amplitudes associated with indistinguishable processes, the probability of coincident two-photon detection is

$$\begin{aligned}
P_{\text{coin}} &= \frac{1}{4}|1 + e^{i\Phi}|^2 = \frac{1}{2}[1 + \cos\Phi] \\
&= \frac{1}{2}\left[1 + \cos\left(\frac{\omega_p}{2c}(\Delta L_1 + \Delta L_2)\right)\right].
\end{aligned} \tag{9.57}$$

This result exhibits 100% visibility, meaning that the minimum probability of coincident detection is zero. This, in turn, means that the photons entering the two interferometers become *anti-correlated*: if one takes the short path then the other takes the long path, and vice versa. The meaning of the maximum is that the photons become *correlated* in the interferometers: either both take the short path or both take the long path. No classical or hidden-variable model can predict a visibility greater that 50%. Fringes of visibility greater than 70.7% violate a Bell-type inequality. The experiments of Kwiat *et al.* [27] exhibit a visibility of 80.4 ± 0.6%. Other experimental realizations of the Franson experiment are those of Ou *et al.* [28], Brendel *et al.* [29], and Shih *et al.* [30].

9.8 Applications of Down-Converted Light to Metrology without Absolute Standards

Finally in this chapter, we briefly discuss the application of down-converted light to problems of practical interest. These applications are possible because of the tight correlations of the photons that are spontaneously emitted during the down-conversion process, as was originally shown by Burnham and Weinberg [3]. The virtue of using down-converted light is, as we shall see, that measurements can be performed to yield absolute results without the need for a calibrated standard.

We first discuss the determination of the absolute calibration of a photon detector, essentially measuring its absolute quantum efficiency. A schematic of the measurement is given in Fig. 9.15. Two identical photon detectors are placed in the outputs of a type I down-converter. We regard detector A as the detector to be calibrated and detector B as the trigger, though which is which is quite arbitrary. Because the down-converter creates photons in pairs, coincident photons would be detected in both arms for ideal detectors. The measurement of the quantum efficiency is quite simple. Suppose that N is the number of pairs of photons produced by the down-converter. Whenever detector B "clicks," the experimenter

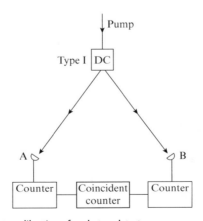

Fig. 9.15   Schematic for the absolute calibration of a photon detector.

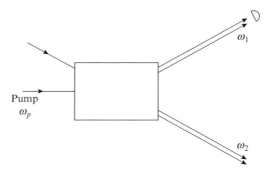

 Fig. 9.16 Schematic for the measurement of absolute radiance.

checks to see if detector A also "clicks." Then the fraction of detections by B for which there is a coincident detection by A is the measured quantum efficiency of detector A. The method does not require knowledge of the quantum efficiency of detector B. If there is no detection by B, the experimenter simply disregards detector A altogether. More quantitatively, if we let the quantum efficiencies of detectors A and B be represented respectively by η_A and η_B, then the number of photons detected by each is $N_A = \eta_A N$ and $N_B = \eta_B N$. The number of coincident counts will be given by $N_C = \eta_A \eta_B N = \eta_A N_B$. Thus $\eta_A = N_B/N_C$, which is obviously independent of the quantum efficiency of detector B. The idea behind this sort of measurement is implicit in the work of Burnham and Weinberg [3]. Migdall *et al.* [31] at NIST performed a test of the method in 1995 – comparing the method with a more conventional method involving a standard – and found good agreement.

Another metrologic application, the only other one we discuss, is to the measurement of absolute radiance. This application was proposed by Klyshko [32] and was first demonstrated in the laboratory in 1979 [33]. It is a variation of the process of SPDC, where a light beam whose radiance is to be measured (the unknown) is aligned to overlap with one of the output beams, say the signal beam, of the down-converter – as indicated in Fig. 9.16. The setup constitutes a parametric amplifier where the presence of the unknown input beam stimulates the production of photon pairs (they must always be produced in pairs) into the output beams. This is indicated in Fig. 9.16 by the double lines at the outputs. The idea is to monitor the output of the idler beam and determine the increase in strength of the radiation in going from spontaneous down-conversion, where the unknown beam is blocked, to the case of stimulated emission, where the unknown input beam is unblocked. The ratio of the strengths of the stimulated and spontaneous output signals yields the radiance of the unknown beam. No standard is required in such a measurement.

The applications discussed here have not involved entanglement. However, applications that do involve entanglement, and thus require type II down-conversion, have been proposed. A specific example is the application to ellipsometry [34]. For reviews with many references, see the papers listed in the bibliography at the end of this chapter.

Problems

1. In the perturbation expansion of Eq. (9.8), in first order the output of a type I down-converter is the pair state $|1\rangle_s|1\rangle_i$. (a) Show that in second order, the output is the pair state $|2\rangle_s|2\rangle_i$. (b) Supposing that the two photons within each of the states, and both pairs of the photons, simultaneously fall onto the inputs of a 50:50 beam splitter, what will be the output state? (c) Show that your result does not agree with that which would be obtained classically using a simple binomial distribution. (d) Devise an experiment able to discriminate between the quantum mechanical and classical predictions for the output of the beam splitter. (*After* working through this problem you may wish to consult a paper by Ou, Rhee, and Wang [35].)

2. Derive a CHSH–Bell inequality starting from the Bell state

$$|\Psi^+\rangle = \frac{1}{\sqrt{2}}(|H\rangle_1|V\rangle_2 + |V\rangle_1|H\rangle_2).$$

3. As mentioned in the text and in Refs. [23, 24], using back-to-back type I down-converters, where one is rotated 90° around the normal to the other and where both are pumped by light polarized at 45°, the Bell state

$$|\Phi^+\rangle = \frac{1}{\sqrt{2}}(|H\rangle_1|H\rangle_2 + |V\rangle_1|V\rangle_2)$$

can be produced. Analyze the possible violations of Bell's inequalities (in the form of CHSH) with respect to this state.

4. For the state $|\Psi^-\rangle$, consider the quantity

$$S_{\text{Bell}} \equiv |C(\theta, \phi) - C(\theta, \phi')| + |C(\theta', \phi') + C(\theta', \phi)|.$$

Show, for a local realistic theory, where we replace $C(\theta, \phi)$ by $C_{\text{HV}}(\theta, \phi)$, that $S_{\text{Bell}} \leq 2$. Find an example where quantum mechanics violates the inequality. (This is the inequality originally discussed by Bell.)

References

[1] J. S. Bell, *Physics (NY)* **1**, 195 (1964). See also J. S. Bell, *Speakable and Unspeakable in Quantum Mechanics* (Cambridge: Cambridge University Press, 1987).
[2] D. Bohm, *Phys. Rev.* **85**, 166 (1952).
[3] D. C. Burnham and D. L. Weinberg, *Phys. Rev. Lett.* **25**, 84 (1970).
[4] A. Joobeur, B. Saleh, and M. Teich, *Phys. Rev. A* **50**, 3349 (1994).
[5] See, for example, R. W. Boyd, *Nonlinear Optics*, 2nd edition (San Diego, CA: Academic Press, 2003).

[6]   C. K. Hong, Z. Y. Ou, and L. Mandel, *Phys. Rev. Lett.* **59**, 2044 (1987). A related experiment was later performed by J. G. Rarity, P. R. Tapster, E. Jakeman, T. Larchuk, R. A. Campos, M. C. Teich, and B. E. A. Saleh, *Phys. Rev. Lett.* **65**, 1348 (1990).

[7]   P. G. Kwiat, A. M. Steinberg, and R. Y. Chiao, *Phys. Rev. A* **45**, 7729 (1992).

[8]   D. M. Greenberger, M. A. Horne, and A. Zeilinger, *Phys. Today* **Aug**, 22 (1993).

[9]   X. Y. Zou, L. J. Wang, and L. Mandel, *Phys. Rev. Lett.* **67**, 318 (1991).

[10] R. Y. Chiao and A. M. Steinberg in E. Wolf (Ed.), *Progress in Optics XXXVII* (Amsterdam: Elsevier, 1997), p. 345; G. Nimtz and W. Heitmann, *Prog. Quant. Electr.* **21**, 81 (1997).

[11] A. M. Steinberg, P. G. Kwiat, and R. Y. Chiao, *Phys. Rev. Lett.* **71**, 708 (1993).

[12] See H. Winful, *Phys. Rev. Lett.* **90**, 023901 (2003); M. Büttiker and S. Washburn, *Nature* **422**, 271 (2003); and the exchange between Winful and Büttiker and Washburn in *Nature* **424**, 638 (2003).

[13] See any good optics text, such as F. L. Pedrotti and L. S. Pedrotti, *Introduction to Optics*, 2nd edition (Englewood Cliffs, NJ: Prentice Hall, 1993). For a discussion of polarized single-photon states and the use of calcite crystals to detect them, along with a discussion of their use in elucidating a number of puzzling features in quantum mechanics, see A. P. French and E. F. Taylor, *An Introduction to Quantum Physics* (New York: W.W. Norton, 1978), chapter 6.

[14] See W. Tittel and G. Weihs, *Quant. Inf. Comp.* **1**, 3 (2001); Y. Shih, *Rep. Prog. Phys.* **66**, 1009 (2003).

[15] A. Einstein, B. Podolsky, and N. Rosen, *Phys. Rev.* **47**, 777 (1935).

[16] D. Bohm, *Quantum Theory* (New York: Prentice Hall, 1951), chapter 22.

[17] J. F. Clauser, M. A. Horne, A. Shimony, and R. A. Holt, *Phys. Rev. Lett.* **23**, 880 (1969).

[18] S. J. Freedman and J. F. Clauser, *Phys. Rev. Lett.* **28**, 938 (1972).

[19] A. Aspect, P. Grangier, and G. Roger, *Phys. Rev. Lett.* **49**, 91 (1982).

[20] A. Aspect, J. Dalibard, and G. Roger, *Phys. Rev. Lett.* **49**, 1804 (1982).

[21] Z. Y. Ou and L. Mandel, *Phys. Rev. Lett.* **61**, 50 (1988).

[22] Y. H. Shih and C. O. Alley, *Phys. Rev. Lett.* **61**, 2921 (1988).

[23] D. Dehlinger and M. W. Mitchell, *Am. J. Phys.* **70**, 898 (2002).

[24] P. G. Kwiat, K. Mattle, H. Weinfurter, A. Zeilinger, A. V. Sergienko, and Y. H. Shih, *Phys. Rev. Lett.* **75**, 4337 (1995); P. G. Kwiat, E. Waks, A. G. White, I. Appelbaum, and P. H. Eberhard, *Phys. Rev. A* **60**, R773 (1999).

[25] J. G. Rarity and P. R. Tapster, *Phys. Rev. Lett.* **64**, 2495 (1990).

[26] J. D. Franson, *Phys. Rev. Lett.* **62**, 2205 (1989).

[27] P. G. Kwiat, A. M. Steinberg, and R. Y. Chiao, *Phys. Rev. A* **47**, R2472 (1993).

[28] Z. Y. Ou, X. Y. Zou, L. J. Wang, and L. Mandel, *Phys. Rev. Lett.* **65**, 321 (1990).

[29] J. Brendel, E. Mohler, and W. Martienssen, *Euro. Phys. Lett.* **20**, 575 (1992).

[30] Y. H. Shih, A. V. Sergienko, and M. H. Rubin, *Phys. Rev. A* **47**, 1288 (1993).

[31] A. Migdall, R. Datla, A. Sergienko, J. S. Orszak, and Y. H. Shih, *Metrologica* **32**, 479 (1995/6).

[32] D. N. Klyshko, *Sov. J. Quantum Electron.* **7**, 591 (1977).

[33] G. Kitaeva, A. N. Penin, V. V. Fadeev, and Yu. A. Yanait, *Sov. Phys. Dokl.* **24**, 564 (1979).

[34] A. F. Abouraddy, K. C. Toussaint, A. V. Sergienko, B. E. A. Saleh, and M. C. Teich, *J. Opt. Soc. Am. B* **19**, 656 (2002).

[35] Z. Y. Ou, J.-K. Rhee, and L. J. Wang, *Phys. Rev. Lett.* **83**, 959 (1999).

Bibliography

D. N. Klyshko, *Photons and Nonlinear Optics* (New York: Gordon and Breach, 1988).

R. Y. Chiao, P. G. Kwiat, and A. M. Steinberg, "Faster than light?", *Sci. Am.* **Aug**, 52 (1993).

P. Harihan and B. C. Sanders, "Quantum phenomena in optical interferometry" in E. Wolf (Ed.), *Progress in Optics XXXVI* (Amsterdam: Elsevier, 1996), p. 49.

R. Y. Chiao and A. M. Steinberg, "Tunneling times and superluminality" in E. Wolf (Ed.), *Progress in Optics XXXVII* (Amsterdam: Elsevier, 1997), p. 345.

L. Mandel, "Quantum effects in one-photon and two-photon interference," *Rev. Mod. Phys.* **71**, S274 (1999).

The discussion of the Bell inequalities above follows closely that of

L. Hardy, "Spooky action at a distance in quantum mechanics," *Contemp. Phys.* **39**, 419 (1998).

The early experiments using atomic cascades to generate polarization-entangled photons are reviewed in

J. F. Clauser and A. Shimony, "Bell's theorem: experimental tests and implications," *Rep. Prog. Phys.* **41**, 1881 (1978).

Two articles that review applications of down-converted light to metrology are

A. Migdall, "Correlated-photon metrology without absolute standards," *Phys. Today* **Jan**, 41 (1999).

A. V. Sergienko and G. S. Sauer, "Quantum information processing and precise optical measurements with entangled photon pairs," *Contemp. Phys.* **44**, 341 (2003).

A recent review article on the Hong–Ou–Mandel effect is

F. Bouchard, A. Sit, Y. Zhang, R. Fickler, F. M. Miatto, Y. Yao, *et al.*, "Two-photon interference: The Hong–Ou–Mandel effect," *Rep. Prog. Phys.* **84**, 012402 (2021).

Experiments in Cavity QED and with Trapped Ions

In this chapter, we discuss two more experimental realizations of quantum optical phenomena, namely the interaction of an effective two-level atom with a quantized electromagnetic field in a high-Q microwave cavity, the subject usually referred to as cavity QED (or sometimes CQED) and in the quantized motion of a trapped ion. Strictly speaking, these experiments are not optical, but they do realize interactions of exactly the type that are of interest in quantum optics, namely the Jaynes–Cummings interaction between a two-level system (an atom) and a bosonic degree of freedom, a single-mode cavity field in the case of a microwave cavity, and a vibrational mode of the center-of-mass motion of a trapped ion, the quanta being phonons in this case. We shall begin with a description of the useful properties of the so-called Rydberg atoms that are used in the microwave CQED experiments, proceed to discuss some general considerations of the radiative behavior of atoms in cavities, the CQED realization of the Jaynes–Cummings model, and then discuss the use of the dispersive, highly off-resonant, version of the model to generate superpositions of coherent states (i.e. the Schrödinger-cat states of the type discussed in Chapters 7 and 9), but this time for a cavity field. Finally, we discuss the realization of the Jaynes–Cummings interaction in the vibrational motion of a trapped ion.

10.1 Rydberg Atoms

A Rydberg atom is an ordinary atom, where one of its electrons, the valence electron in an alkali atom, is excited to a state of very high principal quantum number [1], a Rydberg state. In the experiments to be discussed, the Rydberg atoms used are those of rubidium, with the principal quantum number n (not to be confused with the photon number of a field state, as should be clear by the context) of the valence electron in the vicinity of $n \sim 50$. The electronic binding energy of a Rydberg state is given by

$$E_{nl} = -\frac{R}{(n - \delta_l)^2} = -\frac{R}{n^{*2}}, \tag{10.1}$$

where $n^* = n - \delta_l$ and where δ_l is the "quantum defect" due to the hydrogenic "core," which corrects for the deviation of the binding potential from a purely hydrogenic situation. The parameter R is the Rydberg constant having value $R = 13.6$ eV. The quantum defect does depend on the angular momentum quantum number l and has a value of unity for low l states, decreasing for higher l. As we are interested mainly in high l where δ_l is small, we may characterize our Rydberg states with the usual principal quantum number n. States for which l takes on the highest value allowed for a given n, $l = n - 1$, and with $|m| = n - 1$ (where m is the magnetic quantum number), are known as circular Rydberg states, as in the classical limit they describe an electron in a circular orbit [2]. It is the circular Rydberg states that are used in the CQED experiments. Nussenzweig et al. [3] have described how such states may be prepared in the case of rubidium.

So, what are the special properties of circular Rydberg atoms that make them suitable for CQED experiments? They are numerous. To begin, only one electric dipole transition is allowed, $n \leftrightarrow n - 1, |m| \leftrightarrow |m| - 1$, so that such states closely approximate a two-level system. The "classical" radius of a Rydberg atom scales as $n^2 a_0$, where $a_0 = 0.5$ Å is the Bohr radius. For $n \sim 100$, the atom is about the size of a virus. (Rydberg atoms making transitions from states with principal quantum numbers as high as $n \sim 733$ have been observed astrophysically.) The matrix element of the electric dipole operator between states, between two circular Rydberg states of principal quantum numbers n and n' for $\Delta n = n - n' = 1$, is

$$d = \langle n|\hat{d}|n'\rangle \sim qn^2 a_o, \qquad (10.2)$$

where $\hat{d} = qr$. For $n = 50$ we obtain $d \sim 1390$ A.U., where $qa_0 = 1$ A.U., a value for the dipole moment close to 300 times those of typical optical transitions for low-lying hydrogenic states. The large dipole moments of the circular Rydberg atoms make them particularly attractive for CQED experiments as their coupling to a single-mode cavity field can be quite large. For a Rydberg atom in large-n state, the frequency of the radiation emitted in a transition for which $\Delta n = 1$ is close to the Bohr orbital frequency itself:

$$\omega_0 = \frac{E_{nl} - E_{n'l'}}{\hbar} \simeq \frac{2R}{\hbar n^3}. \qquad (10.3)$$

For $n \sim 50$, the frequency of the emitted radiation is $v_0 = \omega_0 / 2\pi \sim 36$ GHz, which corresponds to a wavelength of $\lambda_0 = c/v_0 \sim 8$ mm. This last number gives a sense of the dimensions required of a cavity to support a standing microwave field.

The rate of spontaneous emission is given by

$$\Gamma = \frac{d^2 \omega_0^3}{3\pi\varepsilon_0 \hbar c^3}. \qquad (10.4)$$

Using the previous results for circular Rydberg transitions, $d^2 \sim n^4$ and $\omega_0^3 \sim n^{-9}$, we find that the decay rate for these transitions is

$$\Gamma \sim \Gamma_0 n^{-5}, \tag{10.5}$$

where $\Gamma_0 = c\alpha^4/a_0$, α here being the fine-structure constant $\alpha = e^2/\hbar c = 1/137$. The rate Γ_0 is that for the usual strongly allowed lower-level spontaneous emissions (we normally deal with optical transitions) and typically has an order of magnitude $\Gamma_0 \sim 10^9 \mathrm{s}^{-1}$, with corresponding excited-state lifetimes $\tau_0 = 1/\Gamma_0 \sim 10^{-9}\mathrm{s}$. The circular Rydberg transitions will have a lower rate and longer excited-state lifetimes given by $\tau = 1/\Gamma = \tau_0 n^5$. For $n \sim 50$ we obtain $\tau \sim 10^{-1}\mathrm{s}$, a very long lifetime indeed compared to the "ordinary" spontaneous emission lifetimes of around $\tau_0 \sim 10^{-9}\mathrm{s}$.

Finally, we discuss the issue of atomic-state detection. For a Rydberg atom, the electron is typically far from the hydrogenic core and its binding energy is rather small, so this allows relatively easy ionization by an external applied field. The electric field normally felt by the electron, with no applied field, from Coulomb's law is

$$E_{\text{Coulomb}} \approx \frac{e}{4\pi\varepsilon_0 (n^2 a_0)^2}, \tag{10.6}$$

and thus the ionization rate with an applied constant uniform field $E \sim E_{\text{Coulomb}}$ goes as n^{-4}. The ionization rate of the Rydberg state with principal quantum number n is smaller than one with $n - 1$ by a factor of $(1 + 4/n)$. Though the difference in the rates for adjacent Rydberg levels is small in comparison to the case for low-lying levels, it is sufficiently large for state-selective measurements to be performed by field ionization. The idea is quite simple: one allows the atom to pass through a set of two ionization detectors, each having different field strengths (see Fig. 10.1). The atom must first encounter the one with the field set to detect the highest

Field ionization detectors

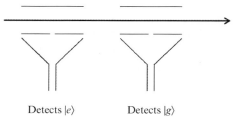

Detects $|e\rangle$ Detects $|g\rangle$

Fig. 10.1   Field ionization of an atomic beam of highly excited Rydberg atoms allows for selective state detection: an electric field is applied that ramps up so that the first detector monitors the excited-state population whereas the stronger field downstream monitors the more strongly bound ground state.

of two states, say the state with quantum number n. The second detector has a stronger applied field to detect the state with quantum number $n - 1$. The strength of the applied fields must be carefully set so that the first detector will ionize the atom only if the electron is in the upper level, while that of the second must be strong enough to ionize the lower. Of course, it is the detection of the field-ionized electron in one of the detectors that constitutes selective atomic state detection.

In summary, circular atomic Rydberg states have the following desirable properties: (1) the transitions are restricted to other circular states, thus setting the stage for them to act as "two-level" atoms; (2) the dipole moments of the allowed transitions are large; (3) the states have long lifetimes; and (4) they may be selectively ionized by applied fields to achieve selective atomic-state detection.

10.2 Rydberg Atom Interacting with a Cavity Field

As we have shown, the wavelengths of the radiation emitted as a result of transitions between neighboring circular Rydberg states is in the range of a few millimeters. This corresponds to microwave radiation. A Rydberg atom in free space then should undergo, in addition to spontaneous emissions, stimulated emissions and absorptions as a result of ambient blackbody radiation in the microwave part of the spectrum. We label an adjacent pair of Rydberg states $|e\rangle$ and $|g\rangle$, which typically could correspond to states of principal quantum numbers $n = 50$ and $n = 49$, respectively. If $U_{\mathrm{BB}}(\omega)$ is the black-body energy density, then the probability rate of stimulated absorption taking the atom from $|g\rangle$ to $|e\rangle$ is given by [4]

$$W_{ge}^{\mathrm{BB}} = \frac{\pi d^2}{\varepsilon_0 \hbar^2} \overline{(\mathbf{e}_a \cdot \mathbf{e})^2} U_{\mathrm{BB}}(\omega_0), \qquad (10.7)$$

where $\mathbf{e}_a$ and $\mathbf{e}$ are the polarization unit vectors of the atomic transition and the radiation field, respectively, and the bar represents an average over all possible polarization and propagation directions. One can show that $\overline{(\mathbf{e}_a \cdot \mathbf{e})^2} = 1/3$. Further, for a thermal field in free space, we can write the energy density as $U_{\mathrm{BB}}(\omega_0) = \rho_{\mathrm{fs}}(\omega_0)\bar{n}(\omega_0)$, where $\rho_{\mathrm{fs}}(\omega_0)$ is the mode density of free space, $\rho_{\mathrm{fs}}(\omega_0) = \omega_0^2/(\pi^2 c^3)$ [see Eq. (2.75)], and

$$\bar{n}(\omega_0) = [\exp(\hbar\omega_0/k_B T) - 1]^{-1}. \qquad (10.8)$$

Thus, we can write

$$W_{ge}^{\mathrm{BB}} = B\bar{n}, \qquad (10.9)$$

where

$$B = \frac{d^2 \omega_0^3}{3\pi \varepsilon_0 \hbar c^3}. \tag{10.10}$$

The probability rate of stimulated emission from the $|e\rangle$ to $|g\rangle$ transition is given by

$$W_{eg}^{\mathrm{BB}} = B\,\bar{n}(\omega_0) = W_{ge}^{\mathrm{BB}}. \tag{10.11}$$

Spontaneous emission from state $|e\rangle$ can be taken into account by writing

$$W_{eg}^{\mathrm{BB+SpE}} = B\,[\bar{n}(\omega_0) + 1]. \tag{10.12}$$

From Eq. (10.4) we notice that $B = \Gamma$. From this equivalence, we can write

$$\Gamma = \frac{\pi d^2}{3\varepsilon_0 \hbar^2} \rho_{\mathrm{fs}}(\omega_0)\hbar\omega_0. \tag{10.13}$$

If $\bar{n} = 0$, the rate of spontaneous emission appears to be related to the vacuum fluctuations of all the modes around the atom. For a Rydberg transition with wavelengths in the millimeter range, $\hbar\omega_0 \ll k_B T$ even at room temperatures, so we may use the Rayleigh–Jeans limit of the Planck radiation law for $\bar{n} \approx k_B T/\hbar\omega_0 \gg 1$ to obtain

$$W_{eg}^{\mathrm{BB}} \approx \frac{d^2 \omega_o^2 k_B T}{3\pi\varepsilon_0 \hbar^2 c^3}. \tag{10.14}$$

Because for circular Rydberg states $d^2 \sim n^4$ and $\omega_0^2 \sim n^{-6}$, we have $W_{eg}^{\mathrm{BB}} \sim n^{-2}$. This needs to be compared with the *total* spontaneous emission rate of the state $|e\rangle$, which is $\Gamma \sim n^{-5}$. Because $W_{eg}^{\mathrm{BB}}/\Gamma \sim n^3$, the stimulated emission rate due to the black-body radiation surrounding the atom is much greater than the rate of spontaneous emission.

The rates W_{eg}^{BB} and Γ are essentially irreversible rates: the atom radiates a photon into a continuum of modes, never to be reabsorbed by the atom. Our goal has been to obtain the conditions under which the atom–field dynamics will be reversible. We can do this by placing the atom in a cavity of appropriate dimensions such that the density of modes is much, much less than in free space. Ideally, we would wish to have a cavity constructed to support just one mode of a frequency close to the resonance with the atom transition frequency ω_0. This will reduce the rate of irreversible *spontaneous* emissions as only one cavity mode is available into which the atom can radiate. At the same time, we wish to suppress the transitions due to the presence of black-body radiation of wavelength corresponding to the atomic transition frequency. For a cavity with walls of low enough temperature such that $\hbar\omega_0 \gg k_B T$, we have $\bar{n} \approx \exp(-\hbar\omega_0/k_B T) \ll 1$, and thus the stimulated transition rate due to thermal photons will be small.

As a simple example of an atom in such a cavity, we consider the case of an atom in the excited state $|e\rangle$ placed in a cavity supporting a single-mode field at resonance with atomic transition frequency ω_0. We assume that the field is initially in the vacuum, so that our initial state of the atom–field system is $|1\rangle = |e\rangle|0\rangle$. When the atom undergoes a spontaneous emission, the system goes to the state $|2\rangle = |g\rangle|1\rangle$. These atom+field states have identical total energies; $E_1 = E_2 = \hbar\omega_0/2$ (at resonance). If there are now other interactions involved, then the system would simply oscillate period- ically between the states $|1\rangle$ and $|2\rangle$ at the vacuum Rabi frequency $\Omega_0 = \Omega(0) = 2\lambda$ (see Section 4.5). But there is one other possibility: the photon could leak out of the cavity before the atom is able to reabsorb it. No matter how low the temperature of the cavity walls, they interact with the atom–field system and have two important effects: the energy field decays as a result of the loss of photons, and there is a decoherence of the quantum dynamics of the system. So, there is a third state that we need to consider in our description of the interaction of an excited atom with a single-mode cavity field: the $|3\rangle = |g\rangle|0\rangle$ is decoupled from the reversible dynamics but appears as a result of the irreversible dynamics associated with losses. State $|3\rangle$ has energy $E_3 = -\hbar\omega_0/2$, lower than E_1 and E_2 by an amount $\hbar\omega_0$, as indicated in Fig. 10.2. This energy goes into the excitation of the reservoir. The issue of incoherent losses is the subject of Chapter 8.

Here, as there, we will introduce the appropriate master equation to model the losses.

If we let the interaction Hamiltonian be represented by

$$\hat{H}_{\mathrm{I}} = \hbar\lambda(\hat{a}\hat{\sigma}_+ + \hat{a}^\dagger\hat{\sigma}_-), \tag{10.15}$$

as we did in Chapter 3, then we can write the master equation for the evolution of the density operator, derived and discussed in Chapter 8, as

$$\frac{d\hat{\rho}}{dt} = -\frac{i}{\hbar}[\hat{H}_{\mathrm{I}}, \hat{\rho}] - \frac{\kappa}{2}(\hat{a}^\dagger\hat{a}\hat{\rho} + \hat{\rho}\hat{a}^\dagger\hat{a}) + \kappa\hat{a}\hat{\rho}\hat{a}^\dagger, \tag{10.16}$$

where we shall take $\kappa = \omega_0/Q$ and where Q is a characteristic quality factor parameter of the cavity describing the rate of loss. The time $T_{\mathrm{r}} = 1/\kappa$ is the cavity relaxation time. For Q very large, the rate of losses will be (desir- ably) small. With the states $|i\rangle$, $i = 1, 2, 3$, such that $\rho_{ij} = \langle i|\hat{\rho}|j\rangle$, Eq. (10.16) is equivalent to the set of coupled first-order differential equations

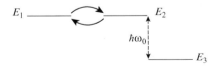

Fig. 10.2 Energy-level diagram for the atom–field system. The states $|1\rangle = |e\rangle|0\rangle$ and $|2\rangle = |g\rangle|1\rangle$ have the same energy and without loss of a photon, the system oscillates between these two states. But if the photon is lost, the system state becomes $|3\rangle = |g\rangle|0\rangle$ with energy lower by an amount $\hbar\omega_0$.

$$\frac{d\rho_{11}}{dt} = \frac{i}{2}\Omega_0(\rho_{12} - \rho_{21}) \equiv \frac{i}{2}\Omega_0 V,$$

$$\frac{d\rho_{22}}{dt} = -\frac{\omega_0}{Q}\rho_{22} - \frac{i}{2}\Omega_0 V,$$

$$\frac{dV}{dt} = i\Omega_0 W - \frac{1}{2}\left(\frac{\omega_0}{Q}\right)V,$$

$$\frac{d\rho_{33}}{dt} = \frac{\omega_0}{Q}\rho_{22},$$

(10.17)

where we have set $V = \rho_{12} - \rho_{21}$ and $W = \rho_{11} - \rho_{22}$. The first three of these equations are precisely of the form of the Bloch equations familiar from magnetic resonance [5] and of optical resonance with a classical field [6]. We refer to Eqs. (10.17) as "one-photon" Bloch equations, although Ω_0 is the vacuum Rabi frequency. In matrix form, the first three of these read

$$\frac{d}{dt}\begin{pmatrix} \rho_{11} \\ \rho_{22} \\ V \end{pmatrix} = \begin{pmatrix} 0 & 0 & i\Omega_0/2 \\ 0 & -\omega_0/Q & -i\Omega_0/2 \\ i\Omega_0 & -i\Omega_0 & -\omega_0/2Q \end{pmatrix}\begin{pmatrix} \rho_{11} \\ \rho_{22} \\ V \end{pmatrix}. \qquad (10.18)$$

The initial conditions are $\rho_{11}(0) = 1$, $\rho_{22}(0) = 0$, and $\rho_{12}(0) = 0$. Of course, we also have $\rho_{33}(0) = 0$. The eigenvalue problem of the matrix in Eq. (10.18) leads to the equation

$$\left(\Lambda + \frac{\omega_0}{2Q}\right)\left(\Lambda^2 + \frac{\omega_0}{Q}\Lambda + \Omega_0^2\right) = 0, \qquad (10.19)$$

whose solutions are

$$\Lambda_0 = -\frac{\omega_0}{2Q},$$

$$\Lambda_\pm = -\frac{\omega_0}{2Q} \pm \frac{\omega_0}{2Q}\left(1 - \frac{4\Omega_0^2 Q^2}{\omega_0^2}\right)^{1/2}. \qquad (10.20)$$

If the cavity decay rate is weak, such that $\omega_0/Q < 2\Omega_0$, then the eigenvalues $\Lambda_\pm$ will be complex and we obtain damped oscillations of frequency Ω_0 in the probability of finding the atom in the excited state $P_e(t) = \rho_{11}(t)$, as given in Fig. 10.3. The oscillations, which here are the vacuum Rabi oscillations, reflect the fact that the spontaneous emission is reversible for weak field decay. The oscillations are damped at the rate $\omega_0/2Q$. But if there is strong cavity damping, such that $\omega_0/Q > 2\Omega_0$, the eigenvalues $\Lambda_\pm$ are real and so there will now be irreversible spontaneous emission as indicated in Fig. 10.4. The largest time constant, and thus the smallest eigenvalue, in this case is Λ_+, which we may approximate as

$$\Lambda_+ \simeq -\frac{\Omega_0^2 Q}{\omega_0} = -\frac{4d^2 Q}{\hbar\varepsilon_0 V}, \qquad (10.21)$$

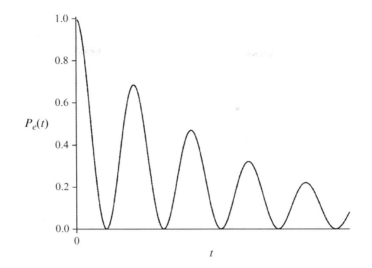

$P_e(t)$

Fig. 10.3 Damped excited-state probability for an atom in a high-Q cavity, where many vacuum Rabi oscillations are visible within a cavity damping time.

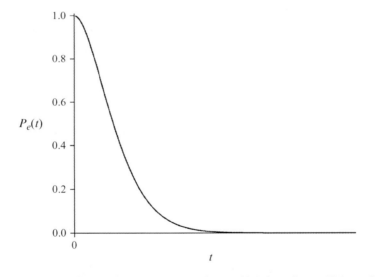

$P_e(t)$

Fig. 10.4 Decay of excited state of an atom in a low-Q cavity, with no visible Rabi oscillations. The atom decays at a rate that depends on the cavity Q.

where the right-hand side has been obtained using the relations $\Omega_0 = 2\lambda$, $\lambda = dg/\hbar$, and where g is given by Eq. (4.89), V being the effective mode volume of the cavity. We have assumed that the atom is sitting near an anti-node of the standing wave cavity field such that $\sin^2(kz) \approx 1$. Therefore we may define the rate of irreversible decay as

$$\Gamma_{\text{cav}} = \frac{4d^2Q}{\hbar \varepsilon_0 V}.$$

(10.22)

Note that the decay rate increases with increasing Q and thus we have a cavity-enhanced rate of spontaneous emission. If we write the decay rate after the form in Eq. (10.13) as

$$\Gamma_{\text{cav}} = \frac{\pi d^2}{\varepsilon_0 \hbar^2} \rho_{\text{cav}}(\omega_0) \hbar \omega_0,$$

(10.23)

then we find that the number of modes per unit volume of a cavity is

$$\rho_{\text{cav}}(\omega) = \frac{4Q}{\pi \varepsilon_0 V},$$

(10.24)

and thus we have

$$\frac{\Gamma_{\text{cav}}}{\Gamma} = \frac{3\rho_{\text{cav}}(\omega)}{\rho_{\text{fs}}(\omega)} = \frac{4Q\lambda_0^3}{3\pi \varepsilon_0 V}.$$

(10.25)

This result, predicting enhanced spontaneous emission in a cavity, was derived by Purcell [7] in the context of magnetic resonance. Of course, the result assumes that the cavity supports a mode of the proper wavelength into which the atom can radiate. But if the cavity is so small that the wavelength λ_0 cannot be supported, spontaneous emission will be inhibited [8], and this has been observed experimentally [9].

10.3 Experimental Realization of the Jaynes–Cummings Model

The experimental realization of cavity-enhanced and reversible transition dynamics has been pioneered by groups at the Max Planck Institute for Quantum Optics and the Ecole Normale Supérieure (ENS).

In the experiments performed by Haroche and collaborators at ENS in Paris, the cavity used consists of a Fabry–Perot resonator constructed out of superconducting niobium spherical mirrors of diameter 50 mm and radius of curvature 40 mm, separated by 27 mm. The measured Q for the cavity is $Q = 3 \times 10^8$, though higher values are not out of the question. The storage time for photons for such a cavity is about 1 ms, shorter than the atomic decay time for a circular Rydberg atom with $n \sim 50$ as estimated above, but much longer than the atom–field interaction time which, at the speeds atoms are injected into the cavity, is on the order of a few microseconds. The cavity walls are cooled to about 1 K, leaving an average of 0.7 microwave photons

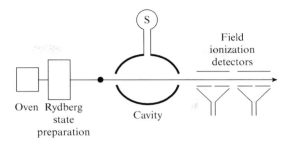

Fig. 10.5 Experimental setup for the realization of the Jaynes–Cummings model. The cavity field is prepared in a coherent state obtained via waveguide from the source S, which itself is driven by classical currents. The atoms are prepared in the excited state, velocity selected, and then directed into the cavity. The selective ionization detectors yield the statistics for the inversion as a function of interaction time, which itself is controlled by the velocity selection.

in the cavity. But this number can be further reduced by injecting a sequence of ground-state atoms through the cavity in order to absorb the photons, effectively "cooling" the cavity further. The procedure reduces the average photon number to 0.1. In Fig. 10.5 we sketch the setup for a typical cavity QED experiment. The microwave cavity, which is open on the sides to allow the passage of atoms, has attached to it a source of classical microwaves (a klystron), whose output can be piped into the cavity to form a coherent state of low amplitude. In addition, there are other sources of classical resonant microwaves that can be used to manipulate the atomic states before and after the atom leaves the cavity. These fields essentially perform Ramsey interferometry [10], as will be demonstrated below, and are thus referred to as Ramsey fields R_1 and R_2. The fields have very rapid decay (cavities of low Q are used with small photon numbers) and thus may be treated classically as no entanglement is generated [11]. Finally, selective ionization detectors are placed after the second Ramsey field.

As for the circular Rydberg atoms, the states used are those of principal quantum numbers $n = 49$, 50, and 51, which we denote respectively as $|f\rangle$, $|g\rangle$, and $|e\rangle$. The energy-level diagram for the atomic states is given in Fig. 10.6. The transition frequency of the $|e\rangle \leftrightarrow |g\rangle$ transition is 51.1 GHz and that of the $|g\rangle \leftrightarrow |f\rangle$ transition is 54.3 GHz. The latter transition frequency is far enough off-resonance with the cavity field mode such that there are no transitions between these levels. However, that does not mean that the cavity has no effect, as we shall see in Section 10.4. The atomic states can be tuned in and out of resonance with the cavity field by the application of a small static electric field (a Stark shift) if necessary.

An early experimental design to observe collapse and revival phenomena in atom–field interactions in CQED was reported in 1987 by Rempe, Walther, and Klein [12]. This involved a thermal field in a cavity and, indeed, an oscillation was observed in the atomic-state statistics, but the experiment was not able to manifest wide enough range interaction times to explore collapse and revival. In 1996, Haroche and collaborators [13] were able to

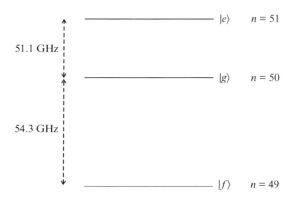

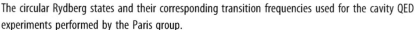 The circular Rydberg states and their corresponding transition frequencies used for the cavity QED experiments performed by the Paris group.

perform an experiment with the requisite control over the interaction times and, furthermore, were able to inject quantized coherent cavity fields of up to $\bar{n} = 1.77(\pm 0.15)$ photons. In this particular experiment, the cavity used had a relaxation time of $T_r = 220$ μs, corresponding to $Q = 7 \times 10^7$. The relaxation time is longer than the atom–field interaction times t, which range from $40 < t < 90$ μs. The corresponding range of atomic velocities v is $110 < v < 250$ m/s. The results of the experiment are given in Fig. 10.7, reprinted from Ref. [13]. The first (A) shows the oscillations obtained with no injected field but with a thermal field containing $\bar{n} = 0.06(\pm 0.01)$ photons. Several oscillations are clearly present. For the coherent fields of increasing amplitude, collapse followed by a revival is evident. The largest coherent field used had an average photon number of $\bar{n} = 1.77(\pm 0.15)$.

10.4 Creating Entangled Atoms in CQED

The experimental arrangements used for the CQED experiments described above, with some modification, can be used to generate entanglement between successive *atoms* injected through the cavity. In Fig. 10.8 we show a setup with two atoms crossing the cavity, where atom 1 is prepared in the excited state with atom 2 initially in the ground state. The cavity field is assumed to be initially in the vacuum. While the atom is in the cavity, the field and atom evolve according to

$$|e\rangle_1 |0\rangle_{cav} \rightarrow \cos(\lambda t_1)|e\rangle_1|0\rangle - i\sin(\lambda t_1)|g\rangle_1|1\rangle. \qquad (10.26)$$

If the speed of the atom is chosen such that $\lambda t_1 = \pi/4$, then we have

$$|e\rangle_1 |0\rangle_{cav} \rightarrow \frac{1}{\sqrt{2}}(|e\rangle_1|0\rangle - i|g\rangle_1|1\rangle). \qquad (10.27)$$

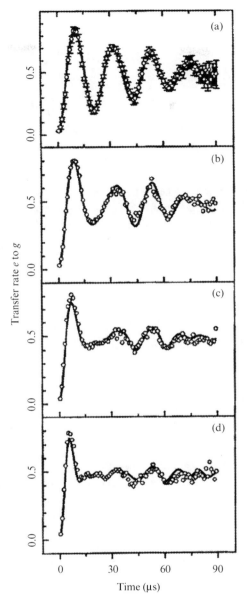

 Fig. 10.7 The $|e\rangle$ to $|g\rangle$ transition rates for various initial cavity field states. (A) is for no injected field with an average of $0.06(\pm0.01)$ thermal photons, whereas (B), (C), and (D) are for injected coherent states with average photon numbers $0.040(\pm0.02)$, $0.85(\pm0.04)$, and $1.77(\pm0.15)$, respectively. Reprinted by permission from Ref. [13]. Copyright 1996 by the American Physical Society.

Sending in the second atom we have

$$|e\rangle_1|g\rangle_2|0\rangle_{\text{cav}} \rightarrow \frac{1}{\sqrt{2}}(|e\rangle_1|g\rangle_2|0\rangle - i|g\rangle_1|g\rangle_2|1\rangle), \qquad (10.28)$$

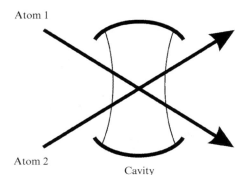

Atom 1

Atom 2

Cavity

Fig. 10.8 Proposed method for creating an entanglement of two atoms. The first atom, prepared in the excited state, is velocity selected such that it emits a photon upon passing through the cavity. Subsequently, the second atom, prepared in the ground state, is velocity selected to absorb the photon. As a result, both atoms, though distant, become entangled.

and using that

$$|g\rangle_2|1\rangle \rightarrow \cos(\lambda t_2)|g\rangle_2|1\rangle - i\sin(\lambda t_2)|e\rangle_2|0\rangle, \tag{10.29}$$

and with the velocity of the second atom chosen so that $\lambda t_2 = \pi/2$, we finally obtain

$$|e\rangle_1|g\rangle_2|0\rangle_{\text{cav}} \rightarrow |\Psi^+\rangle|0\rangle, \tag{10.30}$$

where

$$|\Psi^+\rangle = \frac{1}{\sqrt{2}}(|e\rangle_1|g\rangle_2 + |g\rangle_1|e\rangle_2) \tag{10.31}$$

is just one of the maximally correlated Bell states, this time given in terms of the states of a two-level atom. Using classical resonant fields and selective ionization detection, one could use such a state to test local realist theories by seeking violations of Bell's inequalities. However, as the atom-state measurements are not likely to have space-like separations, this type of test retains a loophole in regard to the criterion of locality. The generation of entangled atoms via the method discussed here was proposed by Cirac and Zoller [14]. Experimental realization was achieved by the Paris group [15].

10.5 Formation of Schrödinger-Cat States with Dispersive Atom–Field Interactions and Decoherence from the Quantum to the Classical

In Section 4.8 we discussed the Jaynes–Cummings model in the case of large detuning. The interaction in that case becomes, as shown in Appendix C, represented by the effective Hamiltonian

$$\hat{H}_{\text{eff}} = \hbar\chi[\hat{\sigma}_+\hat{\sigma}_- + \hat{a}^\dagger\hat{a}\hat{\sigma}_3], \tag{10.32}$$

where $\hat{\sigma}_+\hat{\sigma}_- = |e\rangle\langle g|g\rangle\langle e| = |e\rangle\langle e|$ and $\hat{\sigma}_3 = |e\rangle\langle e| - |g\rangle\langle g|$. The large detuning is between the cavity field frequency and the frequency for the $|e\rangle\leftrightarrow|g\rangle$ transition. The lower-level state $|f\rangle$ is, of course, still dynamical – disconnected as it is very far from resonance with the cavity field. If the atomic state $|e\rangle$ is not to become populated, the effective interaction Hamiltonian can be written as

$$\hat{H}_{\text{eff}} = -\hbar\chi\hat{a}^\dagger\hat{a}|g\rangle\langle g|. \tag{10.33}$$

Let us suppose now that the setup pictured in Fig. 10.9 is used where the Ramsey zone R_1, whose field we now take to be resonant with the $|f\rangle\leftrightarrow|g\rangle$ transition (as is also assumed to be the case for R_2), is used to prepare an atom in a superposition of the form

$$|\psi_{\text{atom}}\rangle = \frac{1}{\sqrt{2}}(|g\rangle + |f\rangle). \tag{10.34}$$

The atom now enters the cavity, which itself has been prepared in a coherent state $|\alpha\rangle$. Thus the initial state of the atom–field system is

$$|\Psi(0)\rangle = |\psi_{\text{atom}}\rangle|\alpha\rangle = \frac{1}{\sqrt{2}}(|g\rangle + |f\rangle)|\alpha\rangle. \tag{10.35}$$

Using the effective Hamiltonian given by Eq. (10.33), this state evolves to

$$|\Psi(t)\rangle = \exp[-i\hat{H}_{\text{eff}}t/\hbar]|\Psi(0)\rangle = \frac{1}{\sqrt{2}}[|g\rangle|\alpha e^{i\chi t}\rangle + |f\rangle|\alpha\rangle]. \tag{10.36}$$

We now suppose that in the second Ramsey zone, R_2, a classical field performs the transformations

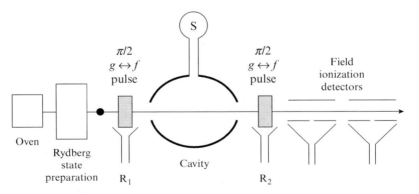

Fig. 10.9   A sketch of the experiment designed to generate the Schrödinger-cat states, as discussed in the text. R_1 and R_2 are the Ramsey zones that effect $\pi/2$-pulses on the $|g\rangle\leftrightarrow|f\rangle$ transition.

$$|g\rangle \rightarrow \cos(\theta/2)|g\rangle + \sin(\theta/2)|f\rangle,$$
$$|f\rangle \rightarrow \cos(\theta/2)|f\rangle - \sin(\theta/2)|g\rangle, \tag{10.37}$$

where the Rabi angle $\theta = \Omega_R T$ and where Ω_R is the semiclassical Rabi frequency and T is the interaction time. Equation (10.36) is transformed into

$$
\begin{aligned}
|\Psi(t),\theta\rangle &= \frac{1}{\sqrt{2}} \left[\left(\cos(\theta/2)|g\rangle + \sin(\theta/2)|f\rangle \right) |\alpha e^{i\chi t}\rangle \right. \\
&\quad \left. + \left(\cos(\theta/2)|f\rangle - \sin(\theta/2)|g\rangle \right) |\alpha\rangle \right] \\
&= |g\rangle \frac{1}{\sqrt{2}} \left[\cos(\theta/2)|\alpha e^{i\chi t}\rangle - \sin(\theta/2)|\alpha\rangle \right] \\
&\quad + |f\rangle \frac{1}{\sqrt{2}} \left[\sin(\theta/2)|\alpha e^{i\chi t}\rangle + \cos(\theta/2)|\alpha\rangle \right].
\end{aligned}
\tag{10.38}
$$

With the interaction time chosen such that $\chi t = \pi$, and with the Rabi angle θ chosen as $\theta = \pi/2$, we have

$$|\Psi(\pi/\chi),\pi/2\rangle = \frac{1}{2}[|g\rangle(|-\alpha\rangle - |\alpha\rangle) + |f\rangle(|-\alpha\rangle + |\alpha\rangle)]. \tag{10.39}$$

With the ionization detectors set to detect the states $|g\rangle$ and $|f\rangle$, a detection of $|f\rangle$ projects the cavity field into an even Schrödinger-cat state

$$|\psi_e\rangle = \mathcal{N}_e(|\alpha\rangle + |-\alpha\rangle), \tag{10.40}$$

while if $|g\rangle$ is detected, the cavity field is projected into the odd cat state

$$|\psi_o\rangle = \mathcal{N}_o(|\alpha\rangle - |-\alpha\rangle), \tag{10.41}$$

states first given in Eqs. (7.115)–(7.117). The atomic velocities required to achieve the requisite phase shift of $\chi t = \pi$ for generating the even and odd cat states is about 100 m/s.

The generation of the even and odd cat states, as just discussed, can (in fact, must) be seen as a consequence of interference. We illustrate this in Fig. 10.10. The Ramsey zones act as beam splitters, the second effectively creating ambiguity in the choice of two "paths."

One of the motivations for creating the Schrödinger-cat states was to provide an opportunity to attempt to monitor the decoherence of a mesoscopic superposition state into a statistical mixture. If we assume that the atom–field interaction time is short enough so that dissipative effects can be ignored while the atom is in the cavity, then after the atom exits we can take the *field* state – assumed to be unentangled with the atom state after the atom is ionized – to evolve according to the master equation

$$\frac{d\hat{\rho}_F}{dt} = -\frac{\kappa}{2}(\hat{a}^\dagger\hat{a}\hat{\rho}_F + \hat{\rho}_F\hat{a}^\dagger\hat{a}) + \kappa\hat{a}\hat{\rho}_F\hat{a}^\dagger, \tag{10.42}$$

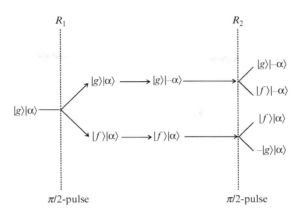

R_1 R_2

$|g\rangle|\alpha\rangle \longrightarrow |g\rangle|-\alpha\rangle \cdots\cdots\cdots\cdots \qquad |g\rangle|-\alpha\rangle$

$|f\rangle|-\alpha\rangle$

$|g\rangle|\alpha\rangle \longrightarrow$

$|f\rangle|\alpha\rangle$

$|f\rangle|\alpha\rangle \longrightarrow |f\rangle|\alpha\rangle \cdots\cdots\cdots\cdots$

$-|g\rangle|\alpha\rangle$

$\pi/2$-pulse $\qquad\qquad\qquad \pi/2$-pulse

Fig. 10.10 A graphical representation of the interference involved in the generation of the Schrödinger-cat states. The Ramsey zones act like beam splitters to create "path" ambiguity. On the far right, the field states associated with like atomic states become superposed upon selective atomic-state ionization.

where we have $\hat{H}_1 = 0$ as the atom and the field are no longer interacting and where we have made the notational replacement $\gamma \to \kappa$. As we showed in Chapter 8, the solution to the master equation for the field initially in an even cat state is

$$\hat{\rho}_F(t) = \mathcal{N}_e^2 \{ |\alpha e^{-\kappa t/2}\rangle \langle \alpha e^{-\kappa t/2}| + |-\alpha e^{-\kappa t/2}\rangle \langle -\alpha e^{-\kappa t/2}|$$
$$+ e^{-2|\alpha|^2(1-e^{-\kappa t})}[|\alpha e^{-\kappa t/2}\rangle \langle -\alpha e^{-\kappa t/2}| + |-\alpha e^{-\kappa t/2}\rangle \langle \alpha e^{-\kappa t/2}|]\}.$$
$$(10.43)$$

Notice, again, that as $t \to \infty$, the last two terms (the "coherences") decay faster than energy itself decays provided that $|\alpha|^2$ is not too small. Under this condition, we find that the initial cat state (a quantum mechanical superposition state) decoheres into a statistical mixture of coherent states

$$\hat{\rho}_F(t) \xrightarrow[t \text{ large}]{} \frac{1}{2}[|\beta(t)\rangle \langle \beta(t)| + |-\beta(t)\rangle \langle -\beta(t)|], \qquad (10.44)$$

where $\beta(t) = \alpha e^{-\kappa t/2}$. Of course, as $t \to \infty$, we obtain the (pure) vacuum state $\hat{\rho}_F(t) \xrightarrow[t \to \infty]{} |0\rangle \langle 0|$.

How can this decoherence be monitored? Davidovich et al. [16] proposed the following method. A second atom (2) prepared in the state $(|g\rangle_2 + |f\rangle_2)/\sqrt{2}$ is injected into the cavity a time T after the first atom and acts as a quantum "mouse" to detect the presence of the cat. The new "initial" state (i.e. just before the second atom is injected) is, for the atom–field system,

$$\hat{\rho}_{AF}(T) = \frac{1}{2}(|g\rangle_2 + |f\rangle_2)(_2\langle g| +\, _2\langle f|)\hat{\rho}_F(T), \qquad (10.45)$$

where $\hat{\rho}_F(T)$ is given by Eq. (10.43). Assuming the velocity of atom 2 is adjusted to that once again, we have $\chi t_2 = \pi$, where t_2 is the interaction time of the second atom with the cavity field, then the atom–field state just after atom 2 exits the cavity is given by the density operator

$$\hat{\rho}_{AF} = \frac{1}{2}[|g\rangle_{2\,2}\langle g|e^{i\pi\hat{a}^\dagger\hat{a}}\hat{\rho}_F(T)e^{-i\pi\hat{a}^\dagger\hat{a}} + |f\rangle_{2\,2}\langle f|\hat{\rho}_F(T)$$
$$+ |g\rangle_{2\,2}\langle f|e^{i\pi\hat{a}^\dagger\hat{a}}\hat{\rho}_F(T) + |f\rangle_{2\,2}\langle g|\hat{\rho}_F(T)e^{-i\pi\hat{a}^\dagger\hat{a}}]. \qquad (10.46)$$

The second Ramsey zone performs a $\pi/2$-pulse on the atom so that just before the atom encounters the ionization detectors, the density operator takes the form

$$\hat{\rho}_{AF}(T) = \frac{1}{4}\Big[(|g\rangle_2 + |f\rangle_2)(_2\langle g|+_2\langle f|)e^{i\pi\hat{a}^\dagger\hat{a}}\hat{\rho}_F(T)e^{-i\pi\hat{a}^\dagger\hat{a}}$$
$$+ (-|g\rangle_2 + |f\rangle_2)(-_2\langle g|+_2\langle f|)\hat{\rho}_F(T)$$
$$+ (|g\rangle_2 + |f\rangle_2)(_2\langle g|+_2\langle f|)e^{i\pi\hat{a}^\dagger\hat{a}}\hat{\rho}_F(T)$$
$$+ (-|g\rangle_2 + |f\rangle_2)(_2\langle g|+_2\langle f|)\hat{\rho}_F(T)e^{-i\pi\hat{a}^\dagger\hat{a}}\Big]. \qquad (10.47)$$

The reduced-density operator for the atom is found by tracing over the field according to

$$\hat{\rho}_A(T) = \mathrm{Tr}_F\hat{\rho}_{AF}(T), \qquad (10.48)$$

from which we find (the details are left as an exercise) that the probability of detecting the state $|g\rangle_2$ is

$$P_{g_2}(T) = {}_2\langle g|\hat{\rho}_A(T)|g\rangle_2 = \frac{1}{2}\left[1 - \frac{e^{-2|\alpha|^2 e^{-\kappa T}} + e^{-2|\alpha|^2(1-e^{-\kappa T})}}{1 + e^{-2|\alpha|^2}}\right], \qquad (10.49)$$

and the probability of detecting $|f\rangle_2$ is

$$P_{f_2}(T) = {}_2\langle f|\hat{\rho}_A(T)|f\rangle_2 = \frac{1}{2}\left[1 + \frac{e^{-2|\alpha|^2 e^{-\kappa T}} + e^{-2|\alpha|^2(1-e^{-\kappa T})}}{1 + e^{-2|\alpha|^2}}\right], \qquad (10.50)$$

where we must have $P_{f_2}(T) + P_{g_2}(T) = 1$. Recall that the initial state of the field, the even cat state, was projected by the detection of the *first* atom to be in the state $|f\rangle$. Well, for $T = 0$ we have that $P_{f_2}(0) = 1$, meaning that if there has not yet been any dissipation of the cavity field, the second atom must also be detected in the state $|f\rangle$. But for increasing T, the probability of detecting the atom in the state $|f\rangle$ starts to slide toward $1/2$. Thus, by increasing the delay time between the first and second atoms, one can monitor the decoherence of the initial even cat state into a statistical mixture.

In the preceding, the Ramsey zones were taken to be resonant with the $|g\rangle\leftrightarrow|f\rangle$, but where the state $|f\rangle$ was too far from resonance with the cavity field to give rise to any shift or the phase of the field. But suppose now that the Ramsey zones are tuned resonant with the $|e\rangle\leftrightarrow|g\rangle$ transition and that the first zone produces the superposition

$$|\psi_{\text{atom}}\rangle = \frac{1}{\sqrt{2}}(|e\rangle + |g\rangle). \qquad (10.51)$$

In this case, we must use the full effective interaction Hamiltonian of Eq. (10.32). With the same initial field state, we would have at time t the state

$$|\Psi(t)\rangle = \frac{1}{\sqrt{2}}[e^{-i\chi t}|e\rangle|\alpha e^{-i\chi t}\rangle + |g\rangle|\alpha e^{i\chi t}\rangle], \qquad (10.52)$$

where we notice that the coherent-state components are counter-rotating in phase space. Maximal entanglement will occur at the maximal phase space separation, which occurs when $\chi t = \pi$, giving

$$|\Psi(\pi/\chi)\rangle = \frac{1}{\sqrt{2}}[-i|e\rangle|-i\alpha\rangle + |g\rangle|i\alpha\rangle]. \qquad (10.53)$$

With the second Ramsey zone effecting the transformations

$$|e\rangle \rightarrow \frac{1}{\sqrt{2}}(|e\rangle + |g\rangle),$$

$$|g\rangle \rightarrow \frac{1}{\sqrt{2}}(|g\rangle - |e\rangle), \qquad (10.54)$$

we obtain for arbitrary interaction time t

$$|\Psi(t), \pi/2\rangle = \frac{1}{2}[|e\rangle(e^{-i\chi t}|e^{-i\chi t}\alpha\rangle - |e^{i\chi t}\alpha\rangle) + |g\rangle(e^{-i\chi t}|e^{-i\chi t}\alpha\rangle + |e^{i\chi t}\alpha\rangle)]. \qquad (10.55)$$

With the detectors set to distinguish between $|e\rangle$ and $|g\rangle$, state reduction produces the cavity field states

$$|\psi_\pm\rangle \sim e^{-i\chi t}|\alpha e^{-i\chi t}\rangle \pm |\alpha e^{i\chi t}\rangle. \qquad (10.56)$$

States of exactly this type have, in fact, been generated experimentally and their decoherence has been observed [17]. However, the atomic velocities were rather high in this experiment, so that phase shifts of $\chi t = \pi$ were not achieved. In fact, the greatest phase shift achieved was with $\chi t < \pi/4$. Nevertheless, this – along with the fact that the initial coherent state injected into the cavity had an average photon $\bar{n} \simeq 10$ – afforded a large enough initial separation of the component coherent states so that observation of the progressive decoherence of the field state was possible.

10.6 Quantum Non-demolition Measurement of Photon Number

In the standard approach to measurement as discussed in Appendix B, the quantity being measured is typically destroyed as a result of the measurement. For example, in the context of quantum optics, if one wishes to measure the photon number of a quantized field state, the photodetectors used will absorb photons, thus removing them from the field state. Here we show, in the context of CQED, how photon number can be measured without the absorption of any photons. This will provide an example of what is known as a quantum non-demolition (QND) measurement [18].

Suppose the cavity field contains a definite, but unknown, number of photons n. A Rydberg atom is then prepared in a superposition state $(|e\rangle + |g\rangle)/\sqrt{2}$ and injected into the cavity. The initial state is thus

$$|\Psi(0)\rangle = \frac{1}{\sqrt{2}}(|e\rangle + |g\rangle)|n\rangle = \frac{1}{\sqrt{2}}(|e\rangle|n\rangle + |g\rangle|n\rangle). \tag{10.57}$$

After a time t, which is the time the atom exits the cavity, the state vector is

$$|\Psi(t)\rangle = \exp[-i\hat{H}_{eff}t/\hbar]|\Psi(0)\rangle$$

$$= \frac{1}{\sqrt{2}}[e^{-i\chi t(n+1)}|e\rangle + e^{i\chi t n}|g\rangle]|n\rangle. \tag{10.58}$$

If a Ramsey field subjects the atom to a resonant $\pi/2$-pulse, thus effecting the transformations of Eq. (10.54), then the state vector becomes

$$|\Psi(t), \pi/2\rangle = \frac{1}{2}\left[|e\rangle\left(e^{-i\chi t(n+1)} - e^{+i\chi t n}\right) + |g\rangle\left(e^{-i\chi t(n+1)} + e^{+i\chi t n}\right)\right]|n\rangle. \tag{10.59}$$

The probabilities of detecting the atom in the excited and ground states, respectively, are then

$$P_e(t) = \frac{1}{2}\{1 + \cos[(2n + 1)t]\},$$

$$P_g(t) = \frac{1}{2}\{1 - \cos[(2n + 1)t]\}, \tag{10.60}$$

which clearly exhibit oscillations (Ramsey fringes) with time. The frequency of the oscillations yields the photon number n. Note that there will be oscillations even when $n = 0$. These Ramsey fringes have been observed by Brune *et al.* [19] for the cases $n = 0$ and $n = 1$. The cavity is prepared in a single-photon state by injecting into the cavity an excited atom of appropriate speed with the cavity tuned into resonance.

10.7 Quantum State Engineering in the Resonant Jaynes–Cummings Model

The discussion in Section 10.5 on the generation of Schrödinger-cat-like states in cavity QED is, of course, an example of quantum state engineering (see Section 7.11), though this time accomplished by the manipulation and measurements performed on an atom that has interreacted with a quantized cavity field. The atom–field interaction in the above was assumed to be dispersive, but this assumption can be relaxed so that one can consider quantum state engineering using the resonant Jaynes–Cummings model to generate a great variety of nonclassical states of single-mode cavity fields.

Recall the solution to the resonant Jaynes–Cummings model for the case of the atom initially in the excited state as given by Eq. (4.120), which we reproduce here:

$$|\psi(t)\rangle = |\psi_g(t)\rangle|g\rangle + |\psi_e(t)\rangle|e\rangle, \tag{10.61}$$

where

$$|\psi_g(t)\rangle = -i\sum_{n=0}^{\infty} C_n \sin(\lambda t\sqrt{n+1})|n+1\rangle,$$

$$|\psi_e(t)\rangle = \sum_{n=0}^{\infty} C_n \cos(\lambda t\sqrt{n+1})|n\rangle. \tag{10.62}$$

The coefficients C_n are from the initial state of the cavity field:

$$|\psi(0)\rangle_{\text{field}} = \sum_{n=0}^{\infty} C_n|n\rangle. \tag{10.63}$$

Now suppose the field and atom interact for a time $t = T$ after which the atom, which is assumed to have left the cavity, is detected, by selective ionization in, say, the excited state $|e\rangle$ such that the cavity field is projected into the state

$$|\Phi(T), e\rangle_f \equiv \frac{|\psi_e(T)\rangle}{\sqrt{\langle\psi_e(T)|\psi_e(T)\rangle}}. \tag{10.64}$$

If the atom is detected in the ground state $|g\rangle$, the field is projected into the state

$$|\Phi(T), g\rangle_f \equiv \frac{|\psi_g(T)\rangle}{\sqrt{\langle\psi_g(T)|\psi_g(T)\rangle}}. \tag{10.65}$$

It is important to understand that the time T does not represent continuous time evolution. It is the time at which the field and the atom cease interacting. Of course, that time will depend on the velocity of the atom as it passes through the cavity, and so can be adjusted by adjusting the atomic velocity. But once the atom leaves the cavity and the atomic measurement is made, the time T is fixed.

Of course, more field states are possible if one assumes that a resonant classical radiation field performs the Ramsey-like transformations

$$|g\rangle \to \cos(\theta/2)|g\rangle + \sin(\theta/2)|e\rangle,$$

$$|e\rangle \to \cos(\theta/2)|e\rangle - \sin(\theta/2)|g\rangle, \tag{10.66}$$

such that just before the atomic state is detected, one has

$$|\psi(t)\rangle \rightarrow |\psi(T), \theta\rangle \equiv |\psi_g(T)\rangle[\cos(\theta/2)|g\rangle + \sin(\theta/2)|e\rangle]$$
$$+ |\psi_e(T)\rangle[\cos(\theta/2)|e\rangle - \sin(\theta/2)|g\rangle]. \tag{10.67}$$

Detection of the excited state projects the field into the state

$$|\Phi(T), \theta, e\rangle_f \equiv \frac{\cos(\theta/2)|\psi_e(T)\rangle + \sin(\theta/2)|\psi_g(T)\rangle}{\| \cos(\theta/2)|\psi_e(T)\rangle + \sin(\theta/2)|\psi_g(T)\rangle \|}, \tag{10.68}$$

while if the ground state is detected, the field is projected into the state

$$|\Phi(T), \theta, g\rangle_f \equiv \frac{\cos(\theta/2)|\psi_g(T)\rangle - \sin(\theta/2)|\psi_e(T)\rangle}{\| \cos(\theta/2)|\psi_g(T)\rangle + \sin(\theta/2)|\psi_e(T)\rangle \|}. \tag{10.69}$$

The introduction of the angle θ provides one more parameter that can be changed, in addition to α and T, in order to adjust the properties of the projected states. Obviously, in the limit $\theta \rightarrow 0$ we recover the results of Eqs. (10.64) and (10.65).

Problem 6 of Chapter 7 deals with the prospect of generating squeezed states of radiation in the Jaynes–Cummings model. For initial coherent states of modest average photon number $\bar{n} = |\alpha|^2 = 5$, the amount of quadrature squeezing obtained is also modest, that being about 20% of the allowed maximum below the vacuum noise level, as was shown by Meystre and Zubairy [20], whereas the squeezing obtained for average photon number $\bar{n} = 100$ can be very large, as was shown by Kuklinski and Madajczyk [21]. On the other hand, it has been shown by Gerry and Ghosh [22] that squeezing can be enhanced even for low photon number upon the implementation of selective atomic measurements, as described in the paragraphs above. It was shown that even with $\bar{n} = 5$, squeezing in excess of 50% below the vacuum noise is possible. That this is possible is understandable, because the projected field states given above are pure states, whereas in the continuously evolving Jaynes–Cummings model, the atom and field are generally entangled. See also Weiher *et al.* [23].

10.8 Realization of the Jaynes–Cummings Interaction in the Motion of a Trapped Ion

We conclude this chapter with a discussion of what amounts to a realization of the Jaynes–Cumming-type interaction, but where the quantized cavity field is replaced by the vibrational motion of the center of mass of a trapped ion. In other words, the *photons* are replaced by *phonons*, the quanta of mechanical vibrations. The realization of such an interaction is the result of significant advances in the laser cooling of ions (and neutral atoms) over the past decade. We do not discuss the method of laser cooling here, but refer the reader to the articles listed in the bibliography at the end of this chapter.

We assume that an ion of mass M has been laser cooled and is contained in some form of electromagnetic trap. Two types of traps are in general use. One is the Penning trap, which uses a combination of a uniform magnetic and quadrupole electric fields. The other is the Paul trap, also known as the r.f. trap, which uses radio frequency electric fields to trap ions. It is, of course, not possible to trap ions with static electric fields on account of Earnshaw's theorem [24]. We shall not go into the details of the traps (but see Horvath *et al.* [25]). In a typical experiment of the type we have in mind involving a single trapped ion, it is the Paul trap that is used. A sketch of such a device is given in Fig. 10.11.

We take as our starting point a two-level ion, with internal electronic energy levels, usually hyperfine levels, denoted $|e\rangle$ and $|g\rangle$, contained within the trap. A laser of frequency ω_L is directed along the axis, denoted the x-axis, of the trap. This laser field is assumed to be tunable. The trap itself is approximated as a harmonic oscillator of frequency v, not to be confused with the velocity v earlier. The value of this frequency is determined by the various trap parameters. The position of the center of mass of the ion is given by the operator $\hat{x}$, which we shall write below in terms of annihilation and creation operators, respectively $\hat{a}$, $\hat{a}^\dagger$, which are now taken to be the annihilation and creation operators of the vibration motion of the center of mass of the ion, not those of a quantized field. (There are *no*

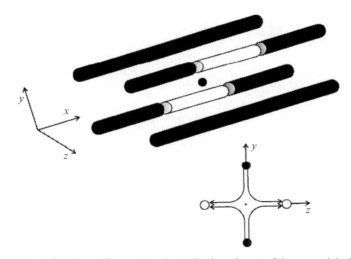

Fig. 10.11   Sketch of a linear r.f. Paul trap. The x-axis is taken to be along the axis of the trap and the black dot represents a single trapped ion. The smaller figure represents an end view. The common r.f. field is applied to the dark electrodes, while the others are held at constant electric potentials.

quantized radiation fields involved with this problem.) The Hamiltonian for our system is given by

$$\hat{H} = \hat{H}_0 + [\mathcal{D}E^{(-)}(\hat{x}, t)\hat{\sigma}_- + h.c.], \tag{10.70}$$

where

$$\hat{H}_0 = \frac{\hbar\omega_0}{2}\hat{\sigma}_3 + \hbar v\left(\hat{a}^\dagger\hat{a} + \frac{1}{2}\right) \tag{10.71}$$

is the interaction-free Hamiltonian, $\mathcal{D}$ is the dipole moment of the $|g\rangle\leftrightarrow|e\rangle$ transition, and $E^{(-)}(\hat{x}, t)$ is the negative frequency part of the laser field, given by

$$E^{(-)}(\hat{x}, t) = E_0 \exp[i(\omega_L t - k_L\hat{x} + \phi)], \tag{10.72}$$

where E_0 is the amplitude of the laser field, $k_L = 2\pi/\lambda_L$ is the wave vector of the field, and ϕ is just some phase. We write the position operator in the usual way as

$$\hat{x} = \sqrt{\frac{\hbar}{2vM}}(\hat{a} + \hat{a}^\dagger). \tag{10.73}$$

Substitution into Eq. (10.72) yields

$$E^{(-)}(\hat{x}, t) = E_o e^{i(\phi+\omega_L t)}e^{-i\eta(\hat{a}+\hat{a}^\dagger)}, \tag{10.74}$$

where $\eta \equiv k_L(\hbar/2vM)^{1/2}$ is the so-called Lamb–Dicke parameter. This parameter is typically small, $\eta \ll 1$, this range being known as the Lamb–Dicke regime. In this regime, the vibrational amplitude of the ion is much smaller than the wavelength of the laser light field. With $\hat{U} = \exp(-i\hat{H}_0 t/\hbar)$, we now transform to the interaction picture, obtaining the interaction Hamiltonian

$$\hat{H}_I = \hat{U}^\dagger\hat{H}\hat{U} + i\hbar\frac{d\hat{U}^\dagger}{dt}\hat{U} \tag{10.75}$$

$$= \mathcal{D}E_0 e^{i\phi}e^{i\omega_L t}\exp[-i\eta(\hat{a}e^{ivt} + \hat{a}^\dagger e^{-ivt})]\hat{\sigma}_- e^{-i\omega_0 t} + h.c.,$$

as the reader can verify. As η is small, we expand to first order

$$\exp[-i\eta(\hat{a}e^{ivt} + \hat{a}^\dagger e^{-ivt})] \approx 1 - i\eta(\hat{a}e^{ivt} + \hat{a}^\dagger e^{-ivt}), \tag{10.76}$$

and thus we have to this order

$$\hat{H}_I \approx \mathcal{D}E_0 e^{i\phi}\left[e^{i(\omega_L-\omega_0)t} - i\eta\left(\hat{a}\, e^{i(\omega_L-\omega_0+v)t} + \hat{a}^\dagger\, e^{i(\omega_L-\omega_0-v)t}\right)\right]\hat{\sigma}_- + h.c. \tag{10.77}$$

Suppose now that the laser is tuned such that $\omega_L = \omega_0 + v$. We shall then have

$$\hat{H}_I \approx \mathcal{D}E_0 e^{i\phi}[e^{i\nu t} - i\eta(\hat{a}\, e^{i2\nu t} + \hat{a}^\dagger)]\hat{\sigma}_- + h.c. \tag{10.78}$$

The terms rotate at frequencies ν and 2ν very rapidly compared with the remaining term, and average to zero. Dropping these terms (we are essentially making a rotating wave approximation), we obtain

$$\hat{H}_I \approx - i\hbar\eta\Omega e^{i\phi}\hat{a}^\dagger\hat{\sigma}_- + h.c., \tag{10.79}$$

which is the Jaynes–Cummings interaction, where $\Omega = \mathcal{D}E_0/\hbar$ is the Rabi frequency associated with the semiclassical laser–atom interaction. But with regard to the Jaynes–Cummings interaction, it is important to note that the internal states of the ion are coupled with the vibrational motion of its center of mass. Further, note that this interaction is not the only one possible. If the laser is tuned according to $\omega_L = \omega_0 - \nu$, we obtain the interaction

$$\hat{H}_I \approx - i\,\hbar\eta\Omega e^{i\phi}\hat{a}\hat{\sigma}_- + h.c., \tag{10.80}$$

which contains the non-energy-conserving interaction of the form $\hat{a}\hat{\sigma}_-$. This model is sometimes known as the anti-Jaynes–Cummings model, an interaction that simply cannot be realized in the case of an atom interacting with a quantized field.

By keeping more terms in the expansion (10.76), other higher-order Jaynes–Cummings-type interactions can be generated. For example, suppose we retain terms to second order and set $\omega_L = \omega_0 \pm 2\nu$. It should be easy to see that the interactions obtained will be of the form

$$\begin{aligned} \hat{H}_I &\sim \eta^2\hat{a}^{\dagger 2}\hat{\sigma}_- + h.c., \text{ for } \omega_L = \omega_0 + \nu, \\ \hat{H}_I &\sim \eta^2\hat{a}^2\hat{\sigma}_- + h.c., \text{ for } \omega_L = \omega_0 - \nu. \end{aligned} \tag{10.81}$$

By taking the laser to be tuned to the frequencies $\omega_L = \omega_0 \pm l\nu, l = 0, 1, 2, \ldots$, it will be evident that interactions of the form

$$\begin{aligned} \hat{H}_I &\sim \eta^l\hat{a}^{\dagger l}\hat{\sigma}_- + h.c., \text{ for } \omega_L = \omega_0 + l\nu, \\ \hat{H}_I &\sim \eta^l\hat{a}^l\hat{\sigma}_- + h.c., \text{ for } \omega_L = \omega_0 - l\nu \end{aligned} \tag{10.82}$$

are possible, though the η^l for $|l| > 2$ may be too small in practice. The frequencies $\omega_L = \omega_0 \pm l\nu$ are known as the "side-band" frequencies and are illustrated in Fig. 10.12 for the cases $l = 0$ and $l = 1$.

Various kinds of nonclassical motional states of a single trapped ion, and various types of ion–phonon couplings, have been studied experimentally. Meekhof et al. [26], in 1996, were able to create thermal, Fock, coherent, and squeezed states in the motion of a trapped $^9Be^+$ ion initially laser cooled to the zero point. They were able to engineer the Jaynes–Cummings interaction and to observe the collapse and revival phenomena in this system. In fact, the collapse and revivals were more pronounced in this system than in the CQED experiment realizing the original model of

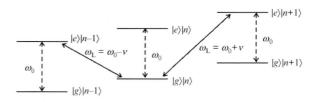

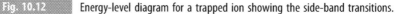

Fig. 10.12 Energy-level diagram for a trapped ion showing the side-band transitions.

a two-level atom interacting with a single-mode field [13]. In addition, they were able to construct the anti-Jaynes–Cummings and the "two-phonon" interactions. In later work [27], this group created Schrödinger-cat-like states consisting of a single trapped ion where the ion was in a superposition of two Gaussian wave packets separated by a few nanometers – not exactly a macroscopic separation but still large on an atomic scale.

10.9 Concluding Remarks

This chapter has mainly been devoted to cavity quantum electrodynamic realizations of the interaction of a quantized radiation field with an effective two-level atom. However, the cavity fields have been taken as microwave fields, not optical fields. This is because – as a practical matter – it is easier to realize the two-level approximation for the atom by using circular Rydberg states whose transitions are in the microwave. The main source of losses in the microwave experiments is the absorption of photons by the walls of the cavity. But cavity QED within the optical spectrum is also an area of active research. An essential difference between microwave and optical cavity QED is that in the latter, one must take into account (assuming the cavity is open) losses from spontaneous emission to modes external to the cavity. This requires an extra dissipative term in the master equation where the field operators $\hat{a}$ and $\hat{a}^{\dagger}$ are replaced by the atomic transition operators $\hat{\sigma}_{-}$ and $\hat{\sigma}_{+}$, respectively. The effects of these additional losses on the collapse and revival phenomenon have been studied by Quang, Knight, and Bužek [28].

Finally, another related problem we have ignored is the case where there is more than one atom in the cavity. Dicke [29] studied (without the field quantized) the collective behavior of a collection of two-level atoms all within a volume $V < \lambda^3$. The atoms do not directly interact with each other, but they can act collectively through their coupling to a common field mode. The phenomenon of "super-radiance" was predicted whereby, with a properly prepared state of a collection of N atoms, the atoms collectively radiate with an intensity proportional to N^2, whereas the intensity would go only as N if the atoms are radiating independently. The extension of the model to the case of a quantized field was first considered by Tavis and Cummings [30]. The mathematics of the operators describing a collection of identical two-level atoms is, as shown originally by Dicke [29], identical to that of the angular momentum algebra and the collective atomic states may be mapped onto angular momentum states $|J, M\rangle$, where $J = N/2$. The state $|J, -J\rangle = |g\rangle_1 |g\rangle_2 \cdots |g\rangle_N$, just the product state of all the atoms with each in the ground state. Similarly, $|J, J\rangle = |e\rangle_1 |e\rangle_2 \cdots |e\rangle_N$. The angular momentum states for other values of M, however, are *not* product states [31]. Upon this basis it is possible to construct a set of *atomic coherent states* [31], also known as the *spin coherent states* [32], analogous in some ways to the field coherent states. Consideration of multi-atom problems is beyond the scope of this book, but the interested reader should consult the references cited and the bibliography at the end of the chapter.

Finally, we mention a more or less newly emergent field of the past 20 years that has much in common with cavity QED. We speak of circuit QED, where the circuits in question are superconducting microwave circuits and where the roles of the atoms are played by Josephson junctions. Most of the ideas behind cavity QED carry over to circuit QED. In fact, phenomena in circuit QED even go quite beyond that to high degree. For example, giant Kerr effects become possible in this regime. An excellent and extensive review article on this field, with many references, has been published by Gu *et al.* [35]. The Les Houches 2011 volume [36] contains many chapters on circuit QED.

Problems

1. Justify the claim in Eq. (10.2) that, for adjacent circular Rydberg states of principal quantum numbers n and $n - 1$, the dipole moment goes as $\sim n^2$.
2. Carry through the calculation of the eigenvalues of the matrix in Eq. (10.18).
3. Suppose that an atom prepared in the superposition state

$$|\psi_{\text{atom}}\rangle = \frac{1}{\sqrt{2}}(|e\rangle + e^{i\varphi}|g\rangle) \tag{10.83}$$

    is injected into a cavity whose field is initially in a vacuum state. Obtain the time evolution for the excited-state population assuming both high and low-Q cavities. What is the effect, if any, upon the relative phase φ?

4. For a cavity field initially in a coherent state of average photon number $\bar{n} = 5$ and with the atom initially in the excited state, numerically integrate the set of differential equations associated with the Jaynes–Cummings model master Eq. (10.16) for the cases where (a) $\kappa = 0$, (b) $\kappa = 0.01\lambda$, and (c) $\kappa = 0.03\lambda$. For each case, plot the atomic inversion as a function of the scaled time λt. Also plot the average photon number as a function of the scaled time. Compare your results with those in Ref. [30].

5. Consider the cat-like superposition state

$$|\text{sup}\rangle \sim |\alpha e^{i\phi}\rangle + |\alpha e^{-i\phi}\rangle. \tag{10.84}$$

Normalize the state. How does the initial, short-time, decoherence rate of this state vary with the angle ϕ?

6. Consider the state for the field as given by Eq. (10.64) which is obtained from the Jaynes–Cummings model upon detection of the atom in the excited states after the atom and field have evolved for a time T. Assuming the field is initially in a coherent state of average photon number $\bar{n} = 5$, determine any nonclassical properties (squeezing, sub-Poissonian statistics) the field might possess for different interaction times T. Is this a Gaussian state of the field?

7. Investigate the possibility of performing a QND measurement of photon number for optical fields. Use the cavity QED method described above as a guide and keep in mind the cross-Kerr interaction described in Chapter 7.

8. For a single trapped ion, obtain the interaction for the case where the laser is tuned to resonance with the internal states of the ion, the $|g\rangle \leftrightarrow |e\rangle$ transition such that $\omega_L = \omega_0$, and where terms of second order in η are retained. Further, if the internal state of the ion is initially the ground state $|g\rangle$, and if the center-of-mass motion is prepared in a coherent state, investigate the evolution of the system. Do the internal and vibrational degrees of freedom become entangled? *Hint*: Obtain the dressed states of the effective interaction Hamiltonian.

References

[1] S. Haroche, in G. Grynberg and R. Stora (Eds.), *New Trends in Atomic Physics* (Amsterdam: Elsevier, 1984), p. 193.

[2] R. G. Hulet and D. Kleppner, *Phys. Rev. Lett.* **51**, 1430 (1983).

[3] P. Nussenzweig, F. Bernardot, M. Brune, J. Hare, J. M. Raimond, S. Haroche, and W. Gawlik, *Phys. Rev. A* **48**, 3991 (1993).

[4]   C. Cohen-Tannoudji, B. Diu, and F. Laloë, *Quantum Mechanics*, Vol. 2 (New York: Wiley Interscience, 1977).

[5]   F. Bloch, *Phys. Rev.* **70**, 460 (1946).

[6]   See L. Allen and J. H. Eberly, *Optical Resonance and the Two-Level Atom* (New York: Wiley Interscience, 1975; Mineola, NY: Dover, 1987), chap. 2.

[7]   E. M. Purcell, *Phys. Rev.* **69**, 681 (1946).

[8]   D. Kleppner, *Phys. Rev. Lett.* **47**, 233 (1981).

[9]   R. G. Hulet, E. S. Hilfer, and D. Kleppner, *Phys. Rev. Lett.* 55, 2137 (1985).

[10] N. F. Ramsey, *Rev. Mod. Phys.* **62**, 541 (1990); *Molecular Beams* (New York: Oxford University Press, 1985).

[11] I. I. Kim, K. M. Fonseca Romero, A. M. Horiguti, L. Davidovich, M. C. Nemes, and A. F. R. de Toledo Piza, *Phys. Rev. Lett.* **82**, 4737 (1999).

[12] G. Rempe, H. Walther, and N. Klein, *Phys. Rev. Lett.* **58**, 353 (1987).

[13] M. Brune, F. Schmidt-Kaler, A. Maali, J. Dreyer, E. Hagley, J. M. Raimond, and S. Haroche, *Phys. Rev. Lett.* **76**, 1800 (1996).

[14] J. I. Cirac and P. Zoller, *Phys. Rev. A* **50**, R2799 (1994). See also S. J. D. Phoenix and S. M. Barnett, *J. Mod. Opt.* **40**, 979 (1993); I. K. Kudryavtsev and P. L. Knight, *J. Mod. Opt.* **40**, 1673 (1993); M. Freyberger, P. K. Aravind, M. A. Horne, and A. Shimony, *Phys. Rev. A* **53**, 1232 (1996); A. Beige, W. J. Munro, and P. L. Knight, *Phys. Rev. A* **62**, 052102 (2000).

[15] E. Hagley, X. Maître, G. Nogues, C. Wunderlich, M. Brune, J. M. Raimond, and S. Haroche, *Phys. Rev. Lett.* **79**, 1 (1997).

[16] L. Davidovich, M. Brune, J. M. Raimond, and S. Haroche, *Phys. Rev. A* **53**, 1295 (1996).

[17] M. Brune, E. Hagley, J. Dreyer, X. Maître, A. Maali, C. Wunderlich, *et al. Phys. Rev. Lett.* **77**, 4887 (1996).

[18] See the review articles P. Grangier, J. A. Levensen, and J.-P. Poizat, *Nature* **396**, 537 (1998); V. B. Braginsky, V. B. Vorontsov, and K. S. Thorne, *Science* **209**, 547 (1980). An example of an optical application can be found, for example, in F. X. Kärtner and H. A. Haus, *Phys. Rev. A* **47**, 4585 (1993).

[19] M. Brune, P. Nussenzveig, F. Schmidt-Kaler, F. Bernardot, J. M. Raimond, and S. Haroche, *Phys. Rev. Lett.* **72**, 3339 (1994).

[20] P. Meystre and M. S. Zubairy, *Phys. Lett. A* **89**, 390 (1982).

[21] J. R. Kuklinski and J. L. Madajczyk, *Phys. Rev. A* **37**, R3175 (1988).

[22] C. C. Gerry and H. Ghosh, *Phys. Lett. A* **229**, 17 (1997).

[23] K. Weiher, E. Agudelo, and M. Bohmann, *Phys. Rev. A* **100**, 043812 (2019).

[24] See P. Lorrain, D. R. Corson, and F. Lorrain, *Fundamentals of Electromagnetic Phenomena* (New York: W.H. Freeman, 2000), p. 57.

[25] G. Sz. K. Horvath, R. C. Thompson, and P. L. Knight, *Contemp. Phys.* **38**, 25 (1997).

[26] D. M. Meekhoff, C. Monroe, B. E. King, W. M. Itano, and D. J. Wineland, *Phys. Rev. Lett.* **76**, 1796 (1996).

[27] C. Monroe, D. M. Meekhoff, B. E. King, and D. J. Wineland, *Science* **272**, 1131 (1996).

[28] T. Quang, P. L. Knight, and V. Bužek, *Phys. Rev. A* **44**, 6092 (1991).

[29] R. H. Dicke, *Phys. Rev.* **93**, 99 (1954). See also N. E. Rehler and J. E. Eberly, *Phys. Rev. A* **3**, 1735 (1971); J. H. Eberly, *Am. J. Phys.* **40**, 1374 (1972).

[30] M. Tavis and F. W. Cummings, *Phys. Rev.* **170**, 379 (1968).

[31] For an elementary account of the mapping of the atomic states onto angular momentum states, see M. Sargent III, M. O. Scully, and W. E. Lamb Jr., *Laser Physics* (Reading, MA: Addison-Wesley, 1974), Appendix G.

[32] F. T. Arecchi in Ref. 32, E. Courtens, R. Gilmore, and H. Thomas, *Phys. Rev. A* **6**, 2211 (1972).

[33] J. M. Radcliffe, *J. Phys. A: Gen. Phys.* **4**, 313 (1971).

[34] S. M. Barnett and P. L. Knight, *Phys. Rev. A* **33**, 2444 (1986); R. R. Puri and G. S. Agarwal, *Phys. Rev. A* **33**, R3610 (1986); **35**, 3433 (1997).

[35] X. Gu, A. F. Kockum, A. Miranowicz, Y. Liu, and F. Nori, *Phys. Rep.* **718/719**, 1 (2017).

[36] M. Devoret, B. Huard, R. Schoelkopf, and L. F. Cugliandolo (Eds.), *Les Houches 2011, Quantum Machines: Measurement and Control of Engineered Quantum Systems* (Oxford: Oxford University Press, 2014).

Bibliography

Two books on Rydberg atoms:

T. F. Gallagher, *Rydberg Atoms* (Cambridge: Cambridge University Press, 1994).

J. P. Connerade, *Highly Excited Atoms* (Cambridge: Cambridge University Press, 1998).

Some review articles on cavity QED with emphasis on microwave cavities are

E. A. Hinds, "Cavity quantum electrodynamics," in D. Bates and B. Bederson (Eds.), *Advances in Atomic, Molecular, and Optical Physics*, Vol. **28** (Amsterdam: Elsevier, 1990), p. 237.

D. Meschede, "Radiating atoms in confined spaces: From spontaneous emission to micromasers," *Phys. Rep.* **211**, 201 (1992).

P. Meystre, "Cavity quantum optics and the quantum measurement process," in E. Wolf (Ed.), *Progress in Optics XXX* (Amsterdam: Elsevier, 1992).

S. Haroche, "Entanglement experiments in cavity QED," *Fortschr. Phys.* **51**, 388 (2003).

J. M. Raimond, M. Brune, and S. Haroche, "Manipulating quantum entanglement with atoms and photons," *Rev. Mod. Phys.* **73**, 565 (2003).

A collection of articles covering most aspects of cavity QED is

P. Berman (Ed.), *Cavity Quantum Electrodynamics* (New York: Academic Press, 1994).

Some references on ion traps and the physics of trapped ions:

P. K. Ghosh, *Ion Traps* (Oxford: Oxford University Press, 1995).

J. I. Cirac, A. S. Parkins, R. Blatt, and P. Zoller, "Nonclassical states of motion in trapped ions," in B. Bederson and H. Walther (Eds.), *Advances in Atomic, Molecular, and Optical Physics*, Vol. **37** (Amsterdam: Elsevier, 1996), p. 237.

D. Liebfried, R. Blatt, C. Monroe, and D. Wineland, "Quantum dynamics of single trapped ions," *Rev. Mod. Phys.* **75**, 281 (2003).

Two reviews covering all aspects of cavity QED, including optical cavities and collective behavior in multi-atom systems, are

S. Haroche, "Cavity quantum electrodynamics," in J. Dalibard, J. M. Raimond, and J. Zinn-Justin (Eds.), *Fundamental Systems in Quantum Optics, Les Houches Session LIII* (Amsterdam: Elsevier, 1992), p. 767.

S. Haroche and J.-M. Raimond, *Exploring the Quantum, Atoms, Cavities, and Photons* (Oxford: Oxford University Press, 2006).

For an extensive discussion of super-radiance, see

M. G. Benedict, A. M. Ermolaev, V. A. Malyshev, I. V. Sokolov, and E. D. Trifonov, *Superradiance* (Bristol: Institute of Physics, 1996).

A volume with several chapters on the topics of this chapter, including early work on circuit QED, with connections to quantum information science, is

D. Estève, J.-M. Raimond, and J. Dailbard (Eds.), *Les Houches 2003, Quantum and Information Processing* (Amsterdam: Elsevier, 2004).

In 2012, Nobel Prizes were awarded to Serge Haroche and David Wineland for their experimental work in cavity QED and trapped ions, respectively. Their Nobel lectures are

S. Haroche, "Nobel Lecture: Controlling photons in a box and exploring the quantum to classical boundary," *Rev. Mod. Phys.* **85**, 1083 (2013).

D. J. Wineland, "Nobel Lecture: Superposition, entanglement, and the raising of Schrödinger's cat," *Rev. Mod. Phys.* **85**, 1103 (2013).

11 Applications of Entanglement: Heisenberg-Limited Interferometry and Quantum Information Processing

"All information is physical," the slogan advocated over many years by Rolf Landauer of IBM, has recently led to some remarkable changes in the way we view communications, computing, and cryptography. By employing quantum physics, several objectives that were thought impossible in a classical world have now proven to be possible. Quantum communications links, for example, make it impossible to eavesdrop without detection. Quantum computers (were they to be realized) could turn some algorithms that are labeled "difficult" for a classical machine, no matter how powerful, into ones that become "simple." The details of what constitutes "difficult" and "easy" are the subject of mathematical complexity theory, but an example here will illustrate the point and the impact that quantum information processors will have on all of us. The security of many forms of encryption is predicated on the difficulty of factoring large numbers. Finding the factors of a 1024-digit number would take longer than the age of the universe on a computer designed according to the laws of classical physics, and yet can be done in the blink of an eye on a quantum computer were it to have a comparable clock speed. But only if we can build one, and that's the challenge! No one has yet realized a quantum register of the necessary size, or quantum gates with the prerequisite accuracy. Yet it is worth the chase, as a quantum computer with a modest-sized register could outperform any classical machine. And in addition, quantum mechanics will allow us to construct a number of novel nonclassical technologies, some requiring already realizable resources, as we will see.

Quantum information processing offers a qualitatively different way in which to think about manipulating information. We may well be forced to adopt this new way as we exhaust classical resources and push traditional information technology into the quantum regime. One of Moore's laws [1], formulated by the founder of INTEL, using historical data, states that the number of transistors per chip (i.e. the complexity of computers) grows exponentially with time; more precisely, it doubles every one and a half years (see Fig. 11.1).

This law has been obeyed almost precisely over the past 50 years. If this exponential growth is extrapolated into the near future we see that, at the rate suggested by Moore's law, a bit of information will be

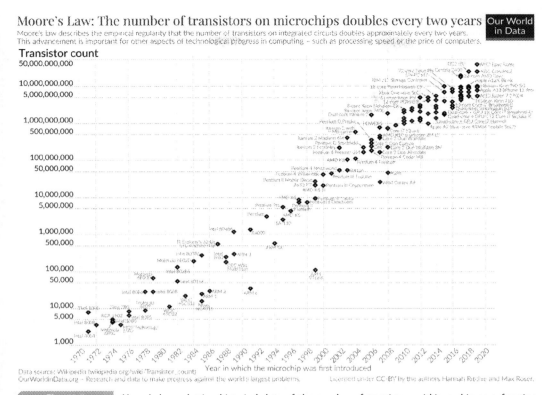

Data source: Wikipedia (wikipedia.org/wiki/Transistor_count)
OurWorldinData.org – Research and data to make progress against the world's largest problems.   Licensed under CC-BY by the authors Hannah Ritchie and Max Roser.

Fig. 11.1 Moore's law, plotting historical data of the number of transistors within a chip as a function of date, demonstrating exponential growth. Data source: Wikipedia (wikipedia.org/wiki/Transistor_count), OutWorldinData.org. Licensed under CC-BY by the authors Hannah Ritchie and Max Roser.

encoded into a single-atom size object by the time of writing (as far as we know, yet to be actually realized). In fact, quantum effects may well begin to be important due to the shrinking size and cross-talk across the chip, and may have a non-negligible effect upon computation. So not only our theoretical curiosity, but also technological progress requires that we study and understand quantum information processing. In this chapter we are interested in what quantum mechanics allows us to do, which is substantially different from what classical physics allows. The implication of Moore's law is that soon there will be about one electron per transistor and that this lone electron will be confined to a region small enough that it will act as a quantum mechanical particle, and not as a classical charged billiard ball. Such devices have been realized in low-dimensional structures. Fortunately, at this point, quantum physics not only becomes more important in designing classical computers, but also offers the whole new field of quantum computing with an entirely different mindset.

11.1 The Entanglement Advantage

We have already discussed in previous chapters, and in Appendix A, the nature of entanglement and some of its consequences – such as the violation of Bell's inequalities. Entanglement can loosely be described as a kind of correlation. But it is quite a bit more than that and, besides, classical systems can also exhibit correlations. A correlation between two systems is simply the statement that if a measurement of one system yields the result A, then a measurement on the second system will yield the result B with some probability. Perfect correlation occurs when the second result is certain, given the outcome of the first. Some readers may be familiar with John Bell's paper "Bertlmann's socks and the nature of reality" [2]. Eccentric Professor Bertlmann always wears socks of two different colors, one of which is always pink. If an observer sees only one of Bertlmann's feet and notices a pink sock, then he knows that the other is certainly not pink. On the other hand, if he observes, say, a green sock, then he knows that the other *is* pink. The correlations in this case are entirely of a classical nature and there is nothing mysterious in this example. Products of spin states of the form $|\uparrow\rangle_1|\uparrow\rangle_2$ and $|\downarrow\rangle_1|\downarrow\rangle_2$ have obvious correlations and, though the spin states of each particle are quantum mechanical, the correlations between the states are purely classical. For Bell states of the form

$$|\psi^\pm\rangle = \frac{1}{\sqrt{2}}(|\uparrow\rangle_1|\downarrow\rangle_2 \pm |\downarrow\rangle_1|\uparrow\rangle_2), \tag{11.1}$$

there are also classical-like correlations in the sense that a measurement of the z-component of the spin of one particle also determines the outcome of a measurement of the z-component of the spin of the other, although in the case of the Bell states there is an element of probability involved: 50% of the experimental runs give $|\uparrow\rangle_1|\downarrow\rangle_2$ and 50% $|\downarrow\rangle_1|\uparrow\rangle_2$, and thus outcomes of measurements of spin along the z-direction are always correlated. But when we consider measurements of spin along other directions, the product states exhibit no correlations between measurements on the different particles, whereas the Bell states do exhibit certain kinds of correlations, correlations that lead to violations of Bell-type inequalities. Bell's inequality, as should be clear from the discussion in Chapter 9, can be violated only by states that are entangled.

As we have previously said, so far, most (if not all) of the experimental tests of Bell's inequalities support quantum mechanics. But entanglement is now viewed not merely as an esoteric feature of quantum mechanics suitable only for metaphysical contemplation and tests of local hidden-variable theories, but rather it serves as the basis for a new technology. The strong nonclassical correlations possessed by entangled states are precisely what we may exploit in a number of quantum technologies that we now describe.

11.2 Two No-Go Theorems: No-Signaling and No-Cloning

Before we get into the specifics of quantum information processing, we first consider two no-go theorems that put limits on what is possible with quantum entanglement.

Bell's inequality, and its violation by quantum mechanical systems, is generally understood to be a manifestation of nonlocality displayed by multiparticle systems where the particles (which could be field modes) are prepared in an entangled state and are then made to have space-like separations, meaning that measurements performed on each of the particles cannot be connected by a light signal. This has led to speculation that quantum mechanics could be used to perform faster-than-light (super-luminal) communication if some system and/or procedure can be found to implement it. However, it is possible to show that such super-luminal signaling is prohibited within standard quantum mechanics. This is the no-signaling theorem of Ghirardi *et al.* [3]. We shall prove this theorem here. After that, we shall prove another no-go theorem, the no-cloning theorem.

11.2.1 The No-Signaling Theorem

We start by assuming two parties, Alice (A) and Bob (B), want to communicate. They each have access to one part of a system described, in the most general way possible, by the density operator $\hat{\rho}_{AB}$, where A and B represent the parts of the system accessible to Alice and Bob, respectively. We also assume that the density operator cannot be factored into a product of density operators for the individual parts of the systems, that is

$$\hat{\rho}_{AB} \neq \hat{\rho}_A \otimes \hat{\rho}_B. \tag{11.2}$$

Alice selects a local operation (an operation that acts on her particle only) from a set of unitary operations $\{\hat{U}_A^{(k)} : k = 1, 2, ..., N\}$, such that the resulting density operator for the combined systems must be

$$\hat{\rho}_{AB}^{(k)} = \hat{U}_A^{(k)} \hat{\rho}_{AB} \hat{U}_A^{(k)\dagger}, \tag{11.3}$$

where we now have label k on the new combined density operator, indicating it is generated by the action of the kth unitary operator acting on the state of particle A. Our objective is to answer the question: Does the reduced density operator for particle B depend on the label k? If it does not, then Bob cannot perform any measurement to determine k, hence there can be no signaling via quantum mechanics.

From the density operator of Eq. (11.3) we obtain the reduced density operator for particle B as given by

$$\hat{\rho}_B^{(k)} = \mathrm{Tr}_A[\hat{U}_A^{(k)}\hat{\rho}_{AB}\hat{U}_A^{(k)\dagger}] = \sum_l {}_A\langle l|\hat{U}_A^{(k)}\hat{\rho}_{AB}\hat{U}_A^{(k)\dagger}|l\rangle_A, \tag{11.4}$$

where we have assumed the existence of a complete set of orthonormal states of some observable associated with particle A. Resolving unity on both sides of $\hat{\rho}_{AB}$, we have

$$\hat{\rho}_B^{(k)} = \sum_n \sum_m \sum_l {}_A\langle l|\hat{U}_A^{(k)}|m\rangle_A \langle m|\hat{\rho}_{AB}|n\rangle_A \langle n|\hat{U}_A^{(k)\dagger}|l\rangle_A$$

$$= \sum_n \sum_m \sum_l \langle n|\hat{U}_A^{(k)\dagger}|l\rangle_A \langle l|\hat{U}_A^{(k)}|m\rangle_A \langle m|\hat{\rho}_{AB}|n\rangle_A, \tag{11.5}$$

from which, because $\hat{U}_A^{(k)\dagger}\hat{U}_A^{(k)} = 1$, it follows that

$$\hat{\rho}_B^{(k)} = \sum_n {}_A\langle n|\hat{\rho}_{AB}|n\rangle_A = \hat{\rho}_B, \tag{11.6}$$

which is independent of the label k. Therefore, Bob cannot perform a measurement on his particle B to determine Alice's choice of unitary transformation to act on particle A, and hence there can be no signaling via quantum mechanics. Note that the formalism used in the proof does not explicitly incorporate the space-like separation of particles A and B. The proof is general in that respect. The point is that if one cannot use quantum entanglement for signaling, then one cannot use quantum entanglement for faster-than-light signaling.

11.2.2 The No-Cloning Theorem

For a classical system, which can be examined at will, an arbitrary number of copies of the system can be made. For quantum systems, the situation is quite different. If we have access to only one system in an arbitrary quantum state, it is impossible to copy it. That this is so can be seen as follows.

Suppose we have two systems denoted A and B (not to be confused with Alice and Bob of the previous subsection). Further, suppose system A is in the state $|\varphi\rangle_A$, which is assumed to be arbitrary and unknown. To copy it into system B means we have to make the transformation $|\varphi\rangle_A|i\rangle_B \rightarrow |\varphi\rangle_A|\varphi\rangle_B$, where $|i\rangle_B$ is itself an arbitrary input state of system B (possibly the ground state). This is what we mean by cloning. To implement the cloning, we require a unitary transformation acting on both systems. We denote this $\hat{U}_{AB}$ and it is supposed to have the effect

$$\hat{U}_{AB}|\varphi\rangle_A|i\rangle_B = |\varphi\rangle_A|\varphi\rangle_B. \tag{11.7}$$

Now, suppose we take two states to copy, $|\varphi_1\rangle_A$ and $|\varphi_2\rangle_A$:

$$\hat{U}_{AB}|\varphi_1\rangle_A|i\rangle_B = |\varphi_1\rangle_A|\varphi_1\rangle_B,$$
$$\hat{U}_{AB}|\varphi_2\rangle_A|i\rangle_B = |\varphi_2\rangle_A|\varphi_2\rangle_B. \tag{11.8}$$

We take the inner products of the respective right and left sides of these equations to obtain

$$_A\langle\varphi_2|\varphi_1\rangle_A{}_B\langle\varphi_2|\varphi_1\rangle_B = {}_A\langle\varphi_2|_B\langle i|\hat{U}^\dagger_{AB}\hat{U}_{AB}|\varphi_1\rangle_A|i\rangle_B$$
$$= {}_A\langle\varphi_2|\varphi_1\rangle_A{}_B\langle i|i\rangle_B = {}_A\langle\varphi_2|\varphi_1\rangle_A. \tag{11.9}$$

This shows that we must have $_A\langle\varphi_2|\varphi_1\rangle_A = 0$ or 1. If $_A\langle\varphi_2|\varphi_1\rangle_A = 1$, the states are identical. If $_A\langle\varphi_2|\varphi_1\rangle_A = 0$, the states are orthogonal and thus they are not arbitrary. They would have been selected from an orthonormal basis and are not arbitrary. Thus, we conclude that with the operator $\hat{U}_{AB}$ it is not possible to copy a second state which is not orthogonal to the first. The no-cloning theorem can be stated as follows: There is no universal copier for pure quantum states [4].

Another way to see that cloning is generally impossible is to consider a superposition state of the form

$$|\psi\rangle_A = c_1|\varphi_1\rangle_A + c_2|\varphi_2\rangle_A, \tag{11.10}$$

and then consider the operation $\hat{U}_{AB}|\psi\rangle_A|i\rangle_B$. Cloning requires that $\hat{U}_{AB}|\psi\rangle_A|i\rangle_B = |\psi\rangle_A|\psi\rangle_B$.

However, if we presume the validity of the transformations in Eqs. (11.8), we arrive at

$$\hat{U}_{AB}|\psi\rangle_A|i\rangle_B = c_1|\varphi_1\rangle_A|\varphi_1\rangle_B + c_2|\varphi_2\rangle_A|\varphi_2\rangle_B, \tag{11.11}$$

which is clearly not of the form

$$|\psi\rangle_A|\psi\rangle_B = (c_1|\varphi_1\rangle_A + c_2|\varphi_2\rangle_A)(c_1|\varphi_1\rangle_B + c_2|\varphi_2\rangle_B). \tag{11.12}$$

The no-cloning theorem is of critical importance in quantum information processing and underlies the security of quantum communications.

11.3 Entanglement and Interferometric Measurements

The key to harnessing the potential of entanglement turns out to be related to our ability to count the number of particles in a system. We can illustrate this by considering the problem of quantum-enhanced phase measurement. Let us say that you wish to determine the phase shift induced by some optical element on a beam. One way to do this is to put the element into one arm of an interferometer and illuminate the interferometer input.

Then a measurement of the difference photocurrent from two detectors looking at the output ports of the interferometer will exhibit a sinusoidal fringe pattern as the phase is changed. A particular phase can be measured by comparing the difference photocurrent with the known maximum and minimum photocurrents. The accuracy of this phase measurement will be set by the photocurrent fluctuations. If the input light is classical, these will be at the shot noise limit. That is, if the input radiation has a mean photon number $\bar{n}$, the sensitivity of the phase measurement will be $\Delta\theta = 1/\sqrt{\bar{n}}$, as we saw in Chapter 6. A sensitivity given by $\Delta\theta = 1/\sqrt{\bar{n}}$ is called shot noise (SN) limited. It is also known as the standard quantum limit (SQL).

Phase-measurement accuracy can be improved if nonclassical light is used to illuminate the interferometer. Recall that when light in a coherent state (classical light) is injected into an interferometer, the light inside is always describable by separable states; the states of light along the two beam paths never become entangled. This is all very classical. But it is possible to improve the measurements of phase shifts by using quantum mechanical states of light [5]. As a simple example, consider the case where exactly one photon is put into each of the input ports a and b of the interferometer, as indicated in Fig. 11.2. (Note that the labeling of the modes has been modified from that used in Chapter 6. Here, the beam along the counterclockwise path is labeled a and that along the clockwise path is labeled b.) The state of the system at the input can be denoted $|1\rangle_a|1\rangle_b$. After the beam splitter, the system is in an entangled state

$$|\psi_2\rangle = \frac{1}{\sqrt{2}}(|2\rangle_a|0\rangle_b + |0\rangle_a|2\rangle_b), \tag{11.13}$$

as given by Eq. (6.45), apart from an irrelevant overall phase. If the phase shifter is in the b-arm of the interferometer, as pictured in Fig. 11.2, then just before the second beam splitter, the state is

$$|\psi_2(\theta)\rangle = \frac{1}{\sqrt{2}}(|2\rangle_a|0\rangle_b + e^{i\theta}|0\rangle_a|2\rangle_b), \tag{11.14}$$

and using the methods of Chapter 6, after the second beam splitter the state is

$$|\text{out}\rangle = \frac{1}{2\sqrt{2}}(1 - e^{2i\theta})(|2\rangle_a|0\rangle_b - |0\rangle_a|2\rangle_b) + \frac{i}{2}(1 + e^{2i\theta})|1\rangle_a|1\rangle_b. \tag{11.15}$$

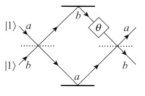

Schematic of an interferometric measurement of a phase shift with twin single-photon states as inputs. Parity measurements must be performed on one of the output beams.

Now to measure the phase shift we might try to follow the procedure used in the case of the coherent state input, or even for the case when the input is $|1\rangle_a|0\rangle_b$, as we discussed in Chapter 6. In those cases, we obtained an oscillatory function of the phase shift θ. But in the present situation there is a problem. If we calculate the average photon number in each of the output beams, that is $\langle \hat{a}_{out}^{\dagger}\, \hat{a}_{out}\rangle$ and $\langle \hat{b}_{out}^{\dagger}\, \hat{b}_{out}\rangle$, we find that they both have a value of unity and do not depend on θ. Obviously their difference will vanish as well. So, the standard measures to obtain θ will not work.

On the other hand, if one has available photon detectors that can resolve counts at the level of one photon, one could do coincident measurements for detecting, say, one photon in each output mode (i.e. detecting the state $|1\rangle_a|1\rangle_b$), the probability of which is, from Eq. (11.15):

$$P_{1,1}(\theta) = \left| \frac{i}{2}(1 + e^{2i\theta}) \right|^2 = \frac{1}{2}[1 + \cos(2\theta)]. \qquad (11.16)$$

One could instead perform coincidence measurements to determine the probability of detecting two photons in one output beam and none in the other:

$$P_{2,0}(\theta) = P_{0,2}(\theta) = \left| \frac{1}{2\sqrt{2}}(1 - e^{2i\theta}) \right| = \frac{1}{4}[1 - \cos(2\theta)]. \qquad (11.17)$$

These probabilities depend on 2θ, not just θ. Thus we can say that probabilities exhibit twofold super-resolution in the phase shift θ.

A generalization of the state of Eq. (11.13) is the so-called N00N [6] state given by

$$|\psi_N\rangle = \frac{1}{\sqrt{2}}(|N\rangle_a|0\rangle_b + |0\rangle_a|N\rangle_b). \qquad (11.18)$$

N00N states are obviously highly entangled states of a two-mode field. These are two-mode Schrödinger-cat-like states, being superpositions of states where all the photons are in one mode and all in the other. For $N > 2$, such a state cannot be generated by beam splitting (but see below). But suppose such a state has somehow been made available. We emphasize that the first beam splitter of the interferometer pictured in Fig. 11.2 must be replaced by some device or process that generates the N00N states. Measurements of the phase shift are to be performed *after* the remaining beam splitter. We assume as before that a phase shift is encoded onto the b-mode, so that we have

$$|\psi_N(\theta)\rangle = \frac{1}{\sqrt{2}}(|N\rangle_a|0\rangle_b + e^{iN\theta}|0\rangle_a|N\rangle_b), \qquad (11.19)$$

just before the beam splitter. Using the methods of Chapter 6, the beam splitter transformation results in the output state

$$|\text{out}\rangle = \left(\frac{1}{\sqrt{2}}\right)^{N+1} \sum_{k=0}^{N} \binom{N}{k}^{1/2} [i^k + e^{iN\theta}i^{N-k}]|N-k\rangle_a|k\rangle_b. \qquad (11.20)$$

From this, the probability of finding $N - k$ photons in mode a and k in mode b is

$$P_{N-k,k}(\theta) = \left(\frac{1}{2}\right)^N \binom{N}{k} |i^k + e^{iN\theta}i^{N-k}|^2, \qquad (11.21)$$

which clearly has a dependence on $N\theta$ and thus there will be interference fringes with an N-fold super-resolution.

As mentioned, it is not possible to generate a $N00N$ state for an arbitrary number of photons N using a beam splitter. However, it is possible to generate, to a good approximation, a superposition of $N00N$ states, that is a superposition of the form

$$|\psi\rangle = \frac{1}{\sqrt{2}} \sum_N C_N (|N\rangle_a|0\rangle_b + |0\rangle_a|N\rangle_b), \qquad (11.22)$$

by mixing at a beam splitter coherent light and squeezed vacuum light of equal, but low, average photon number [7]. States like this are sometimes referred to as "corner states" for reasons that are obvious in Fig. 11.3 Afek et al. [8] have performed such an experiment wherein they performed

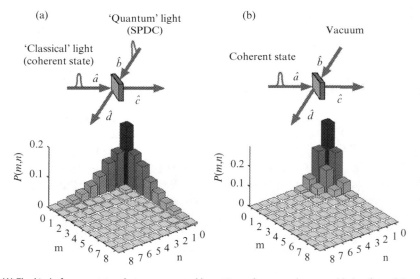

Fig. 11.3 (A) The kind of corner states that are generated by mixing coherent and squeezed light of equal, but low, average photon number. In this case the average photon number per pulse for both beams was 1.2. (B) The same as (A) but with a vacuum state in place of the squeezed vacuum state. Used with the permission of American Association for the Advancement of Science, from Ref. [8]; copyright 2010 permission conveyed through Copyright Clearance Center, Inc.

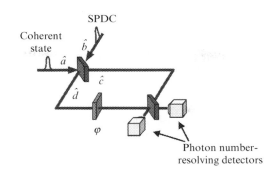

A schematic of the Mach–Zehnder interferometer with photon number-resolving detectors placed at exit ports of the interferometer. Used with the permission of American Association for the Advancement of Science, from Ref. [8]; copyright 2010 permission conveyed through Copyright Clearance Center, Inc.

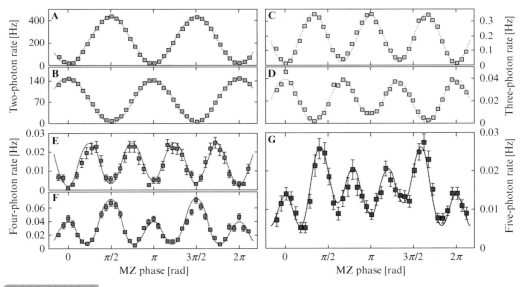

Experimental results of coincidence detections as a function of phase (in radians) for totals 2, 3, 4, and 5 photons as indicated. What is shown are the count rates of the two detectors for given total photon numbers. Used with the permission of American Association for the Advancement of Science, from Ref. [8]; copyright 2010 permission conveyed through Copyright Clearance Center, Inc.

coincidence counting on the output beams of a second beam splitter of a Mach–Zehnder interferometer (Fig. 11.4). That is, they counted N_a photons in mode a coincident with counting N_b photons in mode b such that the total number detected was $N = N_a + N_b$. The experiment demonstrated N-fold super-resolution for total photon numbers $N = 2, 3, 4$, and 5 (see Fig. 11.5).

As mentioned, these coincidence measurements exhibit super-resolution, but it is unclear as to whether or not these coincidence measurements lead to super-sensitivity (i.e. the uncertainty in the phase measurements satisfies the relation $\Delta\theta < \Delta\theta_{\text{SQL}} = 1/\sqrt{\bar{n}}$).

Another approach, one that leads to super-sensitivity as well as super-resolution, is to consider a different observable altogether, one that has no classical analog. This is the photon number parity operator that was introduced in Chapter 3. The idea of using measurements of parity first appeared in the context of spectroscopy, as discussed by Bollinger et al. [9] in the context of trapped ions, and was later adapted for the purpose of optical interferometry [10].

We consider the parity operators of both a and b output modes, these being

$$\hat{\Pi}_a = (-1)^{\hat{a}^\dagger \hat{a}} = \exp(i\pi\hat{a}^\dagger\hat{a}), \quad \hat{\Pi}_b = (-1)^{\hat{b}^\dagger \hat{b}} = \exp(i\,\pi\,\hat{b}^\dagger\hat{b}). \quad (11.23)$$

The expectation values of these operator with respect to the state of Eq. (11.15) are easily found to be

$$\langle\hat{\Pi}_a\rangle = \langle\text{out}|\hat{\Pi}_a|\text{out}\rangle = -\cos(2\theta),$$

$$\langle\hat{\Pi}_b\rangle = \langle\text{out}|\hat{\Pi}_b|\text{out}\rangle = -\cos(2\theta),$$

(11.24)

which depend not just on θ but also on 2θ. The results are identical because of the symmetry in the output state of Eq. (11.15). If we use the calculus of error propagation for the a mode, we will obtain, rather generally, the uncertainty in the phase measurements (the sensitivity of the measurement) as

$$\Delta\theta = \Delta\Pi_a \Big/ \left|\frac{\partial\langle\hat{\Pi}_a\rangle}{\partial\theta}\right|, \quad (11.25)$$

where $\Delta\Pi_a = \sqrt{1 - \langle\hat{\Pi}_a\rangle^2}$ owing to the fact that $\hat{\Pi}_a^2 = \hat{I}$, the identity operator. In our particular case, we obtain $\Delta\theta = 1/2$. If we compare this result with the optimal coherent state result for the *same average photon number* at the input, $\Delta\theta = 1/\sqrt{2}$ for $\bar{n} = 2$, we see an improvement in the phase sensitivity, the resolution, by a factor of $1/\sqrt{2}$. This improvement is the result of the entanglement of the state inside the interferometer.

For the N00N state of Eq. (11.19), one can show that the expectation value of the parity operator is

$$\langle\hat{\Pi}_a\rangle = \begin{cases} (-1)^{N/2}\cos(N\theta) & N \text{ even}, \\ (-1)^{(N+1)/2}\sin(N\theta) & N \text{ odd}. \end{cases} \quad (11.26)$$

The demonstration of this is left as an exercise (see problem 3 of this chapter). From the error propagation formula, we obtain the sensitivity of the phase shift measurement to be

$$\Delta\theta = \frac{\sqrt{1 - \langle\hat{\Pi}_a\rangle^2}}{|(\partial\langle\hat{\Pi}_a\rangle/\partial\theta)|} = \frac{1}{N}, \tag{11.27}$$

which represents an improvement in the sensitivity of the standard quantum limit, that being $\Delta\theta_{\text{SQL}} \equiv 1/\sqrt{N}$, by a factor of $1/\sqrt{N}$. The right-hand side of Eq. (11.27) represents, in fact, the highest level of sensitivity allowed by quantum mechanics for the interferometric measurement of phase shifts [5], and is referred to as the Heisenberg limit (HL): $\Delta\theta_{\text{HL}} \equiv 1/N$.

Of course, there is still the issue of how to detect photon number parity. The obvious approach is to somehow count photons with a resolution at the level of a single photon and raise minus one to that power. This is generally impossible with the standard avalanche detectors. However, new detectors, the so-called superconducting transition edge sensors (TES), are capable of such detection up to a certain number of photons. This will be described later in Section 11.8. Ideally, one might want to detect the parity of the photon number without detecting the number first. In principle, this could be done via quantum non-demolition measurements [11]. However, the large cross-Kerr interactions required are not available in the optical regime, although they are available in the context of circuit QED [12]. As we show below, direct parity detection may not even be necessary.

It turns out that parity measurements in optical interferometry have value even if coherent states are being used in the interferometer. The sensitivity at best is still at the standard quantum limit, but there is an improvement in the *resolution* of such measurements. This was shown to be the case by Gao *et al.* [13].

We refer back to the coherent state interferometry discussed in Chapter 6. We use the output states of the interferometer as given by Eq. (6.85), which we reproduce here for easy reference:

$$|\text{out}\rangle = |i(1 + e^{i\theta})\alpha/2\rangle_a|(e^{i\theta} - 1)\alpha/2\rangle_b. \tag{11.28}$$

We find that

$$\langle\hat{\Pi}_a\rangle = \exp[-\bar{n}(1 + \cos\theta)],$$
$$\langle\hat{\Pi}_b\rangle = \exp[-\bar{n}(1 - \cos\theta)], \tag{11.29}$$

where $\bar{n} = |\alpha|^2$. Note that in this case, the parity expectation values of the two output modes are not identical. In the vicinity of a very small phase shift (i.e. with $\theta \to 0$) we can see that with $\cos\theta \simeq 1 - \theta^2/2$, we have

$$\langle\hat{\Pi}_a\rangle \simeq \exp[-2\bar{n} + \bar{n}\,\theta^2/2],$$
$$\langle\hat{\Pi}_b\rangle \simeq \exp[-\bar{n}\,\theta^2/2]. \tag{11.30}$$

Evidently, in the limit of small θ and large $\bar{n}$, $\langle\hat{\Pi}_a\rangle \to 0$, whereas $\langle\hat{\Pi}_b\rangle$ becomes narrowly peaked around $\theta = 0$. In fact, $\langle\hat{\Pi}_b\rangle$ is a Gaussian centered at $\theta = 0$ with a width given by $\delta\theta = 1/\sqrt{\bar{n}}$.

Obviously the greater $\bar{n}$ the narrower the width, and this is what we mean by super-resolution in parity measurement-based interferometry with coherent states. Obviously, for small phase shifts, parity measurements should be performed on the output b mode.

But suppose the phase shift is large. Suppose θ in the vicinity of π. If we set $\theta = \pi - \delta$, we then have

$$\langle \hat{\Pi}_a \rangle = \exp[-\bar{n}(1 - \cos \delta)],$$
$$\langle \hat{\Pi}_b \rangle = \exp[-\bar{n}(1 + \cos \delta)],$$

(11.31)

where the identity $\cos(\beta - \gamma) = \cos \beta \cos \gamma + \sin \beta \sin \gamma$ was used. Clearly, the parities of the two modes have essentially been reversed so that the a-mode parity should be monitored as it will be narrowly peaked in the limit $\delta \to 0$. Again we have super-resolution. The case for $\theta \to \pi$ is relevant to an experiment to be described below.

But first, we examine the level sensitivity of the phase measurements for parity-based coherent state interferometry. For ease of manipulation, we consider the square of the phase uncertainty, which for a generic parity operator $\hat{\Pi}$ will be given by

$$(\Delta \theta)^2 = \frac{1 - \langle \hat{\Pi} \rangle^2}{(\partial \langle \hat{\Pi} \rangle / \partial \theta)^2}.$$

(11.32)

It is straightforward to show, for the two output modes undergoing parity measurements, that the corresponding phase uncertainties using Eqs. (11.31) and (11.32) are given by

$$(\Delta \theta)_a^2 = \frac{e^{2\bar{n}(1+\cos \theta)} - 1}{\bar{n}^2 \sin^2 \theta},$$
$$(\Delta \theta)_b^2 = \frac{e^{2\bar{n}(1-\cos \theta)} - 1}{\bar{n}^2 \sin^2 \theta}.$$

(11.33)

For the limiting case of $\theta \to 0$, one can use L'Hôpital's rule to show that $\lim_{\theta \to 0} (\Delta \theta)_b^2 = 1/\bar{n}$ or that $\lim_{\theta \to 0} (\Delta \theta)_b = 1/\sqrt{\bar{n}}$, which is the SQL. Similarly, we have $\lim_{\theta \to \pi} (\Delta \theta)_a^2 = 1/\sqrt{\bar{n}}$. Thus, in these limiting cases we obtain super-resolution at the standard quantum limit.

Super-resolution of this form by parity measurements was demonstrated experimentally by Cohen et al. [14]. They detected a relative phase shift of π between the arms of a Mach–Zehnder interferometer by doing photon number counting using a silicon photomultiplier (SiPM, Hamamatsu Photonics, S10362-11-100U) which employs a grid of single-photon avalanche detectors. With this technique, the authors of Ref. [14] were able to perform measurements of photon counts of up to 26 photons with a resolution at the level of a single photon. They used this capability to construct the photon number probability distribution P_n for phase shifts

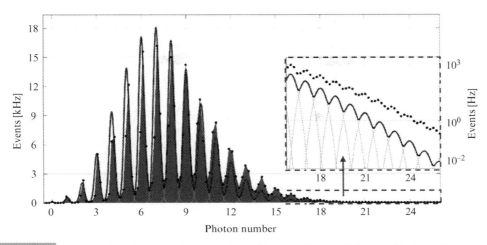

Fig. 11.6   Photon number distribution from experiment fitted to a Poisson distribution. (Courtesy of Lior Cohen. Used with permission.)

set in a narrow range around the value of $\theta = \pi$, a typical example of which is shown in Fig. 11.6. The parity is not determined directly. Rather, the average parity is calculated from the experimentally determined photon number distribution by the formula

$$\langle \hat{\Pi} \rangle = \sum_{n=0}^{\infty} (-1)^n P_n. \tag{11.34}$$

In this way, the average parity is obtained without direct measurements of the parity.

These authors also performed an experiment wherein the phase shift was determined by on–off "parity" detection: the detection of no photons versus the detection of any number of photons, represented by the operators (for the a mode)

$$\hat{Z} = |0\rangle_{a\,a}\langle 0| \quad \text{and} \quad \hat{\mathbf{I}} - \hat{Z} = \sum_{n=1}^{\infty} |n\rangle_{a\,a}\langle n|, \tag{11.35}$$

respectively. The expectation value of the vacuum-state projection operator $\hat{Z}$ with respect to the output state given by Eq. (11.28) is given by

$$P_0 = |_a\langle 0|i(1 + e^{i\theta})\alpha/2\rangle_a|^2 = \exp\left[-\frac{\bar{n}}{2}(1 + \cos\theta)\right], \tag{11.36}$$

which is very similar to $\langle \hat{\Pi}_a \rangle$ as given in Eq. (11.31). The results obtained by this "parity" method were very close to those obtained by the even/odd parity method. The sensitivity in this case was also SN limited.

In Fig. 11.7 are plotted the expectation values of the parity and the probability P_0 of detecting the vacuum state as functions of the phase shift θ in the vicinity of $\theta = \pi$, for coherent states of various average photon

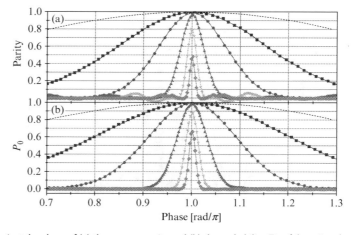

Fig. 11.7 Plots against the phase of (a) the average parity and (b) the probability P_0 of detecting the vacuum with input laser light of average photon numbers 4.6 (squares), 25 (circles), 200 (triangles), 1190 (inverted triangles), and 4150 (rhombuses). Reprinted with permission from Ref. [14] © The Optical Society.

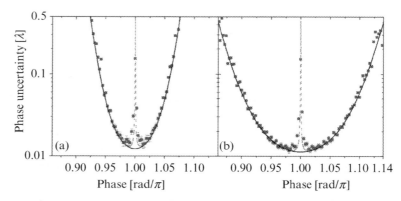

Fig. 11.8 Plots of the phase uncertainty versus the phase for the case of coherent light with an average of 200 photons obtained from (a) the average parity and (b) P_0. Reprinted with permission from Ref. [14] © The Optical Society.

numbers. For the very large average photon numbers indicated, results are possible only very near $\theta = \pi$, where the amplitude of the output a-mode coherent state is significantly reduced, as can be seen from Eq. (11.28), where we notice that $|i(1 + e^{i\theta})\alpha/2\rangle_a \rightarrow |0\rangle_a$ as $\theta \rightarrow \pi$. The narrowing of the parity expectation and of the probability of detecting the vacuum state as $\theta \rightarrow \pi$ constitutes super-resolution.

In Fig. 11.8 are plotted the measured phase uncertainties for both methods, for 200 photons on average, as functions of the phase. They show the expected behavior, but with spikes around $\theta = \pi$ due to imperfect visibility.

The parity measurement approach to interferometry is reviewed, with many references, by Gerry and Mimih [15] and by Birrittella *et al.* [16]. More about photon number parity detection, but in a different setting, can be found in Section 11.8.

11.4 Quantum Teleportation

Another application of entanglement, this one much more in the context of quantum information processing, is quantum teleportation, whereby an unknown quantum state is transferred from one point to another, these points possibly being widely separated. We wish to emphasize that it is the unknown quantum *state* that is to be teleported, not the particle or particles in such states. We shall restrict ourselves to a simple and rather basic description of teleportation, as originally given by Bennett *et al.* [17]. We shall use states denoted $|0\rangle$ and $|1\rangle$ as usual, but we will not necessarily think of them in terms of photon number states but rather as states of some sort of "two-level" system, such as the polarization states of single photons. Furthermore, in the quantum information business, the symbols $|0\rangle$ and $|1\rangle$ and superpositions of these states represent the quantum bits of information, the *qubits*. Qubits are the quantum information-theoretic name for the states of any "two-level" system, and a *single qubit* refers to one system whose states can be represented, in general, by superpositions of the states $|0\rangle$ and $|1\rangle$. Qubits could be realized also by two-level atoms. As this text is concerned in the main with photons, one might wish to interpret the states $|0\rangle$ and $|1\rangle$ as vertically and horizontally polarized states, $|V\rangle$ and $|H\rangle$, respectively, but the procedures described are rather general.

We require two participants in the teleportation procedure (or protocol), usually given the names Alice (A) and Bob (B). We suppose that Alice is given a photonic mode in the quantum state $|\Psi\rangle = c_0|0\rangle + c_1|1\rangle$ and that she wishes to teleport the state to Bob, so that he can recreate it in a photonic mode in his possession. The state is unknown to Alice (i.e. she does not know the coefficients c_0 and c_1). In fact, if she did know the state, the protocol about to be described would be unnecessary as she could simply convey the information to Bob over a classical channel (e.g. via telephone). We further suppose that some source of light can produce an entangled state of the form

$$|\Psi_{AB}\rangle = \frac{1}{\sqrt{2}}(|0\rangle_A|0\rangle_B + |1\rangle_A|1\rangle_B), \qquad (11.37)$$

which is shared between Alice and Bob as indicated by the subscripts. So, Alice has in her possession the state to be teleported, $|\psi\rangle$, and part of the shared state, $|\psi_{AB}\rangle$, whereas Bob has only his part of the shared state, so far. Thus, we may write the total state for Alice and Bob, at this point, as

$$|\Phi_{AB}\rangle := |\Psi\rangle||\Psi_{AB}\rangle = (c_0|0\rangle + c_1|1\rangle)(|0\rangle_A|0\rangle_B + |1\rangle_A|1\rangle_B)/\sqrt{2}. \quad (11.38)$$

We can expand this as

$$|\Phi_{AB}\rangle = \frac{1}{\sqrt{2}}(c_0|0\rangle|0\rangle_A|0\rangle_B + c_0|0\rangle|1\rangle_A|1\rangle_B + c_1|1\rangle|0\rangle_A|0\rangle_B + c_1|1\rangle|1\rangle_A|1\rangle_B)$$

$$= \frac{1}{2}[|\Phi^+\rangle(c_0||0\rangle_B + c_1|1\rangle_B) + |\Phi^-\rangle(c_0||0\rangle_B - c_1|1\rangle_B)$$

$$+ |\Psi^+\rangle(c_0||1\rangle_B + c_1|0\rangle_B) + |\Psi^-\rangle(c_0||1\rangle_B - c_1|0\rangle_B)],$$

$$(11.39)$$

where we have introduced the Bell states

$$|\Phi^+\rangle = \frac{1}{\sqrt{2}}(|0\rangle|0\rangle_A + |1\rangle|1\rangle_A), \quad (11.40)$$

$$|\Phi^-\rangle = \frac{1}{\sqrt{2}}(|0\rangle|0\rangle_A - |1\rangle|1\rangle_A), \quad (11.41)$$

$$|\Psi^+\rangle = \frac{1}{\sqrt{2}}(|0\rangle|1\rangle_A + |1\rangle|0\rangle_A), \quad (11.42)$$

$$|\Psi^-\rangle = \frac{1}{\sqrt{2}}(|0\rangle|1\rangle_A - |1\rangle|0\rangle_A), \quad (11.43)$$

which are mutually orthogonal and thus constitute a basis in a four-dimensional Hilbert space. These states are constructed from the basis states of the unknown state to be teleported and that part of the entangled state shared by Alice. Each Bell state in Eq. (11.9) is correlated with a different superposition of the states in Bob's share of the entangled state provided. Perhaps it is important to point out that nothing physical has happened here yet: we have merely rewritten the direct product of the state to be teleported and the provided, shared, entangled state. The next step in the protocol is for Alice to perform projective measurements onto the Bell basis. Each of the four Bell states $|\Phi^\pm\rangle, |\Psi^\pm\rangle$ will occur randomly with equal probability of $1/4$. Suppose Alice obtains the state $|\Phi^\pm\rangle$ and she *knows* that she has obtained that state. Bob's photonic system is then projected onto the state $c_0|0\rangle_B + c_1|1\rangle_B$. Over a classical channel, Alice then tells Bob that she has detected the Bell state $|\Psi^\pm\rangle$ and thus both know that Bob already has in possession the teleported state and thus he needs to do nothing. Note that neither Alice nor Bob knows what that state is. On the other hand, if Alice reports that she has detected the state $|\Phi^-\rangle$, then Bob's system has been projected onto the state $c_0|0\rangle_B - c_1|1\rangle_B$, and he knows that his state differs from the original by the sign of the second term, so that he can perform the transformation $|0\rangle_B \rightarrow |0\rangle_B, |1\rangle_B \rightarrow -|1\rangle_B$ to obtain the original state. If Alice detects $|\Psi^+\rangle$, Bob has $c_0|1\rangle_B + c_1|0\rangle_B$ and he needs to perform the "flip" operation $|1\rangle_B \rightarrow |0\rangle_B, |0\rangle_B \rightarrow |1\rangle_B$, the equivalent of a logical NOT operation. Finally, if Alice detects $|\Psi^-\rangle$, then Bob has $(c_0|1\rangle_B - c_1|0\rangle_B)$ and must perform the transformation $|1\rangle_B \rightarrow |0\rangle_B, |0\rangle_B \rightarrow -|1\rangle_B$. This completes the description of the teleportation protocol.

Within the two-dimensional basis $(|0\rangle_B, |1\rangle_B)$, the above operations can be represented by a set of 2×2 matrices $\{I_2, \sigma_x, \sigma_y, \sigma_z\}$, where the first is the two-dimensional identity matrix and the rest are the familiar Pauli matrices, as the reader can easily check. Finally, notice that at no time did anyone know the original state being teleported. In the process of performing the Bell state measurements, this original state is itself destroyed.

Quantum teleportation may seem to have an element of magic to it and, if so, this is because the state shared by Alice and Bob is entangled. If we tried teleporting a state by allowing Alice and Bob to share the statistical mixture

$$\hat{\rho}_{AB} = \frac{1}{2}(|0\rangle_A|0\rangle_B \, _A\langle 0| \, _B\langle 0| + |1\rangle_A \, |1\rangle_B \, _A\langle 1| \, _B\langle 1|), \tag{11.44}$$

teleportation would be impossible, as the reader can check. Notice, though, that the mixed state of Eq. (11.14) does, in fact, exhibit correlations; but these correlations are completely of a classical nature.

Quantum teleportation is not merely a theoretical curiosity but is an experimental reality. The initial experiments teleporting qubit states, where the qubits were represented by photon polarization states, were performed by groups in Rome [18] and Innsbruck [19]. Subsequently, a group at Caltech teleported a state with continuous variables [20], one made from squeezed states of light.

11.5 Cryptography

Secure communication is an area of great importance as we move into an information age. Such communication is built on crypto-systems, whose security is based on assumptions concerning the computational difficulty of solving various problems, especially of factoring. An alternative is quantum cryptography, which is based on a fundamental description of nature, namely quantum mechanics, and where security is ensured by the nature of measurements in quantum physics.

The rapid rise of electronic communication and commerce has led to increasing concern with the security and authentication of electronic messages. This is, of course, not a new problem. Since the earliest records of humanity there has been a need and requirement to pass secrets between different parties. For millennia, communicating parties have devised schemes whereby messages can be authenticated (the signature) and secured from unauthorized access (cryptography). Modern methods for secure communication always involve the prior exchange of a random number or binary string, called the key. If the communicating parties share this number with each other and no one else, messages can be securely encrypted and decoded. The method, however, is vulnerable to a third party acquiring access to the key. Quantum mechanics enables two communicating parties to arrive at a shared secure key and yet be certain that no eavesdropper has acquired the key. Quantum cryptography will be reviewed in this section.

Quantum mechanics has the potential to revolutionize information processing in at least two almost complementary ways. Firstly, if a quantum computer can be realized, it could be utilized to factor large numbers beyond the capability of any classical computer. Since (as we will see) the difficulty of factoring underpins secure classical communication, such a quantum computer threatens all current cryptographic security. But while quantum mechanics can undermine cryptographic security, it can also save it. We will show that the basic laws of quantum mechanics can allow us to construct quantum cryptographic protocols secure against interception.

This section is structured as follows. We begin by providing an introduction to classical private and public key crypto-systems. We then consider several of the quantum protocols and their experimental realization in quantum optics. However, before we proceed further, let us define a few essential terms and concepts.

- *Alice and Bob*: Alice is an individual who wants to send a private message to Bob.
- *Eve*: an eavesdropper who wants to intercept the message sent by Alice to Bob and read it. We assume that Eve has full access to the communications channel used to transmit the message.
- *Public channel*: a communication channel through which information is transmitted from one endpoint to another. The eavesdropper has full access to all information passed through this channel.
- *Private channel*: a secure channel to which the eavesdropper has no access. It is believed to be impossible to eavesdrop on information transmitted through this channel. This channel is often referred to as a trusted channel.
- *Key material*: a sequence of preferably random numbers known only to Alice and Bob.
- *Symmetric algorithm*: the symmetric key gets its name from the fact that the encryption key can be calculated from the decryption key and vice versa. In most cases, the encryption key and the decryption key are the same.
- *Public key algorithm*: public key algorithms are asymmetric in nature and are designed such that the key used for encryption is different from the key used for decryption. Furthermore, the decryption key cannot be calculated from the encryption key in a reasonable amount of time.

11.6 Private Key Crypto-systems

Until about 1970, all crypto-systems generally operated on a private key principle. In this situation, the two parties (Alice and Bob) wishing to communicate must have established a shared key beforehand. If Alice wishes to send a message to Bob, then Alice encrypts the message with an encoding key. This encrypted message is then transmitted to Bob in the open (via a public channel). Bob then uses his decoding key (which was established with Alice beforehand) to decrypt the transmitted signal and hence receive the original message from Alice, as indicated in Fig. 11.9.

One of the earliest examples is the so-called *Caesar crypto-system*, which involves translating our normal alphabet into numbers. One simple mechanism is to translate our normal 26-character alphabet (A–Z) into the numerical equivalents $0, \ldots, 25$ (e.g. $A = 0, \ldots, Z = 25$) and add a constant number, 10 say. To perform this operation, one can define a function such that

$$f(x) = \begin{cases} x + 10, & x < 15, \\ x - 15, & x \geq 15, \end{cases} \tag{11.45}$$

or, in the more compact form

$$f(x) = x + 10 \pmod{26}. \tag{11.46}$$

As an example, consider that Alice wants to send Bob the word "HELLO." She begins by translating the word "HELLO" into the numerical equivalents "7 5 11 11 14." She then applies the above function to get her encrypted message "17 15 21 21 24," which can be transmitted to Bob over an open channel. Having previously established the nature of the encoding, Bob knows the inverse function $f^{-1}(x)$ and hence decodes the encrypted message "17 15 21 21 24" to "7 5 11 11 14." The message is then simply converted back into the word "HELLO."

The above protocol requires that effectively, both Alice and Bob know the form of the encoding function (this has to be established previously by some secure means). Alice uses an encoding function and Bob uses the

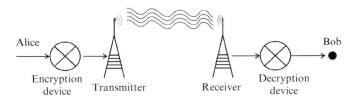

Fig. 11.9　Schematic of a crypto-system.

decoding function that is the inverse of the encoding function. A more complicated version of the encoding could be given by

$$f(x) = ax + b(\operatorname{mod} N), \tag{11.47}$$

where a and b are constants. The problem with this kind of encoding is that once an eavesdropper (Eve) knows the form of the function, she can simply decode the message the same way Bob does. If the same encoding is used repeatedly, then Eve can attack the encoding and it can be broken.

A modification to the encoding scheme proposed above does allow a secure encryption scheme. A very well-known private key crypto-system is the *Vernam cipher* (also called the *one-time pad*). This scheme works by Alice and Bob initially sharing a secret key string. This must be done privately and securely. When Alice and Bob want to pass a message, Alice encodes her message by adding the message and key string together. This is then transmitted over a public channel to Bob. Bob recovers the message by subtracting the same key string from the encoded message. The one-time pad is a well-used, provably secure private key crypto-system. As its name indicates, the one-time pad is secure only if the key material is not reused. In fact, the key material must be at least as large as the size of the message for absolute security.

One of the major difficulties with the one-time pad is that it requires a huge amount of key material that must be distributed to the parties wishing to communicate. This cannot be done over a public channel, which makes it impractical for general usage, but it is still used today in situations requiring provable security. Certain forms of banking still use the Vernam cipher. Today, most of the current crypto-systems use a public key arrangement.

11.7 Public Key Crypto-systems

The private key crypto-systems we have considered previously have relied on the distribution of key material. There is another approach that comes under the heading of public key crypto-systems, which does not require the prior secure communication of key material. Perhaps the most widely used one is the RSA system (Rivest, Shamir, and Adleman [21]). The RSA crypto-system is based on the fact that in order to factorize a number with N digits, a classical computer needs to perform a number of steps that at least grows faster than polynomially with N. Consider the factorizing of the numbers 21 and 1073 by hand: the number 21 is easy but 1073 takes a little more time. Computers have the same problem. As N becomes large, factoring becomes a hard problem on a classical computer and, in fact, the number of steps required becomes exponential with N. This is how the security of the key arises. If these large numbers can be factored, then an eavesdropper can recover the message sent from Alice to Bob. Let us examine the RSA protocol which allows Bob to announce publicly a key (the public key) such that Alice can use the key, encrypt a message, and send it publicly back to Bob. The protocol only allows Bob to decipher it. The RSA protocol is established as follows.

- It begins by Bob taking two very large prime numbers p and q (of size greater than 10^{1000}). He then calculates the quantities $N = pq$ [where obviously $N > 10^{2000}$ and $pq(N) = (p - 1)(q - 1)$]. Bob then takes an integer $e < N$ such that the greatest common divisor between e and $pq(N)$ is 1, in which case e and $pq(N)$ are said to be *coprime* (i.e. they have no common factors other than 1). Once this is achieved, Bob computes $d = e^{-1}(\text{mod } pq(N))$.
- Bob then sends the public key (e, N) to Alice via a public channel.
- Alice now encodes her plain text message m according to the rule $c = m^e$ (mod N) and sends the result c to Bob.
- Once Bob receives the encrypted message c from Alice, he computes c^d (mod N) and recovers the plain text message m.

It is interesting to observe that most of the work in this protocol is done by Bob (not Alice). Alice's main job is to encode the message m she wishes to send to Bob with the public key (e, N).

Given that all the communication is done only via a public channel, what is an effective mechanism for an eavesdropper to break this crypto-system? Eve can attack this crypto-system quite easily. To do this, Eve must simply solve $d = e^{-1}(\text{mod } pq(N))$, where she knows only e and N. Effectively, this just requires Eve to factor N. Once she factors N, she obtains p and q and hence can calculate $pq(N)$. While this may sound like a relatively simple attack, it is not practical to factorize numbers of size $N > 10^{2000}$ on a classical computer with current technology in a reasonable time (in fact, to factor the numbers currently used in the RSA encryption system would take a very large supercomputer many, many years). However, a recent proposal for a new kind of computer architecture, the quantum computer, opens the possibility for factoring these very large numbers in a short time.

The quantum computer at scale, of course, has yet to be realized, but has the potential to revolutionize information processing through its ability to access a vastly larger state space than a classical machine. This, in turn, changes the complexity classification of how difficult computational tasks are, so that some tasks which are known to be extremely difficult (i.e. require exponential resources) on a classical machine become easier (need only polynomial resources) on a quantum machine. The gain comes from using two features of quantum physics: entanglement and quantum interference, allowing massively parallel processing capacity.

In 1985 David Deutsch [22] (with a generalization in 1992 co-authored with Richard Jozsa [23]) proved that in quantum mechanics the complexity level of some problems can change dramatically. This paved the way for a number of advances in quantum algorithms, in particular Peter Shor's for factoring large numbers. In 1994, Peter Shor [24] discovered a quantum algorithm that allows one to factor large numbers in polynomial time so that, provided one had a quantum computer, factoring would become essentially as easy as multiplying. The realization of quantum factoring using various platforms has been successful only for very small numbers up to present. Some years ago, the quantum factoring of the number 15 into its prime factors (5 and 3) using a small-scale nuclear magnetic resonance (NMR) device was reported [25]. Unfortunately, things have not moved much beyond that in the intervening years. The NMR approach cannot be scaled up, and devices with the required number of qubits to factor large numbers are extremely vulnerable to decoherence. Nevertheless, the central point here is that if quantum computing schemes can be scaled up to the point where very large numbers can be factored, RSA crypto-systems become vulnerable to deciphering.

11.8 The Quantum Random Number Generator

Classical computers are frequently called upon to generate strings of random numbers. Random numbers (random bits) play a role in many areas of science and engineering, including secret communication and Monte Carlo simulations. But the numbers generated by classical computers are only pseudo-random, as a deterministic algorithm is always involved. In many of the private key cryptography protocols it is necessary to have local access to a good source of true (and not pseudo-) random numbers. This is also true for the quantum cryptographic schemes we will describe below. While deterministically generated pseudo-random numbers can appear quite random, if one knows the algorithm and the seed, there is the possibility that the string could be reproduced by an eavesdropper. There are also a number of hardware-based random generators based on the complexity of thermal noise fluctuations that produce a chaotic behavior in the system. The difficulty in predicting the result of the chaotic process is attributed to randomness. Under ideal conditions this produces a good source of random numbers. A potential problem, however, is that the thermal noise could be influenced and potentially controlled by someone who tampers with the external environment.

In recent years, however, many proposals have been put forward that take advantage of the irreducible probabilistic nature of the quantum world. Many of these proposals have recently been reviewed by Herrero-Collantes and Garcia-Escartin [26] and in a volume edited by Kollmitzer *et al.* [27]. In fact, in this book we have already discussed a simple quantum optical device that, in principle, can be used to generate strings of random bits (i.e. sequences of 0s and 1s) that can be generated randomly by a quantum mechanical process. We are speaking here of the simple 50:50

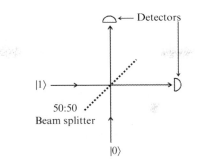

Fig. 11.10 Schematic setup of a quantum random number generator. The apparatus consists of a single-photon state arriving at a 50:50 beam splitter. A vacuum enters the second port of the beam splitter. Assuming that the beam splitter is 50:50, the single photon has a 50% chance of being transmitted through the beam splitter and a 50% chance of being reflected. A random bit is obtained by measuring whether the photon was reflected or transmitted.

beam splitter, discussed in Chapter 6, wherein a single photon is incident in one port with the vacuum in the other, as pictured in Fig. 11.10. In fact, this is the prototype of the optical quantum random number generator. Recall from Chapter 6 that a single photon fallen on a 50:50 beam splitter results in

$$|0\rangle_0 |1\rangle_1 \xrightarrow{\text{BS}} \frac{1}{\sqrt{2}} (i|1\rangle_2 |0\rangle_3 + |0\rangle_2 |1\rangle_3). \tag{11.48}$$

Detectors placed at the outputs yield which-path information. We assign 0 if the photon takes the reflected path, and 1 if it takes the transmitted path. As there is a 50% chance of it taking each path, we have a 50% chance of getting 0 or 1. This is our random bit. By performing this experiment repeatedly, a sequence of truly random bits can be generated.

Well, at least that's what should happen in theory. In reality, there are some problems with this simple idea. One problem is the inability to reliably, and rapidly, produce single photons on demand. Thus, in practice, pulsed weak laser beams are often used. But as we have seen from Chapter 3, a laser beam described by a weak coherent state is dominated by the vacuum, with a significantly smaller probability of the one-photon state appearing in the beam. For a coherent state mixed with a vacuum state, we had

$$|0\rangle_0 |\alpha\rangle_1 \rightarrow \left|\frac{i\alpha}{\sqrt{2}}\right\rangle_2 \left|\frac{\alpha}{\sqrt{2}}\right\rangle_3. \tag{11.49}$$

For a small value of $|\alpha|$ we can expand our states to find

$$\left|\frac{i\alpha}{\sqrt{2}}\right\rangle_2 \left|\frac{\alpha}{\sqrt{2}}\right\rangle_3 \approx |0\rangle_2 |0\rangle_3 + \frac{\alpha}{\sqrt{2}} (i|1\rangle_2 |0\rangle_3 + |0\rangle_2 |1\rangle_3) + \cdots, \tag{11.50}$$

so we see that the state on the left side of Eq. (11.48) does occur, but only with probability $|\alpha|^2$. The vacuum states dominate. Increasing $|\alpha|$ will not do because this brings in higher number states, which are unwanted. Another problem is that perfect 50:50 beam splitters are hard to fabricate, which means that classical de-biasing algorithms or post-processing of the generated bit sequence is required.

An alternative approach to QNRG has recently been proposed [28] and implemented [29], which involves the use of moderately intense laser light along with photon number parity measurements. The idea behind this is as follows. The parity operator for, say, a field mode labeled a is given by

$$\hat{\Pi} = \exp(i\pi\hat{a}^\dagger\hat{a}) = (-1)^{\hat{a}^\dagger\hat{a}} = \hat{\Pi}_e - \hat{\Pi}_o, \tag{11.51}$$

where

$$\hat{\Pi}_e = \sum_{m=0}^{\infty} |2m\rangle\langle 2m|, \quad \hat{\Pi}_o = \sum_{m=0}^{\infty} |2m+1\rangle\langle 2m+1| \tag{11.52}$$

are the even and odd photon number state projection operators, respectively. With respect to the coherent state

$$|\alpha\rangle = e^{-\frac{1}{2}|\alpha|^2} \sum_{n=0}^{\infty} \frac{\alpha^n}{\sqrt{n!}}|n\rangle, \tag{11.53}$$

we find the expectation value of the parity operator to be

$$\langle\alpha|\hat{\Pi}|\alpha\rangle = \langle\hat{\Pi}\rangle_\alpha = e^{-\bar{n}} \sum_{n=0}^{\infty} \frac{(-\bar{n})^n}{n!} = e^{-2\bar{n}}. \tag{11.54}$$

The expectation values of the even and odd state projection operators are the probabilities

$$\begin{aligned} P_e &= \langle\alpha|\hat{\Pi}_e|\alpha\rangle = e^{-\bar{n}} \sum_{m=0}^{\infty} \frac{\bar{n}^{2m}}{(2m)!} \\ &= e^{-\bar{n}}\cosh\bar{n} = \frac{1}{2}(1 + e^{2\bar{n}}), \\ P_o &= \langle\alpha|\hat{\Pi}_o|\alpha\rangle = e^{-\bar{n}} \sum_{m=0}^{\infty} \frac{\bar{n}^{2m+1}}{(2m+1)!} \\ &= e^{-\bar{n}}\sinh\bar{n} = \frac{1}{2}(1 - e^{2\bar{n}}). \end{aligned} \tag{11.55}$$

In the limit that $\bar{n}$ is large, we have $\langle\hat{\Pi}\rangle_\alpha \to 0$, P_e, $P_o \to 1/2$ rapidly for increasingly larger values of $\bar{n}$. The photon number does not have to be very large for these limits to be attained to a fairly high degree of accuracy. For $\bar{n} = 16$, we find that P_e, $P_o = 1/2$ to 13 decimal places. The point is that for large enough $\bar{n}$, there will be no bias in the outcomes of the parity measurements, provided that such measurements can be made.

Obviously, it is necessary to be able to perform photon counting with resolution at the level of a single photon, which is also required for photon number parity-based interferometry, as discussed earlier. This is not possible with the readily available avalanche photodiode (APD) detectors. These devices are, in fact, binary detectors in the sense that they can discriminate between there being no light detected and any number of photons detected without being able to resolve the photon number [29].

However, as discussed in Chapter 7, over the past decade, a new kind of detector has become available, namely the so-called TES, which are superconducting devices capable of bolometric measurements that can distinguish infrared photon number counts with a resolution at the level of one photon. In other words, these devices are photon number resolving (PNR) detectors. For a description of these devices, see Ref. [30]. The early detectors of this type could perform PNR detections to about 10 photons, but very recently it has become possible to extend the performance of these devices to perform PNR detection for up to 100 photons. In fact, this capability was demonstrated in a laboratory realization of the photon number parity-based QRNG described above by Eaton *et al.* [31] in an experiment with laser light of average photon number $\bar{n} = 57$.

One drawback of the TES devices is their long "rest" times, of the order of 10 μs, which can limit the speed of random number generation. This can be circumvented to some extent by using modulo 4 measurement binning, which requires higher average photon number as there are now four possible outcomes instead of two. This can be continued to modulo 8, 16, 32, and so on. Binning in this way constitutes a higher-order form of photon number parity [32]. However, higher average photon numbers of the coherent state are required to maintain the unbiased generation of random numbers, and this runs out of steam in terms of producing unbiased random numbers simply because the technology for PNR detection is limited at present to about 100 photons.

In the above, we assumed the availability of pure coherent states. Here we show that a statistical mixture of identical amplitude coherent states will suffice. We consider the phase-averaged coherent state

$$\hat{\rho}_{\text{coh}} = \frac{1}{2\pi} \int\limits_{0}^{2\pi} d\varphi |re^{i\varphi}\rangle \langle re^{i\varphi}|, \tag{11.56}$$

where $r = |\alpha| = \sqrt{\bar{n}}$. It is easy to show that

$$\langle \hat{\Pi} \rangle = \text{Tr}[\hat{\rho}_{\text{coh}} \hat{\Pi}] = \frac{1}{2\pi} \int\limits_{0}^{2\pi} d\varphi \sum_{n=0}^{\infty} (-1)^n |\langle n|re^{i\varphi}\rangle|^2$$

$$= e^{-\bar{n}} \sum_{n=0}^{\infty} \frac{(-\bar{n})^n}{n!} = e^{-2\bar{n}}, \tag{11.57}$$

which is the same as Eq. (11.54).

11.9 Quantum Cryptography

Given that the quantum computer (a device yet to be built) is able to attack current public key crypto-systems (such as RSA), what means of absolute secure communication is possible? We know the private key crypto-system (one-time pad) is not vulnerable to the quantum computing architecture described above. However, one-time pad classical schemes require a large amount of key material being shared between the two parties Alice and Bob. This has had to be done via private channels, to which the eavesdropper has no access. There are a number of practical solutions to this, such as Alice sending Bob a sealed can by courier with the key material within. Here, Alice and Bob must trust the courier not to have read and copied the key material. Another solution based on the principles of quantum mechanics is possible, namely *quantum key distribution* (QKD), and it provides a mechanism for Alice and Bob to create a shared key over public channels. Once both parties have this key material, secure communication can take place. In all QKD proposals it is critical that one has a source of random numbers.

11.9.1 Quantum Key Distribution

Quantum key distribution offers the possibility for two remote parties – Alice and Bob – to exchange a secret key without physically meeting or requiring a trusted party to deliver the key material. The security of QKD is conditioned on the principles of quantum mechanics, a theory well tested and thought to be correct.

The central idea behind quantum key distribution is that it is impossible for an eavesdropper to obtain all the information from the transmitted quantum state. What does this statement mean? Consider a single qubit in the superposition state $|\psi\rangle = c_0|0\rangle + c_1|1\rangle$. (We again use the symbols $|0\rangle$ and $|1\rangle$ to represent in a general manner the basis of our qubits. For a realization in terms of polarization states, we could again take $|0\rangle = |V\rangle$ and $|1\rangle = |H\rangle$.) For a single general measurement with c_0 and c_1 unknown, it is impossible to determine the state $|\psi\rangle$ precisely. For instance, if "0" is obtained on a single measurement, it is impossible to determine what the c_0 coefficient was (see Fig. 11.11). A single measurement yielding a result "0" does not distinguish between $|0\rangle$, $(|0\rangle + |1\rangle)/\sqrt{2}$, or $(|0\rangle + i|1\rangle)/\sqrt{2}$, or many other possible states for that matter. Over many runs with identical copies of the state, one can determine, via a number of measurements, the exact state. However, in QKD, the qubit is never reused and so it is impossible for an eavesdropper to determine the state completely if several non-orthogonal states are used. The key to QKD is the use of these non-orthogonal states.

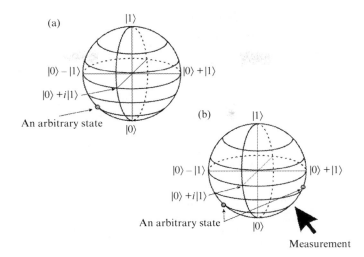

Fig. 11.11 Diagram representing a qubit on the surface of a sphere (a). The ground state $|0\rangle$ is situated at the south pole of the sphere. The excited state $|1\rangle$ is situated at the north pole of the sphere. Superpositions of $|0\rangle$ and $|1\rangle$ lie on the surface of the sphere between the north and south poles. Equal superpositions of $|0\rangle$ and $|1\rangle$ lie on the equator. An arbitrary state can lie anywhere on the sphere. In (b) we show a measurement of a point on the sphere. On this sphere we have indicated two unknown states. These states are not orthogonal and so a single measurement cannot confirm which one of the two states we have. It is this key point we exploit in quantum key distribution.

There are a number of QKD protocols that we will examine in detail below. We begin with the first one discovered, the BB84 protocol.

11.9.2 The BB84 Protocol

The BB84 protocol (Bennett and Brassard [33]) begins by Alice creating and sending to Bob a random set of qubits, where the choice of qubits comes from the set

$$
\begin{aligned}
|\psi_0\rangle &= |0\rangle, \\
|\psi_1\rangle &= |1\rangle, \\
|\psi_+\rangle &= \frac{1}{\sqrt{2}}(|0\rangle + |1\rangle), \\
|\psi_-\rangle &= \frac{1}{\sqrt{2}}(|0\rangle - |1\rangle).
\end{aligned}
\tag{11.58}
$$

Note that $\langle\psi_0|\psi_1\rangle = 0$ and $\langle\psi_+|\psi_-\rangle = 0$. The first two states form a basis and the second two states form another. In terms of single-photon polarization states, the second two states are oriented at $\pm 45°$ to the first two. Clearly, the four states of Eq. (11.58) are not *all* orthogonal to each other: $\langle\psi_{0,1}|\psi_{+,-}\rangle \neq 0$. Hence, there is no measurement protocol that allows us with 100% certainty to determine the states Alice sends to Bob. Bob measures randomly in one of two bases, either "0" and "1" or "+" and "−" [here "+" and "−" represent measuring $(|0\rangle + |1\rangle)/\sqrt{2}$ and $(|0\rangle - |1\rangle)/\sqrt{2}$]. After the measurements are complete, Alice and Bob communicate via a public channel as to certain properties of the states sent and measured. More precisely, Alice announces whether the qubit was sent in the 0, 1 basis or the +, − basis but not which exact state. Bob then announces which basis he measured in, but not the result. They then keep between them the results only if Alice's sending and Bob's measurement basis are the same. This gives a smaller subset of qubits (half the size) where Alice and Bob know what was sent and measured without having communicated exactly which state was sent. A key has then been established.

When Alice sends the qubits to Bob, it is possible that an eavesdropper (Eve) may intercept some or all of the transmitted qubits and interfere with them. Now what is the action of the eavesdropper? Basically, the eavesdropper must perform some action between the time that Alice sends her quantum state and the time Bob receives it. This could be any form of measurement or no measurement. Let us consider the case where Eve performs a definite projective measurement. Eve does not know which state has been sent (and cannot determine this at the time of her measurement), and so must choose whether to measure in the 0, 1 basis or the +, − basis. If Eve guesses the right basis (and she will not know this until after Alice and Bob have communicated), then she can transmit the right state to Bob (so that he does not know that it has been intercepted). If, however, Eve chooses the wrong basis to measure in, then the state she transmits is wrong and not what Bob should get. If Bob performs some form of error checking, then he will see that the wrong state has been sent and hence that Eve is present.

Alice and Bob − in the second half of the protocol − publicly disclose the basis of the state sent and the measurement basis. From this they generate an initial amount of shared key material. However, as we mentioned above, Eve may have interfered with this and hence know some of the key. Also, because there is a 50% chance that she measured in the wrong basis and hence transmitted the wrong state to Bob, there will not be agreement between all the bits in Alice's and Bob's shared key. Alice and Bob can, however, see if an eavesdropper has been present by selecting a subset of the key bits and publicly telling each other what they are. By checking the error rate (cases where their bits disagree), they can see the presence of the eavesdropper. If this error rate is not too high, Alice and Bob can perform classical privacy amplification on the remaining bits to create a secure shared key. We summarize this protocol in Fig. 11.12 and describe an implementation in Fig. 11.13.

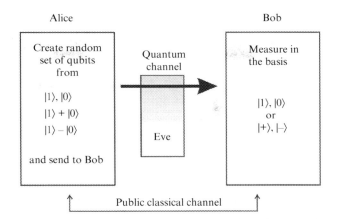

Fig. 11.12 Summary of the procedure to implement the BB84 protocol. After Bob's measurement, Alice and Bob communicate whether they sent and measured a state in $|0\rangle$, $|1\rangle$ or $|+\rangle$, $|-\rangle$ basis. They keep the results only when the sent and measured bases agree. Then they use a subset of this basis-agreed data to check for the presence of Eve.

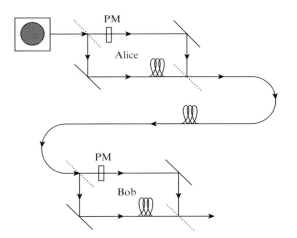

Fig. 11.13 Schematic representation of a realization of the BB84 protocol using the interferometer. In this realization, the shorter and longer paths through the interferometer define the 0 and 1 states (instead of using polarization, which is hard to preserve in commercial fiber). Phase modulators (PM) are positioned within the upper arms of both Bob's and Alice's interferometer.

11.9.3 The B92 Protocol

The BB84 protocol has been generalized to use other bases and states. One of the most well-known is the B92 protocol [34], which uses two non-orthogonal states rather than four. This protocol is established as follows. Alice creates a random bit a (either 0 or 1). She then transmits to Bob the qubit

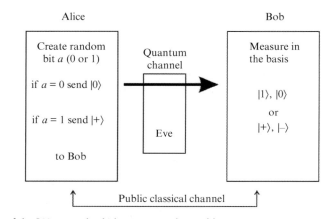

Summary of the B92 protocol, which uses non-orthogonal bases.

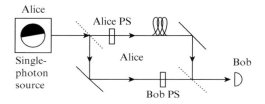

Schematic representation of an optical realization of the B92 protocol: here the whole transmission between Bob and Alice takes place within the arms of an interferometer.

$$|\psi\rangle = \begin{cases} |0\rangle, & \text{if } a = 0, \\ |+\rangle, & \text{if } a = 1. \end{cases} \qquad (11.59)$$

Now Bob generates a random bit a' (again either 0 or 1). If the bit is 0, then Bob measures in the 0, 1 basis. If the bit is 1, he measures in the +, − basis. From Bob's measurement b, he obtains either a 0 or a 1, which he then announces over a public channel to Alice. Both Alice and Bob keep their random bits a and a' secret. Their initial key material is formed from the subset of results where the measurement result b was 1. This occurs only if Alice had $a = 0$ and Bob had $a' = 1 - a$ or vice versa, with a probability of 1/2. The key is a for Alice and a' for Bob. Now because they wish to estimate the effects of the potential eavesdropper, they can use some of the key material to check for errors and then perform appropriate privacy amplification if the error rate is not too high. This then establishes secure shared key material. We summarize this protocol in Fig. 11.14 and show an implementation in Fig. 11.15.

11.9.4 The Ekert Protocol

In the protocols we have discussed so far, we have relied on Alice sending single photons prepared in polarization or interferometric bases. Artur Ekert in 1991 [35] proposed a new quantum cryptographic protocol, which exploits quantum entanglement and Bell's inequality.

Suppose, during a period of time, a quantum source emits two photons in an entangled state: the photons can be direction or polarization entangled, for example. We can use these in a secure communication system, as we shall now describe. The basic idea of the Ekert protocol over the BB84 or B92 protocol consists of replacing the single quantum channel carrying qubits from Alice to Bob with a channel that carries two entangled qubits (one to Alice and one to Bob) from a central common source. We will describe the use of this novel quantum resource next.

Consider that Alice and Bob share a set of n entangled qubit pairs, where each pair is of the form

$$|\psi\rangle = \frac{1}{\sqrt{2}}(|0\rangle|0\rangle + |1\rangle|1\rangle). \tag{11.60}$$

The state on the left belongs to Alice and that on the right to Bob. Let us examine the establishment of a shared bit. Alice and Bob share the entangled pair (one qubit each); each now creates a random number (with values 0, 1) independently – called, say, a (by Alice) and a' (by Bob). If Alice's random number is 0, she measures her qubit in the 0, 1 basis, otherwise she uses the $+, -$ basis. She then records the measurement result b. Bob does the same

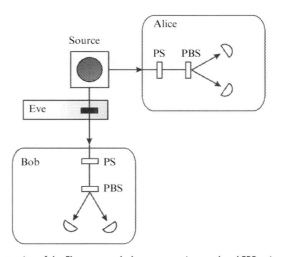

Fig. 11.16 Schematic representation of the Ekert protocol: the source emits correlated EPR pairs, one of which is transmitted to Alice and the other to Bob. Polarizing beam splitters (PBS) route the correlated photons to detectors.

thing with his qubit; if Bob's random number is 0, he measures in the 0, 1 basis, otherwise he uses the $+, -$ basis. He records the result b'. Alice and Bob now communicate their results (b, b') publicly. They keep the results only if $b = b'$, which establishes the key bit (a, a'). Performing this on the n entangled pairs establishes the required key. In Fig. 11.16, we sketch the layout of the Ekert protocol for a quantum cryptography system.

The experimental realizations of these schemes are reviewed at length by Gisin *et al.* [22].

11.10 Future Prospects for Quantum Communication

Quantum physics, as we have described, offers the world of communications both opportunities and threats. Opportunities, from the fundamental nature of indivisible quanta and our inability to copy them, and threats because of the enormous potential of quantum computing – provided the decoherence obstacle can be overcome so that large-scale quantum registers can be manipulated. We have concentrated here on secure key distribution for cryptography as it represents the most developed quantum technology. In QKD, classical information is transmitted using a quantum resource. However, quantum information itself can also be transmitted: for example, the quantum state of a system can be transferred from Alice to Bob if they share the quantum resource of entanglement. This is the basis of quantum teleportation as we have already discussed. This area has developed into one of the major growth fields of modern physics and is likely to generate new insights, new applications, and new technologies over the coming decade.

11.11 Gates for Quantum Computation

A full discussion of quantum computing lies way beyond the scope of this book. But it does seem worthwhile nevertheless to discuss some aspects of quantum computing, in particular the notion of quantum registers and quantum gates and their quantum optical realizations. This we shall do in what follows.

We begin with a discussion of quantum registers. The qubits we again represent in the generic basis $|0\rangle$ and $|1\rangle$, also known as the computational basis. The most general pure state for a single qubit is the superposition state

$$|\psi_1\rangle = c_0|0\rangle + c_1|1\rangle, \quad |c_0|^2 + |c_1|^2 = 1. \tag{11.61}$$

A quantum register is a collection of qubits, say, N qubits. For example, the 3-qubit register in the state

$$|1\rangle \otimes |0\rangle \otimes |1\rangle \equiv |101\rangle = |5\rangle \tag{11.62}$$

provides a binary representation of the decimal number 5, $(5)_{10} = (101)_2$, as indicated on the right-hand side. But now suppose that the first qubit of the register is in the balanced superposition state $(|0\rangle + |1\rangle)/\sqrt{2}$. The state of the register is then

$$
\begin{aligned}
\frac{1}{\sqrt{2}}(|0\rangle + |1\rangle) \otimes |0\rangle \otimes |1\rangle &= \frac{1}{\sqrt{2}}(|001\rangle + |101\rangle) \\
&= \frac{1}{\sqrt{2}}(|1\rangle + |5\rangle).
\end{aligned}
\tag{11.63}
$$

This means that the 3-qubit quantum register *simultaneously* represents the decimal numbers 1 and 5. If all three qubits are in balanced superposition states, then the register is in the state

$$
\begin{aligned}
\frac{1}{\sqrt{2}}(|0\rangle + |1\rangle) \otimes &\frac{1}{\sqrt{2}}(|0\rangle + |1\rangle) \otimes \frac{1}{\sqrt{2}}(|0\rangle + |1\rangle) \\
&= \frac{1}{2^{3/2}}(|000\rangle + |001\rangle + |010\rangle + |011\rangle + |100\rangle + |101\rangle + |110\rangle + |111\rangle) \\
&= \frac{1}{2^{3/2}}(|0\rangle + |1\rangle + |2\rangle + |3\rangle + |4\rangle + |5\rangle + |6\rangle + |7\rangle),
\end{aligned}
\tag{11.64}
$$

and thus a 3-qubit register can represent the eight decimal numbers 0–7 simultaneously. For any decimal number a given by $a = a_0 2^0 + a_1 2^1 + \cdots + a_{N-1} 2^{N-1}$, where the coefficients a_j, $j = 0, 1, \ldots, N$ are restricted to 0 or 1, we can write

$$
|a\rangle = |a_{N-1}\rangle \otimes |a_{N-2}\rangle \otimes \ldots \otimes |a_1\rangle \otimes |a_0\rangle \equiv |a_{N-1} a_{N-2} \cdots a_0\rangle.
\tag{11.65}
$$

The most general N-qubit quantum register state is

$$
|\psi_N\rangle = \sum_{a=0}^{2^N - 1} c_a |a\rangle.
\tag{11.66}
$$

With an N-qubit register, all the decimal numbers from 0 to $2^N - 1$, a total of 2^N numbers, can be represented simultaneously in a state of the form of Eq. (11.66).

To do any kind of processing, we need logic gates just as in a classical computer. The important difference between the quantum logic gates and classical ones is that the former are always reversible, in that the outputs uniquely depend on the inputs, which does not hold for classical gates. It holds for the quantum gates because they are implemented by unitary transformations. We begin with 1-qubit gates and then proceed to 2-qubit gates.

The first gate we consider is the Hadamard gate. This is a gate that creates balanced superpositions of the basis states $|0\rangle$ and $|1\rangle$ and is represented by the letter H. With $x \in \{0, 1\}$, the Hadamard gate, described by the unitary operator $\hat{U}_H$, effects the transformation

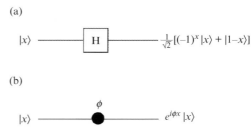

(a)

(b)

Fig. 11.17 (a) Circuit symbol of a Hadamard gate. (b) Circuit symbol of the 1-qubit phase gate.

$$\hat{U}_H|x\rangle = \frac{1}{\sqrt{2}}[(-1)^x|x\rangle + |1 - x\rangle]. \tag{11.67}$$

Thus, $\hat{U}_H|0\rangle = (|0\rangle + |1\rangle)/\sqrt{2}$ and $\hat{U}_H = (|0\rangle - |1\rangle)/\sqrt{2}$. Note that the left-hand side of Eq. (11.24) may be obtained from Hadamard transformations on three bits each in a state $|0\rangle$. If the qubits are realized by the ground and excited states of a two-level atom, then the Hadamard transformation is just a $\pi/2$-pulse. The circuit symbol for the Hadamard gate is given in Fig. 11.17a.

A second 1-qubit gate, the phase gate represented by $\hat{U}_{PG}(\phi)$, effects the transformation

$$\hat{U}_{PG}(\phi)|x\rangle = e^{ix\phi}|x\rangle, \tag{11.68}$$

and thus $\hat{U}_{PG}(\phi)|0\rangle = |0\rangle$ and $\hat{U}_{PG}(\phi)|1\rangle = e^{i\phi}|1\rangle$. The circuit symbol for this gate is given in Fig. 11.17b. It turns out that a combination of Hadamard and phase gates can produce various superposition states of a single qubit. The sequence

$$\hat{U}_{PG}\left(\phi + \frac{\pi}{2}\right)\hat{U}_H\hat{U}_{PG}(2\theta)\hat{U}_H|0\rangle = \cos\theta|0\rangle + e^{i\phi}\sin\theta|1\rangle, \tag{11.69}$$

as represented in Fig. 11.18, yields the most general 1-qubit pure state.

So far, we have not encountered registers in entangled states. One possible entangled 2-qubit register state is $(|00\rangle + |11\rangle)/\sqrt{2}$. How can these, as a matter of principle, be made? We have already encountered ways of generating entangled states in the earlier parts of this book through some very specific interactions. At the moment, we are only interested in formulating a rather general procedure, without regard to any specific realization. Thus, we introduce a 2-qubit gate called the controlled not (C-NOT), denoted by the unitary operator $\hat{U}_{C\text{-}NOT}$. If we let the first qubit be the control qubit and the second be the target, then we can represent the action of this gate symbolically by

$$\hat{U}_{C\text{-}NOT}|x\rangle|y\rangle = |x\rangle|\mathrm{mod}_2(x + y)\rangle, \tag{11.70}$$

where $x, y \in \{0, 1\}$. The gate is represented diagrammatically by the circuit in Fig. 11.19a. If the control is in state $|0\rangle$ and the target is also in state $|0\rangle$,

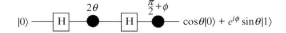

$$|0\rangle \overset{2\theta}{\boxed{H}} \overset{\frac{\pi}{2}+\phi}{\boxed{H}} \cos\theta|0\rangle + e^{i\phi}\sin\theta|1\rangle$$

Fig. 11.18 Circuit for the generation of the state of Eq. (11.29) using phase and Hadamard gates.

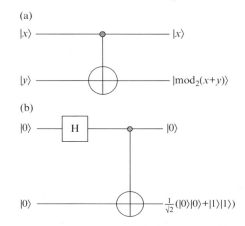

(a)

$|x\rangle$ $|x\rangle$

$|y\rangle$ $|mod_2(x+y)\rangle$

(b)

$|0\rangle$ H $|0\rangle$

$|0\rangle$ $\frac{1}{\sqrt{2}}(|0\rangle|0\rangle+|1\rangle|1\rangle)$

Fig. 11.19 (a) Circuit symbol for a C-NOT. (b) Circuit for creating a 2-qubit entangled state using a Hadamard gate and a C-NOT.

the target is unchanged. But if the control is in $|1\rangle$ then the target "flips" from $|0\rangle$ to $|1\rangle$. But suppose the control qubit is prepared, by the Hadamard transformation, in the superposition state $(|0\rangle + |1\rangle)/\sqrt{2}$. Then we shall have, for the target initially in $|0\rangle$ and $|1\rangle$, respectively

$$\hat{U}_{\text{C-NOT}}\hat{U}_{\text{H1}}|0\rangle|0\rangle = \hat{U}_{\text{C-NOT}}\frac{1}{\sqrt{2}}(|0\rangle + |1\rangle)|0\rangle = \frac{1}{\sqrt{2}}(|00\rangle + |11\rangle),$$

$$\hat{U}_{\text{C-NOT}}\hat{U}_{\text{H1}}|0\rangle|1\rangle = \hat{U}_{\text{C-NOT}}\frac{1}{\sqrt{2}}(|0\rangle + |1\rangle)|1\rangle = \frac{1}{\sqrt{2}}(|01\rangle + |10\rangle),$$

$$(11.71)$$

where H1 means that the Hadamard gate acts only on the first qubit. In both cases we end up with entangled states. The entire process, for the target initially in state $|0\rangle$, is represented by the diagram in Fig. 11.19b.

Another 2-qubit gate is the controlled phase gate $\hat{U}_{\text{CPG}}(\phi)$ acting according to

$$\hat{U}_{\text{CPG}}(\phi)|x\rangle|y\rangle = e^{i\phi xy}|x\rangle|y\rangle, \qquad (11.72)$$

represented graphically in Fig. 11.20. The phase factor $e^{i\phi xy}$ appears only when $x = 1 = y$.

Out of these gates, essentially any quantum computational algorithm can be implemented with the proper concatenation. Networks can be

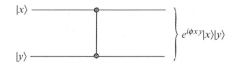

Fig. 11.20 Circuit symbol for a controlled phase gate.

devised to bring about certain kinds of output. One, very simple, circuit has already been discussed and others will be addressed in the problems.

A function in a computer is a mapping from one set of numbers to another:

$$f : \{0, 1, \ldots, 2^{m-1}\} \rightarrow \{0, 1, \ldots, 2^{n-1}\}, \tag{11.73}$$

where m and n are integers. A classical computer evaluates the function at each input $0, 1, \ldots, 2^m - 1$ to obtain the set of numbers $f(0), f(1), \ldots, f(2^m - 1)$. But the computation is generally irreversible and thus it is not possible to associate a given output with a particular input. Quantum mechanically, it is not possible to evolve an input register $|x\rangle$ into an output register $|f(x)\rangle$ by a unitary transformation. However, by using two registers, one to contain the output and another to retain the input, we can implement the quantum mechanical evaluation of a function as a unitary transformation according to

$$\hat{U}_f |x\rangle |0\rangle = |x\rangle |f(x)\rangle, \tag{11.74}$$

where different input and output states will be orthogonal: $\langle x|x'\rangle = \delta_{xx'}, \langle f|f'\rangle = \delta_{ff'}$.

As an example of a quantum computation, simple enough to describe here yet able to illustrate the potential power of a quantum computer taking full advantage of quantum entanglement, we now describe the quantum algorithm known as Deutsch's problem [22], the first quantum algorithm proposed, and one of the simplest. The problem proposed by Deutsch is this. Consider a function f: $\{0, 1\} \rightarrow \{0, 1\}$. There are only four *possible* outcomes, and these are $f(0) = 0$, $f(0) = 1$, $f(1) = 0$, and $f(1) = 1$. Given an unknown function f, determine, *with only one operation*, whether or not the function is constant, meaning that $f(0) = f(1)$, or is varying, $f(0) \neq f(1)$. The quantum computer will be required to implement the transformation

$$|x\rangle |y\rangle \rightarrow |x\rangle |\mathrm{mod}_2 \left(y + f(x)\right)\rangle, \tag{11.75}$$

where $|x\rangle$ is the input qubit and $|y\rangle$ represents the qubit of the quantum computer hardware. The idea is to prepare these qubits, using Hadamard gates, in the product state

$$|\Psi_{\mathrm{in}}\rangle = |x\rangle |y\rangle = \frac{1}{\sqrt{2}}(|0\rangle + |1\rangle)\frac{1}{\sqrt{2}}(|0\rangle - |1\rangle)$$
$$= \frac{1}{2}(|0\rangle |0\rangle - |0\rangle |1\rangle + |1\rangle |0\rangle - |1\rangle |1\rangle). \tag{11.76}$$

The minus signs in this last equation are important. If we now allow this state to undergo the transformation of Eq. (11.35), we obtain the output state

$$|\Psi_{\text{out}}\rangle = \frac{1}{2}\Big(|0\rangle|\overline{f}(0)\rangle - |0\rangle|\overline{f}(0)\rangle + |1\rangle|f(1)\rangle - |1\rangle|\overline{f}(1)\rangle\Big)$$
$$= \frac{1}{2}\Big[|0\rangle\Big(|f(0)\rangle - |\overline{f}(0)\rangle\Big) + |1\rangle\Big(|f(1)\rangle - |\overline{f}(1)\rangle\Big)\Big], \tag{11.77}$$

where the bar means to invert: $\overline{0} = 1$, $\overline{1} = 0$. Notice that Eq. (11.77) is the result of simultaneous, parallel, processing on both states of the first qubit in the superposition $|x\rangle \sim |0\rangle + |1\rangle$. This displays the potential power of a quantum computer: the possibility of massive parallel processing. If f is a constant function, then the output is

$$|\Psi_{\text{out}}\rangle_{f\text{const}} = \frac{1}{2}\Big[(|0\rangle + |1\rangle)\Big(|f(0)\rangle - |\overline{f}(0)\rangle\Big)\Big], \tag{11.78}$$

whereas if f is not constant

$$|\Psi_{\text{out}}\rangle_{f\text{not const}} = \frac{1}{2}\Big[(|0\rangle - |1\rangle)\Big(|f(0)\rangle - |\overline{f}(0)\rangle\Big)\Big]. \tag{11.79}$$

Note that in the two cases the first qubit states are orthogonal. If we apply a Hadamard transformation to this first qubit we have

$$\hat{U}_{H1}|\Psi_{\text{out}}\rangle_{f\text{const}} = |0\rangle\Big(|f(0)\rangle - |\overline{f}(0)\rangle\Big),$$
$$\hat{U}_{H1}|\Psi_{\text{out}}\rangle_{f\text{not const}} = |1\rangle\Big(|f(0)\rangle - |\overline{f}(0)\rangle\Big). \tag{11.80}$$

Thus, a single measurement on the first qubit determines whether or not f is constant.

The Deutsch algorithm is, obviously, very simple. A slight generalization of this algorithm by Deutsch and Jozsa has been implemented in an ion-trap quantum computer containing two trapped ions [37].

A much more involved algorithm is that of Shor for factoring large numbers into primes [24]. We shall not review that algorithm here, but refer the reader to specialized reviews of quantum computing. We point out again though that Shor's algorithm has been implemented, on an NMR quantum computer [18], in order to find the prime factors of $15 = 3 \times 5$, a demonstration of the algorithm and obviously not a demonstration of the power of a quantum computer. Unfortunately, the NMR approach appears not to be scalable. Another important quantum algorithm is Grover's search algorithm [38].

Grover's search algorithm can be implemented as a quantum random walk – another important algorithm. An introductory overview of quantum random walks, which includes a discussion of the Grover algorithm in this context, has been given by Kempe [39]. A more comprehensive tutorial article on quantum random walks has been presented by Reitzner *et al.* [40]. Grover's search algorithm could be implemented as a quantum walk algorithm [39].

11.12 An Optical Realization of Some Quantum Gates

There have been many proposals for realizing the quantum gates in quantum optical systems such as polarized photons, cavity QED, and in systems of trapped ions. Here we shall discuss all-optical realizations of some gates based on beam splitters, Kerr interactions, and optical phase shifters. Two photonic modes are required and these form the basis of a "dual-rail" realization of a quantum computer discussed by Chuang and Yamamoto [29]. Only single-photon states are involved and, to make clear the mapping of the dual-photon states onto qubit bases $|0\rangle$ and $|1\rangle$, we start with the realization of the Hadamard gate in terms of a 50:50 beam splitter, as pictured in Fig. 11.21a. The input mode operators we take as $\hat{a}$ and $\hat{b}$ and the beam splitter is chosen to be of the type whose output mode operators $\hat{a}'$ and $\hat{b}'$ are related to the input ones according to the transformation

$$\hat{a}' = \frac{1}{\sqrt{2}}(\hat{a} + \hat{b}), \quad \hat{b}' = \frac{1}{\sqrt{2}}(\hat{a} - \hat{b}), \tag{11.81}$$

or, when inverted

$$\hat{a} = \frac{1}{\sqrt{2}}(\hat{a}' + \hat{b}'), \quad \hat{b} = \frac{1}{\sqrt{2}}(\hat{a}' - \hat{b}'). \tag{11.82}$$

We assign the computational state $|0\rangle$ to the input product state $|0\rangle_a|1\rangle_b$, that is we take $|0\rangle \equiv |0\rangle_a|1\rangle_b$, and similarly we take $|1\rangle \equiv |1\rangle_a|0\rangle_b$. From the

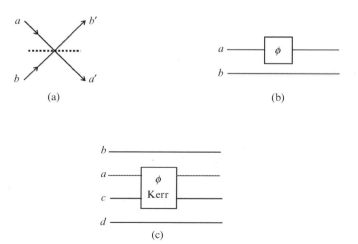

(a) Designation of the input and output modes of the 50:50 beam splitter in a dual-rail optical implementation of the Hadamard transformation. (b) Optical dual-rail implementation of a phase gate. (c) Optical implementation of a dual-rail 2-qubit controlled phase gate.

rules developed in Chapter 6, and letting $\hat{U}_{BS}$ represent the beam-splitter transformation operator, it follows that

$$
\begin{aligned}
\hat{U}_{BS}|0\rangle &= \hat{U}_{BS}|0\rangle_a|1\rangle_b \\
&= \frac{1}{\sqrt{2}}(|0\rangle_a|1\rangle_b + |1\rangle_a|0\rangle_b) = \frac{1}{\sqrt{2}}(|0\rangle + |1\rangle), \\
\hat{U}_{BS}|1\rangle &= \hat{U}_{BS}|1\rangle_a|0\rangle_b \\
&= \frac{1}{\sqrt{2}}(|0\rangle_a|1\rangle_b - |1\rangle_a|0\rangle_b) = \frac{1}{\sqrt{2}}(|0\rangle - |1\rangle),
\end{aligned}
\tag{11.83}
$$

where we have dropped the primes for the states after the beam splitter. Clearly, this device provides a realization of the Hadamard gate. We wish to make it clear that, despite the appearance of two photonic modes containing either the vacuum or a single photon, those are not the basis of the qubits. It is the product of the states of the two modes that map onto the computational basis. We are still dealing with a 1-qubit gate.

The 1-qubit phase gate can be realized by placing a phase shifter in the a-beam, the operator for this device being $\hat{U}_{PG} = \exp(i\phi\hat{a}^\dagger\hat{a})$. It is easy to see that

$$
\begin{aligned}
\hat{U}_{PG}|0\rangle &= \exp(i\phi\hat{a}^\dagger\hat{a})|0\rangle_a|1\rangle_b = |0\rangle, \\
\hat{U}_{PG}|1\rangle &= \exp(i\phi\hat{a}^\dagger\hat{a})|1\rangle_a|0\rangle_b = e^{i\phi}|1\rangle.
\end{aligned}
\tag{11.84}
$$

The device is represented graphically in Fig. 11.21b.

For 2-qubit gates we must have two sets of two photonic modes, say (a, b) and (c, d). Following the convention we have been using, we have the identifications

$$
\begin{aligned}
|0\rangle|0\rangle &= |0\rangle_a|1\rangle_b|0\rangle_c|1\rangle_d, \\
|0\rangle|1\rangle &= |0\rangle_a|1\rangle_b|1\rangle_c|0\rangle_d, \\
|1\rangle|0\rangle &= |1\rangle_a|0\rangle_b|0\rangle_c|1\rangle_d, \\
|1\rangle|1\rangle &= |1\rangle_a|0\rangle_b|1\rangle_c|0\rangle_d.
\end{aligned}
\tag{11.85}
$$

The controlled phase gate (CPG) can, in principle, be realized by the cross-Kerr interaction between the modes a and c, as indicated in Fig. 11.21c. The interaction is represented by the operator $\hat{U}_{CPG} = \exp(i\phi\hat{a}^\dagger\hat{a}\hat{c}^\dagger\hat{c})$. It is easy to verify that, for 2-qubit states, only the state $|1\rangle|1\rangle$ acquires the phase shift $e^{i\phi}$. However, there is a problem in generating arbitrary, especially large, phase shifts in this manner. Large phase shifts ϕ will require high $\chi^{(3)}$ nonlinear susceptibilities, which are generally not available.

Lastly, we discuss a possible optical realization of the C-NOT gate. This can be done, at least in principle, with a Mach–Zehnder interferometer coupled to an external mode, which we take to be the c-mode as pictured in

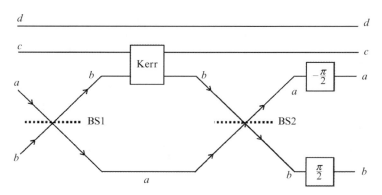

Fig. 11.22 An optical realization of a C-NOT using a Fredkin gate. The state modes a, b and c, d represent the two qubits. The phase shifters in the outputs of the Mach–Zehnder interferometer are to adjust the phase factors of the outputs.

Fig. 11.22, via a cross-Kerr interaction. The Kerr "evolution" is described by the unitary operator

$$\hat{U}_{\text{Kerr}}(\eta) = \exp(i\eta \hat{b}^\dagger \hat{b} \hat{d}^\dagger \hat{d}), \qquad (11.86)$$

where the parameter η is proportional to the nonlinear susceptibility $\chi^{(3)}$. The first beam splitter will be taken as a type described by the transformation

$$\hat{U}_{\text{BS1}} = \exp\left[i\frac{\pi}{4}(\hat{a}^\dagger \hat{b} + \hat{a}\hat{b}^\dagger)\right], \qquad (11.87)$$

and the second beam splitter we take to be conjugate to the first such that $\hat{U}_{\text{BS2}} = \hat{U}^\dagger_{\text{BS1}}$. Thus, the operator describing the evolution just after the second beam splitter must be given by

$$\hat{U}_{\text{F}}(\eta) = \hat{U}^\dagger_{\text{BS1}} \hat{U}_{\text{Kerr}}(\eta) \hat{U}_{\text{BS1}}, \qquad (11.88)$$

where the label F stands for Fredkin. The gadget represented by this operator is the quantum optical form of what is known as the Fredkin [40] gate. A quantum optical realization of the Fredkin gate was first discussed by Milburn [41]. After a bit of algebra, which we leave to the reader, we arrive at

$$\hat{U}_{\text{F}}(\eta) = \exp\left[i\frac{\pi}{2}\hat{c}^\dagger \hat{c}(\hat{a}^\dagger \hat{a} + \hat{b}^\dagger \hat{b})\right]\exp\left[\frac{\eta}{2}\hat{c}^\dagger \hat{c}(\hat{a}^\dagger \hat{b} - \hat{b}^\dagger \hat{a})\right]. \qquad (11.89)$$

The first term will just yield a phase factor depending on whether or not mode c contains a photon. The second term though is a conditional transformation. If the c-mode is in the vacuum, no transformation will occur. But if there is a single photon in the c-mode, a transformation must occur. For our purposes we must take $\eta = \pi$. Using the identifications in

Eq. (11.45), and taking the qubits of the c, d modes to be the control and those of the a, b as the target, reversed in order compared with the above discussion, we can easily show that

$$
\begin{aligned}
\hat{U}_{\mathrm{F}}(\eta)|0\rangle|0\rangle &= |0\rangle|0\rangle, \\
\hat{U}_{\mathrm{F}}(\eta)|1\rangle|0\rangle &= |1\rangle|0\rangle, \\
\hat{U}_{\mathrm{F}}(\eta)|0\rangle|1\rangle &= i|1\rangle|1\rangle, \\
\hat{U}_{\mathrm{F}}(\eta)|1\rangle|1\rangle &= -i|0\rangle|1\rangle,
\end{aligned}
\tag{11.90}
$$

where we have used the transformation

$$
\exp\left[\frac{\pi}{2}(\hat{a}^{\dagger}\hat{b} - \hat{b}^{\dagger}\hat{a})\right]
\begin{cases}
|0\rangle_a|1\rangle_b = i|1\rangle_a|0\rangle_b, \\
|1\rangle_b|0\rangle_b = -i|0\rangle_a|1\rangle_b,
\end{cases}
\tag{11.91}
$$

as the reader can verify. To remove the factors i and $-i$ we can, if necessary, insert appropriate phase shifters into the output beams. Thus, the unitary operator describing a quantum optical C-NOT is

$$
\hat{U}_{\text{C-NOT}} = \exp\left(-i\frac{\pi}{2}\hat{a}^{\dagger}\hat{a}\right)\exp\left(-i\frac{\pi}{2}\hat{b}^{\dagger}\hat{b}\right)\hat{U}_{\mathrm{F}}(\pi).
\tag{11.92}
$$

Unfortunately, as we have already remarked, Kerr media with large non-linearities would be required to realize these gates, but they are not available. A C-NOT gate has been realized experimentally with polarized photons [41]. Knill *et al.* [42] have proposed a different approach using linear optics, but where state reduction is an essential ingredient during the computation, not just at the outputs.

There are many other possible interactions that can be used to make a quantum C-NOT gate, including cavity QED [43], laser-cooled trapped ions [44], atoms in optical lattices interacting via collisions [45], and so on. All of them have the feature that the interaction of one qubit with another is conditional on the state of the partner qubit. For example, if an atom in an optical lattice trap is brought close to a second atom in an adjacent well of the lattice, then the energy levels of the valence electrons in each of the atoms are shifted owing to the fluctuations of the dipole moment of the other atom. The shifts are greater for higher levels. Thus, an electron in an excited state picks up phase more slowly than one in the ground state. This conditional phase shift (i.e. the amount of phase that depends on the state of the control electron) forms the basis of this kind of conditional state change performed by a quantum C-NOT gate.

The atoms must be controlled very carefully, though. It turns out that the fidelity of gate operation has to be about one part in a thousand or better in order to make a quantum computer realistic. This is but one of the hurdles to harnessing the undoubted quantum potential.

11.13 Decoherence and Quantum Error Correction

Quantum coherence is fragile. For the reasons we have discussed in Chapter 8 and elsewhere, it is difficult to maintain superposition states of many particles in which each particle may be physically separated from all the others. Entanglement is delicate. The reason for this is that all systems, quantum or classical, are not isolated. They interact with everything around them – local fluctuating electromagnetic fields, the presence of impurity ions, coupling of unobserved degrees of freedom to the system containing the qubit, and so on. These fluctuations destroy quantum interference. A simple, but useful, analogy is the interference of optical waves in Young's double-slit experiment. In that apparatus, waves from two spatially separated portions of a beam are brought together. If the two parts of the beam have the same phase, then the fringe pattern remains stable. But if the phase of one part of the beam is drifting with respect to the other, then the fringe pattern will be washed out. And the more slits you have in the screen, the lower the visibility for the same amount of phase randomization per pair of slits.

It might appear that the situation is hopeless: how could one hope to control the phases and amplitudes of superpositions involving many qubits? But amazingly, quantum mechanics provides a way to solve this problem, through even higher levels of entanglement. In classical information processing, inevitable environmental noise is dealt with by error correction. In its simplest form, this involves repeating the message transmission or calculation until a majority result is obtained. But there are more efficient ways, for example through the use of a parity check on a block of bits. A similar notion of encoding logical "1s" and "0s" into blocks of qubits and performing a parity check to reveal errors can be applied to a quantum register [46]. But there is one difficulty: if you measure the register, you would destroy the superposition state encoded in it. So how do we determine what might be wrong with the register qubits without looking at them? Simple: entangle them with an ancillary register and measure the ancilla! Because the two registers are correlated, the results of the measurement of the ancilla will tell you how to fix the errors in the processing register, without destroying any coherent superpositions in the processing register itself.

Another way to retain register coherence is to know a little about the sort of noise that is acting on it [47]. If the noise has some very slow components (or those with very long wavelengths), then it is sometimes possible to find certain combinations of qubit states for which the noise on one qubit exactly cancels the noise on another. These qubit states live in a "decoherence-free subspace," or DFS [48]. If you can use only computational states that lie in this DFS, then your computer will be immune to environmental perturbations.

It is the possibility of combining these tools to combat noise that leads researchers to believe that a quantum computer can be built even though decoherence lurks around every corner.

Problems

1. Given the output state in Eq. (11.15) of the Mach–Zehnder interferometer of Fig. 11.2 with twin single photons as inputs, verify Eq. (11.24).

2. Suppose that the inputs to the Mach–Zehnder interferometer are the twin two-photon states $|2\rangle_a|2\rangle_b$. Obtain the uncertainty of the phase measurement in this case.

3. Assuming the production of an $N00N$ state and the encoding of a phase shift as given by Eq. (11.19), verify Eqs. (11.26) and (11.27).

4. Examine the question: Is photon number parity-based QRNG possible with thermal states of light?

5. Verify the statement made in Section 11.3 to the effect that teleporting a state with the shared statistical mixture

$$\hat{\rho}_{AB} = \frac{1}{2}\left(|0\rangle_A|0\rangle_B \,_A\langle 0|\,_B\langle 0| + |1\rangle_A|1\rangle_B \,_A\langle 1|\,_B\langle 1|\right)$$

    is impossible.

6. Suppose that Alice, Bob, and Claire each share one of the "particles" A, B, or C prepared in the three-particle entangled state

$$|\Psi\rangle = \frac{1}{\sqrt{2}}\left(|0\rangle_A|0\rangle_B|0\rangle_C - |1\rangle_A|1\rangle_B|1\rangle_C\right)$$

    as indicated by the labels. Suppose Alice has the unknown state $|\psi\rangle = c_0|0\rangle + c_1|1\rangle$. Investigate the prospects for teleporting the unknown state to both Bob and Claire.

7. Show that the unitary operator for the Hadamard gate can be represented as

$$U_H = \frac{1}{\sqrt{2}}\left[|0\rangle\langle 0| + |1\rangle\langle 0| + |0\rangle\langle 1| - |1\rangle\langle 1|\right].$$

8. The 1-qubit operator $\hat{X} = |0\rangle\langle 1| + |1\rangle\langle 0|$ represents a logical NOT gate (also known as an inverter) because $\hat{X}|0\rangle = |1\rangle$ and $\hat{X}|1\rangle = |0\rangle$. Show that the controlled-not (C-NOT) gate is represented by the 2-qubit operator $\hat{U}_{C\text{-}NOT} = |0\rangle\langle 0| \otimes \hat{I} + |1\rangle\langle 1| \otimes \hat{X}$, where $\hat{I} = |0\rangle\langle 0| + |1\rangle\langle 1|$ is the identity operator and where the first qubit is the control.

9. The 3-qubit gate known as the Toffoli gate is given by

$$\hat{U}_{\text{TG}}|x_1\rangle|x_2\rangle|y\rangle = |x_1\rangle|x_2\rangle|\text{mod}_2(x_1 x_2 + y)\rangle.$$

Show that this gate is a controlled-controlled-not gate. Draw a circuit diagram to represent the gate. Find the representation of the operator $\hat{U}_{\text{TG}}$ in terms of the projection and inversion operator of the three qubits.

10. Design an optical realization of the Toffoli gate described in the previous problem.

11. Let $\hat{U}_{\text{Cl}}$ represent a 2-qubit operator that "clones" any quantum state according to $\hat{U}_{\text{Cl}}|\psi\rangle|0\rangle = |\psi\rangle|\psi\rangle$. Show that such an operator does not, in fact, exist.

References

[1] G. E. Moore, *Electronics* **38** (1965), April 19 issue.

[2] Reprinted in J. S. Bell, *Speakable and Unspeakable in Quantum Mechanics* (Cambridge: Cambridge University Press, 1987), p. 139.

[3] G. C. Ghirardi, A. Rimini, and T. Weber, *Lett. Nuovo Cimento* **27**, 293 (1980); G. C. Ghirardi, R. Grassi, and T. Weber, *Europhys. Lett.* **6**, 95 (1988).

[4] J. Audretsch, *Entangled Systems* (Weinheim: Wiley-VCH, 2007), p. 161.

[5] See, for example, J. A. Dunningham, *Contemp. Phys.* **47**, 257 (2006).

[6] See J. P. Dowling, *Contemp. Phys.* **49**, 127 (2008) and references cited therein.

[7] H. F. Hofmann and T. Ono, *Phys. Rev. A* **76**, 031806 (2004).

[8] I. Afek, O. Ambar, and Y. Silberberg, *Science* **328**, 879 (2010).

[9] J. J. Bollinger, W. M. Itano, D. J. Wineland, and D. J. Heinzen, *Phys. Rev. A* **54**, R4649 (1996).

[10] C. C. Gerry, *Phys. Rev. A*, 61, 043811 (2000).

[11] C. C. Gerry, A. Benmoussa, and R. A. Campos, *Phys. Rev. A* **72**, 053818 (2005).

[12] See X. Gu, A. F. Kockum, A. Miranowicz, Y. Liu, and F. Nori, *Phys. Rep.* **718/719**, 1 (2017) and references cited therein.

[13] Y. Gao, P. Anisimov, C. F. Wildfeuer, J. Luine, H. Lee, and J. P. Dowling, *J. Opt. Soc. Am. B* **27**, A170 (2010).

[14] L. Cohen, D. Istrati, L. Dovrat, and H. S. Eisenberg, *Opt. Express* **22**, 11945 (2014).

[15] C. C. Gerry and J. Mimih, *Contemp. Phys.* **51**, 497 (2010).

[16] R. J. Birrittella, P. M. Alsing, and C. C. Gerry, *AVS Quant. Sci.* **3**, 014701 (2021).

[17] C. H. Bennett, G. Brassard, C. Crepeau, R. Jozsa, A. Peres and W. K. Wootters, *Phys. Rev. Lett.* **70**, 1895 (1993).

[18] D. Boschi, S. Branca, F. De Martini, and L. Hardy, *Phys. Rev. Lett.* **80**, 1121 (1998).

[19] D. Bouwmeester, J.-W. Pan, K. Mattle, M. Eibl, H. Weinfurter, and A. Zeilinger, *Nature* **390**, 575 (1997).

[20] A. Furuwasa, J. Sørensen, S. L. Braunstein, C. A. Fuchs, H. J. Kimble, and E. S. Polzik, *Science* **282**, 706 (1998).

[21] R. Rivest, A. Shamir, and L. Adleman, *On Digital Signatures and Public-Key Cryptosystems*. MIT Laboratory for Computer Science Technical Report MIT/LCS/TR-212 (January 1979).

[22] D. Deutsch, *Proc. R. Soc. Lond.* A **400**, 97 (1985).

[23] D. Deutsch and R. Jozsa, *Proc. R. Soc. Lond.* A **439**, 553 (1992).

[24] P. W. Shor, in S. Goldwasser (Ed.), *Proceedings of the 35th Symposium on Foundations of Computer Science, Los Alamitos* (Washington, DC: IEEE Computer Society Press, 1994), p. 124.

[25] L. M. K Vandersypen, M. Steffen, G. Breyta, C. S. Yannoni, M. H. Sherwood, and I. L. Chuang, *Nature* **414**, 883 (2001).

[26] M. Herrero-Collantes and J. C. Garcia-Escartin, *Rev. Mod. Phys.* **89**, 015004 (2017).

[27] C. Kollmitzer, S. Schauer, S. Rass, and B. Rainer (Eds.), *Quantum Random Number Generation, Theory and Practice* (Cham: Springer, 2020).

[28] C. C. Gerry, R. J. Birrittella, P. M. Alsing, A. Hossameldin, M. Eaton, and O. Pfister, *J. Opt. Soc. Am.* **39**, 1068 (2022).

[29] See C. Silberhorn, *Contemp. Phys.* **48**, 143 (2007).

[30] See A. E. Lita, A. J. Miller, and S. W. Nam, *Opt. Express* **16**, 3032 (2008).

[31] M. Eaton, A. Hossanmeldin, R. J. Birrettella, P. M. Alsing, C. C. Gerry, C. Cuevas, *et al. Nature Photonics* (in press).

[32] R. A. Campos and C. C. Gerry, *Phys. Lett.* A **343**, 27 (2005).

[33] C. H. Bennett and G. Brassard, in *Proceedings of the IEEE Conference on Computers, Systems and Signal Processing* (New York: IEEE Press, 1984). See also C. H. Bennett, F. Besette, G. Brassard, L. Salvail, and J. Smolin, *J. Cryptol.* **5**, 3 (1992).

[34] C. H. Bennett, *Phys. Rev. Lett.* **68**, 3121 (1992).

[35] A. K. Ekert, *Phys. Rev. Lett.* **67**, 661 (1991).

[36] N. Gisin, G. Ribordy, W. Tittel, and H. Zbinden, *Rev. Mod. Phys.* **74**, 145 (2002).

[37] S. Guide, M. Riebe, G. P. T. Lancaster, C. Becher, J. Eschner, H. Häffner, *et al.*, *Nature* **421**, 48 (2003).

[38] L. K. Grover, *Phys. Rev. Lett.*, **79**, 325 (1997).

[39] J. Kempe, *Contemp. Phys.* **44**, 307 (2003).

[40] D. Reitzner, D. Nagaj, and V. Bužek, arXiv:1207.7283 (2013).

[39] I. L. Chuang and Y. Yamamoto, *Phys. Rev. A* **52**, 3489 (1995).

[40] E. Fredkin and T. Toffoli, *Int. J. Theor. Phys.* **21**, 219 (1982).

[41] G. J. Milburn, *Phys. Rev. A* 62, 2124 (1989).

[42] E. Knill, R. Laflamme, and G. Milburn, *Nature* **409**, 46 (2001).

[43]   See T. Sleator and H. Weinfurter, *Phys. Rev. Lett.* **74**, 4087 (1995);
       P. Domokos, J. M. Raimond, M. Brune, and S. Haroche, *Phys. Rev.
       A* **52**, 3554 (1995).

[44]   See C. Monroe, D. M. Meekhof, B. E. King, W. M. Itano, and
       D. J. Wineland, *Phys. Rev. Lett.* **75**, 4714 (1995).

[45]   See D. R. Meacher, *Contemp. Phys.* **39**, 329 (1998); I. H. Deutsch and
       G. K. Brennen, *Forts. Phys.* **48**, 925 (2000).

[46]   A. Calderbank and P. Shor, *Phys. Rev. A* **52**, R2493 (1995); *Phys.
       Rev. A* **54**, 1098 (1996); A. M. Steane, *Phys. Rev. Lett.* **77**, 793 (1995).

[47]   G. M. Palma, K.-A. Suominen, and A. K. Ekert, *Proc. R. Soc. Lond.
       A* **452**, 567 (1996).

[48]   D. Lidar, I. L. Chuang, and B. Whaley, *Phys. Rev. Lett.* **81**, 2594
       (1998).

Bibliography

A book covering many aspects of cryptography is

P. Garrett, *Making and Breaking Codes: An Introduction to Cryptology*
(Upper Saddle River, NJ: Prentice Hall, 2001).

A fun read on the history of cryptography is

S. Singh, *The Code Book: The Science of Secrecy from Ancient Eygpt to
Quantum Cryptography* (New York: Anchor Books, 1999).

*Useful books that have been published on all aspects of quantum informa-
tion are*

M. Nielsen and I. L. Chuang, *Quantum Information and Quantum
Computation* (Cambridge: Cambridge University Press, 2001).

V. Vedral, *Introduction to Quantum Information Science* (Oxford: Oxford
University Press, 2006).

J. Audretsch, *Entangled Systems* (Weinheim: Wiley-VCH, 2007).

N. D. Mermin, *Quantum Computer Science* (Cambridge: Cambridge
University Press, 2007).

S. M. Barnett, *Quantum Information* (Oxford: Oxford University Press,
2009).

J. A. Jones and D. Jaksch, *Quantum Information, Computation and
Communication* (Cambridge: Cambridge University Press, 2012).

W.-H. Steeb and Y. Hardy, *Problems and Solutions in Quantum Computing
and Quantum Information*, 4th edition (Singapore: World Scientific,
2018).

J. Bergou, M. Hillery, and M. Saffman, *Quantum Information Processing,
Theory and Implementation* (New York: Springer, 2021).

A book that is particularly relevant to the subject of the present book is

P. Kok and B. W. Lovett, *Introduction to Optical Quantum Information Processing* (Cambridge: Cambridge University Press, 2010).

Below is a list of review articles we have found to be accessible and with many references to the original literature

D. P. DiVincenzo, "Quantum computation," *Science* **270**, 255 (1995).

R. J. Hughes, D. M. Alde, P. Dyer, G. G. Luther, G. L. Morgan, and M. Schauer, "Quantum cryptography," *Contemp. Phys.* **36**, 149 (1995).

S. J. D. Phoenix and P. D. Townsend, "Quantum cryptography: How to beat the code breakers using quantum mechanics," *Contemp. Phys.* **36**, 165 (1995).

A. Barenco, "Quantum physics and computers," *Contemp. Phys.* **37**, 375 (1996).

M. B. Plenio and V. Vedral, "Teleportation, entanglement and thermodynamics in the quantum world," *Contemp. Phys.* **39**, 431 (1998).

J. Kempe, "Quantum random walks: An introductory overview," *Contemp. Phys.* **44**, 307 (2003).

E. Gerjuoy, "Shor's factoring algorithm and modern cryptography: An illustration of the capabilities inherent in quantum computers," *Am. J. Phys.* **73**, 521 (2005).

T. Spiller, W. J. Munro, S. D. Barrett, and P. Kok, "An introduction to quantum information processing: Applications and realizations," *Contemp. Phys.* **46**, 207 (2005).

A. Montanaro, "Quantum algorithms: An overview," *npj Quant. Inf.* **2**, 15023 (2016).

A. Acin and L. Masanes, "Certified randomness in quantum physics," *Nature* **540**, 213 (2016).

F. W. Strauch, "Resource letter QI-1: Quantum information," *Am. J. Phys.* **84**, 495 (2016).

D. J. Bernstein and T. Lange, "Post-quantum cryptography," *Nature* **549**, 188 (2017).

A. W. Harrow and A. Montanaro, "Quantum computational supremacy," *Nature* **549**, 203 (2017).

Quantum key distribution can now be demonstrated in the undergraduate laboratory:

A. Bista, B. Sharma, and E. J. Galvez, "A demonstration of quantum key distribution with entangled photons for the undergraduate laboratory," *Am. J. Phys.* **89**, 111 (2021).

Appendix A The Density Operator, Entangled States, the Schmidt Decomposition, and the Von Neumann Entropy

A.1 The Density Operator

Quantum mechanical state vectors $|\psi\rangle$ convey the maximal amount of information about a system allowed by the laws of quantum mechanics. Typically, the information consists of quantum numbers associated with a set of commuting observables. Furthermore, if $|\psi_1\rangle$ and $|\psi_2\rangle$ are two possible quantum states, then so is their coherent superposition

$$|\psi\rangle = c_1|\psi_1\rangle + c_2|\psi_2\rangle \tag{A.1}$$

if the coefficients c_1 and c_2 are known. If the states $|\psi_1\rangle$ and $|\psi_2\rangle$ are orthogonal ($\langle\psi_2|\psi_1\rangle = 0$) then we must have $|c_1|^2 + |c_2|^2 = 1$. But there are frequently (in fact, more often than not) situations where the state vector is not precisely known. There are, for example, cases where the system of interest is interacting with some other system, possibly with some very large system (e.g. a reservoir, with which it becomes entangled). It may be possible to write state vectors for the multicomponent system but not for the subsystem of interest. For example, for a system of two spin-1/2 particles with the eigenstates of, say, the z-component of the spin denoted $|\uparrow\rangle$ for spin up and $|\downarrow\rangle$ for spin down, a possible state vector of the combined system is

$$|\psi\rangle = \frac{1}{\sqrt{2}}[|\uparrow\rangle_1|\downarrow\rangle_2 - |\downarrow\rangle_1|\uparrow\rangle_2], \tag{A.2}$$

the so-called singlet state (total angular momentum zero), also known as one of the "Bell" states. Equation (A.2) is an example of an entangled state. An entangled state cannot be factored, in any basis, into a product of states of the two subsystems, that is

$$|\psi\rangle \neq |\text{spin 1}\rangle|\text{spin 2}\rangle \text{(for an entangled two-spin state).} \tag{A.3}$$

Entanglement is, apart from the superposition principle itself, an essential mystery of quantum mechanics, as was pointed out in 1935 by Schrödinger himself. Note though that entanglement follows from the superposition principle and is not something imposed on the theory. So, Feynman's dictum that the superposition principle contains "the only mystery" is still correct.

Quantum states described by state vectors are said to be *pure* states. States that cannot be described by state vectors are said to be in *mixed* states. Mixed states are described by the density operator

$$\hat{\rho} = \sum_i |\psi_i\rangle p_i \langle\psi_i| = \sum_i p_i |\psi_i\rangle\langle\psi_i|, \tag{A.4}$$

where the sum is over an ensemble (in the sense of statistical mechanics), with p_i the probability of the system being in the ith state of the ensemble $|\psi_i\rangle$, where $\langle\psi_i|\psi_i\rangle = 1$. The probabilities satisfy the obvious relations

$$0 \le p_i \le 1, \quad \sum_i p_i = 1, \quad \sum_i p_i^2 \le 1. \tag{A.5}$$

For the special case where all the p_i vanish except, say, the jth one, $p_i = \delta_{ij}$, we obtain

$$\hat{\rho} = |\psi_j\rangle\langle\psi_j|, \tag{A.6}$$

the density operator for the pure state $|\psi_j\rangle$. Note that the density operator for this case is just the projection operator onto the state $|\psi_j\rangle$ and that, for the more general case of Eq. (A.4), the density operator is a sum of the projection operators over the ensemble, weighted with the probabilities of each member of the ensemble.

We now introduce a complete, orthonormal, basis $\{|\varphi_n\rangle\}$ $(\sum_n |\varphi_n\rangle\langle\varphi_n| = \hat{I})$ of eigenstates of some observable. Then, for the ith member of the ensemble, we may write

$$|\psi_i\rangle = \sum_n |\varphi_n\rangle\langle\varphi_n|\psi_i\rangle = \sum_n c_n^{(i)}|\varphi_n\rangle, \tag{A.7}$$

where $c_n^{(i)} = \langle\varphi_n|\psi_i\rangle$. The matrix element of $\hat{\rho}$ between the n and n' eigenstates is

$$\langle\varphi_n|\hat{\rho}|\varphi_{n'}\rangle = \sum_i \langle\varphi_n|\psi_i\rangle p_i \langle\psi_i|\varphi_{n'}\rangle = \sum_i p_i c_n^{(i)} c_{n'}^{(i)*}. \tag{A.8}$$

The quantities $\langle\varphi_n|\hat{\rho}|\varphi_{n'}\rangle$ form the elements of the density matrix. Taking the trace of this matrix, we have

$$\mathrm{Tr}\hat{\rho} = \sum_n \langle \varphi_n | \hat{\rho} | \varphi_n \rangle = \sum_i \sum_n \langle \varphi_n | \psi_i \rangle p_i \langle \psi_i | \varphi_n \rangle$$

$$= \sum_i \sum_n p_i \langle \psi_i | \varphi_n \rangle \langle \varphi_n | \psi_i \rangle = \sum_i p_i = 1. \tag{A.9}$$

Since $\hat{\rho}$ is Hermitian [as is evident from its construction in Eq. (A.4)], the diagonal elements $\langle \varphi_n | \hat{\rho} | \varphi_n \rangle$ must be real, and it follows from Eq. (A.9) that

$$0 \le \langle \varphi_n | \hat{\rho} | \varphi_n \rangle \le 1. \tag{A.10}$$

Now let us consider the square of the density operator: $\hat{\rho}^2 = \hat{\rho} \cdot \hat{\rho}$. For a pure state where $\hat{\rho} = |\psi\rangle\langle\psi|$, it follows that

$$\hat{\rho}^2 = |\psi\rangle\langle\psi|\psi\rangle\langle\psi| = |\psi\rangle\langle\psi| = \hat{\rho}, \tag{A.11}$$

and thus

$$\mathrm{Tr}\hat{\rho}^2 = \mathrm{Tr}\hat{\rho} = 1. \tag{A.12}$$

For a statistical mixture

$$\hat{\rho}^2 = \sum_i \sum_j p_i p_j |\psi_i\rangle\langle\psi_i|\psi_j\rangle\langle\psi_j|. \tag{A.13}$$

Taking the trace, we have

$$\mathrm{Tr}\hat{\rho}^2 = \sum_n \langle \varphi_n | \hat{\rho}^2 | \varphi_n \rangle$$

$$= \sum_n \sum_i \sum_j p_i p_j \langle \varphi_n | \psi_i \rangle \langle \psi_i | \psi_j \rangle \langle \psi_j | \varphi_n \rangle$$

$$= \sum_i \sum_j p_i p_j |\langle \psi_i | \psi_j \rangle|^2 \tag{A.14}$$

$$\le \left[\sum_i p_i \right]^2 = 1.$$

Equality holds only if $|\langle \psi_i | \psi_j \rangle|^2 = 1$ for *every* pair of states $|\psi_i\rangle$ and $|\psi_j\rangle$. This is possible only if all the $|\psi_i\rangle$ are collinear in Hilbert space (i.e. equivalent up to an overall phase factor). Thus, we have the following criteria for pure and mixed states:

$$\begin{aligned} \mathrm{Tr}\hat{\rho}^2 &= 1 \text{ for a pure state,} \\ \mathrm{Tr}\hat{\rho}^2 &< 1 \text{ for a mixed state.} \end{aligned} \tag{A.15}$$

Perhaps a simple example is warranted at this point. Consider a superposition of, say, the vacuum and one-photon number states

$$|\psi\rangle = \frac{1}{\sqrt{2}}(|0\rangle + e^{i\phi}|1\rangle), \tag{A.16}$$

where ϕ is just some phase. The density operator associated with this state is given by

$$\hat{\rho}_\psi = |\psi\rangle\langle\psi| = \frac{1}{2}\left[|0\rangle\langle0| + |1\rangle\langle1| + e^{i\phi}|1\rangle\langle0| + e^{-i\phi}|0\rangle\langle1|\right]. \tag{A.17}$$

On the other hand, the density operator for an equally populated mixture of vacuum and one-photon states is

$$\hat{\rho}_M = \frac{1}{2}\left[|0\rangle\langle0| + |1\rangle\langle1|\right]. \tag{A.18}$$

The two density operators differ by the presence of the "off-diagonal" or "coherence" terms in the former, the terms being absent in the case of the mixture. The absence of the coherence terms in the mixture is, of course, what makes the distinction between a state exhibiting full quantum mechanical behavior and one that does not. It is easy to check that $\mathrm{Tr}\hat{\rho}_M^2 = 1/2$.

For one of the states of the ensemble $|\psi_i\rangle$, by itself pure, the expectation value of some operator $\hat{O}$ is given by

$$\langle\hat{O}\rangle_i = \langle\psi_i|\hat{O}|\psi_i\rangle. \tag{A.19}$$

For the statistical mixture, the ensemble average is given by

$$\langle\hat{O}\rangle = \sum_i p_i\langle\psi_i|\hat{O}|\psi_i\rangle, \tag{A.20}$$

which is just the average of the quantum mechanical expectation values weighted with the probabilities p_i. More formally, we may write

$$\langle\hat{O}\rangle = \mathrm{Tr}(\hat{\rho}\hat{O}) \tag{A.21}$$

since

$$\begin{aligned}
\mathrm{Tr}(\hat{\rho}\hat{O}) &= \sum_n \langle\varphi_n|\hat{\rho}\hat{O}|\varphi_n\rangle \\
&= \sum_n \sum_i p_i\langle\varphi_n|\psi_i\rangle\langle\psi_i|\hat{O}|\varphi_n\rangle \\
&= \sum_i \sum_n p_i\langle\psi_i|\hat{O}|\varphi_n\rangle\langle\varphi_n|\psi_i\rangle \\
&= \sum_i p_i\langle\psi_i|\hat{O}|\psi_i\rangle.
\end{aligned} \tag{A.22}$$

A.2 Two-State System and the Bloch Sphere

For a two-state system, be it a spin-1/2 particle, a two-level atom, or the polarizations of a single photon, there always exists a description in terms of the Pauli operators $\hat{\sigma}_1, \hat{\sigma}_2, \hat{\sigma}_3$ satisfying the commutation relations

$$[\hat{\sigma}_i, \hat{\sigma}_j] = 2i\varepsilon_{ijk}\hat{\sigma}_k. \tag{A.23}$$

In a basis where $\hat{\sigma}_3$ and $\hat{\sigma}^2 = \hat{\sigma}_1^2 + \hat{\sigma}_2^2 + \hat{\sigma}_3^2$ are diagonal, these operators can be written in matrix form as

$$\sigma_1 = \begin{pmatrix} 0 & 1 \\ 1 & 0 \end{pmatrix}, \quad \sigma_2 = \begin{pmatrix} 0 & -i \\ i & 0 \end{pmatrix}, \quad \sigma_3 = \begin{pmatrix} 1 & 0 \\ 0 & -1 \end{pmatrix}. \tag{A.24}$$

Any Hermitian 2×2 matrix can be expressed in terms of the Pauli matrices and the 2×2 identity matrix $\hat{I}_2$, and this includes, of course, the density operator. That is, we can write

$$\rho = \begin{pmatrix} \rho_{11} & \rho_{12} \\ \rho_{21} & \rho_{22} \end{pmatrix} = \frac{1}{2}\begin{pmatrix} 1 + s_3 & s_1 + i s_2 \\ s_1 - i s_2 & 1 - s_3 \end{pmatrix} = \frac{1}{2}(\hat{I}_2 + \mathbf{s} \cdot \sigma), \tag{A.25}$$

where the vector $\mathbf{s} = \{s_1, s_2, s_3\}$ is known as the Bloch vector. For a pure state $\hat{\rho} = |\Psi\rangle\langle\Psi|$, the Bloch vector has unit length, $\sum_i |s_i|^2 = 1$, and points in some direction specified by the spherical coordinate angles θ and ϕ in a three-dimensional Euclidean space. The associated quantum state can be represented in terms of these angles as

$$|\Psi\rangle = \cos\left(\frac{\theta}{2}\right) e^{-i\phi/2}|\uparrow\rangle + \sin\left(\frac{\theta}{2}\right) e^{i\phi/2}|\downarrow\rangle. \tag{A.26}$$

In general, and including the case of mixed states where $|\mathbf{s}| \le 1$, the density operator of the form of Eq. (A.25) has two eigenvalues

$$\begin{aligned} g_1 &= \frac{1}{2}\left[1 + \sqrt{s_1^2 + s_2^2 + s_3^2}\right] = \frac{1}{2}[1 + |\mathbf{s}|], \\ g_2 &= \frac{1}{2}\left[1 - \sqrt{s_1^2 + s_2^2 + s_3^2}\right] = \frac{1}{2}[1 - |\mathbf{s}|], \end{aligned} \tag{A.27}$$

and its eigenvectors are determined by the two vectors $\mathbf{u}$ and $-\mathbf{u}$ shown in the Bloch sphere in Fig. A.1. For pure states, where $|\mathbf{s}| = 1$, $\mathbf{u}$ coincides with $\mathbf{s}$ and its tip lies on the surface of the Bloch sphere. For a mixed state where $|\mathbf{s}| < 1$, vector $\mathbf{u}$ points in the same direction as $\mathbf{s}$ but unlike $\mathbf{s}$ maintains unit length so that its tip always lies on the surface of the Bloch sphere. Equation (A.26) can be used to express $\mathbf{u}$ and $-\mathbf{u}$ in terms of the vectors in state space.

A.3 Entangled States

Let us now consider a two-particle (or two-mode) system (known as a bipartite system) and, for simplicity, let us assume that each particle can be in either of two one-particle states $|\psi_1\rangle$ or $|\psi_2\rangle$. Using the notation

$$|\psi_1^{(1)}\rangle \text{ (particle 1 in state 1)},$$
$$|\psi_2^{(1)}\rangle \text{ (particle 1 in state 2)},$$
$$|\psi_1^{(2)}\rangle \text{ (particle 2 in state 1)},$$

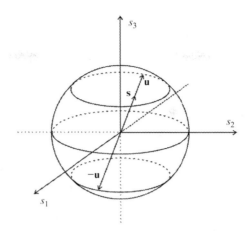

Representation of the density matrix of a two-state system in terms of the Bloch sphere and Bloch vector. The three components of the Bloch vector $\mathbf{s} = \{s_1, s_2, s_3\}$ specify the density operator according to the parameterization of Eq. (A.25). The two eigenvalues are $(1 \pm |\mathbf{s}|)/2$ and the two eigenvectors are specified by $\mathbf{u}$ and $-\mathbf{u}$.

$$|\psi_2^{(2)}\rangle \ \text{(particle 2 in state 2)},$$

we consider a pure two-particle superposition state (in general, an entangled state)

$$|\Psi\rangle = C_1|\psi_1^{(1)}\rangle \otimes |\psi_2^{(2)}\rangle + C_2|\psi_2^{(1)}\rangle \otimes |\psi_1^{(2)}\rangle, \qquad (A.28)$$

an example of which is given in Eq. (A.2). (We have inserted the direct product symbol here for emphasis, though we generally assume it to be understood and clear from the context throughout this book.) Clearly, this can be extended for multiparticle (multipartite) systems. For such multipartite systems, we can define reduced density operators for each of the subsystems by tracing multipartite density operators over the states of all the other systems. In the present case, with the density operator of the two-particle system given by $\hat{\rho} = |\Psi\rangle\langle\Psi|$, the reduced density operator for particle 1 is

$$\begin{aligned}\hat{\rho}^{(1)} &= \mathrm{Tr}_2\hat{\rho} = \langle\psi_1^{(2)}|\hat{\rho}|\psi_1^{(2)}\rangle + \langle\psi_2^{(2)}|\hat{\rho}|\psi_2^{(2)}\rangle \\ &= |C_1|^2|\psi_1^{(1)}\rangle\langle\psi_1^{(1)}| + |C_2|^2|\psi_2^{(1)}\rangle\langle\psi_2^{(1)}|.\end{aligned} \qquad (A.29)$$

This has the form of a mixed state for particle 1 as long as $C_i \neq 0, i = 1, 2$. Similarly for particle 2:

$$\hat{\rho}^{(2)} = \mathrm{Tr}_1\hat{\rho} = |C_1|^2|\psi_1^{(2)}\rangle\langle\psi_1^{(2)}| + |C_2|^2|\psi_2^{(2)}\rangle\langle\psi_2^{(2)}|. \qquad (A.30)$$

Evidently, when only one of the particles is considered without regard to the other, it is generally in a mixed state. Thus, one may characterize the degree of entanglement according to the degree of purity of either of the subsystems. If $\mathrm{Tr}[\hat{\rho}^{(2)}]^2 = 1$, the state $|\Psi\rangle$ is not an entangled state; but if

$\text{Tr}[\hat{\rho}^{(2)}]^2 < 1$, we may conclude that $|\Psi\rangle$ describes an entanglement between subsystems 1 and 2.

A.4 Schmidt Decomposition

There is another convenient way to approach the problem of characterizing entanglement, at least for cases where there are only two subsystems. We refer to the use of the von Neumann entropy, which we introduce in Section A.5. But first, as a preliminary, we proceed to introduce the Schmidt decomposition [1, 2]. To keep the discussion general, we do not restrict the dimensions of the Hilbert spaces of the subsystems.

Suppose we let $\{|a_i\rangle, i = 1, 2, 3, ...\}$ and $\{|b_j\rangle, j = 1, 2, 3, ...\}$ form orthonormal bases for subsystems $\mathcal{U}$ and $\mathcal{V}$, respectively. If the Hilbert spaces of these systems are denoted $\mathcal{H}_\mathcal{U}$ and $\mathcal{H}_\mathcal{V}$, respectively, then any state $|\Psi\rangle \in \mathcal{H} = \mathcal{H}_\mathcal{U} \otimes \mathcal{H}_\mathcal{V}$ can be written as

$$|\Psi\rangle = \sum_{i,j} c_{ij} |a_i\rangle \otimes |b_j\rangle. \tag{A.31}$$

In these bases, the density operator for the composite system is

$$\begin{aligned}\hat{\rho} = |\Psi\rangle\langle\Psi| &= \sum_{i,j,k,l} c_{ij} c_{kl}^* |a_i\rangle\langle a_k| \otimes |b_j\rangle\langle b_l| \\ &= \sum_{i,j,k,l} \rho_{ij,kl} |a_i\rangle\langle a_k| \otimes |b_j\rangle\langle b_l|,\end{aligned} \tag{A.32}$$

where $\rho_{ij,kl} = c_{ij} c_{kl}^*$. The density operators of each of the subsystems are given by the partial traces

$$\begin{aligned}\hat{\rho}_u = \text{Tr}_v\hat{\rho} &= \sum_j \langle b_j|\hat{\rho}|b_j\rangle = \sum_{i,j,k} \rho_{ij,kj} |a_i\rangle\langle a_k|, \\ \hat{\rho}_v = \text{Tr}_u\hat{\rho} &= \sum_i \langle a_i|\hat{\rho}|a_i\rangle = \sum_{i,j,k} \rho_{ij,il} |b_j\rangle\langle b_l|.\end{aligned} \tag{A.33}$$

But it is possible, for any bipartite system in a *pure* state, to write that state in the form

$$|\Psi\rangle = \sum_i g_i |u_i\rangle \otimes |v_i\rangle, \tag{A.34}$$

where $\{|u_i\rangle\}$ and $\{|v_i\rangle\}$ are orthonormal bases in $\mathcal{H}_\mathcal{U}$ and $\mathcal{H}_\mathcal{V}$, respectively, and where $\sum_i |g_i|^2 = 1$. The density operator for the composite system is then

$$\hat{\rho} = \sum_{i,k} g_i g_k^* |u_i\rangle \langle u_k| \otimes |v_i\rangle \langle v_k|. \tag{A.35}$$

The proof of this is as follows. Let us choose $\{|u_i\rangle\}$ such that $\hat{\rho}_u$ is diagonal in this basis and the dimension of $\mathcal{H}_\mathcal{U}$ is not greater than the dimension of $\mathcal{H}_\mathcal{V}$. Let $\{|v'_i\rangle\}$ be an arbitrary basis in $\mathcal{H}_\mathcal{V}$. Then we can write

$$\hat{\rho} = |\Psi\rangle\langle\Psi| = \sum_{i,j,k,l} c_{ij} c_{kl}^* |u_i\rangle \langle u_k| \otimes |v'_k\rangle \langle v'_l|. \tag{A.36}$$

However, $\hat{\rho}_u = \sum_{i,k,j} c_{ij} c_{kj}^* |u_i\rangle\langle u_k|$ must be diagonal, therefore $\sum_j c_{ij} c_{kj}^* = |g_i|^2 \delta_{ik}$. This condition allows us to switch to another orthogonal basis in $\mathcal{H}_\mathcal{U}$. For each $|g_i| \neq 0$ and for each $|u_i\rangle$, we can define the relative state

$$|v_i\rangle = \sum_j \left(\frac{c_{ij}}{g_i}\right) |v'_j\rangle, \tag{A.37}$$

where $i = 1, 2, ..., N$ and N is not greater than the dimension of $\mathcal{H}_\mathcal{U}$. The density operator of the composite system may now be written as

$$\begin{aligned}\hat{\rho} &= \sum_{i,j,k,l} g_i g_k^* |u_i\rangle \langle u_k| \otimes \left(\frac{c_{ij}}{g_i}\right) |v'_j\rangle \langle v'_l| \left(\frac{c_{kl}^*}{g_k^*}\right) \\ &= \sum_{i,k} g_i g_k^* |u_i\rangle \langle u_k| \otimes |v_i\rangle \langle v_k|,\end{aligned} \tag{A.38}$$

where only N basis vectors $\{|v_i\rangle\}$ from $\mathcal{H}_\mathcal{V}$ are sufficient to express the state vector or the density operator of the composite system. This completes the proof.

Before proceeding, some remarks are in order. (i) The summation of the single index in the Schmidt decomposition goes to the smaller of the dimensions of the two Hilbert spaces $\mathcal{H}_\mathcal{U}$ and $\mathcal{H}_\mathcal{V}$. (ii) The decomposition is not unique and cannot, in general, be extended to multipartite systems with greater than two subsystems. (iii) In the bases chosen for the Schmidt decomposition, both $\hat{\rho}_u$ and $\hat{\rho}_v$ are diagonal and have the same positive spectrum:

$$\begin{aligned}\hat{\rho}_u &= \sum_i |g_i|^2 |u_i\rangle\langle u_i|, \\ \hat{\rho}_v &= \sum_i |g_i|^2 |v_i\rangle\langle v_i|.\end{aligned} \tag{A.39}$$

It is this feature that facilitates the use of the von Neumann entropy, introduced in Section A.5, to characterize entanglement in a bipartite system. Finally, the summation over the single index in the Schmidt decomposition goes to the smaller of the dimensionalities of the two Hilbert spaces $\mathcal{H}_\mathcal{U}$ and $\mathcal{H}_\mathcal{V}$. This means that if a two-state system is entangled with a system of N dimensions, where $N > 2$, then the Schmidt decomposition has only two terms. An example of this for a two-level atom interacting with a single-mode field is discussed at the end of Chapter 4.

But first we give a simple example of the Schmidt decomposition for a bipartite system. We consider a spin-singlet (SS) state of the form

$$|SS\rangle = \frac{1}{\sqrt{2}}(|\uparrow_z\rangle_1|\downarrow_z\rangle_2 - |\downarrow_z\rangle_1|\uparrow_z\rangle_2), \tag{A.40}$$

where $|\uparrow_z\rangle_1$ is the state for particle 1 of spin up along the z-axis, and so on. Clearly, this state already possesses a Schmidt decomposition. But because the spin-singlet state is rotationally invariant, it is particularly easy to find other Schmidt decompositions. For example, we can write

$$|SS\rangle = \frac{1}{\sqrt{2}}(|\uparrow_x\rangle_1|\downarrow_x\rangle_2 - |\downarrow_x\rangle_1|\uparrow_x\rangle_2), \tag{A.41}$$

or more generally

$$|SS\rangle = \frac{1}{\sqrt{2}}(|\uparrow_\mathbf{n}\rangle_1|\downarrow_\mathbf{n}\rangle_2 - |\downarrow_\mathbf{n}\rangle_1|\uparrow_\mathbf{n}\rangle_2), \tag{A.42}$$

where $\mathbf{n}$ is a unit vector for a quantization axis in some arbitrary direction.

A.5 Von Neumann Entropy

The concept of entropy is familiar from thermodynamics and is commonly understood as a measure of disorder: the greater the disorder within a system, the greater the entropy. From the point of view of statistical mechanics and information theory, entropy can be thought of as a measure of missing information, the information that would be gained if a complete measurement of a system could be performed [3].

The von Neumann definition of entropy closely parallels the statistical mechanical one. For a density operator $\hat{\rho}$, the von Neumann entropy is defined as [4]

$$S(\hat{\rho}) = -\mathrm{Tr}[\hat{\rho} \ln \hat{\rho}]. \tag{A.43}$$

For a pure state $S(\hat{\rho}_{\mathrm{pure}}) = 0$ and thus repeated measurements of the state in question yield no new information. But for a mixed state we shall have $S(\hat{\rho}_{\mathrm{mixed}}) > 0$. In general, the entropy is not easy to calculate. However, in a basis for which the density operator is diagonal, such as the Schmidt basis, the entropy may be evaluated from the diagonal elements according to

$$S(\hat{\rho}) = -\sum_k \rho_{kk} \ln \rho_{kk}. \tag{A.44}$$

Because all of the numbers ρ_{kk} must be real and $0 \le \rho_{kk} \le 1$, it follows that $S(\hat{\rho})$ must be positive semi-definite.

As an example, consider now the bipartite state

$$|\psi\rangle = \frac{1}{\sqrt{1 + |\xi|^2}} (|0\rangle_1 |0\rangle_2 + \xi |1\rangle_1 |1\rangle_2), \tag{A.45}$$

which clearly has the form of a Schmidt decomposition. The density operators for each of the subsystems are

$$\hat{\rho}_1 = \frac{1}{(1 + |\xi|^2)} [|0\rangle_{11}\langle 0| + |\xi|^2 |1\rangle_{11}\langle 1|],$$
$$\hat{\rho}_2 = \frac{1}{(1 + |\xi|^2)} [|0\rangle_{22}\langle 0| + |\xi|^2 |1\rangle_{22}\langle 1|]. \tag{A.46}$$

Then it follows that

$$S(\hat{\rho}_1) = -\left\{ \frac{1}{(1 + |\xi|^2)} \ln \left[\frac{1}{(1 + |\xi|^2)} \right] + \frac{|\xi|^2}{(1 + |\xi|^2)} \ln \left[\frac{|\xi|^2}{(1 + |\xi|^2)} \right] \right\} = S(\hat{\rho}_2). \tag{A.47}$$

For the case where $\xi = 0$, we have the non-entangled product state $|0\rangle_1 |0\rangle_2$ and $S(\hat{\rho}_1) = S(\hat{\rho}_2) = 0$, as expected. But for $|\xi| = 1$ we have $S(\hat{\rho}_1) = S(\hat{\rho}_2) = \ln 2$, which represents maximal entanglement for this state, indicating maximal *quantum* correlations. Information about these correlations is discarded upon tracing over one of the subsystems.

Finally, in the case of a pure state of a bipartite system, where the subsystems are of different dimensions (as is easy to check), the entropies of the subsystems are equal. This is a consequence of the Schmidt basis enforcing the same dimensionality on each of the subsystems, as mentioned above. An example of such is provided, in the case of the Jaynes–Cummings model, at the end of Chapter 4.

A.6 Dynamics of the Density Operator

Finally, in the absence of dissipative interactions and in the absence of an explicitly time-dependent interaction, the density operator evolves unitarily according to

$$\frac{d\hat{\rho}}{dt} = \frac{i}{\hbar} [\hat{\rho}, \hat{H}], \tag{A.48}$$

easily proved by using the fact that each of the $|\psi_i\rangle$ of the ensemble satisfies the Schrödinger equation

$$i\hbar\frac{d|\psi_i\rangle}{dt} = \hat{H}|\psi_i\rangle. \qquad (A.49)$$

Alternatively, we may write

$$\hat{\rho}(t) = \hat{U}(t,0)\hat{\rho}(0)\hat{U}^{\dagger}(t,0), \qquad (A.50)$$

where the unitary evolution operator $\hat{U}(t,0)$ satisfies the equation

$$i\hbar\frac{d\hat{U}}{dt} = \hat{H}\hat{U}. \qquad (A.51)$$

Equation (A.26) is known as the von Neumann equation, the quantum mechanical analog of the Liouville equation associated with the evolution phase-space probability distributions in statistical mechanics. The equation is sometimes written as

$$i\hbar\frac{d\hat{\rho}(t)}{dt} = [\hat{H},\hat{\rho}(t)] \equiv \mathcal{L}\hat{\rho}(t), \qquad (A.52)$$

where $\hat{\mathcal{L}}$ is known as the Liouvillian superoperator.

References

[1] E. Schmidt, *Math. Annal.* **63**, 433 (1907). For an early application in quantum mechanics, see the chapter by H. Everett III in B. S. DeWitt and N. Graham (Eds.), *The Many Worlds Interpretation of Quantum Mechanics* (Princeton, NJ: Princeton University Press, 1973), p. 3.

[2] A. Ekert and P. L. Knight, *Am. J. Phys.* **63**, 415 (1995).

[3] See, for example, the discussion in J. Machta, *Am. J. Phys.* **67**, 1074 (1999).

[4] J. von Neumann, *Gott. Nachr.* **1**, 245 (1927).

Bibliography

Two useful review articles on the density operator that we recommend are

U. Fano, "Description of states in quantum mechanics by density matrix and operator techniques," *Rev. Mod. Phys.* **29**, 74 (1957).

D. Ter Haar, "Theory and application of the density matrix," *Rep. Prog. Phys.* **24**, 304 (1961).

See also

M. Weissbluth, *Atoms and Molecules* (New York: Academic Press, 1978), chap. 13.

A review on entanglement and its characterization is

D. Bruß, "Characterizing entanglement," *J. Math. Phys.* **43**, 4237 (2003).

A review of entropy in physics and information theory can be found in

A. Wehrl, "General properties of entropy," *Rev. Mod. Phys.* **50**, 221 (1978).

Appendix B Quantum Measurement Theory in a (Very Small) Nutshell

The following is in no way meant to be an extensive review of the subject of quantum measurement theory. For more details, the reader should consult the bibliography at the end of this appendix. Here we present only those aspects of the theory necessary to understand the results of state reductive measurements, particularly for situations involving entangled states.

Suppose the operator $\hat{Q}$ is a Hermitian operator representing some observable Q. Further, we let the states $\{|q_n\rangle; n \text{ integer}\}$ be the eigenstates of $\hat{Q}$ with the eigenvalues being q_n: $\hat{Q}|q_n\rangle = q_n|q_n\rangle$. The eigenstates are complete and resolve unity according to

$$\sum_n |q_n\rangle\langle q_n| = \hat{1}. \tag{B.1}$$

Each of the terms $|q_n\rangle\langle q_n|$ form a projection operator onto the state with eigenvalue q_n, $\hat{P}(q_n) := |q_n\rangle\langle q_n|$. A state $|\psi\rangle$ (assumed normalized: $\langle\psi|\psi\rangle = 1$) can be expanded in terms of the eigenstates of $\hat{Q}$ according to

$$|\psi\rangle = \hat{1}|\psi\rangle = \sum_n |q_n\rangle\langle q_n|\psi\rangle = \sum_n c_n|q_n\rangle, \tag{B.2}$$

where the coefficient $c_n = \langle q_n|\psi\rangle$ is, in general, a complex number and is known as a probability amplitude. If $|\psi\rangle$ is the state vector just before a measurement of the observable Q, then the probability of obtaining the outcome q_n, $P(q_n)$, is

$$P(q_n) = \langle\psi|\hat{P}(q_n)|\psi\rangle = |c_n|^2. \tag{B.3}$$

According to the orthodox (or Copenhagen) interpretation of quantum mechanics, the system described by the state vector $|\psi\rangle$ is *not* in one of the eigenstates of $\hat{Q}$ prior to the measurement, the measurement merely revealing which eigenstate; rather, the state vector *collapses*, or *reduces*, to one of the eigenstates upon measurement. This process is sometimes represented by the symbols

$$|\psi\rangle \xrightarrow[\text{of } \hat{Q}]{\text{measurement}} |q_n\rangle. \tag{B.4}$$

Prior to the measurement, the value of the observable associated with the operator $\hat{Q}$ in the system described by the state vector $|\psi\rangle$ is objectively indefinite. It is not merely a matter of only knowing what the value of the observable is with a probability given by Eq. (B.3), as we would if we were talking about statistical mechanics, but rather that one cannot assign definite values to observables for systems in superposition states without leading to conflict with experimental observations. The state reduction process does throw the system into a state with a definite value of the observable and subsequent measurements of that observable will return the same state. The dynamics of state reduction is not described by the Schrödinger equation. The interaction of the detector with the particle whose quantum state is being measured may destroy that particle. For example, when a photodetector "clicks" it does so because the photon itself has been destroyed (absorbed).

It is sometimes possible to arrange experiments to perform selective, or filtered, measurements, sometimes called von Neumann projections. Such measurements are invoked in Chapter 9, where polarization filters are employed, and in Chapter 10, where field ionization is used in the context of cavity quantum electrodynamics. A projective measurement on one of the eigenstates $|q_n\rangle$ requires a filtration of all the other eigenstates of the operator. Mathematically we may represent this projection as $\hat{P}(q_n)|\psi\rangle$, which in normalized form reads

$$|\psi\rangle \xrightarrow[\text{measurement}]{\text{projective}} |q_n\rangle = \frac{\hat{P}(q_n)|\psi\rangle}{\langle\psi|\hat{P}(q_n)|\psi\rangle^{1/2}}. \tag{B.5}$$

The filtering need not project onto an eigenstate of $\hat{Q}$. Suppose we could somehow filter the system so that the superposition state $|\psi_s\rangle = \frac{1}{\sqrt{2}}(|q_1\rangle + |q_2\rangle)$ is measured. The projection operator associated with this state is $\hat{P}_s = |\psi_s\rangle\langle\psi_s|$ and we have the projection

$$|\psi\rangle \xrightarrow[\text{measurement onto } |\psi_s\rangle]{\text{projective}} e^{i\varphi}|\psi_s\rangle = \frac{\hat{P}_s|\psi\rangle}{\langle\psi|\hat{P}_s|\psi\rangle^{1/2}}, \tag{B.6}$$

where $\hat{P}_s|\psi\rangle = |\psi_s\rangle\langle\psi_s|\psi\rangle = |\psi_s\rangle(|c_1 + c_2|e^{i\varphi})/\sqrt{2}$ and $\langle\psi|\hat{P}_s|\psi\rangle = |c_1 + c_2|^2/2$. The phase φ results in an irrelevant overall phase factor.

As a specific example, consider a single photon in the polarization state

$$|\theta\rangle = |H\rangle\cos\theta + |V\rangle\sin\theta, \tag{B.7}$$

where we have used the convention given in Chapter 9. If a Polaroid filter is placed in the beam and oriented along the horizontal (vertical) direction, then the state vector of Eq. (B.7) reduces to $|H\rangle$ ($|V\rangle$). On the other hand, if we try to project onto the state

$$|\theta^\perp\rangle = -|H\rangle\sin\theta + |V\rangle\cos\theta, \tag{B.8}$$

where $\theta^\perp = \theta + \pi/2$, by orienting the filter along that direction, we find that we get no photons at all passing through: $\langle\theta^\perp|\theta\rangle = 0$. But if we orient the filter along some direction ϑ to the horizontal axis, the photon will be projected onto the polarization state

$$|\vartheta\rangle = |H\rangle \cos\vartheta + |V\rangle \sin\vartheta, \tag{B.9}$$

as the reader can easily check.

Let us now turn to the case of two-mode states. If we have two modes labeled 1 and 2, we may write a general two-mode state as

$$|\psi\rangle = \sum_{n,m} c_{nm} |q_n\rangle_1 |s_m\rangle_2, \tag{B.10}$$

where the states $|q_n\rangle_1$ are eigenstates of $\hat{Q}_1$ defined in the Hilbert space of mode 1 and the states $|s_m\rangle$ are eigenstates of some operator $\hat{S}_2$ defined in the Hilbert space of mode 2. Certain choices of the coefficients c_{nm} will render the state an entangled state. In either case, an unfiltered measurement of the observables $\hat{Q}_1$ and $\hat{S}_2$ reduces the state vector to

$$|\psi\rangle \xrightarrow[\text{of } \hat{Q}_1 \text{ and } \hat{S}_2]{\text{measurement}} |q_n\rangle_1 |s_m\rangle_2, \tag{B.11}$$

whereas, say, a projective measurement onto $|q_n\rangle_1$ reduces the state vector of Eq. (B.10) to

$$|\psi\rangle \xrightarrow[\text{onto } |q_n\rangle_1]{\text{projection}} |\psi, q_n\rangle = \frac{\hat{P}_1(q_n)|\psi\rangle}{\langle\psi|\hat{P}_1(q_n)|\psi\rangle^{1/2}}$$

$$= |q_n\rangle_1 \frac{\displaystyle\sum_m c_{nm}|s_m\rangle}{\displaystyle\sum_m |c_{nm}|^2}. \tag{B.12}$$

The projection holds whether or not we have an entangled state, as do the more general projections of the type discussed above. Notice that if the original state *is* entangled, the projective measurement creates a factorized state. As an example, let us consider a state of the form

$$|\psi\rangle = \frac{1}{\sqrt{2}} (|H\rangle_1|V\rangle_2 - |V\rangle_1|H\rangle_2), \tag{B.13}$$

one of the Bell states. Suppose that in mode 1 a Polaroid filter is placed at an angle of $\pi/4$ to the horizontal. This causes a projection onto the state

$$|\pi/4\rangle_1 = \frac{1}{\sqrt{2}} (|H\rangle_1 + |V\rangle_1) \tag{B.14}$$

in mode 1 and reduces the original state vector to

$$|\psi\rangle \xrightarrow[\text{onto } |\pi/4\rangle_1]{\text{projection}} \frac{\hat{P}_1(\frac{\pi}{4})|\psi\rangle}{\langle\psi|\hat{P}_1(\frac{\pi}{4})|\psi\rangle^{1/2}} = |\pi/4\rangle_1 \frac{1}{\sqrt{2}} (|V\rangle_2 - |H\rangle_2). \tag{B.15}$$

Note that a projective measurement on one part of an entangled system projects the other part into a particular state of the other part of the system. The results of measurements are generally probabilistic, but once we know the result of a measurement performed on one part of a system, we know precisely the state in the other part of the system. The correlations exhibited by these projections are of a highly nonclassical nature and are ultimately responsible for the violations of Bell's inequalities as described in Chapter 9.

Bibliography

Excellent modern discussions on the issues raised in this appendix can be found in

C. J. Isham, *Lectures on Quantum Theory* (London: Imperial College Press, 1995).

One should also see

K. Gottfried, *Quantum Mechanics Volume I: Fundamentals* (Reading, MA: Addison-Wesley, 1989), chap. IV.

Of course, one should consult

J. von Neumann, *Mathematical Foundations of Quantum Mechanics* (Princeton, NJ: Princeton University Press, 1955), chap. VI especially.

Appendix C Derivation of the Effective Hamiltonian for Dispersive (Far Off-Resonant) Interactions

Here we derive the effective Hamiltonian for interactions where there is a large detuning between the relevant frequencies involved. In order to keep the discussion rather general, we start with the full Hamiltonian of the form

$$\hat{H} = \hat{H}_0 + \hat{H}_I, \tag{C.1}$$

where $\hat{H}_0$ is the interaction-free Hamiltonian and $\hat{H}_I$ is the interaction Hamiltonian. We take the latter to be of the general form

$$\hat{H}_I = \hbar g(\hat{A} + \hat{A}^\dagger), \tag{C.2}$$

where $\hat{A}$ represents the product of operators describing the interaction, and g is the coupling constant. The operator product $\hat{A}$ is assumed to have no explicit time dependence. In the case of atom–field interactions, we have $\hat{A} = \hat{a}\hat{\sigma}_+$ and $\hat{H}_0 = \hbar\omega\hat{a}^\dagger\hat{a} + \hbar\omega_0\hat{\sigma}_3/2$. This specific case was discussed by Schneider et al. [1]. The time-dependent Schrödinger equation in the Schrödinger picture (SP) is

$$i\hbar\frac{d}{dt}|\psi_{SP}(t)\rangle = (\hat{H}_0 + \hat{H}_I)|\psi_{SP}(t)\rangle, \tag{C.3}$$

where $|\psi_{SP}(t)\rangle$ is the state vector in that picture.

We now transform to the interaction picture (IP) using the interaction-free part of the Hamiltonian and the operator $\hat{U}_0 = \exp(-i\hat{H}_0 t/\hbar)$ [2]. The state vector in the IP is given by

$$|\psi_{IP}(t)\rangle = \hat{U}_0^{-1}|\psi_{SP}(t)\rangle, \tag{C.4}$$

and the Schrödinger equation becomes

$$i\hbar\frac{d}{dt}|\psi_{IP}(t)\rangle = \hat{H}_{IP}(t)|\psi_{IP}(t)\rangle, \tag{C.5}$$

where

$$\begin{aligned}\hat{H}_{IP}(t) &= \hat{U}_0^{-1}\hat{H}\hat{U}_0 - i\hbar\hat{U}_0^{-1}\frac{d\hat{U}_0}{dt} \\ &= \hbar g(\hat{A}e^{i\Delta t} + \hat{A}^\dagger e^{-i\Delta t}),\end{aligned} \tag{C.6}$$

and where Δ is the detuning whose form depends on the form of $\hat{A}$. For the atom–field interaction, we have $\Delta = \omega_0 - \omega$. The detuning will be assumed large in the sense described below.

The solution to Eq. (C.5) can be written formally as

$$|\psi_{\text{IP}}(t)\rangle = \hat{T}\left[\exp\left(-\frac{i}{\hbar}\int_0^t d\,t'\hat{H}_{\text{IP}}(t')\right)\right]|\psi_{\text{IP}}(0)\rangle, \qquad (C.7)$$

where $|\psi_{\text{IP}}(0)\rangle = |\psi_{\text{SP}}(0)\rangle$ and $\hat{T}$ is the time-ordering operator. We make the perturbative expansion

$$\hat{T}\left[\exp\left(-\frac{i}{\hbar}\int_0^t d\,t'\hat{H}_{\text{IP}}(t')\right)\right]$$

$$= \hat{T}\left[\hat{1} - \frac{i}{\hbar}\int_0^t d\,t'\hat{H}_{\text{IP}}(t') - \frac{1}{2\hbar^2}\int_0^t d\,t'\int_0^t d\,t''\,\hat{H}_{\text{IP}}(t')\hat{H}_{\text{IP}}(t'') + \cdots\right] \qquad (C.8)$$

$$= \hat{1} - \frac{i}{\hbar}\int_0^t dt'\hat{H}_{\text{IP}}(t') - \frac{1}{2\hbar^2}\hat{T}\left[\int_0^t dt'\int_0^t dt''\,\hat{H}_{\text{IP}}(t')\hat{H}_{\text{IP}}(t'')\right] + \cdots.$$

The last term, when time ordered, reads

$$\hat{T}\left[\int_0^t dt'\int_0^t dt''\,\hat{H}_{\text{IP}}(t')\hat{H}_{\text{IP}}(t'')\right] = 2\int_0^t dt'\hat{H}_{\text{IP}}(t')\int_0^{t'} dt''\,\hat{H}_{\text{IP}}(t''). \qquad (C.9)$$

The second term in Eq. (C.8) yields

$$\int_0^t dt'\hat{H}_{\text{IP}}(t') = \hbar g\left[\hat{A}\,\frac{e^{i\Delta t'}}{i\Delta}\bigg|_0^t - \hat{A}^{\dagger}\,\frac{e^{-i\Delta t'}}{i\Delta}\bigg|_0^t\right]$$

$$= \frac{\hbar g}{i\Delta}[\hat{A}(e^{i\Delta t} - 1) - \hat{A}^{\dagger}(e^{-i\Delta t} - 1)]. \qquad (C.10)$$

The second-order term now becomes

$$\int_0^t dt'\hat{H}_{\text{IP}}(t')\int_0^{t'} dt''\,\hat{H}_{\text{IP}}(t'') = \frac{\hbar^2 g^2}{i\Delta}\int_0^t dt'[\hat{A}^2 e^{2i\Delta t'} - \hat{A}^2 e^{i\Delta t'} - \hat{A}^{\dagger 2}e^{-2i\Delta t'} + \hat{A}^{\dagger 2}e^{-i\Delta t'}$$

$$+ \hat{A}^{\dagger}\hat{A}(1 - e^{-i\Delta t}) - \hat{A}\hat{A}^{\dagger}(1 - e^{i\Delta t})]. \qquad (C.11)$$

Clearly, the integration will give rise to terms that go as g^2/Δ^2 and these will be small for large detuning. Those terms we drop. This is essentially the rotating-wave approximation as those terms will be rapidly oscillating and will average to zero. What remains gives us

$$\int_0^t dt' \hat{H}_{\mathrm{IP}}(t') \int_0^{t'} dt'' \hat{H}_{\mathrm{IP}}(t'') \approx \frac{i\hbar^2 g^2 t}{\Delta}[\hat{A}, \hat{A}^\dagger]. \tag{C.12}$$

Thus, to second order we have

$$\hat{\mathcal{T}}\left[\exp\left(-\frac{i}{\hbar}\int_0^t dt' \hat{H}_{\mathrm{IP}}(t')\right)\right] \approx \hat{1} - \frac{g}{\Delta}[\hat{A}(e^{i\Delta t} - 1) - \hat{A}^\dagger(e^{-i\Delta t} - 1)]$$
$$- \frac{ig^2 t}{\Delta}[\hat{A}, \hat{A}^\dagger]. \tag{C.13}$$

If the mean "excitation" $\langle \hat{A} \rangle \approx \langle \hat{A}^\dagger \hat{A} \rangle^{1/2}$ is not too large and if

$$\left|\frac{g}{\Delta}\langle \hat{A}^\dagger \hat{A} \rangle^{1/2}\right| \ll 1, \tag{C.14}$$

assumed valid because of the large detuning, then the second term of Eq. (C.13) can be dropped and thus we have

$$\hat{\mathcal{T}}\left[\exp\left(-\frac{i}{\hbar}\int_0^t dt' \hat{H}_{\mathrm{IP}}(t')\right)\right] \approx \hat{1} - \frac{i}{\hbar}t\hat{H}_{\mathrm{eff}}, \tag{C.15}$$

where

$$\hat{H}_{\mathrm{eff}} = \frac{\hbar g^2}{\Delta}[\hat{A}, \hat{A}^\dagger]. \tag{C.16}$$

For the Jaynes–Cummings interaction we have $\hat{A} = \hat{a}\,\hat{\sigma}_+$, so that

$$[\hat{A}, \hat{A}^\dagger] = \hat{\sigma}_+\hat{\sigma}_- + \hat{a}^\dagger\hat{a}\,\hat{\sigma}_3. \tag{C.17}$$

Thus, the effective Hamiltonian in this case is

$$\hat{H}_{\mathrm{eff}} = \hbar\chi(\hat{\sigma}_+\hat{\sigma}_- + \hat{a}^\dagger\hat{a}\,\hat{\sigma}_3), \tag{C.18}$$

where $\chi = g^2/\Delta$, in agreement with the expression given in Eq. (4.185). The term $\hbar\chi\hat{\sigma}_+\hat{\sigma}_-$, present even in the absence of photons, is a kind of cavity-induced atomic Kerr effect giving rise to an energy shift on the bare excited atomic state $|e\rangle$.

Finally, as another example, suppose we consider the case of the degenerate parametric down-converter with a quantized pump field. The full Hamiltonian for this process is given in Eq. (7.84). In our notation we have $\hat{H}_0 = \hbar\omega\hat{a}^\dagger\hat{a} + \hbar\omega_p\hat{b}^\dagger\hat{b}$, where ω_p is the pump frequency and $\hat{A} = i\hat{a}^{\dagger 2}\hat{b}$ with $g = \chi^{(2)}$. The effective interaction for large detuning is then

$$\hat{H}_{\text{eff}} = -\frac{\hbar[\chi^{(2)}]^2}{\Delta}(4\hat{a}^\dagger\hat{a}\,\hat{b}^\dagger\hat{b} + 2\hat{b}^\dagger\hat{b} - \hat{a}^{\dagger 2}\hat{a}^2), \qquad (\text{C.19})$$

where $\Delta = 2\omega - \omega_p$. The dispersive form of the interaction of a Kerr-like term $(\hat{a}^{\dagger 2}\hat{a}^2)$, a cross-Kerr-like term $(4\hat{a}^\dagger\hat{a}\,\hat{b}^\dagger\hat{b})$, and a "frequency-pulling" term $(2\hat{b}^\dagger\hat{b})$. Thus, in case of large detuning between pump and signal fields, the parametric down-converter mimics a Kerr medium. Klimov *et al.* [3] have discussed such interactions, though they have used Lie algebraic methods to arrive at the effective Hamiltonians.

References

[1] S. Schneider, A. M. Herkommer, U. Leonhardt, and W. Schleich, *J. Mod. Opt.* **44**, 2333 (1997).
[2] See, for example, J. J. Sakurai, *Advanced Quantum Mechanics* (Reading, MA: Addison-Wesley, 1967), chap. 4.
[3] See A. B. Klimov, L. L. Sánchez-Soto, and J. Delgado, *Opt. Commun.* **191**, 419 (2001) and references cited therein.

Appendix D Nonlinear Optics and Spontaneous Parametric Down-Conversion

Down-conversion results from a nonlinear interaction of pump radiation with media, where the induced polarization is so strongly affected by the radiation that it deforms beyond the linear response that generates the usual dispersion and absorption. For our purposes, we can expand the nonlinear polarization in a power series in the applied radiation field. Crystals commonly used in nonlinear optics are highly anisotropic and their response is described in tensorial form according to

$$\hat{P}_i = \chi^{(1)}_{i,j}\hat{E}_j + \chi^{(2)}_{i,j,k}\hat{E}_j\hat{E}_k + \chi^{(3)}_{i,j,k,l}\hat{E}_j\hat{E}_k\hat{E}_l + \cdots, \qquad (D.1)$$

where $\chi^{(m)}$ is the mth-order electric susceptibility tensor [1] and where repeated indices imply a sum. The energy density is then $\varepsilon_0 E_i P_i$ and thus the *second-order* contribution to the Hamiltonian, the interaction Hamiltonian, is

$$\hat{H}^{(2)} = \varepsilon_0 \int_V d^3\mathbf{r}\,\chi^{(2)}_{i,j,k}\hat{E}_i\hat{E}_j\hat{E}_k, \qquad (D.2)$$

where the integral is over the interaction volume. We now represent the components of the fields as Fourier integrals of the form

$$\hat{E}(\mathbf{r},t) = \int d^3\mathbf{k}[\hat{E}^{(-)}(\mathbf{k})e^{-i[\omega(\mathbf{k})t - \mathbf{k}\cdot\mathbf{r}]} + \hat{E}^{(+)}(\mathbf{k})e^{i[\omega(\mathbf{k})t - \mathbf{k}\cdot\mathbf{r}]}], \qquad (D.3)$$

where

$$\hat{E}^{(-)}(\mathbf{k}) = i\sqrt{\frac{2\pi\hbar\omega(\mathbf{k})}{V}}\,\hat{a}^\dagger(\mathbf{k}) \quad\text{and}\quad \hat{E}^{(-)}(\mathbf{k}) = i\sqrt{\frac{2\pi\hbar\omega(\mathbf{k})}{V}}\,\hat{a}(\mathbf{k}). \quad (D.4)$$

The operators $\hat{a}(\mathbf{k})$ and $\hat{a}^\dagger(\mathbf{k})$ are the annihilation and creation operators, respectively, with momentum $\hbar\mathbf{k}$.[*] If we substitute field expressions of the above form into Eq. (D.2) and retain only the terms important for the case where the signal and idler modes are initially in vacuum states, we obtain the interaction Hamiltonian

---

[*] The commutation relations for these operators where the wave vector is continuous take the form $[\hat{a}(\mathbf{k}), \hat{a}^\dagger(\mathbf{k}')] = \delta^{(3)}(\mathbf{k} - \mathbf{k}')$.

$$\hat{H}_I(t) = \varepsilon_0 \int_V d^3\mathbf{r} \int d^3\mathbf{k}_s d^3\mathbf{k}_i \chi_{lmn}^{(2)}$$

$$\times \hat{E}_{pl}^{(+)} e^{i[\omega_p(\mathbf{k}_p)t - \mathbf{k}_p \cdot \mathbf{r}]} \hat{E}_{sm}^{(-)} e^{-i[\omega_s(\mathbf{k}_s)t - \mathbf{k}_s \cdot \mathbf{r}]} \hat{E}_{in}^{(-)} e^{-i[\omega_i(\mathbf{k}_i)t - \mathbf{k}_i \cdot \mathbf{r}]} + h.c.$$

$$(D.5)$$

The conversion rates for the process depend on the second-order electric susceptibility $\chi^{(2)}$, but typically have efficiencies in the range 10^{-7} to 10^{-11}, extremely low rates. For this reason, in order to obtain significant output in the signal and idler beams, it is necessary to pump the medium with a very strong coherent field, which we can model as a classical field obtained from a laser as long as we are interested in interactions over a short enough time such that depletion of pump photons can be ignored – the parametric approximation. The pump laser is usually in the ultraviolet, while the photons arising from the down-conversion are usually in the visible spectral range.

From time-dependent perturbation theory, assuming the initial states of the signal and idler modes are in vacuum states, which we denote for the moment as $|\Psi_0\rangle$, we obtain, to first order, $|\Psi\rangle \approx |\Psi_0\rangle + |\Psi_1\rangle$, where [2]

$$|\Psi_1\rangle = -\frac{i}{\hbar} \int dt \hat{H}(t) |\Psi_0\rangle$$

$$= \mathcal{N} \int d^3\mathbf{k}_s d^3\mathbf{k}_i \delta\left(\omega_p - \omega_s(\mathbf{k}_s) - \omega_i(\mathbf{k}_i)\right) \qquad (D.6)$$

$$\times \delta^{(3)}(\mathbf{k}_p - \mathbf{k}_s - \mathbf{k}_i) \hat{a}_s^\dagger(\mathbf{k}_s) \hat{a}_i^\dagger(\mathbf{k}_i) |\Psi_0\rangle,$$

where $\mathcal{N}$ is a normalization factor into which all constants have been absorbed. One sees that the delta functions contain the phase-matching conditions

$$\omega_p = \omega_s + \omega_i,$$

$$\mathbf{k}_p = \mathbf{k}_s + \mathbf{k}_i. \qquad (D.7)$$

In the case of type I phase matching, we end up with the state given by Eq. (9.9), which we arrived at by assuming specific momenta which can be post-selected by the placement of a screen with properly located holes at the down-converter.

References

[1] See R. W. Boyd, *Nonlinear Optics*, 2nd edition (New York: Academic Press, 2003).
[2] See Y. Shih, *Rep. Prog. Phys.* **66**, 1009 (2003).

Index